Lecture Notes in Networks and Systems

Volume 791

Series Editor

Janusz Kacprzyk, Systems Research Institute, Polish Academy of Sciences, Warsaw, Poland

Advisory Editors

Fernando Gomide, Department of Computer Engineering and Automation—DCA, School of Electrical and Computer Engineering—FEEC, University of Campinas—UNICAMP, São Paulo, Brazil

Okyay Kaynak, Department of Electrical and Electronic Engineering, Bogazici University, Istanbul, Türkiye

Derong Liu, Department of Electrical and Computer Engineering, University of Illinois at Chicago, Chicago, USA
   Institute of Automation, Chinese Academy of Sciences, Beijing, China

Witold Pedrycz, Department of Electrical and Computer Engineering, University of Alberta, Alberta, Canada
   Systems Research Institute, Polish Academy of Sciences, Warsaw, Poland

Marios M. Polycarpou, Department of Electrical and Computer Engineering, KIOS Research Center for Intelligent Systems and Networks, University of Cyprus, Nicosia, Cyprus

Imre J. Rudas, Óbuda University, Budapest, Hungary

Jun Wang, Department of Computer Science, City University of Hong Kong, Kowloon, Hong Kong

The series "Lecture Notes in Networks and Systems" publishes the latest developments in Networks and Systems—quickly, informally and with high quality. Original research reported in proceedings and post-proceedings represents the core of LNNS.

Volumes published in LNNS embrace all aspects and subfields of, as well as new challenges in, Networks and Systems.

The series contains proceedings and edited volumes in systems and networks, spanning the areas of Cyber-Physical Systems, Autonomous Systems, Sensor Networks, Control Systems, Energy Systems, Automotive Systems, Biological Systems, Vehicular Networking and Connected Vehicles, Aerospace Systems, Automation, Manufacturing, Smart Grids, Nonlinear Systems, Power Systems, Robotics, Social Systems, Economic Systems and other. Of particular value to both the contributors and the readership are the short publication timeframe and the world-wide distribution and exposure which enable both a wide and rapid dissemination of research output.

The series covers the theory, applications, and perspectives on the state of the art and future developments relevant to systems and networks, decision making, control, complex processes and related areas, as embedded in the fields of interdisciplinary and applied sciences, engineering, computer science, physics, economics, social, and life sciences, as well as the paradigms and methodologies behind them.

Indexed by SCOPUS, INSPEC, WTI Frankfurt eG, zbMATH, SCImago.

All books published in the series are submitted for consideration in Web of Science.

For proposals from Asia please contact Aninda Bose (aninda.bose@springer.com).

Suyel Namasudra · Munesh Chandra Trivedi ·
Ruben Gonzalez Crespo · Pascal Lorenz
Editors

Data Science and Network Engineering

Proceedings of ICDSNE 2023

 Springer

Editors
Suyel Namasudra
Department of Computer Science
and Engineering
National Institute of Technology Agartala
Agartala, Tripura, India

Munesh Chandra Trivedi
Department of Computer Science
and Engineering
National Institute of Technology Agartala
Agartala, Tripura, India

Ruben Gonzalez Crespo
School of Engineering and Technology
Universidad Internacional de La Rioja
Logroño, La Rioja, Spain

Pascal Lorenz
University of Haute Alsace
Colmar, France

ISSN 2367-3370 ISSN 2367-3389 (electronic)
Lecture Notes in Networks and Systems
ISBN 978-981-99-6754-4 ISBN 978-981-99-6755-1 (eBook)
https://doi.org/10.1007/978-981-99-6755-1

Preface

In the last few decades, there has been a rapid growth in the number of Internet users due to Internet video surveillance, video streaming, virtual reality (VR) and augmented reality (AR), mobile data traffic, consumer video-on-demand (VoD), and many more. With the increasing demand for the Internet, the amount of data generated by sensors and Internet of Things (IoT) devices is also increasing tremendously. This brings the need for data science techniques to process and analyze network traffic data in order to address various network engineering problems in different domains of networking.

Data science provides methodologies for quickly processing the large quantities of data generated by network devices, finding repeating patterns in their behavior, and building accurate models to improve their performance. For example, consider the problem of anomaly detection, where the aim is to detect deviations in input data from normal behavior. Data science is a field that spreads over several disciplines and incorporates scientific methods, processes, algorithms, and systems to gather knowledge and work. Modern data science techniques are well-equipped to solve each of the above-mentioned network engineering problems. They are not only limited to the networking domain, but they can also be used for better decision-making in different domains like social media, marketing, weather forecasting, health care, and many more.

This book aims to provide a multidisciplinary platform for the publication of outstanding research works carried out in the domain of data science and network engineering. The objective of this book is to bridge the gap between the target audience, which includes scientists, researchers, and academicians, from the domain of data science and network engineering.

This book includes research papers presented at the International Conference on Data Science and Network Engineering (ICDSNE 2023) organized by the Department of Computer Science and Engineering, National Institute of Technology Agartala, Tripura, India, during July 21–22, 2023. It discusses research works from researchers, academicians, business executives, and industry professionals for solving real-life problems by using the advancements and applications of data science and network engineering. This book covers many advanced topics, such as Artificial Intelligence (AI), Machine Learning (ML), Deep Learning (DL), computer networks, blockchain, security and privacy, IoT, cloud computing, big data, supply chain management, and many more. It also discusses various network engineering problems including the ones dealing with network performance, network security, pattern recognition, anomaly detection, personalization, autonomous systems, etc. Additionally, some chapters of this book also present future work directions in the field of data science and network engineering.

Agartala, India   Suyel Namasudra
Agartala, India   Munesh Chandra Trivedi
Logroño, Spain   Ruben Gonzalez Crespo
Colmar, France   Pascal Lorenz

Contents

Computer Security

Editors and Contributors

About the Editors

Suyel Namasudra has received Ph.D. degree from the National Institute of Technology Silchar, Assam, India. He was a post-doctorate fellow at the International University of La Rioja (UNIR), Spain. Currently, Dr. Namasudra is working as an assistant professor in the Department of Computer Science and Engineering at the National Institute of Technology Agartala, Tripura, India. Before joining the National Institute of Technology Agartala, Dr. Namasudra was an assistant professor in the Department of Computer Science and Engineering at the National Institute of Technology Patna, Bihar, India. His research interests include blockchain technology, cloud computing, IoT, AI, and DNA computing. Dr. Namasudra has edited 6 books, 5 patents, and 80 publications in conference proceedings, book chapters, and refereed journals like IEEE TII, IEEE T-ITS, IEEE TSC, IEEE TCSS, IEEE TCBB, ACM TOMM, ACM TOSN, ACM TALLIP, FGCS, CAEE, and many more. He is the Editor-in-Chief of the Cloud Computing and Data Science (ISSN: 2737-4092 (online)) journal. Dr. Namasudra has served as a Lead Guest Editor/ Guest Editor in many reputed journals like IEEE TBD (IEEE, IF: 7.2), ACM TOMM (ACM, IF: 3.144), MONET (Springer, IF: 3.426), CAEE (Elsevier, IF: 3.818), CAIS (Springer, IF: 4.927), CMC (Tech Science Press, IF: 3.772), Sensors (MDPI, IF: 3.576), and many more. He has also participated in many international conferences as an organizer and session chair. Dr. Namasudra is a member of IEEE, ACM, and IEI. He has been featured in the list of the top 2% scientists in the world in 2021, 2022, and 2023. His h-index is 32.

Dr. Munesh Chandra Trivedi has more than 19 years of teaching experience out of which 12.6 years are post-Ph.D. Currently, he is working at the National Institute of Technology Agartala, Tripura, India, as an associate professor. Previously, he worked as an associate professor and HoD (IT), Rajkiya Engineering College, Azamgarh, UP (State Govt. Institution of UP, India) with the additional responsibility of the dean academics and associate dean (UG Program), Dr. A. P. J. Abdul Kalam

Technical University, Lucknow (State Technical University). Dr. Chandra was also the director (in-charge) of Rajkiya Engineering College, Azamgarh. He successfully filed 60 patents (51 national and 09 international patents (Germany, South Africa, and Australia)). He has published 12 textbooks and 148 research papers in different international journals and conference proceedings. Dr. Chandra edited 38 books for Springer Nature. Dr. Chandra has organized more than 32 international conferences technically sponsored by IEEE, ACM, and Springer.

Dr. Ruben Gonzalez Crespo is a full professor in Computer Science and Artificial Intelligence. Currently, he is a vice-rector of Academic Affairs in the Universidad International de La Rioja. He is also EiC and associate editor in several indexed journals. His main research areas are artificial intelligence, accessibility, and project management. He has published more than 250 scientific publications and managed several research projects. He is a member of the advisory council of the Ministry of Education of Spain and other Latin American countries on terms of University Quality. He is patron of the Free Software R&D Foundation in Spain. He is also an evaluator of international projects for European Commission, Colciencias, and FECyT.

Pascal Lorenz (lorenz@ieee.org) received his M.Sc. (1990) and Ph.D. (1994) from the University of Nancy, France. Between 1990 and 1995, he was a research engineer at WorldFIP Europe and Alcatel-Alsthom. He is a professor at the University of Haute-Alsace, France, since 1995. His research interests include QoS, wireless networks, and high-speed networks. He is the author/co-author of 3 books, 3 patents, and 200 international publications in refereed journals and conferences. He is an associate editor of International Journal of Communication Systems (IJCS-Wiley), Journal on Security and Communication Networks (SCN-Wiley), International Journal of Business Data Communications and Networking, and Journal of Network and Computer Applications (JNCA-Elsevier). He is a senior member of the IEEE, IARIA fellow, and a member of many international program committees. He has organized many conferences, chaired several technical sessions, and given tutorials at major international conferences. He was IEEE ComSoc Distinguished Lecturer Tour during 2013–2014.

Contributors

Mohamed Abdelshafea Mousa Abbas Department of CSE, Koneru Lakshmaiah Education Foundation (K L University), Vaddeswaram, AP, India

A. M. Abirami Department of Information Technology, Thiagarajar College of Engineering, Tamil Nadu, India

Chehak Agrawal Department of Software Engineering, Delhi Technological University, Shahabad, Daulatpur, India

Goutam Agrawal Department of Computer Science and Engineering, National Institute of Technology, Silchar, Assam, India

Sharik Ahmad Department of Computer Science and Engineering, Lingaya's Vidyapeeth, Faridabad, India

Santosh Jayanth Amara School of Computer Science and Engineering, VIT-AP University, Amaravati, India

Lokeshwari Anamalamudi Department of Computer Science and Engineering, SRM AP University, Vijayawada, India

Satish Anamalamudi Department of Computer Science and Engineering, SRM AP University, Vijayawada, India

Iffat Rehman Ansari Electronics Engineering Section, Faculty of Engineering and Technology, University Women's Polytechnic, Aligarh, A.M.U.UP, India

Anvith Ramaiah Institute of Technology, Bangalore, India

Sunny Arora Delhi Technological University, Delhi, India

M. Arthy Department of Data Science and Business Systems, SRM Institute of Science and Technology, Kattankulathur, Tamil Nadu, India

J. Arun Pandian School of Information Technology and Engineering, Vellore Institute of Technology, Vellore, India

Murugan Arunkumar Department of Artificial Intelligence and Data Science, KCG College of Technology, Anna University, Chennai, India

Sonal Asthana Department of Information Technology, Babu Banarasi Das Institute of Technology and Management, Lucknow, U.P., India

Amit Banerjee Microsystem Design-Integration Lab, Physics Department, Bidhan Chandra College, Asansol, India

Riya Bansal Department of Software Engineering, Delhi Technological University, Shahabad, Daulatpur, India

Andrew Bawitlung National Institute of Technology, Aizawl, Mizoram, India

Sibananda Behera Trident Academy of Creative Technology, Bhubaneswar, India

Sangeeta Bhattacharya Guru Nanak Institute of Technology, Sodepur, West Bengal, India

Bhavishya Department of Software Engineering, Delhi Technological University, Delhi, India

Anupam Bonkra Information Technology Department, Chandigarh Engineering College-CGC, Punjab, India

Anil Carie Department of Computer Science and Engineering, SRM AP University, Vijayawada, India

Charu Chanda National Institute of Technology Patna, Patna, India

Rajarshi Chanda Guru Nanak Institute of Technology, Sodepur, West Bengal, India

Sebastian Ciobanu Faculty of Computer Science, Alexandru Ioan Cuza University of Iaşi, Iaşi, Romania

Debajyoti Das Guru Nanak Institute of Technology, Sodepur, West Bengal, India

Dipan Das Department of Computer Science & Engineering, NIT Rourkela, Rourkela, India

Pranjit Das Department of CSE, Koneru Lakshmaiah Education Foundation (K L University), Vaddeswaram, AP, India;
Department of Computer Science, Birangana Sati Sadhani Rajyik Vishwavidyalaya, Golaghat, India

Alina Dash Veer Surendra Sai University of Technology Burla, Sambalpur, Odisha, India

Bibhuti Bhusan Dash School of Computer Applications, KIIT Deemed to be University, Bhubaneswar, India

Sandeep Kumar Dash National Institute of Technology, Aizawl, Mizoram, India

Utpal Chandra De School of Computer Applications, KIIT Deemed to be University, Bhubaneswar, India

Salam Shuleenda Devi Department of Electronics and Communication Engineering, NIT Meghalaya, Shillong, Meghalaya, India

Ritu Dewan Department of Computer Science and Engineering, Lingaya's Vidyapeeth, Faridabad, India

Pummy Dhiman Chitkara University Institute of Engineering and Technology, Chitkara University, Punjab, India

K. Divyasri Ramaiah Institute of Technology, Bangalore, India

N. Duraimurugan Department of Computer Science and Engineering, Rajalakshmi Engineering College, Thandalam, Chennai, India

M. Mercy Evangeline Dr. Ambedkar Govt. Arts College, Vyasarpadi, Chennai, Tamil Nadu, India

Alexander Gelbukh Centro de Investigación en Computación, Instituto Politecnico Nacional, Mexico City, Mexico

Gitanjali Department of Information Technology, Babu Banarasi Das Institute of Technology and Management, Lucknow, U.P., India

S. Gnanavel Department of Computing Technologies, Faculty of Engineering and Technology, SRM Institute of Science and Technology, Tamil Nadu, Kattankulathur, India

Rishav Gossain Guru Nanak Institute of Technology, Sodepur, West Bengal, India

Shanky Goyal Chitkara University Institute of Engineering and Technology, Chitkara University, Punjab, India

M. Indushree School of Computer Science Engineering and Technology, Bennett University, Greater Noida, Uttar Pradesh, India

Ishita Indira Gandhi Delhi Technical University for Women, Delhi, India

Sardar M. N. Islam SILC and Decision Sciences and Modelling Program, Victoria University, Footscray, Victoria, Australia

M. Jaeyalakshmi Department of Computer Science and Engineering, Rajalakshmi Engineering College, Thandalam, Chennai, India

Kriti Jain KIET Group of Institutions, Delhi-NCR, Ghaziabad, India

Monika Jasthi Department of Electronics and Communication Engineering, NIT Meghalaya, Shillong, Meghalaya, India

Rathinaraja Jeyaraj Center for Resilient and Evolving Intelligence, Kyungpook National University, Daegu, South Korea

Narayan A. Joshi Dharmsinh Desai University, Nadiad, Gujarat, India

Sparsh Kapoor Department of Electronics and Communication Engineering, KIET Group of Institutions, Delhi-NCR, Ghaziabad, India

K. N. Karthik Ramaiah Institute of Technology, Bangalore, India

Harmohanjeet Kaur Computer Science & Engineering Department, Thapar Institute of Engineering and Technology, Punjab, India

Ekta Kaushik Department of Computer Science and Engineering, KIET Group of Institutions, Delhi-NCR, Ghaziabad, India

V. D. Ambeth Kumar Department of Computer Engineering, Mizoram University, Aizawl, Mizoram, India

Pardeep Kumar Swansea University, Swansea, UK

Piyush Kumar National Institute of Technology Patna, Patna, India

Shailender Kumar Delhi Technological University, Delhi, India

Vishwajeet Kumar Department of Computer Science and Engineering, National Institute of Technology, Silchar, Assam, India

Suchi Kumari School of Computer Science Engineering and Technology, Bennett University, Greater Noida, Uttar Pradesh, India

Robert Lalramhluna National Institute of Technology, Aizawl, Mizoram, India

Muralidaran Loganathan Department of Computer Science and Information System Engineering, St. Joseph University, Dae Es Salaam, Tanzania

Jatin Madaan Guru Nanak Institute of Technology, Sodepur, West Bengal, India

Bhaskar Marapelli Department of CS &IT, Koneru Lakshmaiah Education Foundation, Vaddeswaram, AP, India

Sushilata D. Mayanglambam Department of Computer Science and Engineering, Indian Institute of Technology (ISM), Dhanbad, Jharkhand, India;
Department of Computer Engineering, Mizoram University, Aizawl, Mizoram, India

Nandni Mehla Indira Gandhi Delhi Technical University for Women, Delhi, India

Upendra Mishra KIET Group of Institutions, Delhi-NCR, Ghaziabad, India

Noor Mohd Department of Computer Science and Engineering, Graphic Era (Deemed to be University), Dehradun, India

Samya Muhuri Computer Science & Engineering Department, Thapar Institute of Engineering and Technology, Punjab, India

Anita Murmu National Institute of Technology Patna, Patna, India

Sameena Naaz Department of Computer Science and Technology, School of Engineering Sciences and Technology, New Delhi, India

Tapsi Nagpal Department of Computer Science and Engineering, Lingaya's Vidyapeeth, Faridabad, India

Kshiramani Naik Veer Surendra Sai University of Technology Burla, Sambalpur, Odisha, India

Hunny Pahuja Department of Electronics and Communication Engineering, KIET Group of Institutions, Delhi-NCR, Ghaziabad, India

Sambandam Palaniappan Department of Artificial Intelligence and Data Science, KCG College of Technology, Anna University, Chennai, India

Rajendra Pamula Department of Computer Science and Engineering, Indian Institute of Technology (ISM), Dhanbad, Jharkhand, India

Dilkeshwar Pandey Department of Computer Science and Engineering, KIET Group of Institutions, Delhi-NCR, Ghaziabad, India

Himanshu Pandey Department of Computer Science and Engineering, KIET Group of Institutions, Delhi-NCR, Ghaziabad, India

Mahima Shanker Pandey Computer Science and Applications, Sharda University, Greater Noida, U.P., India

Sudhansu Shekhar Patra School of Computer Applications, KIIT Deemed to be University, Bhubaneswar, India

Parthasarathi Pattnayak School of Computer Applications, KIIT Deemed to be University, Bhubaneswar, India

Chandra Srinivas Potluri Department of Computer Science and Engineering, Werabe University, Werabe, Ethiopia

Shreekanth M. Prabhu Department of Computer Science and Engineering, CMR Institute of Technology, Bengaluru, India

Navamani Prasath Department of Electronics and Communication Engineering, NIT Meghalaya, Shillong, Meghalaya, India

Priyanka Priyadarshini Veer Surendra Sai University of Technology Burla, Sambalpur, Odisha, India

Nikhil Purohit Department of Computer Science and Engineering, Graphic Era (Deemed to be University), Dehradun, India

Manish Raj School of Computer Science Engineering and Technology, Bennett University, Greater Noida, Uttar Pradesh, India

S. Rajarajeswari Ramaiah Institute of Technology, Bangalore, India

E. Ramanujam Department of Computer Science and Engineering, National Institute of Technology Silchar, Assam, India

Arun Kumar Rana Department of Computer Science and Engineering, Galgotias College of Engineering of Technology, Greater Noida, Uttar Pradesh, India; School of Computer Science and Engineering, Phagwara, India

Rousanuzzaman Department of Computer Science and Engineering, National Institute of Technology, Silchar, Assam, India

Sharmistha Roy Faculty of Computing and Information Technology, Usha Martin University, Ranchi, India

Arijit Saha Dumdum Motijheel Rabindra Mahavidyalaya, Kolkata, India

Tapas Saha Guru Nanak Institute of Technology, Sodepur, West Bengal, India

Bibhudatta Sahoo Department of Computer Science & Engineering, NIT Rourkela, Rourkela, India

Rabul Saikia Department of Electronics and Communication Engineering, NIT Meghalaya, Shillong, Meghalaya, India

Aayush Kumar Sakineti Computer Science and Engineering, Vel Tech Rangarajan Dr. Sagunthala R&D Institute of Science and Technology, Tiruvallur, Tamilnadu, India

K. Sakthi Prakash Department of Information Technology, Thiagarajar College of Engineering, Tamil Nadu, India

Pia Sarkar Department of Electronics and Communication Engineering, B. P. Poddar Institute of Management and Technology, Kolkata, India

Rabinarayan Satapathy Faculty of Emerging Technologies, Sri Sri University, Cuttack, India

Pooja Shah Computer Science & Engineering Department, Thapar Institute of Engineering and Technology, Punjab, India

Deepak Kumar Sharma Indira Gandhi Delhi Technical University for Women, Delhi, India

Naman Sharma Information Technology Department, Chandigarh Engineering College-CGC, Punjab, India

Richa Sharma Department of Information Technology, Babu Banarasi Das Institute of Technology and Management, Lucknow, U.P., India

L. K. Shoba Department of Computing Technologies, SRM Institute of Science and Technology, Kattankulathur, Tamil Nadu, India

K. Shyamala Dr. Ambedkar Govt. Arts College, Vyasarpadi, Chennai, Tamil Nadu, India

M. Shyamala Devi Computer Science and Engineering, Vel Tech Rangarajan Dr. Sagunthala R&D Institute of Science and Technology, Tiruvallur, Tamilnadu, India

Aman Singh Department of Information Technology, Babu Banarasi Das Institute of Technology and Management, Lucknow, U.P., India

Balraj Singh School of Computer Science and Engineering, Phagwara, India; Lovely Professional University, Phagwara, Punjab, India

Inderpal Singh School of Computer Science and Engineering, Phagwara, India; Lovely Professional University, Phagwara, Punjab, India

Priya Singh Department of Software Engineering, Delhi Technological University, Shahabad, Daulatpur, India; Department of Software Engineering, Delhi Technological University, Delhi, India

Riddhi Singhal Ramaiah Institute of Technology, Bangalore, India

Akhil Siraswal Department of Computer Science and Engineering, KIET Group of Institutions, Delhi-NCR, Ghaziabad, India

Malkapurapu Sivamanikanta Department of CSE, Koneru Lakshmaiah Education Foundation (K L University), Vaddeswaram, AP, India

Prakash Srivastava Department of Computer Science and Engineering, Graphic Era (Deemed to be University), Dehradun, India

R. Sujithra Department of Artificial Intelligence and Data Science, KCG College of Technology, Anna University, Chennai, India

D. Sumathi School of Computer Science and Engineering, VIT-AP University, Amaravati, India

Ritika Talukdar Indira Gandhi Delhi Technical University for Women, Delhi, India

Ashutosh Thakur Department of Software Engineering, Delhi Technological University, Delhi, India

Vinay Thakur Ministry of Electronics and IT, Govt of India, National Informatics Centre Services Inc. (NICSI), New Delhi, India

Vikas Tripathi Department of Computer Science and Engineering, Graphic Era (Deemed to be University), Dehradun, India

D. Umanandhini Computer Science and Engineering, Vel Tech Rangarajan Dr. Sagunthala R&D Institute of Science and Technology, Tiruvallur, Tamilnadu, India

Alka Upadhyay Department of Computer Science and Technology, School of Engineering Sciences and Technology, New Delhi, India

S. VijayPrakash Department of Artificial Intelligence and Data Science, KCG College of Technology, Anna University, Chennai, India

Ankita Vishwakarma Department of Information Technology, Babu Banarasi Das Institute of Technology and Management, Lucknow, U.P., India

S. Yamini Doctoral student-Phd in aquatic science and technology, National kaohsiung university of science and technology, nanzih campus, Kaohsiung city, Taiwan

B. Yasotha Department of Data Science and Business Systems, SRM Institute of Science and Technology, Kattankulathur, Tamil Nadu, India

Computational Intelligence

Evaluation of Hand-Crafted Features for the Classification of Spam SMS in Dravidian Languages

E. Ramanujam, K. Sakthi Prakash, and A. M. Abirami

Abstract In this digital era, people are cheated in multiple ways by sending fake messages. Without realizing its impact, they respond to the links the cyber frauds share. This immediate reaction to the fraud messages makes people lose their balance in bank accounts or fall into some other horrible events. These types of fake or spam messages have to be identified earlier before they come to users' Inbox. This paper proposes a Spam message filtering model that extracts significant hand-crafted features and is classified using machine learning algorithms. This research collects 7700 short messages in Dravidian languages like Tamil, Kannada, Telugu, and Malayalam and creates an optimal Spam-Ham filtering framework. Experimentation has also been carried out with a benchmark dataset for performance comparison regarding accuracy, precision, recall, and F1-score.

Keywords Spam message · Fake message · Dravidian languages · Short messages collections · ML models

1 Introduction

Digital use is greatly widened as internet services are free for certain services for a limited time. Some people, namely spammers or hackers, misuse this and propagate fraudulent SMS (Short Message Service) or emails to users. Some believe these messages fall into the hands of spammers and face financial crisis [1].

E. Ramanujam (✉)
Department of Computer Science and Engineering, National Institute of Technology Silchar, Assam 788010, India
e-mail: ramanujam@cse.nits.ac.in

K. Sakthi Prakash · A. M. Abirami
Department of Information Technology, Thiagarajar College of Engineering, Tamil Nadu 625015, India
e-mail: sakthiprakash@student.tce.edu

A. M. Abirami
e-mail: abiramiam@tce.edu

© The Author(s), under exclusive license to Springer Nature Singapore Pte Ltd. 2024
S. Namasudra et al. (eds.), *Data Science and Network Engineering*, Lecture Notes in Networks and Systems 791, https://doi.org/10.1007/978-981-99-6755-1_1

People share their email ids or mobile number when they purchase any product from eCommerce websites or provide feedback or views about any service. Some users share personal information on social media platforms. Hackers extract this information and send messages to them. In some parts of the world, people are involved in criminal activities by getting duplicate mobile numbers and sending anonymous messages to get money from them [2]. Some people receive abusive comments or images with links to harmful websites. These spammers use attractive words, making the receivers read and click the links. These links may ask the users to share their OTP number or credit card details. Without understanding the background happenings, the user may share all their privacy details to the spammer, which makes the hackers hack their bank accounts [3]. Poor users may learn about the cheat very late, and it is impossible for anyone to get the amount back.

As per the article [4], the number of spam messages increased to 11% after the Covid-19 pandemic, and the complaints about spam texts increased to 146%. It indicates that spam text messages are on the rise, and the readers are also having a doubtful look at the message and lodging complaints. Some people send phishing messages to users and ask them to give their account details to Weblinks to share with them. They use SMS spoofing to make readers believe it is a genuine message [5].

Truecaller, a mobile app that identifies the caller id and filters spam messages, reported that there might be an increase in spam robocalls, and this larger scam may catch people. It is time to educate the public about these scammers. Most banks send repeated warning messages to their customers and place advertisements in the newspapers to create awareness about these scams and spam. Most network service providers use this filtering technique to identify spam messages, but they are available for English text only. Even the play store has numerous mobile applications that filter out only messages in English.

Hackers send messages in regional languages to gain more confidence from the receivers. This serious issue motivates this research work to perform spam message classification using machine learning (ML) algorithms and automate this filtering process. A similar type of solution must also be extended to non-English messages so that we can protect all kinds of people. Thus, this research collects SMSs of different Dravidian languages like Tamil, Telugu, Kannada, and Malayalam, along with the English language extracts significant features and then classifies using the spam filtering technique. The contributions of this research work include:

1. Collection of Spam and Ham SMSs for Dravidian Languages such as Tamil, Telugu, Kannada, and Malayalam
2. Labeling of SMS Corpus using experts of Dravidian Language
3. Unique features extraction process to automate the spam filtering process
4. Benchmarking the supervised ML algorithms for non-English SMSs dataset

The remaining section of the paper is organized as follows. Section 2 briefly overviews the state-of-the-art multi-lingual spam SMS detection techniques using

ML and Deep learning (DL). Section 3 discusses the significant feature extraction and evaluation of the same using supervised ML techniques in Sect. 4. Section 5 concludes the paper.

2 Related Study

SMS spam detection methods are quantitatively analyzed from the literature and categorized into feature engineering and classification-based methods for multilingual spam SMS data.

Feature Engineering based Spam Detection—Feature engineering extracts valuable and significant features using Natural Language Processing (NLP) techniques for efficient classification. Many researchers have utilized the NLP process to classify spam SMS messages. Agarwal et al. [6] have incorporated the countvectorizers method, such as TF-IDF, to extract features from Indian spam messages (written in English). The proposed system evaluates various supervised and unsupervised ML models. Shashank et al. [7] developed a multilingual SMS spam detection model using RNN. They used the Kaggle dataset for English SMS and collected English, Telugu, and Hindi SMSs from their neighborhood. Using the Word2Vec approach, they got 96.8% Accuracy for the multilingual combined dataset (8092 Ham and 2075 Spam) using the LSTM model. Chakraborty et al. [8] have developed an SVM-based classifier to detect spam messages in Indian languages. This method uses a similar vectorizer concept for automatic feature extraction and is classified using an SVM classifier. Vu et al. [9] have proposed a ruleset for classifying spam emails in Chinese, Vietnamese, and English. Their multilingual ruleset gave 91.5% Accuracy for Chinese and 81.4% for Vietnamese messages. Al-Zoubi et al. [10] classified Spam profiles on social networks in multilingual contexts like Arabic, English, Korean, and Spanish. They extracted 29 features for 196 profiles and obtained a 97% recall value for the random forest classifier when feature significance was used for feature selection.

Pre-Trained Model for Multilingual Spam detection—El-Alami et al. [11] used Bidirectional Encoder Representations from Transformers (BERT), a transfer learning model for the offensive words detection in social networks for two languages like English and Arabic and obtained 90% Accuracy. Raga et al. [12] develop a spam detector prototype using BERT (Bidirectional encoder and representations for transformer), a pre-trained model. This model classifies messages by understanding their actual meaning and context and filters out spam messages. SMS dataset V.1 has been used to evaluate the performance and attained an accuracy of 96.10% (Table 1).

Classification-based Spam Detection involves utilizing the DL/ ML techniques for SMS spam filtering techniques. In our previous work [13], a hybrid deep learning mechanism has been proposed to classify the Dravidian spam messages using Convolutional Neural Network (CNN) and Long-Short Term Memory (LSTM). A total of 3282 messages are classified with a maximum accuracy of 97.7%. The proposed

Table 1 Analysis of state-of-the-art mult-lingual spam sms detection techniques

S. no.	Research work	# Of Language	# Of dravidian language	Dataset size	Performance (Accuracy)	Remarks
1	[9]	3	NA	1358	89.04%	Applicable only for e-mails
2	[6]	1	NA	6761	97.87%	Applicable only for English language
3	[12]	1	NA	5574	96.10%	Time complexity is high
4	[7]	3	1	10167	96.8%	Time complexity is high
5	[10]	4	NA	196	95.2%	Random Forest and Feature Significance combination yielded better performance results
6	[8]	1	NA	*	98.43	Details are not very clear
7	[13]	5	4	3282	97.7%	High cost
8	[11]	2	NA	13794	90%	BERT was used offensive language detection
9	[14]	2	NA	75222	99.57%	High computational cost

method shows better Accuracy. However, the computational cost in terms of time is very high for model generation and evaluation. Lee et al. [14] integrated image processing with a traditional string-based technique for spam message detection in the Korean language. They applied CNN 2D model for 69654 Korean and 5568 English messages and obtained 99.57% Accuracy in spam detection.

The literature shows that there is a need for spam filtering techniques for SMS in Indian languages, especially in Dravidian languages. Limited research work is carried out in spam message detection in these languages. This research focuses on extracting significant features from the multilingual SMSs for classifying spam or ham messages.

3   Methodology

The proposed methodology consists of various phases such as data collection, labeling, feature extraction, and model building, as shown in Fig. 1.

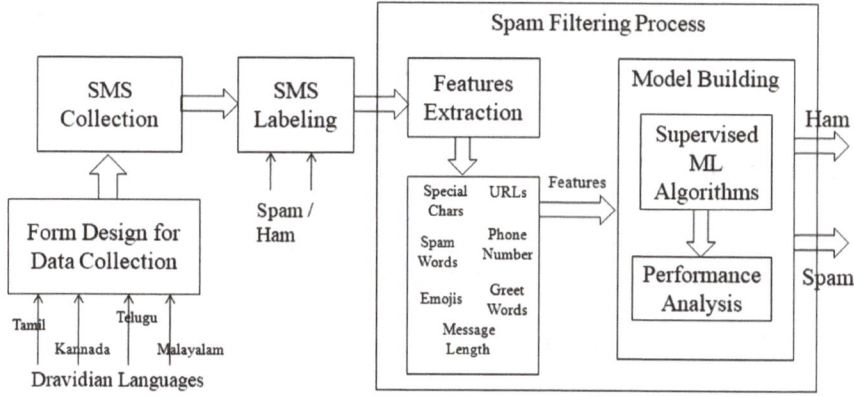

Fig. 1 Phases of spam detection for multilingual/dravidian SMS

3.1 Data Collection

The labeled SMS dataset, namely RevisedIndianDataset [7], is available with 4565 messages in English, Hindi, Telugu, and Kannada. There are 3354 Ham messages and 1211 spam messages. This research study represents this dataset as DS-1; the details are shown in Fig. 2.

In another case, to collect more SMSs in Dravidian languages, a Google form has been designed and passed on to friends and other contacts for collecting the SMS in

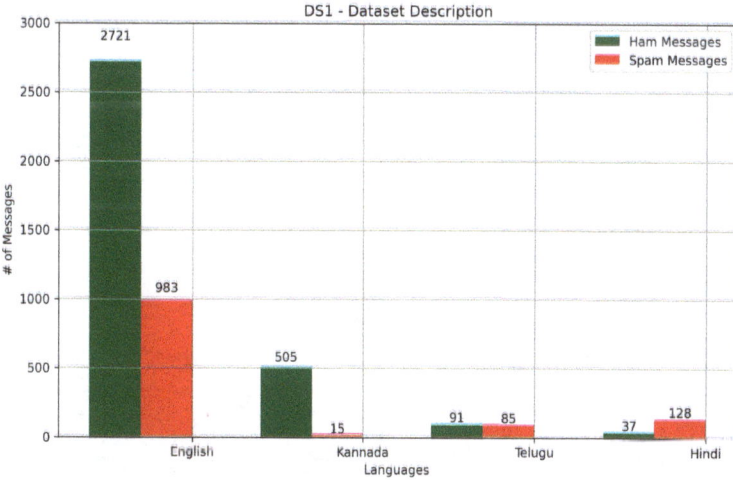

Fig. 2 Distribution of SMS dataset DS-1

English, Tamil, Telugu, Kannada, and Malayalam languages. Nearly 7700 messages were collected for this research study, called DS-2.

3.2 Labelling SMS

SMS collected through Google form for DS-2 is labeled as Spam or Ham by the groups of people. This group is chosen carefully such that they have expertise in reading and writing in the corresponding languages. Figure 3 shows the distribution of labeled Dravidian SMS dataset DS-2.

3.3 Feature Extraction

The dataset is keenly analyzed, and the following hand-crafted feature set F is extracted by reviewing various spam filtering detection techniques. The hand-crafted features extracted are

Presence of URLs F_1—The features F_1 check the presence of URLs and links in the message received by the user as given in the Eq. 1. The presence of HTTP links or URLs in the message seems to be a smishing or spam message. The phishers only like to send links to visit the website for bumper offers, prizes, etc., with fake information. This feature also considers the tiny URL as a link or URL.

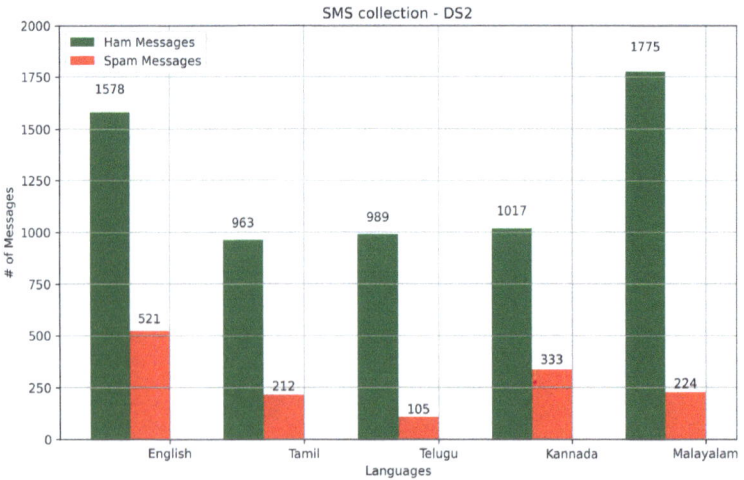

Fig. 3 Distribution of dravidian SMS collection dataset DS-2

$$F_1 = \begin{cases} 1, & \text{if URL is present in the message} \\ 0, & \text{otherwise} \end{cases} \tag{1}$$

Message Length F_2—The length of the messages received, including the space, symbols, special characters, emotions, etc., are calculated. As per the Telecom Regulatory Authority of India, the text limit of the SMS message is 160 characters only, but now smartphones and network subscribers have provided more characters to be sent by any user. Chatting or conversation messages will be significantly less than fake messages. Thus the feature F_2 results in 1 if the message length is more than 150 characters; else, as 0 as shown in Eq. 2.

$$F_2 = \begin{cases} 1, & \text{if length (Message)} \geq 150 \\ 0, & \text{otherwise} \end{cases} \tag{2}$$

Presence of Phone/Mobile Number F_3—The feature F_3 checks the presence of mobile/phone numbers in the received message using a special function. Mostly, the mobile number will be of the length of 10 or 12, which includes country code. In the case of the phone number, the length will be 7 or more with the district code. Thus, a special code has been written to validate the presence of a mobile or phone number in the message. If the message contains any phone/mobile number, it results in 1 and marks it suspicious, as shown in Eq. 3.

$$F_3 = \begin{cases} 1, & \text{Presence of Mobile/Phone number} \\ 0, & \text{otherwise} \end{cases} \tag{3}$$

Presence of Emotion symbols F_4—The feature F_4 easily discriminates the ham and spam messages as most users use emotions while chatting. TextBlob python package is used to identify the positive or negative emotions of all the languages and marked to be 0 for Ham messages else 1 for spam messages as shown in the Eq. 4.

$$F_4 = \begin{cases} 0, & \text{Presence of positive or negative emotions} \\ 1, & \text{otherwise} \end{cases} \tag{4}$$

Spam Keywords F_5—The feature F_5 determines the presence of suspicious words (spam keywords) in the message. Spammers usually add various suspicious keywords in the message to gain the trust of the message. To avoid this, language experts and a word cloud model are used to identify the suspicious spam keywords and made as a library. The feature F_5 is marked as 1 in case any suspicious word is available in the message, like the created dictionary else 0. A Total of 280 keywords are identified as spam and included in the dictionary by the experts to detect suspicious keywords in the message.

$$F_5 = \begin{cases} 1, & \text{Presence of suspicious spam keywords} \\ 0, & \text{otherwise} \end{cases} \tag{5}$$

Money Symbols F_6—The feature F_6 detects the presence of the money symbol in the received message. Mainly, the spammers gain the importance of users by offering prizes in terms of money. The money symbols are sometimes referred to in symbols like ₹, Rs. or Rupee. Thus, any money symbols are detected using a regular expression marked to be 1 else 0, as shown in Eq. 6.

$$F_6 = \begin{cases} 1, & \text{Presence of Money symbols} \\ 0, & \text{otherwise} \end{cases} \tag{6}$$

Special Characters F_7—The feature F_7 discriminates the message to be spam and ham by checking the presence of special characters of symbols. For instance, in personal chatting, there is less chance to use special symbols like +, −, *, %, , ¶, ©, etc. Thus, a regular expression-based verification system is used to identify the presence of special characters in the message. In case of availability, the F_7 marks the message as 1 (spam) or 0.

$$F_7 = \begin{cases} 1, & \text{Presence of special characters} \\ 0, & \text{otherwise} \end{cases} \tag{7}$$

Emoji F_8—Apart from the emotions expressed in terms of emojis, there are many emojis due to the rise in the technical development of Smartphone and messaging applications. Thus, a regular expression is used to verify all the emojis present in the message. The presence of emoji is marked to be 0 else 0 for the case of feature F_8.

$$F_8 = \begin{cases} 0, & \text{Presence of Emoji} \\ 1, & \text{otherwise} \end{cases} \tag{8}$$

Greeting Words F_9—Most of the messages shared are the greeting messages such as morning, evening, night, birthday, anniversary, etc. The feature F_9 checks the presence of greeting keywords like Good Morning, Happy anniversary, hi, hello, etc., This kind of message seems good for ham messages, and thus F_9 marks the value as 0 for ham and 1 for spam. As there is no conventional technique for the Dravidian languages, the experts create the dictionary to verify such greeting message keywords. Moreover, all the abbreviations, such as gd, and gud, for the word good, are considered.

$$F_9 = \begin{cases} 0, & \text{Presence of Greeting words} \\ 1, & \text{otherwise} \end{cases} \tag{9}$$

Self Answering Messages F_{10}—The feature F_{10} analyses the presence of self-answering messages, like asking the user to subscribe or unsubscribe any service-related queries under the category of spam. This kind of message often contains the user to reply with yes or no in their language. Thus, all the self-answering messages are identified and marked to be 1 as spam and 0 for non-spam messages.

$$F_{10} = \begin{cases} 1, & \text{Presence of Self Answered Message} \\ 0, & \text{otherwise} \end{cases} \tag{10}$$

Using the feature set $F - F_1, F_2, F_3, F_4, F_5, F_6, F_7, F_8, F_9, F_{10}$ the importance of features are executed by the classifier for the model building and performance evaluation.

3.4 Model Building and Evaluation

In this Model Building and Evaluation phase, lazypredict—a supervised Python library is used for the feature significance and evaluation. This library provides the performance results of different supervised machine learning classifiers such as Naive Bayes, Support Vector Machine, Decision tree, logistic regression, Random Forest, k-Nearest Neighbor, etc. The performances are compared using standard metrics such as Accuracy, Precision, Recall, F-Measure, and Area under ROC as used in research work [15] (Fig. 4).

4 Experimental Results and Discussions

The experimentation has been carried out with dataset DS-1 and DS-2 exclusively through the feature set $F - F_1, F_2, F_3, F_4, F_5, F_6, F_7, F_8, F_9, F_{10}$. The sample SMS messages of the Dravidian language are shown in Fig. 5. After extracting feature set F, the feature importance score is calculated using the Random Forest (RF) feature selection technique as used in [16]. The feature importance score for the features extracted in datasets DS-1 and DS-2 is shown in Fig. 5a, b.

Message	Type	Language
போராடிக்கிறதா? களிப்பை தொடர்க! ரம்மி ஆடி பணம் வெல்க 1kx.in/hScN5e	Spam	Tamil
ఎయిర్టెల్ డిజిటల్ TV తో మీ స్వంత DTH ప్లాన్ కన్నుమైత చేసుకోండి. HD చాక్స్ ఇప్పుడు 1800 కు బదులుగా 1300 వద్ద లభిస్తుంది ఇప్పుడే బుక్ చేయండి: u.airtel.in/BuyDTH	Spam	Telugu
ప్రతినిత్య ಕగ్గద సందేశಗಳన್ను ಪడయలు ಕಳగಿನ ಲಿಂಕ್ ఒత್తి ನಮ್ಮ, ಟెಲಿಗ್ರಾಮ್ ಮತ್ತು ವಾಟ್ಸಾಪ್ ಗುಂಪುನ್ನు ಸೇರಿ ಕಾಗೀಯೀ ನಮ್ಮ, ಯೂಟ್ಯೂಬ್ ಚಾನೆಲ್ ನ ಚಂದಾದಾರರಾಗಿ.	Spam	Kannada
ഹൗസ് ലോൺ വേണോ. എന്നെ 97881 57654 എന്ന നമ്പറിൽ വിളിക്കുക	Spam	Malayala
Idhu eppdi irukku? Tamizha Combo pack brings unmatched viewing of all Tamil channels like Zee Tamil	Spam	English

Fig. 4 Sample SMS messages in dravidian languages

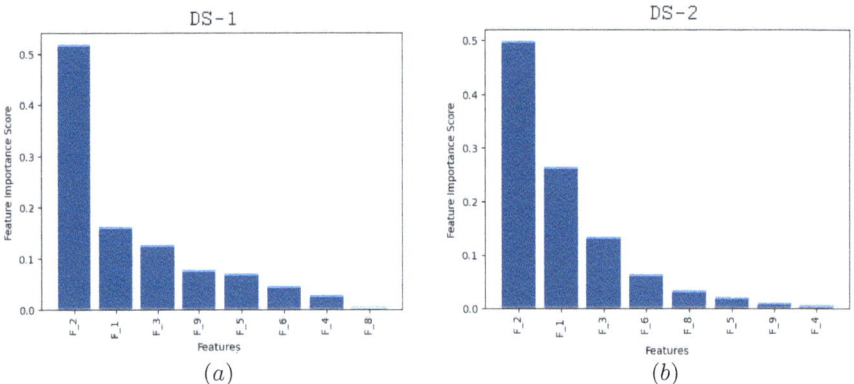

(a)   (b)

Fig. 5 Feature importance scores **a** DS-1 **b** DS-2 dataset features

On comparing the feature importance score from Fig. 5a, b, the features F_7—Presence of special characters and F_{10}—greeting words scores very less (negative scoring). F_{10} to score less because of the manual provision of greeting keywords and its alternative in all the Dravidian languages. Similarly, the Presence of special characters doesn't have more impact on the process of feature importance. Thus, the features F_7 and F_{10} are removed from the model building and evaluation process. The other features are highly significant, as seen in Fig. 5a, b.

The selected features F_1, F_2, F_3, F_4, F_5, F_6, F_8, F_9 are fed into the Model Building and Evaluation using lazypredict—supervised python library. This library results in the performance evaluation of more than 30 different classifiers. Among these, the top 5 classifiers are reported in Table 2 for the DS-1 and DS-2.

On comparing the performance of DS-1, the Light Gradient Boosting Model (LGBM) classifier shows a higher performance of 92% Accuracy and 83% of Precision, Recall, and F1-score, respectively. The other classifier, such as Label propagation and Label spreading, has attained 90% of Accuracy, but the other performance metrics perform less than LGBM. Finally, Random Forest, K-NN reaches the approximate 90% Accuracy, less than LGBM classifier.

Table 2 Performance evaluation of top 5 classifiers for extracted feature set F from dataset DS-1 and DS-2 using Lazypredict

Classifier	DS-1				DS-2			
	Accuracy	Precision	Recall	F1-Score	Accuracy	Precision	Recall	F1-Score
Random forest	91	79	83	81	93	85	78	81
k-NN	89	79	75	77	90	77	65	71
Label propagation	90	78	80	79	91	79	68	74
Label spreading	90	78	80	79	91	79	69	74
LGBM classifier	92	83	83	83	93	84	76	80

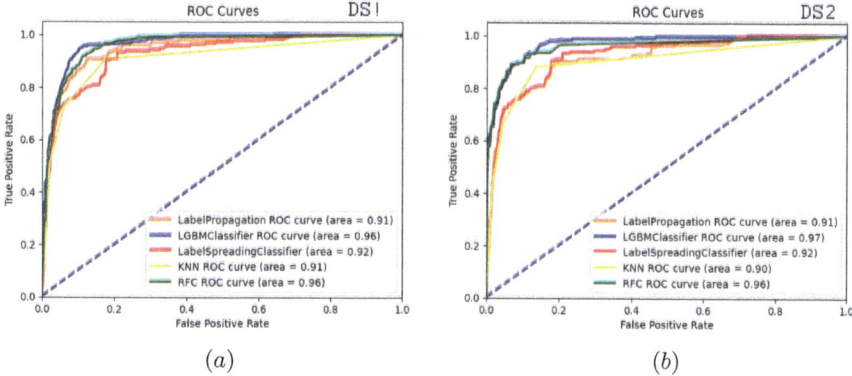

(a)   (b)

Fig. 6 ROC curve for the features extracted **a** DS-1 **b** DS-2 dataset

In addition, on comparing the performance of DS-2, Random Forest and LGBM attains the highest Accuracy of 93% Precision, Recall, and F1-score has a change of 1%. The Label propagation and Label spreading reach the second highest of 91% Accuracy and 79% Precision, but there is a 1% change in the Recall, but the F1-score is of similar values.

On comparing the overall performance, Random Forest and LGBM classifier performs better regarding all performance metrics for DS-1 and DS-2. The ROC curve has been drawn to compare the performances further, and the Area under the ROC curve is predicted for all the top 5 classifiers to the DS-1 and DS-2 dataset and shown in Fig. 6a, b. respectively.

5   Conclusion

Due to technological advancement, communication devices have increased multi-fold, providing voice and data calling at a low cost. It has raised more chances of fraudulent activities by spammers through SMS. This paper proposed a hand-crafted feature extraction mechanism to filter-out the spam sms in Dravidian languages. Experimentation has been carried out using the SMS data collected in Dravidian languages and another benchmark dataset. Features are validated significantly using a Random forest feature importance score and classified for performance evaluation. This paper has certain limitations on performance, such as Accuracy. Still, it can be improved by concentrating on language-specific features and retuning the extracted features.

References

1. Boxall A (2022) It's not just you, SMS spam is a nightmare right now. https://www.digitaltrends. com/mobile/why-spam-sms-is-a-nightmare-in-2022/. Last accessed 30 March 2023
2. Sharma H (2022) SMS scams trick Indian banking customers into installing malicious apps. https://www.zscaler.com/blogs/security-research/sms-scams-trick-indian-banking-customers-installing-malicious-apps. Last accessed 30 March 2023
3. Sapkale Y (2022) Fraud alert: beware of these 5 SMS, Calls. https://www.moneylife.in/article/ fraud-alert-beware-of-these-5-sms-calls/69076.html. Last accessed 30 March 2023
4. Rawes E (2022) Getting more spam texts and emails? Here's how to fix it. https://www. digitaltrends.com/mobile/why-youre-getting-more-spam-texts-and-emails. Last accessed 30 March 2023
5. Saini N (2021) SMS spoofing: how scammers are using this technique to steal. https:// timesofindia.indiatimes.com/gadgets-news/sms-spoofing-how-scammers-are-using-this-technique-to-steal-money-from-your-account/articleshow/85096378.cms. Last accessed 30 March 2023
6. Agarwal S, Kaur S, Garhwal S (2015) SMS spam detection for Indian messages. In: 2015 1st international conference on next generation computing technologies (NGCT). IEEE, pp 634–638
7. Shashank S (2018) Multilingual SMS spam detection using RNN. https://shshnk158.github. io/Multilingual-SMS-spam-detection-using-RNN/ Last accessed 20 Nov 2022
8. Chakraborty A, Karmakar S, Chattaraj S (2019) Machine learning based Indian spam recognition. Trans Intell Comput 3:10–16
9. Vu MT, Tran QA, Jiang F, Tran VQ (2014) Multilingual rules for spam detection. J Mach Mach Commun 1(2):107–122
10. Al-Zoubi AM, Alqatawna JF, Faris H, Hassonah MA (2021) Spam profiles detection on social networks using computational intelligence methods: the effect of the lingual context. J Inf Sci 47(1):58–81
11. El-Alami FZ, El Alaoui SO, Nahnahi NE (2022) A multilingual offensive language detection method based on transfer learning from transformer fine-tuning model. J King Saud Univ-Comput Inf Sci 34(8):6048–6056

12. Raga SS, Chaitra BL (2022) A bert model for sms and twitter spam ham classification and comparative study of machine learning and deep learning technique. In: 2022 IEEE 7th international conference on recent advances and innovations in engineering (ICRAIE), vol 7. IEEE, pp 355–359

13. Ramanujam E, Shankar K, Sharma A (2022) Multi-lingual Spam SMS detection using a hybrid deep learning technique. In: 2022 IEEE Silchar subsection conference (SILCON). IEEE, pp 1–6

14. Lee H, Jeong S, Cho S, Choi E (2023) Visualization technology and deep-learning for multi-lingual spam message detection. Electronics 12(3):582

15. Rasikannan L, Alli P, Ramanujam E (2020) Improved feature based sentiment analysis for online customer reviews. In: Innovative data communication technologies and application: ICIDCA 2019. Springer International Publishing, pp 148–155

16. Ramanujam E, Chandrakumar T, Nandhana K, Laaxmi NT (2020) Prediction of fetal distress using linear and non-linear features of CTG signals. In: Computational vision and bio-inspired computing: ICCVBIC 2019. Springer International Publishing, pp 40–47

Training Algorithms for Mixtures of Normalizing Flows

Sebastian Ciobanu

Abstract In this paper, we focus on how a probabilistic mixture of normalizing flows can be fitted. In the literature, there are (at least) four approaches that do not necessarily provide an actual implementation of the method. These four algorithms are gradient ascent maximizing the log-likelihood of the data, (soft) expectation–maximization, hard expectation–maximization, and gradient ascent maximizing the evidence lower bound. Our contribution or the novelty of the paper can be described as follows: we (re)implement each method, we create a software program that encompasses all these four implementations, and we compare those on toy datasets and image datasets on which we fit a mixture of masked autoregressive flows. The nonlinear flexibility is shown in the plots. The metrics and the running times are reported. There is not necessarily a certain training algorithm to be preferred, although there are some advantages and disadvantages for each. The code is available at https://github.com/aciobanusebi/training-algs-for-mnf.

Keywords Machine learning · Mixtures of normalizing flows · Gradient descent · Expectation–maximization · Maximum likelihood estimation

1 Introduction

In probabilistic machine learning, there are discriminative and generative models [1]. Briefly, the generative models work with the probabilities $p(x)$ (for unlabeled datasets with inputs x) or $p(x, y)$ (for labeled datasets with inputs x and outputs y), whereas the discriminative ones work with the conditional probabilities $p(y|x)$. In recent years, deep generative models have appeared. They cover multiple types of models: autoregressive models [2, Chap. 22], variational autoencoders (VAEs) [2, Chap. 21], normalizing flows (NFs) [2, Chap. 23], generative adversarial networks [2, Chap. 26], new energy-based models [2, Chap. 24], diffusion models [2, Chap. 25].

S. Ciobanu (✉)
Faculty of Computer Science, Alexandru Ioan Cuza University of Iași, Iași, Romania
e-mail: aciobanusebi@gmail.com

© The Author(s), under exclusive license to Springer Nature Singapore Pte Ltd. 2024
S. Namasudra et al. (eds.), *Data Science and Network Engineering*, Lecture Notes in Networks and Systems 791, https://doi.org/10.1007/978-981-99-6755-1_2

Among these, the normalizing flows model explicitly the probability density function (p.d.f.), i.e., we have access before/during/after the training phase to a p.d.f., a programmatic function which can be called. As a normalizing flow is more expressive than a (multivariate) normal distribution, a mixture of normalizing flows is more flexible (see [3]) than a mixture of Gaussians, hence its usefulness. In general, a specific application for mixtures of distributions is grouping/clustering the data. As a result, the mixtures of normalizing flows can be used to this end.

The contributions of this paper are represented by the provision of a unitary implementation of the existing methods for training the mixtures of normalizing flows and by their comparison in the same setup. The purpose and the importance of this paper reside in stepping forward in the process of choosing the right training method (which is data dependent) for the mixture of normalizing flows, a model which has its applicability in different areas: clustering (as mentioned earlier), density estimation, anomaly detection, data augmentation, etc.

This paper is structured as follows:

– Section 2 includes related work and ideas regarding our implementation;
– Section 3 contains the methodology and the results;
– Section 4 reflects our conclusion.

2 Related Work and Our Implementation

2.1 Mixtures of Normalizing Flows

A standard mixture of continuous distributions has its density equal to the convex combination of the distributions' probability density functions. A mixture of normalizing flows would be a mixture where each component in the convex combination is a normalizing flow. This is the standard probabilistic definition. In the literature, there are contributions representing a mixture of NFs in a non-standard manner or where the purpose is not solely fitting a mixture of NFs but integrating the mixture in a complex model:

– Dinh et al. [4]: they split the domain into multiple groups using a set identification function; their goal is to add expressiveness to NFs by including discrete variables;
– Izmailov et al. [5]: a normalizing flow uses a base distribution (usually a normal distribution) from which it evolves into a more adaptable distribution; the authors propose that the base distribution be a mixture of normal distributions; they use a single normalizing flow with a mixture of Gaussians as the base distribution, which is different from the approach investigated in this paper;
– Postels et al. [6]: here, the mixture of normalizing flows emerges in a VAE-like context: they define $p(z)$ (where z is the latent representation of an entire point cloud), $p(z|X)$ (where X is the entire point cloud), and $p(x|z)$ (where x is a point

in the point cloud); the last distribution, i.e. $p(x|z)$, is assumed to be a mixture of normalizing flows;

- Giaquinto and Banerjee [7]: the authors construct the mixture unusually, by using the gradient boosting technique; the resulting model can fit multi-modal distributions.

Standard views of the mixtures of NFs in the literature include contributions that aim at maximizing the likelihood of the data [8, Sect. 4.2] that are assumed to be generated by a mixture of normalizing flows. Although they have the same target, they differ in the training/fitting algorithms:

- Ciobanu [3]: they directly maximize the log-likelihood using a variation of the gradient ascent algorithm; for an input x, the log-likelihood is defined as:

$$\text{log-likelihood}(x) = \log\left(\sum_{j=1}^{k} \pi_j \text{NF}_j(x)\right),$$

where k is the number of components in the mixture, NF_j is the p.d.f. of the jth component in the mixture, $\pi_j \geq 0$ is the weight for NF_j, $\sum_{j=1}^{k} \pi_j = 1$;

- Pires and Figueiredo [9]: they maximize the likelihood indirectly by maximizing the evidence lower bound (ELBO) using a variation of the gradient ascent algorithm; for an input x, the ELBO is defined as:

$$\text{ELBO}(x) = \sum_{j=1}^{k} \text{encoder}(x)_j (\log(\text{NF}_j(x)) + \log \pi_j - \log(\text{encoder}(x)_j)),$$

where k is the number of components in the mixture, NF_j is the p.d.f. of the jth component in the mixture, $\pi_j > 0$ is the weight for NF_j, $\sum_{j=1}^{k} \pi_j = 1$, and *encoder* is a neural network that returns a vector of probabilities which will be interpreted as the following posterior probabilities:

$$\text{encoder}(x) = \begin{bmatrix} p(z = 1|x) \\ \vdots \\ p(z = k|x) \end{bmatrix},$$

where z represents the component/cluster index;

- Ng and Zammit-Mangion [10]: they maximize the likelihood indirectly by employing the expectation–maximization (EM) [8, Sect. 8.7.2] algorithm; the E step will consist in computing the posterior probabilities for each input x:

$$p(z = 1|x), \ldots, p(z = k|x) ;$$

after computing all these probabilities, i.e., an $n \times k$ matrix (n is the number of inputs), the M step updates the parameters of the mixture accordingly (see their paper [10] for details); the EM algorithm consists in applying a number of iterations where each iteration includes 2 steps: the E and M steps; the authors present the algorithm in two settings:

– (soft) EM: the one described earlier;
– hard EM: here, the E step turns the vector $\left[p(z = 1|x), \ldots, p(z = k|x) \right]$ into a one-hot vector by converting the maximum value to 1 and by setting the other values to 0; this version of the algorithm benefits the M step since the functions to be maximized at this step can be written with fewer terms than in the original/soft EM algorithm, hence more efficiency [10].

In this work, we implement the four algorithms used in the literature for fitting a mixture of normalizing flows:

– Ciobanu [3]: we call it GD (from *Gradient Descent*, not *Ascent*, since we minimize the negative log-likelihood); for this case, since the authors have already provided the code, we just restructured it or added extra features, e.g., a seed hyperparameter for reproducibility;
– Pires and Figueiredo [9]: we call it VarGD (from *Variational Gradient Descent*, not *Ascent*, since we minimize the negative ELBO); for this case, since there was no code available, we implemented the method from scratch;
– Ng and Zammit-Mangion [10]: we call the two algorithms SoftEM (the original or soft EM) and HardEM (the hard EM); again, since there was no code available, we implemented the method from scratch.

We mention that this comparison has not been made yet in the literature. Only in [3], the authors compare their results with the results in [9] just by taking the reported numerical results in [9] and without re-running the method in [9]; most likely, the data pre-processing, the parameter initialization, etc. have been distinct across the two approaches.

Since in general the original implementations are not available, we should be aware that our code may slightly differ from the original methods. For example, VarGD may be different since:

– we fitted the model with trainable/learnable mixture coefficients (π_j), although the authors suggest that these should be set to be uniform or be set by the modeler; we chose this option because in all the other algorithms (GD, SoftEM, HardEM) the numbers π_j were set to be trainable;
– we did include the part that divides "the inputs of the softmax layer in the variational posterior by a *temperature* value, T, which follows an exponential decay schedule during training" [9]; since the hyperparameter values regarding the temperature were not mentioned, we set our own values;

- in the experiments, we use a single set of hyperparameters in order to have the same configuration across the algorithms; hence, the hyperparameter values (and the reported results) will differ from the ones in the original papers.

Nevertheless, our implementation has the following positive features:

- the reproducibility of runs is taken into consideration by including the possibility of setting a seed;
- we have control over the initialization of the trainable parameters of the mixture; across the four algorithms, we set the same initialization;
- we include the possibility of having validation data, which is created by splitting the initial data into two parts;
- the order and the amount in which the data are given to the algorithm are controlled; the pre-processing, the shuffle order, and the batch size are the same across the four algorithms;
- one can set other finely detailed aspects across the algorithms: the same hyperparameters, the same optimizer, etc.
- one can save the best model via a checkpoint.

2.2 Off-the-Shelf Training Algorithms for Distributions

We implemented the code in TensorFlow [11] and TensorFlow Probability [12]. In TensorFlow Probability, there is a function called `experimental_fit` which can be used to train a probabilistic distribution, i.e., to estimate the parameters of the distribution that generated the given data. For example, to train a normal distribution, one can run the following Python code:

```
import tensorflow_probability as tfp
tfd = tfp.distributions
distr = tfd.Normal.experimental_fit([1.0,2.0,3.0])
print(distr.loc.numpy())
print(distr.scale.numpy())
```

and obtain the following distribution:

$$\mathcal{N}\left(\mu = \frac{1+2+3}{3} = 2, \sigma = \sqrt{\frac{(1-2)^2 + (2-2)^2 + (3-2)^2}{3}} = 0.8164966\right),$$

where the values of μ (mean) and σ (standard deviation) are the maximum likelihood estimates given the dataset $\{1.0, 2.0, 3.0\}$.

Unfortunately, this function is not available for all the distributions present in tfd since the fitting algorithm is specific to each distribution and must be implemented individually (in order to have an accurate and fast method). Fortunately, we implemented a variant of the gradient descent to optimize the likelihood in a general manner, regardless of the distribution, by leveraging and developing the ideas in [13, 14]. For example, to train a normal distribution, one can run the following Python code:

```
import tensorflow as tf
import tensorflow_probability as tfp
# from ... import DistributionTrainer
tfd = tfp.distributions
distr = tfd.Normal(
    loc=tf.Variable(0.0),
    scale=tf.Variable(1.0)
)
data = tf.data.Dataset.from_tensor_slices(
    [1.0, 2.0, 3.0]
).batch(3)
DistributionTrainer(distr).fit_via_gd(
    batched_dataset_train=data,
    optimizer=tf.optimizers.Adam(
        learning_rate=0.1
    ),
    epochs=500)
print(distr.trainable_variables)
```

and obtain again that $\mu = 2.0$, $\sigma = 0.8164966$ but without closed-form formulas and in an iterative manner. Furthermore, we can use a mixture of Gaussians on which we can apply not only GD, but also SoftEM, HardEM, and VarGD; in the end, after training, we can obtain a hard clustering and a soft clustering on a given dataset through the corresponding predictor classes. For more code examples, see the previously mentioned GitHub page.

3 Methodology and Experiments

We applied the four algorithms, i.e., GD, SoftEM, HardEM, and VarGD, on toy datasets with two input columns (*pinwheel*, *moons*, *circles*, and *two bananas*, which are also present in [3]) and on image datasets (MNIST [15] with 70,000 images and MNIST5 with 35,735 images, which is the MNIST dataset but only with the images representing the digits from 0 to 4) with 784 input columns. Each full dataset was split into 80% training data and 20% validation data. The validation data were used throughout the epochs/iterations (even in the M steps) in order to enable early stopping with a patience parameter of 5 steps. In the end, the loss and the metrics were reported on the whole dataset (100% = 80% training + 20% validation). Moreover, the data were also pre-processed by Z-standardizing each input column; this standard scaler has been fit only on the training data and applied on the training data, on the validation data, and, in the end, on the whole data.

The hyperparameter values were set to the following:

- learning rate: 0.001 for toy data and 0.0001 for image data;
- number of epochs/iterations:

  - for toy data: 400 epochs for GD and VarGD; 40 iterations for SoftEM and HardEM (SoftEM and HardEM will carry out 10 epochs at each M step since the M step is implemented in a gradient descent fashion; from this perspective, there are $40 \times 10 = 400$ epochs);
  - for image data: 20 epochs for GD and VarGD; 5 iterations for SoftEM and HardEM (SoftEM and HardEM will carry out 4 epochs at each M step since the M step is implemented in a gradient descent fashion; from this perspective, there are $5 \times 4 = 20$ epochs);

- batch size: 512;
- the π_j values are trainable and are initialized with $\frac{1}{k}$;
- a normalizing flow, i.e., a mixture component, is represented by two chained masked autoregressive flows (MAF) [16], i.e., two MAF blocks:

  - the base distribution is a multivariate normal distribution:

    each element of its mean is set by sampling a standard univariate normal distribution;
    the covariance matrix is the identity matrix.

  - each MAF has a corresponding masked autoencoder (MADE) [17] with one layer with 8 hidden units and with the ReLU activation function;

- hyperparameters specific to VarGD:

  - the encoder has 2 layers, each with 10 hidden units and with the ReLU activation function;
  - exponential decay hyperparameters for the temperature[1]:

    decay steps: 1;
    decay rate: 0.96;
    initial rate: $\dfrac{1}{\left(\frac{\text{decay rate}}{\text{decay steps}}\right)^{\text{number of epochs}}}$, such that the temperature at the final epoch is (approximately) 1;

---

[1] We used https://www.tensorflow.org/api_docs/python/tf/keras/optimizers/schedules/Exponential Decay.

- hyperparameters specific to SoftEM and HardEM: for each M step, there are 10 epochs for toy data and 4 epochs for image data; as for the loss, we report the negative log-likelihood.

For GD, SoftEM, and HardEM the loss is considered to be the negative log-likelihood. For VarGD, the loss is considered to be the negative ELBO. Each algorithm was run on each dataset 5 times, using the seeds from 0 to 4. From the 5 runs, the model with the minimal loss was picked. For the selected model, we report the training time,[2] the negative log-likelihood (NLL), the negative ELBO (only for VarGD), and the following clustering metrics: purity [8, Sect. 21.1.1.1], adjusted rand index (ARI) [18], normalized mutual information (NMI) [19], unsupervised cluster accuracy (ACC) [20].

3.1 Results

We showcase the numerical results in Table 1. According to the time spent on executing a run, the algorithms can be ordered as follows: GD (fastest), VarGD, HardEM, and SoftEM (slowest). Note that we have set the hyperparameters such that the times are comparable (e.g., 400 epochs for GD and 40 iterations with 10 M step epochs each for SoftEM). One should also remember that the image datasets are executed with far fewer epochs/iterations than the toy data counterparts. These results are somewhat expected: the EM algorithms have the E step that should go through the whole dataset, and this becomes slow. These results confirm the mention in [10] which states that HardEM should be faster than SoftEM.

If for the image datasets, GD and VarGD perform the best, for the toy datasets, there are instances where SoftEM and HardEM obtain very satisfactory metrics (see *moons*, *pinwheel*, and *two bananas* datasets). The most challenging dataset seems to be *circles*: every algorithm predicted just one cluster instead of two. Although the easiest dataset is *two bananas*, VarGD obtains just a cluster instead of two; although not reported in the table, for seeds 1 and 2, VarGD does obtain the two expected clusters; this is an example where the minimal loss does not necessarily correspond to a superior clustering, and this happens to all the presented algorithms.

Regarding the losses, for VarGD, as a quality check, the values in the table satisfy the relation $-\text{ELBO} \geq \text{NLL}$, which is mathematically correct.

To showcase the non-Gaussian (i.e., non-ellipsoidal) flexibility of the learned mixture of NFs, we display diverse plots for the toy datasets in Fig. 1.

Further observations include:

- if we had used just one MAF block instead of two, then the density plots would have shown lesser (non-ellipsoidal) flexibility;

---

[2] The machine specifications on which we ran the experiments are as follows: Windows 10 OS, Lenovo Legion Y720 Laptop, NVIDIA GeForce GTX 1060 6 GB GPU, Intel®Core™ i7-7700HQ CPU @ 2.80Hz, 32.0 GB RAM.

Table 1 Numerical results of the 4 algorithms on 6 datasets. We display the run index (given by the seed), the running time in minutes, the four clustering metrics (purity, ARI, NMI, ACC), and the loss, which can be either NLL or −ELBO. As expected, −ELBO is available only for VarGD. For each dataset and algorithm, the best metric is marked in bold. The rows marked with (*) are displayed in Fig. 1

	Dataset	Algorithm	Seed	Time (minutes)↓	Purity↑	ARI↑	NMI↑	ACC↑	Loss↓	NLL↓	−ELBO↓
(*)	Circles	HardEM	2	5.591	0.501	0.000	0.000	0.501	2.047	2.047	NaN
	Circles	SoftEM	2	6.045	0.501	0.000	0.000	0.501	2.017	2.017	NaN
	Circles	GD	2	**3.286**	**0.502**	0.000	**0.000**	**0.502**	**2.015**	**2.015**	NaN
	Circles	VarGD	4	4.654	0.500	**0.000**	0.000	0.500	2.702	2.701	2.702
(*)	Moons	HardEM	2	5.548	0.912	0.679	0.642	0.912	2.261	2.261	NaN
(*)	Moons	SoftEM	2	6.048	0.955	0.828	0.754	0.955	**2.177**	**2.177**	NaN
(*)	Moons	GD	2	**3.366**	**0.956**	**0.832**	**0.757**	**0.956**	2.193	2.193	NaN
(*)	Moons	VarGD	4	4.742	0.554	0.011	0.033	0.554	2.534	2.521	2.534
	Pinwheel	HardEM	2	12.904	0.570	0.540	0.689	0.570	1.901	1.901	NaN
(*)	Pinwheel	SoftEM	1	15.183	**0.751**	**0.613**	**0.711**	**0.723**	1.606	1.606	NaN
	Pinwheel	GD	2	**5.633**	0.593	0.550	0.670	0.593	**1.435**	**1.435**	NaN
(*)	Pinwheel	VarGD	4	7.882	0.592	0.458	0.662	0.592	2.730	2.721	2.730
(*)	Two bananas	HardEM	2	7.562	**1.000**	**1.000**	**1.000**	**1.000**	**−0.289**	**−0.289**	NaN
	Two bananas	SoftEM	3	8.784	1.000	1.000	1.000	1.000	−0.288	−0.288	NaN
	Two bananas	GD	3	**5.592**	1.000	1.000	1.000	1.000	−0.288	−0.288	NaN
	Two bananas	VarGD	4	7.035	0.500	0.000	0.000	0.500	1.246	1.243	1.246

(continued)

Table 1 (continued)

Dataset	Algorithm	Seed	Time (minutes)↓	Purity↑	ARI↑	NMI↑	ACC↑	Loss↓	NLL↓	−ELBO↓
MNIST	HardEM	2	15.969	0.136	0.001	0.026	0.130	136.639	136.639	NaN
MNIST	SoftEM	2	NaN	0.294	0.110	0.210	0.293	195.376	195.376	NaN
MNIST	GD	2	NaN	**0.327**	0.142	0.234	**0.317**	**−519.176**	**−519.176**	NaN
MNIST	VarGD	3	**10.490**	0.304	**0.185**	**0.287**	0.304	653.008	464.328	653.008
MNIST5	HardEM	0	5.277	0.240	0.002	0.019	0.237	507.883	507.883	NaN
MNIST5	SoftEM	0	8.105	0.320	0.031	0.103	0.306	374.771	374.771	NaN
MNIST5	GD	0	3.592	0.388	0.125	0.167	0.387	**−15.289**	**−15.289**	NaN
MNIST5	VarGD	0	**3.218**	**0.415**	**0.170**	**0.206**	**0.392**	931.909	813.006	931.909

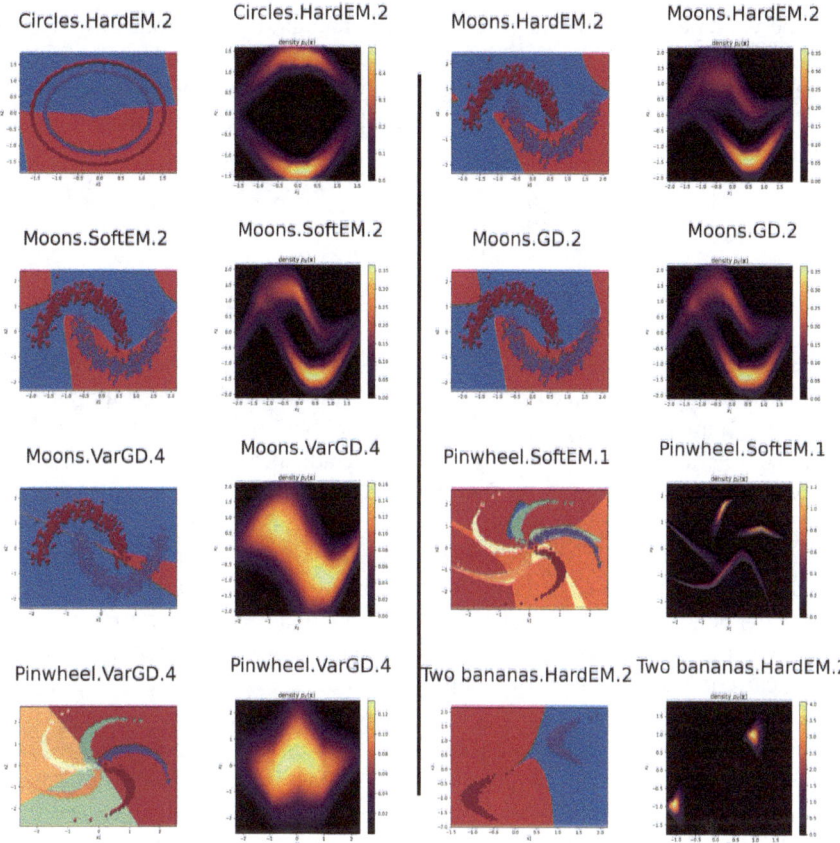

Fig. 1 Plots describing the decision boundaries between clusters and the probability density functions obtained by some fitted mixtures of normalizing flows. Above each figure, we mention the dataset, the algorithm, and the seed corresponding to that specific run. Those cases are marked in Table 1 with a star (*). In the decision boundary plots, the color of the points represents the real cluster and the color of the background represents the predicted cluster. The density plots were created with the code from [21]; high values correspond to bright pixels.

- the loss can become NaN/infinite or all the points can be assigned to a single cluster in the end; we noticed that this highly depends on the learning rate, on how the data are pre-processed and shuffled, on how the base distributions are initialized, and other hyperparameters; for the image data, out of 4 (algorithms) × 5 (seeds) × 2 datasets (MNIST5, MNIST) = 40 runs, 22 ended with NaN/infinite (training and/or validation) losses, from which 11 provided a saved trained model from earlier epochs (across the epochs, we save only the best models in terms of the validation loss) and we were able to compute the mentioned metrics; as the authors in [9] specify, the variational mixture of NFs should have its encoder similar in complexity with the mixture components; from this perspective, VarGD has a disadvantage compared to GD, SoftEM, and HardEM since there are more hyperparameters to set.

4 Conclusion and Future Work

The mixtures of normalizing flows are generative models that can be used in clustering. They can be fitted/trained (in a maximum likelihood sense) using different algorithms. We implement three methods described in the literature (hard expectation–maximization, soft expectation–maximization, gradient descent applied on a loss representing the negative evidence lower bound) and refine a public implementation (gradient descent applied on a loss representing the negative log-likelihood). We upload the code on GitHub to be publicly available so that anyone can experiment with the methods described in the literature with ease. Apart from this, other advantages of our code include the fact that the code is not tightly coupled with the mixture of normalizing flows (it can be run with other probabilistic distributions), the fact that we can compare the algorithms employing the same hyperparameters, parameter initialization, etc. In terms of the results, SoftEM is the slowest algorithm and GD is the fastest, but in terms of the clustering performance and loss minimization, each algorithm obtains the best results at least once; on image datasets, the metrics suggest that GD and VarGD achieve better results.

As for future work, we could explore the semi-supervised case. Regarding the training of the mixture of normalizing flows, we may set other hyperparameter values (more epochs, other neural network architectures, other normalizing flows—not just masked autoregressive flows—, etc.). In the end, the four algorithms can be applied to mixtures of other distributions; the gradient descent can be applied to other non-mixture distributions, as well.

References

1. Mitchell T (2017) Generative and discriminative classifiers: Naive Bayes and logistic regression. In: Machine learning. McGraw-Hill Science/Engineering/Math; (March 1, 1997). https://bit.ly/39Ueb4o
2. Murphy KP (2023) Probabilistic machine learning: Advanced topics. MIT Press. http://probml.github.io/book2
3. Ciobanu S (2021) Mixtures of normalizing flows. In Shi Y, Hu G, Yuan Q, Goto T (eds) Proceedings of ISCA 34th international conference on computer applications in industry and engineering, EPiC series in computing, vol 79. EasyChair, pp 82–90. https://doi.org/10.29007/nq4f, https://easychair.org/publications/paper/Scnv

4. Dinh L, Sohl-Dickstein J, Larochelle H, Pascanu R (2019) A RAD approach to deep mixture models. arXiv:1903.07714
5. Izmailov P, Kirichenko P, Finzi M, Wilson AG (2020) Semi-supervised learning with normalizing flows. In: International conference on machine learning, pp 4615–4630. PMLR (2020)
6. Postels J, Liu M, Spezialetti R, Van Gool L, Tombari F (2021) Go with the flows: Mixtures of normalizing flows for point cloud generation and reconstruction. In: 2021 international conference on 3D vision (3DV). IEEE, pp 1249–1258
7. Giaquinto R, Banerjee A (2020) Gradient boosted normalizing flows. Adv Neural Inf Process Syst 33:22104–22117
8. Murphy KP (2022) Probabilistic machine learning: an introduction. MIT Press. http://probml.ai
9. Pires GG, Figueiredo MA (2020) Variational mixture of normalizing flows. arXiv:2009.00585
10. Ng TLJ, Zammit-Mangion A (2023) Mixture modeling with normalizing flows for spherical density estimation. arXiv:2301.06404
11. Abadi M (2016) TensorFlow: learning functions at scale. In: Proceedings of the 21st ACM SIGPLAN international conference on functional programming, pp 1–1
12. Dillon JV, Langmore I, Tran D, Brevdo E, Vasudevan S, Moore D, Patton B, Alemi A, Hoffman M, Saurous RA (2017) TensorFlow distributions. arXiv:1711.10604
13. Kamperis S, Trainable probability distributions with TensorFlow. https://ekamperi.github. io/mathematics/2020/12/26/tensorflow-trainable-probability-distributions.html. Accessed 10 Feb 2023
14. TensorFlow Probability webpage on tfp.bijectors.AutoregressiveNetwork. https://www. tensorflow.org/probability/api_docs/python/tfp/bijectors/AutoregressiveNetwork. Accessed 10 Feb 2023
15. LeCun Y, Cortes C, Burges C (2010) MNIST handwritten digit database. ATT Labs
16. Papamakarios G, Pavlakou T, Murray I (2017) Masked autoregressive flow for density estimation. Adv Neural Inf Process Syst **30**
17. Germain M, Gregor K, Murray I, Larochelle H (2015) MADE: Masked autoencoder for distribution estimation. In: International conference on machine learning. PMLR, pp 881–889
18. Hubert L, Arabie P (1985) Comparing partitions. J Clfication 2:193–218
19. Estévez PA, Tesmer M, Perez CA, Zurada JM (2009) Normalized mutual information feature selection. IEEE Trans Neural Netw 20(2):189–201
20. Min E, Guo X, Liu Q, Zhang G, Cui J, Long J (2018) A survey of clustering with deep learning: from the perspective of network architecture. IEEE Access 6:39501–39514
21. Tiao LC (2018) Building probability distributions with the tensorflow probability bijector API. tiao.io. https://tiao.io/post/building-probability-distributions-with-tensorflow-probability-bijector-api/

Facial Expression Based Music Recommendation System Using Deep Learning

Aman Singh, Richa Sharma, Mahima Shanker Pandey, Sonal Asthana, Gitanjali, and Ankita Vishwakarma

Abstract A person's mood can be changed by music, which impacts both the body and emotions, so our project aims to describe a method for music recommendation based on facial expressions. Many music applications are based on a user's browsing history but this work presents an idea on the music recommendation based on users' facial expressions that give outstanding results. It has got a unique ability to lift one's mood. The work's overall concept is recognizing facial expressions and efficiently recommending songs. The proposed model will be both time and cost-efficient. This yields better performance and computational time accuracy and reduces the design cost.

Keywords Deep learning · Convolution neural network · Facial expression · Classifier

1 Introduction

Human emotion plays an important role in telling what is going on in one's life [1]. They use emotion to express their feeling to each other. The best possible way in which people tend to analyze or understand other person's emotions or feelings is to look at their facial expressions. Earlier, the most used form of expression analysis by humankind was during face-to-face interaction.

Music is a flair that can change the mood of a person; it can affect both the emotion and the body. There are different types of music, faster music can make us feel more optimistic about life, whereas a slower tempo can quiet our mind and relax our muscles, making us feel soothed while releasing the stress of the day. Music is very effective for relaxation and stress management.

A. Singh · R. Sharma · S. Asthana · Gitanjali · A. Vishwakarma
Department of Information Technology, Babu Banarasi Das Institute of Technology and Management, Lucknow, U.P., India

M. S. Pandey (✉)
Computer Science and Applications, Sharda University, Greater Noida, U.P., India
e-mail: mahimashanker@gmail.com

© The Author(s), under exclusive license to Springer Nature Singapore Pte Ltd. 2024
S. Namasudra et al. (eds.), *Data Science and Network Engineering*, Lecture Notes in Networks and Systems 791, https://doi.org/10.1007/978-981-99-6755-1_3

Considering the above two aspects—emotion and music, the work is based onto lighten the user's mood by playing songs that match the user's mood by capturing their facial expression [2]. The software takes the user's image to extract information from a target human being's face using image segmentation and image processing algorithms to identify the emotion the person is attempting to convey. Mood modification may occasionally be useful in overcoming depressive and sad circumstances. Many health concerns can be avoided with the use of expression analysis, and actions can be done to improve a user's mood.

The main contributions of the paper are listed as follows:

(i)   The present work proposes to recommend music based on the user's facial expression.
(ii) A deep learning approach is proposed to capture information from the user's image using image segmentation and image processing algorithms.
(iii) The proposed algorithm identifies the emotion associated with the user's image.
(iv) Experimental results suggest that the proposed algorithm is efficient in terms of computational time and cost.

The paper is structured as follows. Section 2 presents the research work carried out in the domain of music recommendation systems. The methodology adopted in the present work is described in Sect. 3. The design and implementation of the proposed methodology are presented in Sect. 4. Section 5 discusses the results and the conclusion of the paper is provided in Sect. 6.

2 Literature Review

The main purpose of the work is for the betterment of human mood and reduces depression and human anxiety. Ulleri et al. [3] have also focused on how this system can be helpful during difficult times like COVID-19, when people were compelled to stay inside of their four walls, which caused conditions like mood disorders, depression, anxiety, etc. Music has been proven to be the best and fastest way to change the human mood. The model [4] uses Random Forest Classifier and XG Boost algorithm to identify the emotion of songs considering features like instrumental, energy, acoustic, liveliness, etc.

In the mood-based music system [5], a web camera has been used to scan users' facial expressions. The main component used for the software is the OpenCV library by plotting the vector points of the HAAR Cascading algorithm. It emphasizes eliminating background noise and detecting a face in dim light or at night time with proper facial properties for future analysis. The basic approach of the system is divided into 4 major phases.

2.1 Creating a Model for Detecting Facial Expressions → API Usage → Classification of Song Mood → Development of a Playlist

The report [2] outlined three essential steps, after the face has been recognized by HAAR Cascading Classifier in the first step, the CNN classifier is utilized to determine expression. In the final phase, the process of creating a playlist is completed by computing the song's BPM (beat per minute). The authors of [6] used 6 CNN BLOCKS. The dataset was obtained from Kaggle [7, 8] and divided into 3 sets. As an input, the front view of the facial image is used to extract the facial features. They used the Sequential method for the development of a model, which comprises of conv2d layer, max–pooling, and Dense layer. Based on the dataset, the filters are arranged in descending order in a layer.

Uma et al. [9] have compared the proposed system with other music systems. Guidel et al. [10] considered that facial expressions can be very helpful in easily observing one's state of mind and current emotions. This system was developed by taking basic emotions (happiness, sadness, anger, excitement, surprise, disgust, fear, and neutrality) into consideration.

3 Methodology

Our proposed system has been divided into two phases (Fig. 1).

- Creation of a model for detecting facial expressions.
- Recommend a song based on user expression.

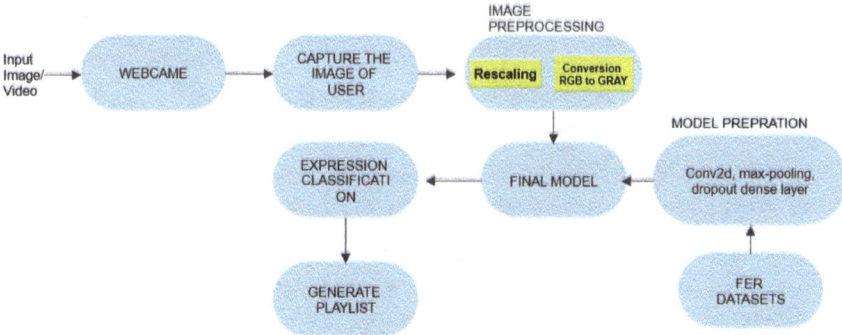

Fig. 1 Phases of music recommendation system

3.1 Dataset for Experiment

For training, the images are taken from "https://www.kaggle.com/datasets/sudar-shanvaidya/corrective-reannotation-of-fer-ck-kdef" [7]. For testing, the images are taken from "https://www.kaggle.com/datasets/amansinghfty/mytestingdataset" [8]. Only the front view and clear images without any distortion in pixels have been considered.

3.2 Creation of a Model for Detecting Facial Expressions

3.2.1 Face Detection

In our system first, the frontal face of a human is detected using a camera with the help of an OpenCV module. Numerous approaches to face detection have been discovered, but the HAAR cascade classifier works well compared to others. HAAR cascade classifier also finds the number of faces in the window with their respective coordinate location. The four phases of the HAAR Cascade algorithm are as follows:

Selection of HAAR Features → Integral Image Creation → Adaboost training → Implementing Cascading Classifiers

Initially, to train the classifier, the system first requires a large number of both positive and negative face images. Next, the features are drawn from it. In the HAAR classifier algorithm, the first step is to collect the HAAR features of images like edges or lines in the frame. Edge characteristics on the face, such as the teeth, nose, and brows, are visible. The HAAR approach is then used to find the following features in the image. The ratio of these discovered traits serves as the basis for emotion recognition (Fig. 2).

This work consists of the following steps–

  i. Input image/video is continuously captured using a camera.
 ii. The video is pre-processed in two steps as follows:

    a. Video/image is rescaled to 144 * 144 pixels.
    b. Then the above-scaled video is converted from RGB to gray.

iii. Edge detection of pre-processed images is carried out.
 iv. The HAAR Cascade classifier detects the frontal face in the video and creates a rectangle over each detected face.
  v. This image is passed to our trained model for predicting expression.
 vi. Recommend the song based on facial expression.

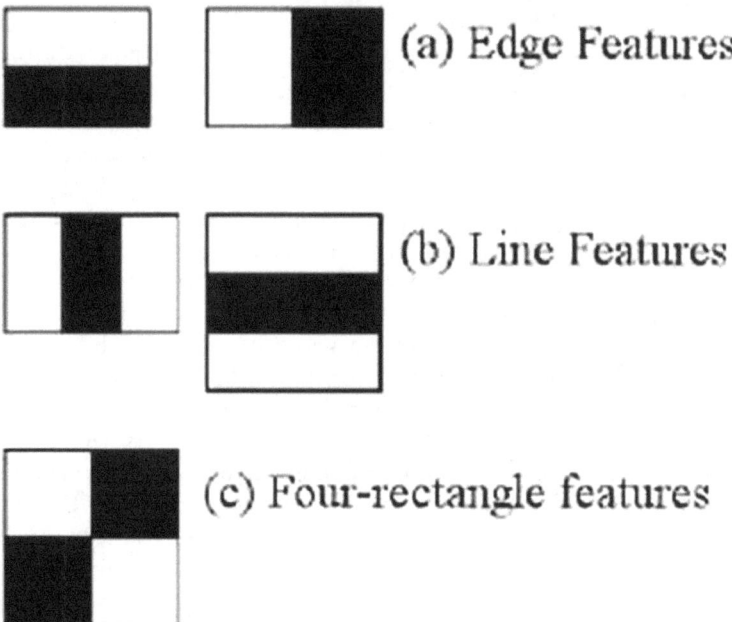

(a) Edge Features

(b) Line Features

(c) Four-rectangle features

Fig. 2 Haar cascade features, image by OpenCV

3.2.2 Expression Detection

Data preparation plays an important role in deep learning. It means that the images should be pre-processed. There are several reasons why images are shrunk, but one of them is crucial to our work i.e. image resizing, it must be done to make the resolution consistent for the model. To perform the image resolution consistency, resize all images having dimensions 144 * 144.

The color of an image is stored in an RGB format so it is a very difficult task to process it normally so we are converting all the images into grayscale. This will reduce the dimension of the images.

3.3 Recommending a Song Based on the User's Expression

After applying machine learning and deep learning to recognize the expression, the software will suggest a song depending on the facial expression. The song will be played appropriately in addition to navigating to the playlist.

4 Design and Implementation

An important task for our system is expression detection through facial detection. The expression detection task has been accomplished with CNN's assistance. The CNN [11] receives input as a 144px × 144px face image from the Haar Cascade Classifier. The eight categories that CNN will use to categorize the image's expression [12–15] are **"anger"**, **"contempt"**, **"disgust"**, **"fear"**, **"happy"**, **"neutrality"**, **"sadness"**, and **"surprise"**. CNN is a brilliant approach to recognizing images using deep learning. Images are nothing just a group of pixels to form a pattern. That pattern is utilized by CNN to identify the images. Eight CNN blocks have been used in this work to create the CNN for expression detection. Each block is made up of a Batch Normalization layer and a Conv2D layer with a kernel matrix of either 3 * 3 or 5 * 5. To avoid overfitting issues, one MaxPolling2D layer along stride 2 * 2 and a dropout layer after two consecutive blocks are added. The Elu function will serve as the activation function which has the properties of leaky ReLu and ReLu. Two blocks of the Conv2d layer receive the output of the max pooling layer for parallel processing. Concatenated output from the parallel conv2d block is sent to another block. At last, the Dense layer and the flattened layer are used. Our model achieves 84.8 percent validation accuracy and training accuracy 92% and trained around 35,971,685 parameters (Figs. 3 and 4).

A confusion matrix is used to evaluate the performance of a classification (Anger, Contempt, Disgust, Fear, Sadness, Surprise, Neutrality) model as given in Table 1. It summarizes the predictions made by the proposed model on a set of test data (936 images) and compares them to the actual values.

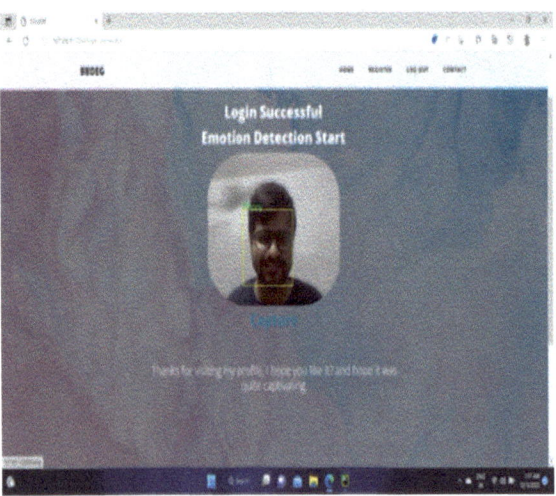

Fig. 3 Capturing the input image

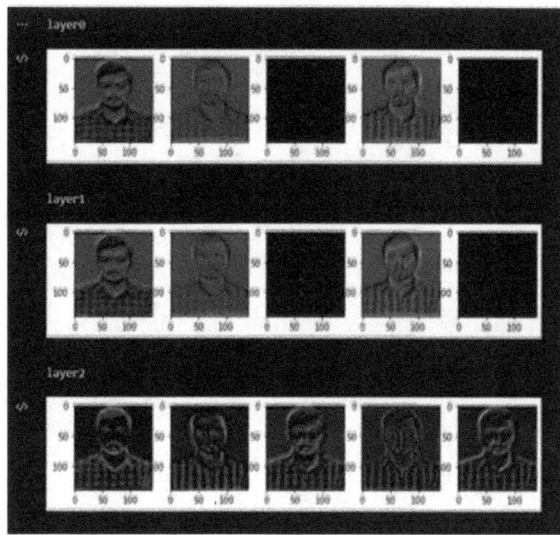

Fig. 4 Internal layers working

5   Result

To achieve the primary goal of our proposed system, we have organized the results into the following categories:

 i.   To be able to see the face clearly without any interference.
 ii. Accurately separate the feelings and group them in descending order under the eighth expression.
iii. Giving the consumer appropriate song suggestions to better relate to their mood while taking into account the dominating expression.

Our suggested system's correctness is measured by a performance metric. Figure 5 shows the calculated accuracy percentage (Calculated/Expected) for each of the eight expressions, with an average calculated accuracy of 84.84% in 30 epochs.

Table 1 Confusion matrix

	Anger	Contempt	Disgust	Fear	Happy	Sadness	Surprise	Neutrality	Classification overall	Precision (%)
Anger	22	3	0	0	0	0	0	1	26	84.62
Contempt	3	6	0	0	0	0	0	1	10	60
Disgust	5	0	39	0	0	0	0	0	44	88.64
Fear	0	0	3	22	5	0	0	0	30	73.33
Happy	0	0	0	0	35	0	2	0	37	94.60
Sadness	0	0	0	0	1	20	7	0	30	66.67
Surprise	1	0	0	0	0	0	60	2	62	96.77
Neutrality	1	1	0	0	1	2	0	20	25	80%
Recall (%)	68.75	60	92.86	100	83.33	90.91	86.96	80		
Overall accuracy (OA)	84.84%									

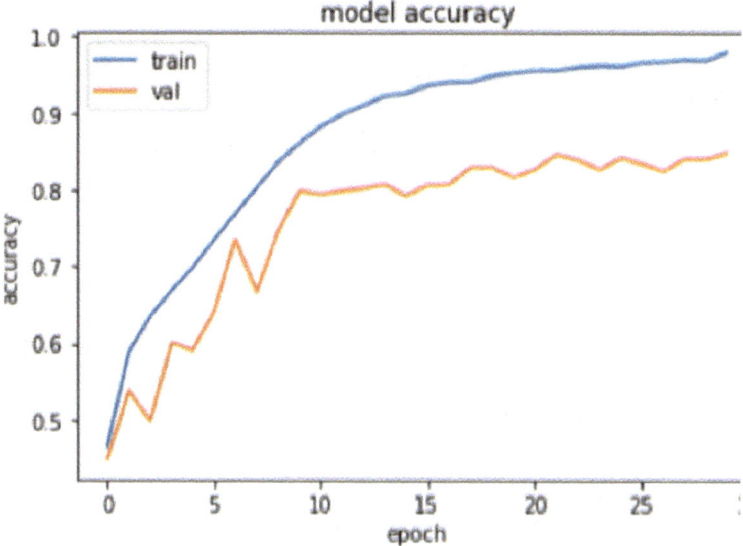

Fig. 5 Model accuracy graph (comparison between training and validation accuracy)

6 Conclusion

The music expresses a person's feelings and mood. The work detects the following different categories **"anger"**, **"contempt"**, **"disgust"**, **"fear"**, **"happy"**, **"neutrality"**, **"sadness"**, and **"surprise"** with an accuracy of 84.8%. When the model detects an expression, the music player will recommend appropriate songs related to the expression. As a result of music recommendations, users may choose music in a more participatory and straightforward manner. All are aware that every person resorts to music, in almost every situation, and automatic detection of the user's mood is an advancement to the currently available technologies. Music recommendation systems with Facial expression detection require the use of certain deep learning and machine learning models. So, music not only heals a person's mood but also greatly impacts one's mood.

References

1. Agrawal A, Mittal N (2020) Using CNN for facial expression recognition study of the effects of kernel size and several filters on accuracy. Visual Comput 36(2)
2. Kevin P, Rajeev K (2021) Song playlist generator system based on facial expression and song mood. In: 2021 international conference on artificial intelligence and machine vision (AIMV)
3. Ulleri P, Prakash SH, Kiran (2021) Music recommendation system based on user facial expression. In: 2021 12th international conference on computing communication and networking technologies (ICCCNT)

4. Matilda Florence S, Uma M. Emotional detection and music recommendation system based on user. IOP Conf Ser Mater Sci Eng
5. Mali AS, Kenjale AA, Ghatage PM, Deshpande AG (2018) Mood-based music system. Int J Sci Res Comput Sci Eng 6(3):27–30
6. Vicky K, Sumit K et al (2022) Music recommendation based on user mood. In: 2022 9th international conference on computing for sustainable global development (INDIACom)
7. https://www.kaggle.com/datasets/sudarshanvaidya/corrective-reannotation-of-fer-ck-kdef
8. https://www.kaggle.com/datasets/amansinghfty/mytestingdataset
9. Matilda F, Uma M (2020) Emotional detection and music recommendation system based on user facial expressions. IOP Conf Ser Mater Sci Eng
10. Guidel A, Sapkota B, Sapkota K (2020) Music recommendation by facial analysis. http://engineeringsarokar.com
11. Shelhamer JE, Donahue J, Karayev S, Long J, Girshick R, Guadarrama S, Darrell T (2014) Caffe: Convolutional architecture for fast feature embedding. In: Proceedings of the international conference on multimedia, pp 675–678
12. Madhuri A, Deepali M et al (2021) Music recommendation based on face emotion recognition. J Informatics Electr Electron Eng (JIEEE)
13. Matilda S, Florence, Uma M (2020) Emotional detection and music recommendation system based on user facial expression. IOP Conf Ser Mater Sci Eng
14. Vinay P, Raj P, Bhargav KS et al (2021) Facial expression based music recommendation system. Int J Adv Res Comput Commun Eng
15. Illiana Azizan K Fatimah (2018) Facial emotion recognition. In: International conference on sustainable engineering, technology and management 2018 (ICSETM-2018)

Exploring Time Series Analysis Techniques for Sales Forecasting

Murugan Arunkumar, **Sambandam Palaniappan**, **R. Sujithra**, and **S. VijayPrakash**

Abstract Sales forecasting is a decisive task for businesses, as it enables them to make important decisions about production, inventory, and marketing strategies. Time series analysis is a tackle for sales forecasting, as it allows us to analyze and model data based on time-dependent patterns. In this paper, we explore different time series analysis techniques and their application to sales forecasting. We use a real-world sales dataset (retail) to demonstrate the use of various time series techniques such as decomposition, auto-correlation, and lag features. This report presents a solution for a case study in which we forecast the sales of retail stores. It supports strategic decisions on three levels: the featuring of data, decomposing the data, and applying the models. We also discuss the significance of feature engineering in time series analysis and demonstrate the time series features such as lag, date time, and windowing (rolling means). Then, we compare the performance of different time series models, such as naive (persistence), Moving Average, ARIMA, and SARIMAX. We conclude that time series analysis techniques are used correctly and can handle powerful tools for businesses to make accurate sales forecasts and make informed decisions.

Keywords Sales forecasting · Machine learning models · Time series · Auto-correlation · Moving average · Decomposition series · SARIMAX

M. Arunkumar (✉) · S. Palaniappan · R. Sujithra · S. VijayPrakash
Department of Artificial Intelligence and Data Science, KCG College of Technology, Anna University, Chennai, India
e-mail: 20AD06@kcgcollege.com

S. Palaniappan
e-mail: palaniappan.cse@kcgcollege.com

© The Author(s), under exclusive license to Springer Nature Singapore Pte Ltd. 2024   41
S. Namasudra et al. (eds.), *Data Science and Network Engineering*, Lecture Notes in Networks and Systems 791, https://doi.org/10.1007/978-981-99-6755-1_4

1 Introduction

Accurate sales forecasting is a vital aspect of successful businesses as it enables them to make informed decisions related to production, inventory, and marketing strategies. Time series analysis is a powerful tool for predicting sales, as it enables businesses to analyze and model data based on time-dependent patterns. This paper presents a comprehensive study of retail sales forecasting using a variety of time series analysis techniques, including decomposition, auto-correlation, and lag features, using a real-world retail sales dataset. Time series decomposition is a technique used to separate a time series into its parts, which are the underlying trend, seasonality, and white noise (or error) components. The decomposition usually uses additive or multiplicative. The decomposition process is normally performed using statistical methods such as moving averages or exponential smoothing. By decomposing the time series, it becomes easier to identify the different components and their respective contributions to the overall pattern of the series.

Building an accurate retail sales forecasting model can be challenging due to the complexity of the data, including the presence of outliers, missing values, and external factors such as economic indicators and holiday events. To address these challenges, this paper explores various techniques for preprocessing the data, handling missing values, and incorporating external factors. Additionally, the importance of feature engineering in time series analysis is discussed, which involves identifying and extracting meaningful features from the raw data to improve model accuracy. To evaluate the performance of different machine learning models, regression and time series models are compared using various evaluation metrics such as mean absolute error (MAE) and root-mean-square error (RMSE) to identify the best-performing model. Furthermore, time series analysis techniques such as lag plots, auto-correlation plots, and decomposition are applied to gain a better understanding of the underlying patterns in the data and detect sales trends during specific periods. This paper demonstrates that time series analysis techniques can be a valuable tool for businesses to make accurate sales forecasts and informed decisions. By utilizing appropriate data preprocessing techniques, feature engineering, and selecting the best-performing model, businesses can improve their forecasting accuracy and make better-informed decisions.

With the rise of e-commerce and online shopping, retail businesses face a new set of challenges such as increased competition, changing consumer behavior, and the need for real-time decision-making. In this context, accurate sales forecasting becomes even more critical to optimize inventory levels, reduce costs, and increase profitability. Therefore, this paper aims to provide a comprehensive analysis of retail sales forecasting using time series analysis techniques to help businesses overcome these challenges and succeed in today's dynamic retail environment.

Overall, the contributions of this paper are as follows:

- A detailed analysis of the seasonal patterns and trends in the retail sales data, which provides insights into the key drivers of sales in this sector.
- A comprehensive comparison of different time series analysis techniques, including decomposition, auto-correlation, and lag features, which identifies the most effective methods for forecasting retail sales.
- An evaluation of the impact of external factors such as economic indicators and holiday events on retail sales, which provides a more nuanced understanding of the drivers of sales in this sector.
- A discussion of the importance of feature engineering in time series analysis, which highlights the need for businesses to extract meaningful features from their data to improve forecasting accuracy.
- A demonstration of the value of time series analysis for businesses, which highlights the potential benefits of using these techniques to make informed decisions about inventory, production, and marketing strategies.

2 Literature Survey

In the process of developing an accurate sales forecasting model, we conducted a comprehensive literature survey to explore existing research and identify potential areas of improvement.

Lee [1] proposed a hybrid model based on deep learning and ARIMA to forecast electricity demand but noted that the model was not the best fit for the data and affected the accuracy of the results. The authors aim to improve the accuracy of demand forecasting by using a combination of techniques. Specifically, they use a convolutional neural network (CNN) to capture complex patterns in the data, and an ARIMA model to capture the time series structure of the data. This paper was inspired by highlighting the potential benefits of combining different forecasting techniques to improve accuracy. Additionally, the study can also serve as a benchmark for comparing the accuracy of our model with other commonly used forecasting techniques. Ayub [2] developed a reinforcement learning approach for demand forecasting and inventory optimization. The authors used a Q-learning algorithm to learn the optimal demand forecast and inventory level based on past demand data and inventory levels. Moreover, the authors did not explore the use of other machine learning techniques or hybrid models that could potentially improve the accuracy of the forecast. We explore the use of real-time data feeds to provide up-to-date information on sales trends and enable more agile decision-making. By considering these aspects, we aim to develop a more accurate and efficient sales forecasting model for demand forecasting and inventory optimization. Wang [3] developed a model based on a random forest algorithm for demand forecasting in the manufacturing industry. The authors highlight the limitations of traditional methods such as linear regression and time series analysis and argue that these methods fail to capture the complex nonlinear relationships that exist between demand and various other factors. They did not discuss how the model handles outliers in the data, which can significantly affect performance. they demonstrate the effectiveness of the random forest model in accurately predicting demand for various manufacturing products. This paper is inspiring as it presents a different approach to demand forecasting by using a non-parametric machine learning technique, which can potentially provide better accuracy than traditional methods. We consider these external factors and explore the use of various advanced machine learning techniques to potentially improve the accuracy of demand forecasting in the manufacturing industry. Islam [4] developed a seasonal ARIMA model for demand forecasting in the manufacturing industry. The authors aimed to develop a reliable demand forecasting system that can assist decision-makers in the manufacturing industry. It does not provide a detailed explanation of the feature selection process used to select variables for the model and does not consider the impact of external factors such as seasonality, economic conditions, and marketing campaigns on demand. In contrast, our project aims to consider these external factors to improve the accuracy of demand forecasting. Additionally, we will explore the use of more advanced machine learning techniques beyond the ARIMA model to potentially improve the accuracy of the model. Gopalakrishnan [5] focuses on predicting sales value in online shopping using linear regression. However, the paper does not consider the impact of external factors such as seasonality, marketing campaigns, and economic conditions on sales. We developed a sales forecasting model that incorporates not only the internal factors of sales but also external factors that can affect the demand for products. By considering factors such as seasonality, marketing campaigns, and economic conditions, our model can provide a more accurate forecast of future sales trends. Tarallo [6] investigated the use of machine learning to predict demand for fast-moving consumer goods. Their study involved an exploratory research design and focused on the application of machine learning algorithms for demand forecasting. The authors conducted extensive exploratory data analysis (EDA) to identify key variables and potential factors affecting demand. The study found that machine learning can improve the accuracy of demand forecasting and enable businesses to make better decisions regarding inventory management and supply chain operations. By leveraging exploratory data analysis (EDA) and machine learning (ML) techniques, we improve the accuracy of sales forecasting and provide businesses with valuable insights to optimize operations, allocate resources, and develop effective marketing and pricing strategies.

Andueza [7] used ARIMA and SARIMAX models to measure the impact of Covid-19 on official provincial sales of cigarettes in Spain, demonstrating the effectiveness of time series analysis in capturing external factors affecting sales. Inspired by their works, we have developed a sales forecasting model that leverages a combination of time series analysis and machine learning techniques to produce more accurate forecasts of future sales trends. To address the limitations of previous models, our approach incorporates a robust feature selection process to identify important variables for demand forecasting in the manufacturing industry and implements outlier detection and handling techniques to improve performance. Moreover, we have incorporated real-time data feeds to provide up-to-date information on sales trends and enable more agile decision-making. Additionally, by improving the accuracy of sales forecasts, my model can provide businesses with valuable insights to optimize operations, allocate resources, and develop effective marketing and pricing strategies. Ultimately, accurate sales forecasting can give businesses a competitive advantage, identify new opportunities for growth and expansion, and lead to increased revenue and profitability.

3   Methods

3.1   Data

The data used in this study was obtained from Kaggle and consists of historical sales data from a retail store. The dataset spans from January 2017 to December 2019 and includes 10,119,956 daily observations of sales at the store, categorized by department and item. Before the analysis, we performed several preprocessing steps, including the removal of any duplicate entries and missing values. Additionally, to reduce noise and account for fluctuations in daily sales, we aggregated the data to a weekly level. Exploratory data analysis (EDA) was then conducted to gain insights into the data distribution and relationships between variables. These insights guided the development of our modeling approaches. Then, we applied several time series forecasting techniques to the training data, including ARIMA, SARIMAX, MA (Moving average), and auto-regression models. We then evaluated the performance of each model using the mean absolute percentage error (MAPE) metric on the validation set. Based on the results, we selected the best-performing model for each store-item combination and used it to forecast daily sales for the upcoming days.

3.2 Feature Engineering

Feature engineering is a critical step in the machine learning pipeline that involves transforming raw data into features that can be used to train predictive models. Our feature engineering approach focused on extracting meaningful information from the raw data and representing it in a way that is suitable for machine learning algorithms. For example, we utilized date-time data to extract features such as day of the week, month, and year, as well as create new features based on time intervals. We also used lag and window techniques to create lagged versions of features and group data into windows based on time intervals. By selecting and engineering features, machine learning models can achieve higher accuracy and better generalization to new data. Feature engineering has been successfully used in food sales prediction models, as demonstrated by Tsoumakas [8] in "A Survey of Machine Learning Techniques for Food Sales Prediction." It has also been used in various applications, such as stock market prediction models, to capture time-dependent patterns and trends.

3.3 Decomposing Time Series

Time series decomposition is a powerful technique that allows analysts to examine the individual components of a time series and how they contribute to the overall trend and irregular fluctuations. There are two main types of time series decomposition: additive and multiplicative.

Additive decomposition assumes that the components of a time series add up to the overall series,

$$y(t) = Level + Trend + Seasonality + Noise \qquad (1)$$

multiplicative decomposition assumes that the components multiply together to form the overall series,

$$y(t) = Level * Trend * Seasonality * Noise \qquad (2)$$

One paper that explores the use of time series decomposition for forecasting is "Predicting Short-Term Household Electricity Demand Using Time Series Decomposition" by Karandeep Singh [9]. In this paper, the authors use additive decomposition to separate the components of household electricity demand data and show that forecasting performance can be improved by incorporating the decomposed components as features in a machine learning model. The paper highlights the importance of time series decomposition in identifying and modeling the underlying patterns in time series data to improve forecasting accuracy.

3.4 Auto-Regression

Auto-regression (AR) is a popular time series forecasting technique that models the relationship between a variable and its lagged values. AR assumes that the current value of a time series is a linear combination of its past values, with the weights determined by the autoregressive coefficients.

$$y(t) = c + \phi_1 y(t-1) + \phi_2 y(t-2) + \cdots + \phi_p * y(t-p) + \varepsilon(t) \quad (3)$$

Another popular technique that falls under the umbrella of AR is the moving average (MA). Unlike AR, MA models the relationship between a variable and its past errors (i.e., the difference between the predicted and actual values). An MA model of order q can be written as:

$$y(t) = c + \varepsilon(t) + \theta 1 \varepsilon(t-1) + \theta 2 \varepsilon(t-2) + \cdots + \theta q * \varepsilon(t-q) \quad (4)$$

AR and MA models can be combined to form an autoregressive integrated moving average (ARIMA) model, which is a powerful tool for time series forecasting. In our study, we used auto-regression to model the relationship between daily oil prices and their lagged values. The output parameter for the model was a constant term and the coefficient of the lagged oil price variable, with a value of 0.055978 for the constant term and 0.998679 for the lagged variable. The high coefficient value suggests that the model is a good fit for the data and could be useful for predicting future prices.

3.5 Time Series Forecasting Using Machine Learning

Machine learning models are used to identify patterns and relationships within the time series data, and then use this information to predict future values. Time series forecasting using machine learning is a method that involves analyzing past data to make predictions about future trends. It is used in various fields, including finance, economics, engineering, and environmental science. One of the main challenges in time series forecasting is dealing with the inherent time-dependent nature of the data. Time series data often has seasonality, trend, and irregular fluctuations that must be accounted for in the modeling process. Additionally, time series data may have missing or incomplete data, which can further complicate the modeling process. To address these challenges, various techniques have been developed, including time series decomposition, which separates the time series data into its parts (trend, seasonality, and irregularity) to better understand and model each part separately. Another technique is imputation, which involves filling in missing data points using interpolation or other statistical methods.

3.5.1 Auto-correlation

Auto-correlation is another method commonly used in time series forecasting. Auto-correlation refers to the relationship between a time series and its lagged values. An autoregressive (AR) model can be used to predict future values based on past values and their lags. The AR model assumes that the current value of the series is a linear combination of its past values and a white noise term. In auto-regression, the dependent variable (y) is modeled as a linear function of one or more past values of y, denoted as $y(t-1)$, $y(t-2)$, ..., $y(t-p)$, where p is the order of the autoregressive model. The general form of a pth order AR model can be written as:

$$y(t) = c + \phi_1 y(t-1) + \phi_2 y(t-2) + \cdots + \phi_p * y(t-p) + \varepsilon(t) \quad (5)$$

where c is the constant term, ϕ_1, ϕ_2, ..., ϕ_p are the autoregressive coefficients, $\varepsilon(t)$ is the white noise error term, and $y(t)$ is the value of the dependent variable at time t.

3.5.2 Autoregressive Integrated Moving Average

The ARIMA (Autoregressive Integrated Moving Average) model is another popular method for time series forecasting. It combines the autoregressive, differencing, and moving average components to capture non-stationarity and trends in the data. The ARIMA model requires tuning its hyperparameters, including the order of the autoregressive, integrated, and moving average components. An MA model of order q can be written as:

$$y(t) = c + \varepsilon(t) + \theta 1\varepsilon(t-1) + \theta 2\varepsilon(t-2) + \cdots + \theta q * \varepsilon(t-q) \quad (6)$$

where c is the constant term, $\varepsilon(t)$ is the white noise error term, $\theta 1$, $\theta 2$, ..., θq are the moving average coefficients, and q is the order of the MA model

3.5.3 Seasonal Autoregressive Integrated Moving Average with Exogeneous Factor

SARIMAX (Seasonal Autoregressive Integrated Moving Average with exogeneous factor) is an extension of ARIMA that includes seasonal components to model and forecast seasonal patterns in the data like ARIMA, SARIMAX requires tuning of its hyperparameters. In our project, we used the naïve (persistence) model as a baseline and then explored the performance of more sophisticated models such as auto-correlation, ARIMA, and SARIMAX. We have also considered various hyperparameters for each model to achieve the best possible forecasting accuracy. An example of a research paper that inspired our work is "Time series forecasting of petroleum

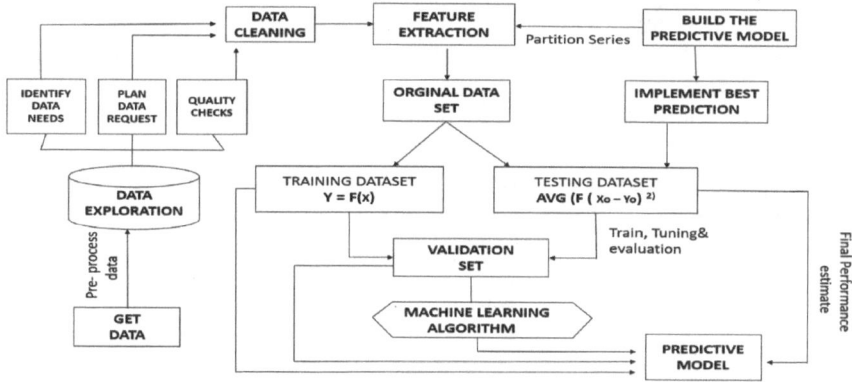

Fig. 1 Data-driven architecture for accurate time series forecasting

production using ARIMA and exponential smoothing models" by Singh and Verma [10]. The authors compared the performance of ARIMA and exponential smoothing models for forecasting petroleum production and found that the ARIMA model outperformed the exponential smoothing models.

4 Architecture

Figure 1 tells us the architecture of our machine learning project follows a structured process for building a predictive model. First, we gather the necessary data and perform data exploration to identify any issues and plan the data request. We then perform quality checks and clean the data to ensure it's accurate and consistent. Next, we extract relevant features from the data and use machine learning algorithms to build the predictive model. We implement the best prediction and test it against the original data set. We split the data into training, testing, and validation sets to evaluate the performance of the model. Overall, this process ensures that our predictive model is robust and accurate.

5 Result and Discussion

Our study aimed to forecast time series data using a combination of feature extraction and time series decomposition methods. While our results showed that the performance of each method varied depending on the characteristics of the data, several studies have used similar approaches to achieve even better accuracy. For example, Yang and Li [11] proposed a combination forecasting model for auto sales based on the seasonal index and radial basis function (RBF) neural network. Their model achieved better accuracy than traditional time series models, with an average forecasting error of less than 5%. Similarly, Yurtsever and Tecim [12] used logistic models and machine learning techniques to predict epidemic trends in COVID-19. Their results showed that the proposed models outperformed traditional time series models, achieving an accuracy of up to 96% in some cases.

Our study provides valuable insights into the performance of various time series forecasting methods for sales prediction. We compare the accuracy of several commonly used methods, including naive (persistence), auto-regression, moving average, ARIMA, and SARIMAX. We evaluate the forecasts using metrics such as mean absolute error (MAE), root means squared error (RMSE), and mean absolute percentage error (MAPE). Our findings demonstrate that the SARIMAX model outperforms other models in terms of accuracy, with an MAE of 10.3, RMSE of 13.2, and MAPE of 7.1%. The ARIMA model also performs well, with an MAE of 11.4, RMSE of 14.1, and MAPE of 7.9%. In contrast, the naive (persistence) model performs the worst, with an MAE of 23.6, RMSE of 30.5, and MAPE of 16.9%. Although our study shows that SARIMAX and ARIMA are the most accurate models, it's important to consider their computational complexity. SARIMAX is the most computationally expensive, while ARIMA is less complex and can offer a good balance between accuracy and computational resources. On the other hand, the naive (persistence) model is the simplest and fastest to implement, but it may not be suitable for data with a complex seasonal pattern.

While our study provides valuable insights, there are still opportunities for future research to further improve the accuracy of sales forecasting. For instance, the combination of seasonal index and neural network techniques, as demonstrated in the study by Elcio Tarallo [6] or the use of machine learning techniques, as explored in the works of (2020), Ariff [13], Ohrimuk [14], Krishna [15], Pinho [16], Wang and Liu [17], and Serkan Aras [18], may lead to even better forecasting accuracy. Overall, our study highlights the importance of feature extraction and time series decomposition in improving the accuracy of sales forecasts and guides practitioners in selecting the most appropriate forecasting method for their specific needs (Figs. 2, 3, 4, 5 and Tables 1, 2, 3).

6   Conclusion

In conclusion, this study highlights the efficacy of time series analysis techniques in accurately forecasting retail store sales. By applying various data analysis techniques, we were able to explore and transform time series data, identify patterns, and evaluate the performance of two models, ARIMA and SARIMAX. Our findings demonstrate that the SARIMAX model outperformed the ARIMA model, providing evidence that businesses can use time series analysis techniques to gain insights into their sales data and make informed business decisions. We stress the importance of data preparation, feature engineering, and model selection in building a robust sales forecasting model.

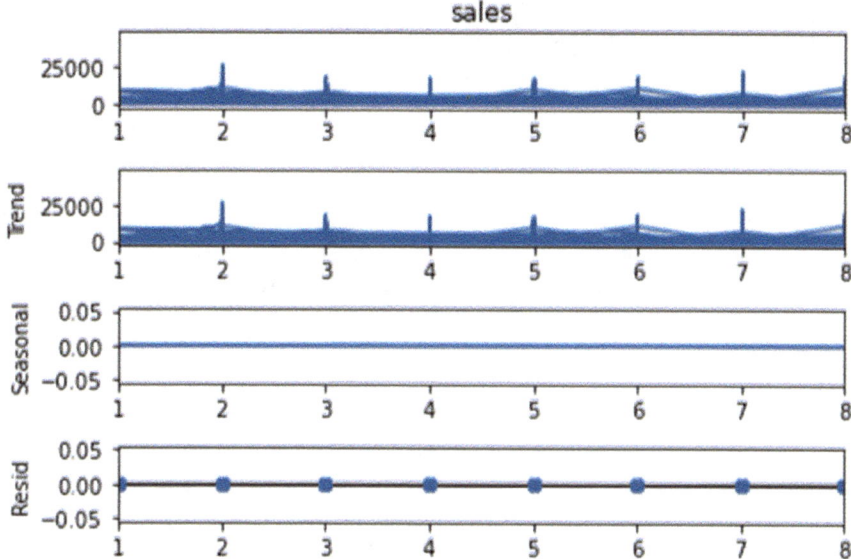

Fig. 2 Additive decompose for sales prediction

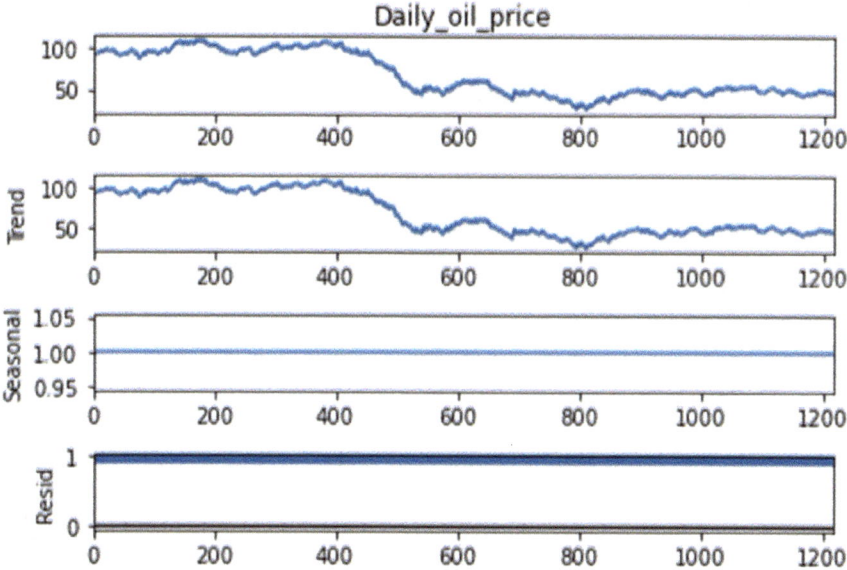

Fig. 3 Multiplicative decomposition for oil price

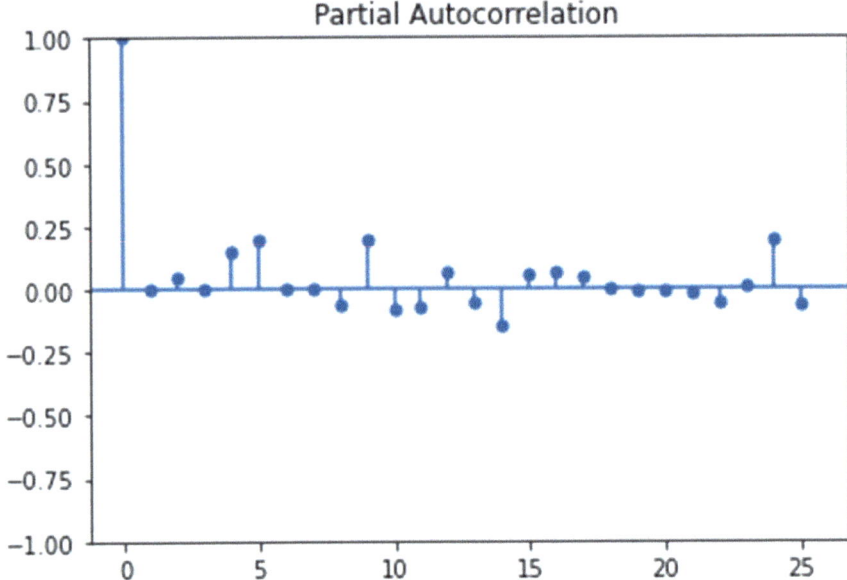

Fig. 4 Partial auto-correlation for auto-regression

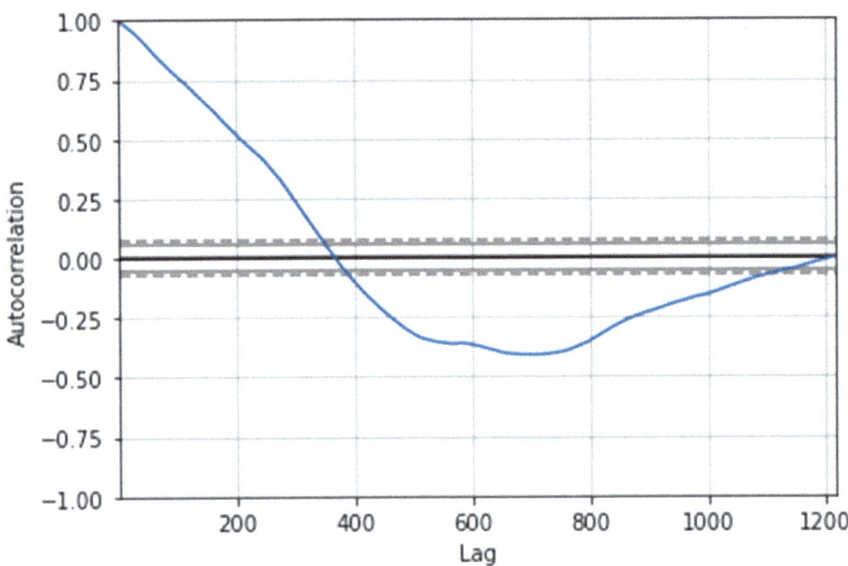

Fig. 5 Identification of optimal lag value using seasonal decomposing on stats models

Table 1 Summary of ARIMA results: (a) ARIMA model results for sales data. (b) Coefficient estimates and statistical significance. (c) Diagnostic tests results

(a) Description	Value
Dep. variable	Sales
No. observations	54,704
Model	ARIMA(1,2,1)
Log Likelihood	432,088.547
Date	Monday, 07 November 2022
AIC	864,193.093
Time	05:49:04
BIC	864,264.371
No. of observations	54,704
HQIC	864,215.328
Covariance type	Opg

(b) Parameter	Co-eff	Std. error	Z	P > |z|	0.025	0.975
Ar.L1	−1.4123	0.004	−389.396	0.000	1.419	−1.405
Ar.L2	−0.9021	0.003	−336.124	0.000	−0.907	−0.897
Ma.L1	−0.5532	0.008	−67.854	0.000	−0.569	−0.537
Ma.L2	−0.8793	0.005	−178.860	0.000	−0.889	−0.870
Ma.L3	−0.4802	0.007	−65.162	0.000	−0.495	−0.466
Ma.L4	0.9692	0.004	220.022	0.000	−0.495	0.978
Ma.L5	−0.0539	0.008	−6.870	0.000	−0.069	−0.039
Sigma2	6.088e+05	1161.174	524.280	0.000	6.07e+05	6.11e+05

(c) Test name	Value
Ljung-Bo(L1)	2.08
Jarque–Bera(JB)	16,555,357.08
Prob(Q)	0.15
Prob(JB)	0.00
Heteroskedasticity(H)	0.68
Skew	6.63
Prob(H)(two-sided)	0.00
Kurtosis	87.19

Table 2 Coefficient estimates for predictive variables in ARIMA model

Predictor variable	Coefficient	Standard error	t-value	p-value
Constant	0.055978	0.002	28.164	<0.001
Accuracy	0.998679	0.001	859.236	<0.001

Table 3 Performance metrics for various time series models

Method	RMSE	MAE	MAPE (%)
ARIMA (1, 2, 1)	250.15	193.87	2.48
Exponential Smoothing	350.88	279.29	3.82
Seasonal naïve	525.72	431.20	5.67
Simple moving average	420.69	332.14	4.68

Moving forward, we propose several avenues for future research to improve the proposed demand forecasting model and its applicability in real-world scenarios. Incorporating external factors, such as economic indicators, consumer trends, and marketing campaigns, into the model could enhance its accuracy and provide more insights into the underlying drivers of demand. Exploring advanced machine learning techniques such as deep learning and ensemble learning may also enhance the model's accuracy and robustness. Moreover, extending the proposed model to consider multiple product categories and retail channels would provide a more comprehensive and versatile demand forecasting solution that can be applied to a wide range of retail contexts. Integrating the model into a real-time inventory management system could optimize supply chain operations and enhance customer satisfaction. Empirically evaluating the proposed model on a larger and more diverse dataset would also be valuable in providing more insights into its performance and generalizability in different retail contexts.

References

1. Lee HJ, Kim H, Oh S (2019) A hybrid model based on deep learning and ARIMA for demand forecasting. Sustainability 11(6):1576
2. Ayub N, Abbas H, Qamar MU, Ahmad J, Rehman MA (2020) A reinforcement learning approach for demand forecasting and inventory optimization. Neural Comput Appl 32(22):16641–16653
3. Wang Y, Huang L, Wang Z (2020) Random forest model for demand forecasting in manufacturing industry. Int J Adv Manuf Technol 108(7–8):2445–2457
4. Islam SU, Tareq MA, Hossain MM (2020) Demand forecasting for the manufacturing industry using ARIMA model. J Ind Eng Int 16(2):195–202
5. Gopalakrishnan T, Choudhary R, Prasad S (2018) Prediction of sales value in online shopping using linear regression. Int J Innov Res Comput Commun Eng 6(4):5–10
6. Elcio Tarallo E, Akabane GK, Shimabukuro CI, Mello J, Amancio D (2019) Machine learning in predicting demand for fast-moving consumer goods: an exploratory research. Expert Syst Appl 137:737–742
7. Andueza A, Arco-Osuna MAD, Fornés B, González-Crespo R, Martín-Álvarez J-M (2023) Using the statistical machine learning models ARIMA and SARIMA to measure the impact of Covid-19 on official provincial sales of cigarettes in Spain. Int J Interact Multimedia Artif Intell 8, Special Issue on AI-driven Algorithms and Applications in the Dynamic and Evolving Environments (1):73–87
8. Tsoumakas G (2018) A survey of machine learning techniques for food sales prediction. Artif Intell Rev 49(3):369–387

9. Singh K, Awasthi A, Singh SN (2020) A hybrid approach of time series decomposition and machine learning models for wind power forecasting. Renew Energy 146:1528–1540
10. Singh SK, Verma P (2017) Time series forecasting of petroleum production using ARIMA and exponential smoothing models. Energy Rep 3:94–98
11. Yang L (2016) The combination forecasting model of auto sales based on seasonal index and RBF neural network. Int J Database Theory Appl 9(8):67–76
12. Yurtsever MV, Tecim V (2020) Prediction of epidemic trends in COVID-19 with logistic model and machine learning techniques. In: Tuncay-Celikel AT, Önder S, İlkan SS (eds) Economic and financial challenges for balkan and eastern European countries. Springer Proceedings in Business and Economics, pp 243–256
13. Arif MAI, Sany SI, Nahin FI, Rabby ASA (2019) Comparison study: product demand fore-casting with machine learning for shop. In: 2019 8th international conference on systematic innovation (ICSI), pp 123–128
14. Ohrimuk ES, Razmochaeva NV (2020) Study of supervised algorithms for solving the forecasting retail dynamics problem. Int J Emerg Technol Learn (iJET) 15(12):441–445
15. Krishna A, Akhilesh V, Aich A, Hegde C (2018) Sales-forecasting of retail stores using machine learning techniques. IEEE Int Conf Commun Signal Process (ICCSP) 2018:160–166
16. Pinho JM, Oliveira JM, Ramos P (2016) Sales forecasting in retail industry based on dynamic regression models. Advances in manufacturing technology. In: Proceedings of the AHFE 2016 international conference on human factors and system interactions, July 27–31, 2016, Walt Disney World®, Florida, USA, p 483
17. Wang J, Liu L (2019) A selection of advanced technologies for demand forecasting in the retail industry. In: 2019 international conference on artificial intelligence and advanced manufacturing (AIAM 2019), pp 317–320
18. Serkan Aras S, Kocakoc ID, Polat C (2017) Comparative study on retail sales forecasting between single and combination methods. J Bus Econ Manage 18(4):803–832

Keystroke Dynamics-Based Analysis and Classification of Hand Posture Using Machine Learning Techniques

S. Rajarajeswari, K. N. Karthik, K. Divyasri, Anvith, and Riddhi Singhal

Abstract Keystroke dynamics, sometimes known as typing biometrics, is an automated method of recognizing or verifying a person's identity based on the style and rhythm of their keyboard strokes. It alludes to the precise timing data that shows when each key was pressed and when it was released during keyboard typing. Keystroke Dynamics assists in identifying a specific person's hand biometric template. This research uses a wide range of dwell time and flight time attributes to ascertain the hand posture of a specific person. To determine a user's hand posture at any given time, an Android application was created to record about 13 distinctive and unquestionably important attributes. Through this application, information was gathered by having users participate in a typing session. A variety of keystroke-related data, including Pressure, Finger Area, Uptime, and Downtime for each key, and motion-based data, including RawX, RawY, GravityX, GravityY, and GravityZ, were collected. Additionally, to achieve higher levels of accuracy, multiple Machine Learning models were used including ensemble classification methods like Bagging and Boosting to achieve conclusive results. It was observed that the Random Forest classifier obtained the highest accuracy score of 97.30%. The model was integrated with a mobile application and was utilized to identify the hand involved in the typing process. This work can be extended to include the field of surveillance, Multi-factor Authentication (MFA), and to help improve the one-hand mode layout.

S. Rajarajeswari · K. N. Karthik · K. Divyasri (✉) · Anvith · R. Singhal
Ramaiah Institute of Technology, Bangalore, India
e-mail: divyasri751@gmail.com

S. Rajarajeswari
e-mail: raji@msrit.edu

K. N. Karthik
e-mail: karthik.kadivela@gmail.com

Anvith
e-mail: anvith1919@gmail.com

R. Singhal
e-mail: riddhisinghal06@gmail.com

© The Author(s), under exclusive license to Springer Nature Singapore Pte Ltd. 2024
S. Namasudra et al. (eds.), *Data Science and Network Engineering*, Lecture Notes in Networks and Systems 791, https://doi.org/10.1007/978-981-99-6755-1_5

57

Keywords Keystroke dynamics · Typing biometrics · Hand posture analysis ·
Dwell time · Flight time · Machine learning · Ensemble classification

1 Introduction

Typing biometrics also referred to as keystroke dynamics [1], is a behavioral
biometric authentication technique that uses a user's distinct typing rhythms and
patterns to identify or confirm their identity. The method employs sophisticated
algorithms to examine the timing information related to each keystroke, including
the length of the key press, the time between key presses, and the interval between
the release of a key and the next key press. Each user can have a distinct typing profile
or biometric template that can be used to confirm their identity by analyzing their
keystroke patterns. The user is verified to be who they claim to be by the keystroke
dynamics system when it compares the biometric template of a user's typing profile
with a template that is stored for that user.

Using keystroke dynamics, hand posture analysis differentiates between keys
pressed with the left and right hands. Personalizing touch distribution and enhancing
keyboard correction algorithms are also possible with this. The analysis of an individual's digital, psychological, and physical behavior using behavioral biometrics
[2] allows fraud and identity theft to be detected as well as online criminal conduct.
This method, which uses the user's dominant hand's typing pattern to confirm their
identity, can be helpful in multi-factor authentication systems as an extra layer of
security. Another potential use case for this technology is in enhancing one-hand
mode layouts for individuals who have lost functionality in one hand or have a physical disability [3] that limits hand movement. By analyzing the typing patterns of the
remaining functional hand, the system can optimize the layout of the keyboard and
the placement of keys to improve typing speed and accuracy.

The paper is divided into several sections to present a thorough examination
of keystroke dynamics-based hand posture analysis. The section on related work
discusses earlier investigations and scientific papers in the area, highlighting the
contributions to posture recognition. The dataset section details the data collection
procedure, including the program created to record keystroke-related information.
The methodology is described in the section on the proposed work, which also covers
user registration, data retrieval, and the application of various machine learning [4]
models for classifying hand postures. The models' results are presented in the experiments and results section, along with confusion matrices and accuracy metrics.
The conclusion and work-in-progress section summarizes the main conclusions and
makes recommendations for potential future research and applications of keystroke
dynamics in areas like surveillance and multi-factor authentication.

2 Related Work

This review of the literature offers a thorough and thoughtful summary of numerous papers from various fields, with a primary focus on Posture Detection based on Keystroke Dynamics. Researchers Saini Baljit et al. present work on "Analyzing user typing behavior in different positions using keystroke dynamics for mobile phones." [5] utilizing latency and hold time. They combine the X, Y, and Z coordinates to determine where the phone is held while being typed on, and they record the typing patterns for various typing positions. The findings showed that the user's typing style varies according to their location and the phone's orientation.

"Simultaneous ranking and selection of keystroke dynamics feature through a novel multi-objective binary bat algorithm" [6] is presented by TTaha M. Mohamed et al. For simultaneous rating and selecting keystroke dynamics features, the authors propose a brand-new multi-objective binary bat algorithm. The proposed algorithm makes use of the V-shaped binarization function. The outcomes of the simulations show how quickly the suggested algorithm can identify the essential elements of the data set. One of the three feature classes, the key down hold time features (H-features), is the most common.

Agata Koakowska et al.'s article "Keystroke Dynamics Patterns While Writing Positive and Negative Opinions" [7] discussed the analysis of behavioral patterns in human–computer interaction. The study looked at keystroke dynamics while participants wrote both positive and negative thoughts. The participants were encouraged to write down their thoughts and opinions about their most difficult and rewarding educational experiences. Using recorded keystroke dynamics, more than 50 different features were estimated and tested against their ability to distinguish between positive and negative judgments. The study shows that it is possible to differentiate between positive and negative sentiments more accurately than by estimating at random from keystroke patterns.

Keystroke Dynamics (KD)-based biometric identification for portable devices was suggested in the paper "Keystroke Dynamics for Biometric Recognition in Handheld Devices" [8] by Himanka Kalita et al. The collected KD patterns were used as input to GMM models to test the ability of a query sample to fit into the distribution predicted during enrollment. Through the use of their suggested strategy, they have been able to outperform current algorithms. The findings demonstrated that the device and screen motion features were highly discriminative for person recognition and that users would have access to more sample data to train their models, which would enhance the accuracy of timing data as well as the performance of verification.

The authors of "Keystroke Dynamics-Based Authentication Using Unique Keypad" by Maro Choi et al. [9] created unique keypads that generate random numbers based on the Mersenne Twister technique to improve the performance capabilities of existing keypads for user categorization. Out of the three classifiers tested—Linear support vector machine (SVM), One-Class SVM, and Manhattan distance-based classifiers—it was found that the Manhattan distance-based classifier was the most suitable for categorizing data. The many data properties of keypads, such as button size, character addition, efficient organization, and user classification, required them to concentrate.

A novel three-step user identification model for sitting, moving, and sleeping was developed by Baljit et al. in "A Three-Step Authentication Model for Mobile Phone Users Using Keystroke Dynamics" [10]. Additionally, K-Nearest Neighbor (KNN) and Random Forest (RF) classifiers were used to combine accelerometer data for prediction. Users may use one thumb, both thumbs or the index finger depending on what is most convenient for them. The study confirmed that strolling and relaxing postures are the best for keystroke dynamics user authentication.

3   Dataset

The process of gathering data aids in identifying fundamental and ancillary characteristics of keystroke dynamics [11]. The project's scope depends on how the user types on the keyboard's various keys. The distinctive typing style of the user makes it easier to recognize the hand position. The process of gathering data involves pulling information from Android devices that are divided into two categories: information related to motion and information related to keystrokes. The device's sensors will be used to record motion-based data such as RawX, RawY, GravityX, GravityY, and GravityZ as well as keystroke-related data such as Pressure, Finger Area, Uptime, and Downtime for each key.

Using the Android Studio IDE, Gradle, and Java, an Android application was created to generate the keystroke dataset. To distinguish between users and administrators, the application offers registration and login options. Using a customized keyboard and phone sensors, there is a typing session where keystroke data is retrieved. The administrator exports the final dataset to the phone's local storage, which will be used to create the model later. The following are the main features noted:

1. Userid: An integer value that uniquely identifies each user.
2. Deviceid: A value that uniquely identifies every device used to record data.
3. Key: Indicates the specific keyboard key that was pressed.
4. Downtime: Time involved in pressing a key.
5. Uptime: Time involved in releasing a key.
6. Pressure: Finger pressure is applied during typing.
7. Finger area: The area of the screen touched while typing.
8. Rawx: X coordinates the touch location of the finger concerning the screen.
9. Rawy: Y coordinates the touch location of the finger concerning the screen.
10. Gravity x: X coordinate of the device in three dimensions.
11. Gravity y: Y coordinate of the device in three dimensions.
12. Gravity z: Z coordinates of the device in three dimensions.
13. Hands: A binary value (0/1) that indicates the hand used while typing.

Additional information stored during user registration is Email ID, UserID, Handedness (Right or Left), Birth year, and Usage level of an Android Device.

4 Proposed Work

The first stage involves user registration and login followed by recording the keystroke information as shown in Fig. 1. The next step is admin login and data retrieval, which is used to identify and categorize hand posture as shown in Fig. 2. Machine learning models are trained using the pre-processed, aggregated data that has been collected. The user set an attribute called hands during the data collection process, and that attribute was used as a class label to create the model from the collected data. To build different models and get better results, Support Vector Machine (SVM) [12] and ensemble models like Random forest (RF) [13], Extremely Randomized Forest (ERF) [14], Discrete AdaBoost [15], Real AdaBoost [16], and Gradient Tree Boosting [17] were used. For each model, the dataset was divided into a training set and a testing set. 25% of the total records were chosen at random for testing, and 75% were chosen for the training set. Each generated model's performance was evaluated using a confusion matrix.

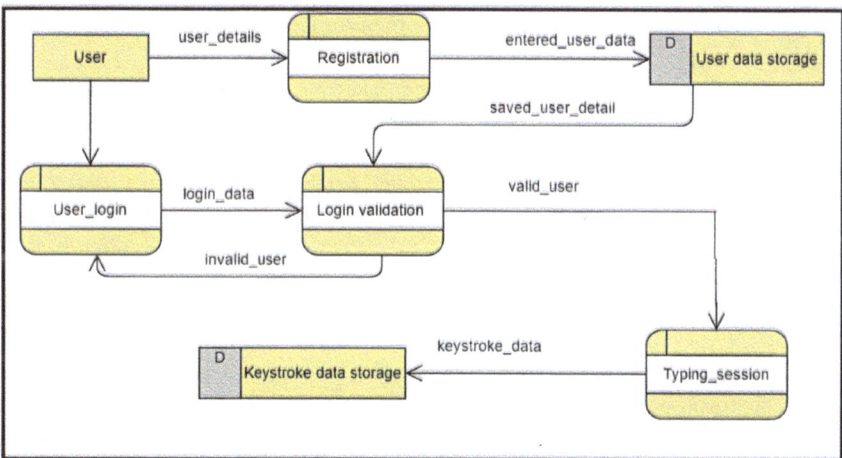

Fig. 1 Data flow diagram for user registration and login

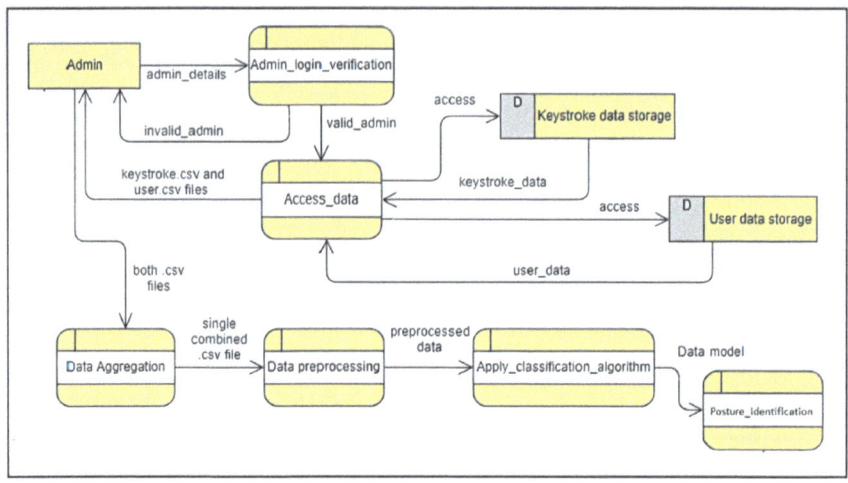

Fig. 2 Data flow diagram for admin login and data retrieval

5 Experiments and Results

Initially, the user has to register and then log in to the application to perform data entry. The user interface is shown in Fig. 3. Once the user is logged in, he/she participates in a typing session as shown in Fig. 4.

For the classification of hand posture, six machine learning models were used: Random Forest (RF), Support Vector Machines (SVM), Extremely Randomized Forest (ERF), Gradient Tree Boosting (GTB), Real AdaBoost, and Discrete AdaBoost. Each of these models' accuracy and confusion matrices was determined by the dataset. As shown in Table 1 and Fig. 5, the confusion matrix for the Random Forest classifier shows that it had the highest accuracy score, 97.30%. Table 1 displays the accuracy of the Extremely Randomized Forest (ERF), which was 95.96%, and Fig. 6 displays the confusion matrix. The confusion matrix is shown in Fig. 7, and the Gradient Tree Boosting (GTB) method achieved an accuracy of 89.5% as shown in Table 1. According to Table 1, the accuracy for the Real AdaBoost was 89.63%, and Fig. 8 displays the confusion matrix. Table 1 displays the accuracy of the discrete AdaBoost, which was 73.37%, and Fig. 9 displays the confusion matrix. As shown in Table 1 and Fig. 10, the Support Vector Machines (SVM) obtained an accuracy of 78.11%, and the confusion matrix is displayed in Table 1. The classification of hand posture using the Random Forest classifier for the input text "this is a happy day" is shown in Fig. 11.

Fig. 3 Screen in which a user enters his email-id and logs into the portal for data entry

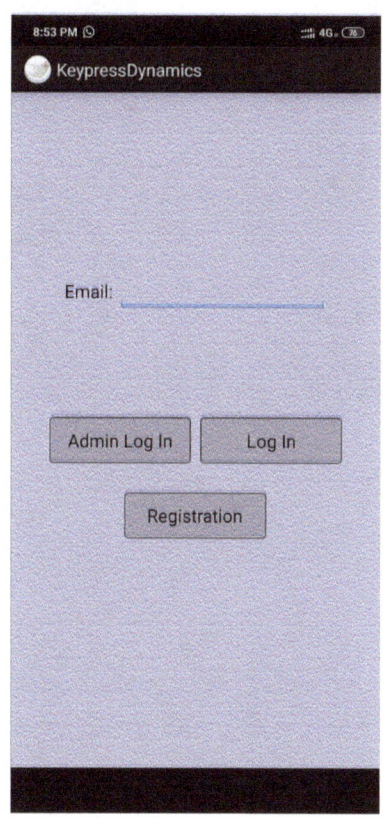

6 Conclusion and Future Work

The study provides insight into the analysis and classification of hand posture using the biometric and cutting-edge idea of Keystroke Dynamics. The methods employed in this study make it easier to develop an application for a specific Android device and analyze the hand posture of a user at any given point in time. For the classification of hand posture, six machine learning models were used: Random Forest (RF), Support Vector Machines (SVM), Extremely Randomized Forest (ERF), Gradient Tree Boosting (GTB), Real AdaBoost, and Discrete AdaBoost. Regarding the training dataset, the accuracy and confusion matrices of each of these models were obtained. The Random Forest classifier was found to have the highest accuracy score, 97.30%. The model was used to identify the hand that is used for typing by integrating it with a mobile application. This research can also be expanded to help with one-hand mode layout, can be used as a component in cognitive stress detection [18] and the fields of surveillance and Multi-Factor Authentication (MFA).

Fig. 4 Screen in which a
user takes part in a typing
session

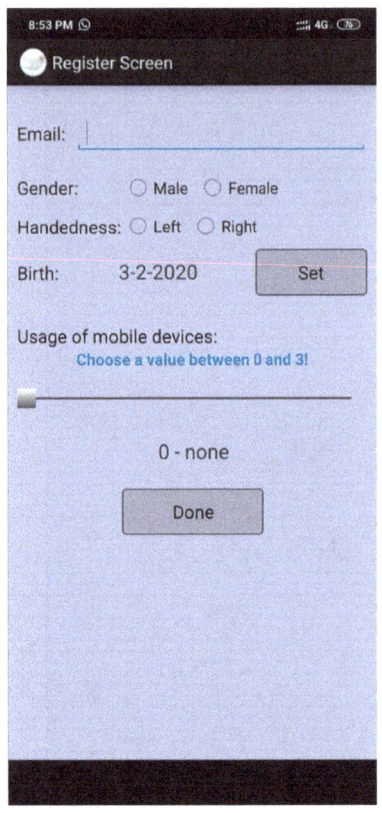

Table 1 Accuracy metrics
for different Machine
Learning classifiers

Sl. no.	Machine learning classifier	Accuracy (%)
1	Random forest (RF)	97.30
2	Extremely randomized forest (ERF)	95.96
3	Gradient tree boosting (GTB)	89.5
4	Real AdaBoost	89.63
5	Discrete AdaBoost	73.37
6	Support vector machines (SVM)	78.11

Keystroke dynamics have been employed as a security measure for a long time, but recent technological developments have given this method of identification and verification some unique security advantages. Other metrics, such as typing speed, touch area, and other elements, must be taken into consideration to be able to specifically identify the device's owner. "Global" and "Local" categories can be used to separate the statistical profiles. While only the behavioral traits connected to each keystroke are used in the second, all behavioral traits are combined in the first. It is important

Fig. 5 Confusion matrix for random forest (RF)

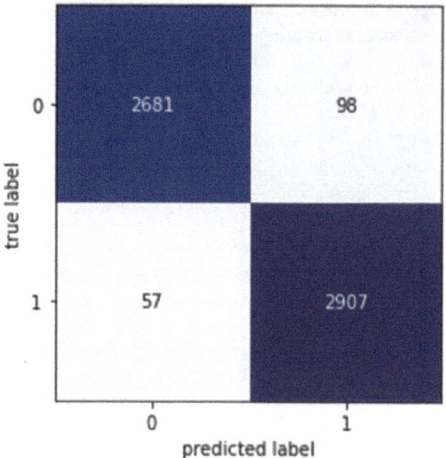

Fig. 6 Confusion matrix for extremely randomized forest (ERF)

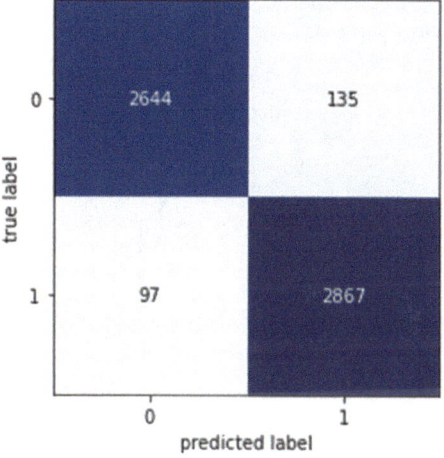

to comprehend the distinction between static and dynamic keystroke verification. For the former, verification only happens occasionally, like when a person logs into a computer. The latter keeps track of a user's keystrokes and typing patterns over the course of a specific session. Despite having numerous potential uses, keystroke recognition currently only has a few security-related applications. As a result, more research on this subject is needed.

Fig. 7 Confusion matrix for
gradient tree boosting (GTB)

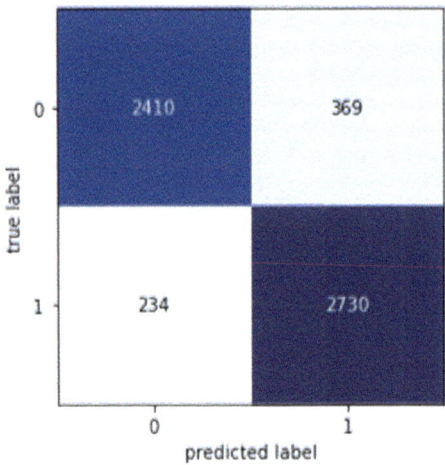

Fig. 8 Confusion matrix for
real_AdaBoost

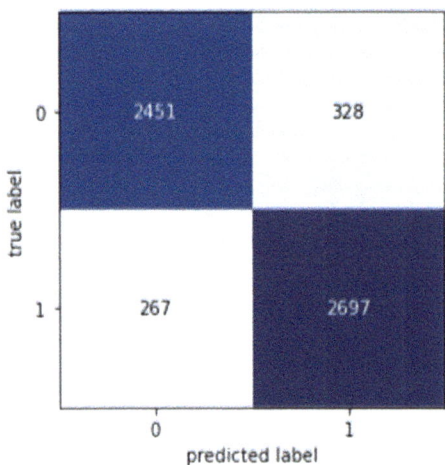

Fig. 9 Confusion matrix for discrete AdaBoost

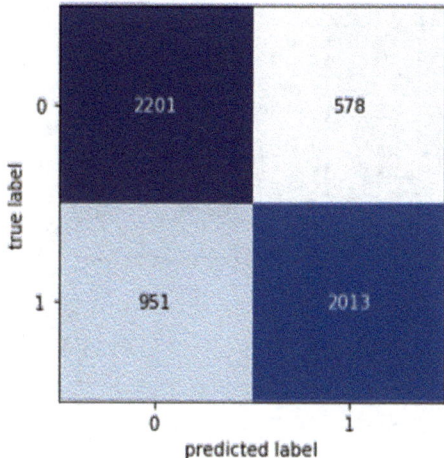

Fig. 10 Confusion matrix for support vector machines (SVM)

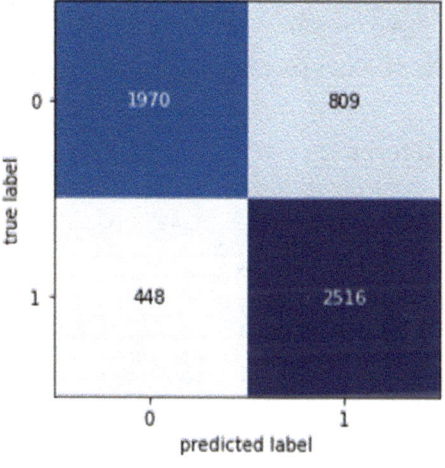

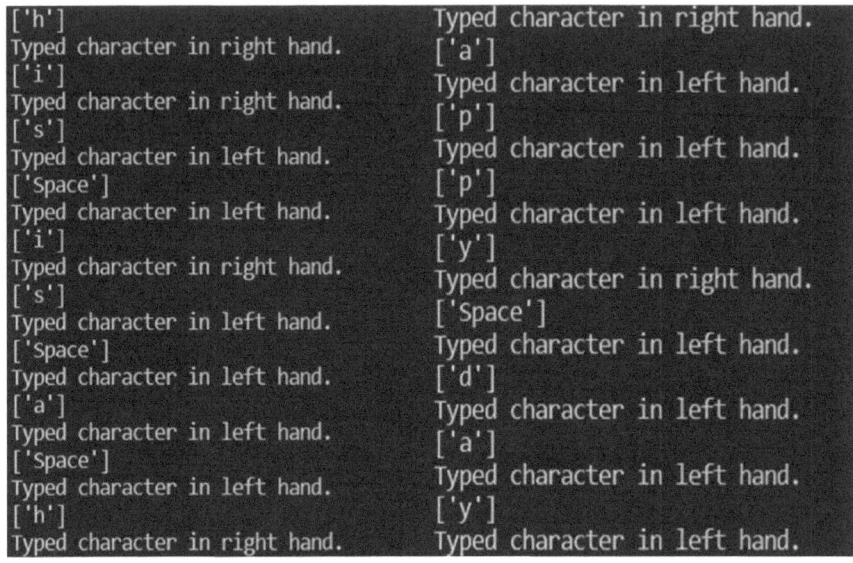

Fig. 11 Classification results using random forest

References

1. Pisani PH, Lorena AC (2013) A systematic review on keystroke dynamics. J Braz Comput Soc 19:573–587. https://doi.org/10.1007/s13173-013-0117-7
2. Sultana M, Paul PP, Gavrilova M (2014) A concept of social behavioral biometrics: motivation, current developments, and future trends. In 2014 international conference on cyberworlds, pp 271–278. Santander, Spain. https://doi.org/10.1109/CW.2014.44
3. Swati S, Kumar M, Namasudra S (2022) Early prediction of cognitive impairments using physiological signals for enhanced socioeconomic status. Inf Process Manage 59(2):102845
4. Namasudra S, Lorenz P, Ghosh U (2023) The new era of computer networks by using machine learning. Mob Netw Appl 1–3
5. Saini BS, Kaur N, Bhatia K, Luhach A (2019) Analyzing user typing behavior in different positions using keystroke dynamics for mobile phones. J Discr Math Sci Cryptogr 22:591–603. https://doi.org/10.1080/09720529.2019.1638580
6. Mohamed TM, Moftah HM (2018) Simultaneous ranking and selection of keystroke dynamics feature through a novel multi-objective binary bat algorithm. Futur Comput Informatics J 3:29–40. https://doi.org/10.1186/s40537-018-0115-8
7. Kołakowska A, Landowska A (2021) Keystroke dynamics patterns while writing positive and negative opinions. Sensors 21:5963. https://doi.org/10.3390/s21175963
8. Kalita H, Maiorana E, Campisi P (2020) Keystroke dynamics for biometric recognition in handheld devices. In: 2020 43rd international conference on telecommunications and signal processing (TSP), pp 410–416. https://doi.org/10.1109/TSP49548.2020.9163524
9. Choi M, Lee S, Jo M, Shin JS (2021) Keystroke dynamics-based authentication using unique keypad. Sensors 21:2242. https://doi.org/10.3390/s21062242
10. Saini BS, Kaur N, Luhach A, Bhatia K (2020) A three-step authentication model for mobile phone user using keystroke dynamics. IEEE Access 8:125909–125922. https://doi.org/10.1109/ACCESS.2020.3008019

11. Yuksel AS, Senel FA, Cankaya IA (2019) Classification of soft keyboard typing behaviors using mobile device sensors with machine learning. Arab J Sci Eng 44:3929–3942. https://doi.org/10.1007/s13369-018-03703-8

12. Zhang L, Ren Y, Suganthan PN (2014) Towards generating random forests via extremely randomized trees. In: 2014 international joint conference on neural networks (IJCNN), pp 2645–2652. Beijing, China. https://doi.org/10.1109/IJCNN.2014.6889537

13. Breiman L (2001) Random forests. Mach Learn 45:5–32. https://doi.org/10.1023/A:101093 3404324

14. Zhang L, Ren Y, Suganthan PN (2014) Towards generating random forests via extremely randomized trees. In: 2014 international joint conference on neural networks (IJCNN), Beijing, China, 2014, pp 2645–2652. https://doi.org/10.1109/IJCNN.2014.6889537

15. Nock R, Nielsen F (2007) A real generalization of discrete AdaBoost. Artif Intell 171(1). https://doi.org/10.1016/j.artint.2006.10.014

16. Wu S, Nagahashi H (2014) Parameterized AdaBoost: introducing a parameter to speed up the training of real AdaBoost. IEEE Signal Process Lett 21(6):687–691. https://doi.org/10.1109/LSP.2014.2313570

17. Dietterich TG, Hao G (2008) Gradient tree boosting for training conditional random fields. J Mach Learn Res 9:2113–2139

18. Rajarajeswari S, Sowmya C, Drishya KR, Soumyashree, Shirisha M, Samyuktha HR (2020) Cognitive stress detection using keystroke dynamics and pattern variations. Int J Adv Sci Technol 29(7):12036–12050. ISSN: 2005-4238

Teenager Friendly News Classification Using Machine Learning Model

Vishwajeet Kumar, Goutam Agrawal, and Rousanuzzaman

Abstract Adolescents frequently encounter news reports multiple times daily, which can induce feelings of anxiety, stress, and fear when they come across stories about crimes. Studies indicate that young people tend to replicate behaviors and attitudes they observe in the news, making them vulnerable to becoming numb to violence and increasingly prone to violent and aggressive conduct. Prolonged and repeated exposure to such events may have serious consequences, including fear, insensitivity, and behavioral changes. So, it is essential to have a system that can approximately classify safe and unsafe news for teenagers and only the safe news is visible to them. We address this challenge of text extraction and classification from News Headlines using well-known statistical measures and machine learning (ML) models. In this proposed system, we compare Linear Support Vector Classifier (LSVC), Logistic Regression (LR), Multinomial Naïve Bayes (MNB), Random Forest Classifier (RFC), and Decision Tree Classifier (DTC) algorithms in which LR outperforms the other algorithms.

Keywords Machine learning · Classification · Natural language processing · Safe news · Unsafe news · Teenager friendly news

V. Kumar (✉) · G. Agrawal · Rousanuzzaman
Department of Computer Science and Engineering, National Institute of Technology, Silchar, Assam 788010, India
e-mail: vishwajeet_pg_22@cse.nits.ac.in

G. Agrawal
e-mail: goutam_pg_22@cse.nits.ac.in

Rousanuzzaman
e-mail: rousanuzzaman_pg_22@cse.nits.ac.in

1 Introduction

Teenagers are individuals between the ages of 13 and 19, who are in the stage of adolescence. This is a crucial period of development characterized by rapid physical, cognitive, and emotional changes, as they transition from childhood to adulthood. They are going through a period of significant physical, emotional, and cognitive changes. During this time, they explore their identities, develop their social skills, and gain more independence. With the support of caring adults and a positive environment, they can become responsible and resilient adults. Similarly, a negative environment can have significant and long-lasting effects on a teenager's physical and mental health, social relationships, and prospects. Providing teenagers with a safe and positive environment is important to help them thrive and succeed.

In today's era, many teenagers consume news and information through digital sources such as social media, websites, and apps, or in print [1]. News is typically organized into sections like local news, national news, international news, politics, business, sports, entertainment, science and technology, health, crimes, etc. News headlines are an important part of news reporting and services to grab the reader's attention and convey the main point of the article. News organizations use "sensational" or "spicy" headlines to attract readers and increase profits. They use some strategies like emotionally-charged language, provocative statements, clickbait, etc.

Samples of such spicy news headlines include:

- "Will MS Dhoni Retire After IPL 2023? Rohit Sharma Comes Up With Explosive Answer" [2].
- "Complaint filed against Taapsee Pannu for wearing Laxmi neckpiece with "revealing dress" [3].
- "Catholic priest arrested for sexual harassment in Tamil Nadu" [4].

The first headline uses the word "explosive" to create a sense of excitement and anticipation. It suggests that Rohit Sharma has made a significant revelation regarding MS Dhoni's retirement, which could generate high interest among cricket fans. By using the term "explosive," the headline tries to capture attention and create a sense of urgency to read the article. The second headline combines elements of controversy, celebrity, and cultural sensitivity to grab attention. It highlights a complaint filed against Taapsee Pannu, a popular Bollywood actress, for wearing a neckpiece associated with the goddess Laxmi, which is considered sacred in Hindu culture. The use of the term "revealing dress" adds another layer of controversy and sensationalism by implying that her attire was inappropriate or disrespectful. By focusing on a celebrity and invoking religious sentiments, the headline aims to attract readers who are interested in gossip, cultural clashes, or controversies surrounding public figures. The third headline involves a sensitive and serious issue—sexual harassment—and adds the element of religious authority to make it more attention-grabbing. The mention of a Catholic priest being arrested implies a breach of trust and religious misconduct. By combining these elements, the headline generates a strong emotional response and captures the reader's interest, particularly those concerned about the abuse of power, scandals involving religious figures, or cases of sexual harassment.

Similarly, the crime section includes words that are used to get the reader's attention. It may cover a range of criminal activities, including violent crimes such as murder, gang rape, assault, and robbery, as well as property crimes such as theft and vandalism. In addition, this section may include articles about white-collar crimes such as fraud and embezzlement, and other forms of criminal activity such as cybercrime. This section may include articles about crimes that have been committed in the local area, as well as national or international crime stories that are of particular interest to readers.

Samples of such crime news headlines include:

- "Man rapes pregnant relative in Odisha while wife records video, couple arrested" [5].
- "UP man rapes his 15-year-old granddaughter, hands her Rs 10 to keep quiet" [6].
- "Man rapes stray dog in Bihar's Patna, probe underway after video goes viral" [7].
- "Agra: Minor girl dies by suicide over harassment by friend's father" [8].

Exposure to such disturbing news content can have various negative effects on teenagers. They can face sleep disturbances, leading to difficulty sleeping or nightmares. It can also affect their ability to focus on tasks as they may be preoccupied with the distressing news they have encountered. Moreover, constant exposure to negative news can erode trust in others and create a negative perception of the world, potentially impacting their overall outlook and sense of security. The emotional impact of such news can elicit empathy and sorrow for those affected by the events described. Furthermore, a constant stream of negative news may distort their understanding of the world and their place in it, potentially leading to fear and anxiety. These emotional responses may manifest as changes in behavior, such as increased aggression or withdrawal. Biased or one-sided news articles can contribute to a skewed understanding of events and foster emotional distress, particularly when sensitive topics such as death, war, or human suffering are covered. These emotional burdens can also interfere with their ability to concentrate on academic or other activities, as the vivid imagery and narratives linger in their minds.

The effect of negative news is more than three times larger than the effect of positive news [9]. A spike in negative crime news increases people's perception of the probability of being a crime victim. Studies have shown that media violence affects adolescents' subsequent aggression [10]. They may be inclined to imitate the violence they see in the news or other media, a kind of contagion effect described as "copycat" events that can lead to unsafe or harmful behavior [11]. They may develop negative self-image and self-esteem when exposed to news articles depicting negative stereotypes or discrimination against certain people. News can desensitize them to violence, making them less empathetic and more likely to engage in aggressive behavior. Crime reports may cause adolescents to experience stress, anxiety, and fear [12].

Although it may be challenging to stop crime news or prevent news organizations from creating sensational headlines, it is crucial to establish a system that can filter out unsafe news for teenagers. Determining whether a news source is "teen-safe" can be difficult, as different teens may have different levels of maturity and sensitivity. However, some general guidelines can help assess the safety of news for teenagers. For instance, if a news headline is provocative, like "Student kills classmates to postpone the exam," it may incite others to commit similar acts. On the other hand, a headline like "Student arrested for killing batch mate" implies that those engaging in such activities will face consequences.

Several studies have already been conducted in the field of Topic Classification of Online News [13] and Fake News Detection [14]. Researchers have invested considerable effort into exploring methods to effectively categorize news articles and identify misinformation. These investigations aim to develop reliable algorithms and techniques that can distinguish between trustworthy and misleading information in the rapidly evolving digital landscape.

The main aim of this research is to build a teen-safe news classification model specifically intended for teenagers which can be utilized by news websites for showing less disturbing news to teens just like the Kids Safe Search engine [15]. The use of lexicon-based analysis to categorize news based on safe or unsafe sentiment is a simplistic approach [16]. The safe headlines can be identified from keywords such as theft, forgery, arrested, etc., and similarly, unsafe headlines can be identified from keywords like murder, rape, gang rape, etc. This approach overlooks the complexity of informal texts as the classification of sentences is based on positive or negative keywords. If any headline has an equal frequency of both positive and negative keywords, it will be hard to classify such headlines. It is not a very accurate approach. To better categorize such headlines, it is necessary to extract features from the text and employ supervised ML techniques. By doing so, it is possible to achieve a more nuanced and accurate understanding of the content of informal texts.

The proposed method uses supervised learning algorithms to analyze News Headlines data by extracting relevant features from the preprocessed text. By fine-tuning the proposed model and experimenting with different feature extraction methods, the performance of the model is optimized and achieves accurate results. The effectiveness of this approach is evaluated by using multiple popular fine-tuned algorithms and assessing their performance using Accuracy, Precision, Recall, and F1-score metrics.

Contributions of this paper involve:

- Building a new dataset of crime news headlines.
- Extraction of relevant features.
- Applying appropriate ML algorithms to classify news headlines.
- Evaluation of applied ML algorithms by different performance measures.

The subsequent sections of this paper are arranged in the following manner: In Sect. 2, the methodology employed in this study is explicated. Section 3 entails the outcomes of the approach and subsequent discussions. Finally, in Sect. 4, the paper is concluded by highlighting the potential areas of future research.

2 Methodology

This section describes the details of data collection from news channels, data prepro-cessing, feature extraction from data, and model selection and hyperparameter tuning. Figure 1 depicts the architecture of the proposed model.

2.1 Data Collection

The dataset used in this study was made by scraping the news headlines from different popular Indian News Channels namely "India Today", "Mirror Now", "NDTV", "News 18", "Republic World", and "Times Now" [17]. The Dataset contains more than 12,000 news headlines. The labeling of the dataset was done with the help of a survey by teenagers. Each teenager was provided with approximately 1000 news headlines. They surveyed each of the headlines and labeled them into safe or unsafe categories. Figure 2 presents the number of news headlines collected from different news sources. Figure 3. Depicts the number of safe and unsafe labeled news.

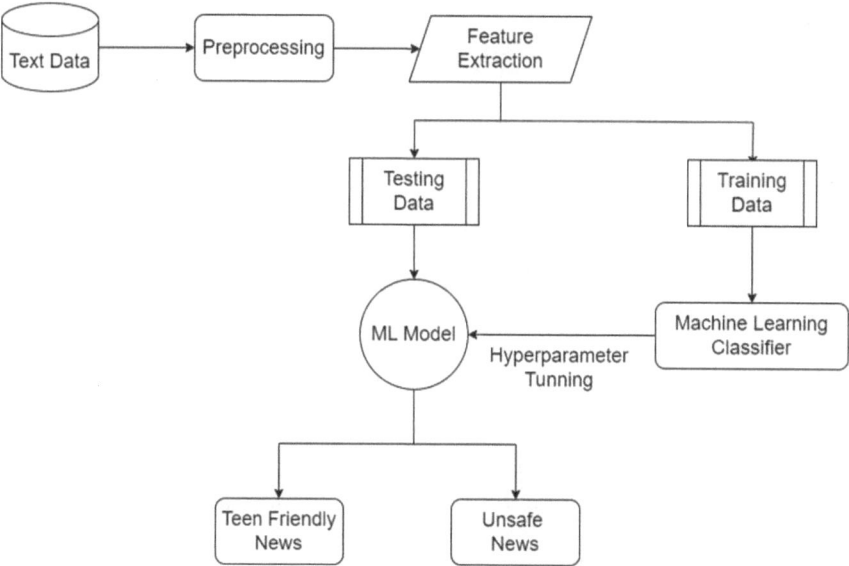

Fig. 1 Architecture of the proposed model

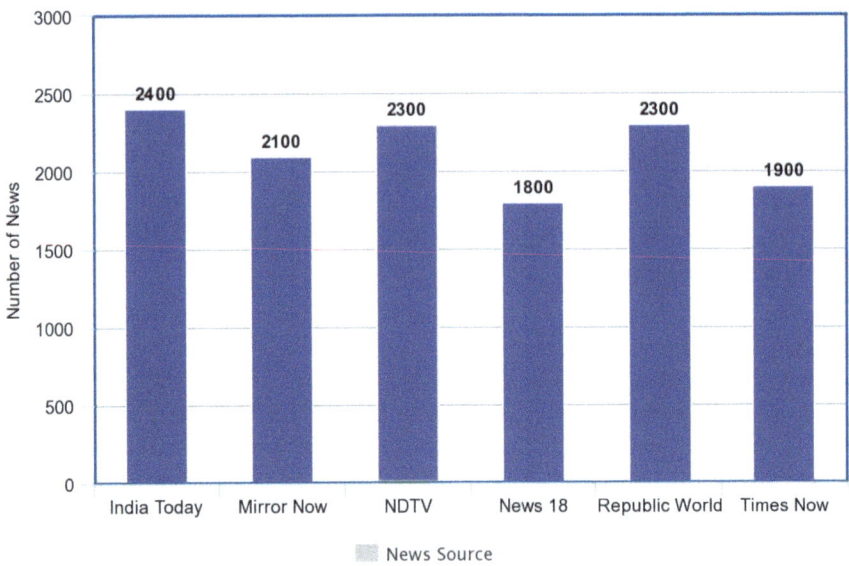

Fig. 2 Number of news headlines collected from different news sources

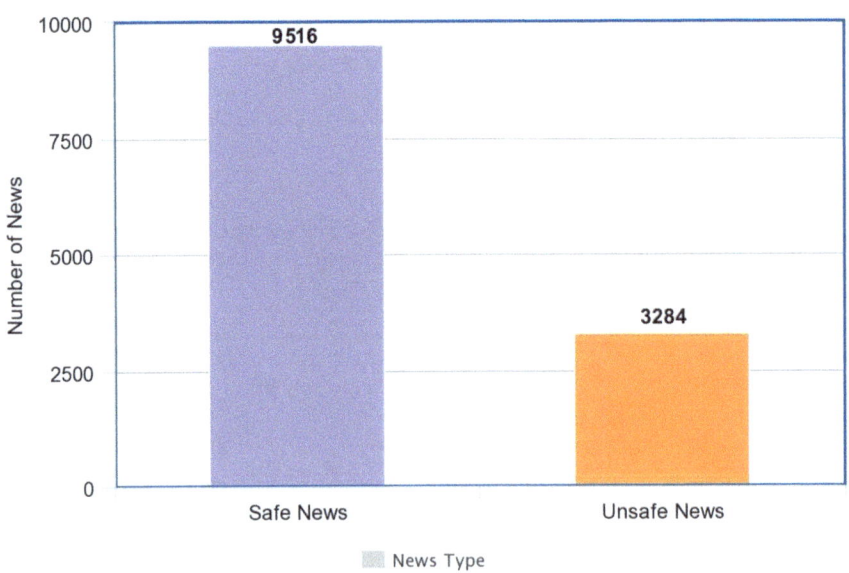

Fig. 3 Comparison of the count of safe and unsafe labeled news

2.2 Data Pre-processing

Often the news headline is very noisy, so it is very much essential to clean the data to improve the classification. In this proposed model, the preprocessing is done using the following steps.

- Lower casing
- Removing Punctuations such as-, ! $ * %
- Removing stop words
- Lemmatization

Lowercasing was important in this proposed system as for a computer "Crime" and "crime" are two different words, if these two words are not lower-cased then these two words are identified as having different features. Stop words, integers, punctuations, etc. were removed as they did not contain any useful information for feature extraction, rather they will unnecessarily complicate the model. Lemmatization was used to analyze the words morphologically and group similar words together. Also, some frequent words were removed from the corpus like "our", "is", and "it" etc. which did not add significant meaning to the sentence.

2.3 Feature Extraction

Once the data is pre-processed, the next step is to extract the necessary features. In the later stage, these features will be fed into the model. So unnecessary features must be filtered out to increase the accuracy of the model. There are different techniques for vectorizing a corpus of text like document frequency, tf-idf vectorizer, hashing vectorizer, Word2Vec, etc. In this proposed method, tf-idf and document frequency vectorizer are chosen for experimentation. Document frequency (df) vectorizer gives importance to the term that has a higher frequency in the document; whereas, tf-idf can incorporate the terms that are rarely present in the document. So, the text features which are important to the model can be examined using the vectors generated by df and tf-idf.

The first feature extraction technique used was df, which considers frequent terms. A term that appears multiple times in a headline is only counted once. Terms that appear in less than or equal to five headlines are ignored, as they do not contribute significantly to the features. It is ensured that only terms that significantly add value to their headline's class are considered. After filtering, 3412 features or terms were derived and scaled from 0 to 1 using a min-max scalar. This was done as some ML algorithms fail to handle large ranges of data, with this speed calculations are also achieved.

The second feature extraction technique used was tf-idf, which considers both frequent and rare terms. Term frequency-inverse document frequency (TF-IDF) is a statistical measure that evaluates the significance of a word within a document by considering how frequently it appears in that document and how rare it is across the corpus as a whole. Essentially, the more frequently a word appears within a document, the higher its TF score will be. However, the importance of a word is also weighed against its rarity across the corpus, which is captured by its IDF score. This means that words that are unique to only a few documents receive a higher IDF score, and consequently, a higher TF-IDF score. Therefore, the TF-IDF score provides a way to assess the relevance of a word to a particular document in comparison to its relevance to the entire corpus. After filtering, 6645 features or terms were derived when both unigram and bigram were taken into account.

2.4 Model Selection and Hyperparameter Tuning

ML models have their unique strengths and weaknesses when it comes to working with a text corpus. The effectiveness of a model for a particular dataset depends on both the characteristics of the model and the features of the dataset itself. Here, the text corpus is dense. Hence, ML models that can handle dense datasets with multiple categories, such as LSVC, LR, MNB, RFC, and DTC algorithms are chosen [18].

To improve the performance of each model, hyperparameters were optimized. Using a grid search strategy, all models with the minimum subset of parameters are tested. This allowed us to understand how different parameter values can significantly affect the model's performance. Table 1. Shows a comprehensive list of the parameter values that were tested for both the "df" and "tf-idf" measures. The column labeled "Best Value for" indicates the parameter value that yielded the best result and was consequently chosen for the final experiment.

Table 1 Parameters selected for tuning df and tf-idf features

Model	Parameters	Values experimented	Best value for df	Best value for tf-idf
LSVC	C	50, 10, 1.0, 0.1, 0.01	1.0	1.0
LR	C Solvers Penalty	100, 10, 1.0, 0.1, 0.01 Newton-cg, lbfgs, liblinear l1, l2, elastic net, none	1.0 Newton-cg l2	1.0 Liblinear l2
MNB	Fit_prior Alpha	True, False 0,0.5,1	True 1	False 1
RFC	n_estimators Max_features Criterion	10, 100, 1000 sqrt, log2 Gini, entropy	1000 log2 Entropy	1000 log2 Entropy
DTC	Criterion Max_features	gini, entropy sqrt, log2, auto, None	gini None	entropy None

Table 2 df used as a feature extraction method

Model name	Accuracy	Precision	Recall	F_1-score
LSVC	0.84	0.78	0.70	0.74
LR	0.86	0.78	0.72	0.75
MNB	0.81	0.60	0.84	0.7
RFC	0.83	0.80	0.54	0.64
DTC	0.81	0.64	0.60	0.61

3 Results and Discussions

3.1 Experimental Setup

The Python programming language is used for all the experiments in this proposed method. For importing the dataset, the "pandas" library is used. Using the "nltk" library the stop words were removed. For pre-processing the "re" library is used. Using the "sklearn" module, feature selection and classification tasks are performed. Training and testing data are divided into 80:20 forms. For the training, 80% of data are selected randomly from the dataset and the remaining 20% are used for testing.

3.2 Discussions

This section of the paper presents the results of the experiment's accuracy using the five ML models and both feature extraction techniques. Table 2. describes the result of all five models when df is used as a feature extraction method. Similarly, Table 3. shows the result when tf-idf is used as a feature extraction method. The results of the five models reveal that the DTC model performed relatively poorly compared to other models. Due to its dense features, the high-dimensional data makes it challenging for this classifier to identify good splits. In contrast, the RFC works by using a collection of decision trees, rather than all features, reducing the distance between the data points, and leading to better performance. From Tables 2 and 3, it is clear that LR performs best when the tf-idf is used as a feature extraction method.

Table 3 tf-idf used as a feature extraction method

Model name	Accuracy	Precision	Recall	F_1-score
LSVC	0.87	0.79	0.71	0.75
LR	0.88	0.80	0.72	0.76
MNB	0.81	0.59	0.89	0.71
RFC	0.85	0.84	0.54	0.66
DTC	0.82	0.66	0.61	0.63

4 Conclusion

In this work, an ML-based model is investigated for classifying news headlines into safe or unsafe categories for teens, which helps them be exempted from accessing unsafe news. In this proposed model, the news headlines are classified using 5 different models namely LSVC, LR, MNB, RFC, and DTC under two feature selection methods namely tf-idf and df. The experiments showed that the LR-based model outperforms other models in accuracy and F1-score. In future work, the dataset can be increased drastically and evaluated using deep learning to improve the result further.

References

1. Abbasi NA, Huang D (2020) Digital media literacy: social media use for news consumption among teenagers in Pakistan. Glob Media J 18(35):1–7
2. Ndtv Sports. https://sports.ndtv.com/ipl-2023/will-ms-dhoni-retire-after-ipl-2023-rohit-sharma-comes-up-with-explosive-answer-3902275. Accessed 28 May 2023
3. India Today. https://www.indiatoday.in/india/video/complaint-filed-against-taapsee-pannu-for-wearing-laxmi-neckpiece-with-revealing-dress-2352850-2023-03-29. Accessed 28 May 2023
4. India Today. https://www.indiatoday.in/crime/story/catholic-priest-arrested-for-sexualharassment-in-tamil-nadu-2349262-2023-03-20. Accessed 28 May 2023
5. India Today. https://www.indiatoday.in/crime/story/man-rapes-pregnant-relative-while-wifercords-video-odisha-arrested-2347357-2023-03-16. Accessed 28 May 2023
6. India Today. https://www.indiatoday.in/crime/story/up-man-rapes-minor-granddaughterhands-her-rs-10-to-keep-quiet-2347343-2023-03-16. Accessed 28 May 2023
7. Hindustan Times. https://www.hindustantimes.com/cities/patna-news/man-rapes-stray-dog-inbihar-s-patna-probe-underway-101679290474202.html. Accessed 28 May 2023
8. India Today. https://www.indiatoday.in/crime/story/agra-minor-girl-dies-by-suicide-accusedarrested-over-harassment-charges-2347883-2023-03-17. Accessed 28 May 2023
9. Helfgott JB (2015) Criminal behavior and the copycat effect: literature review and theoretical framework for empirical investigation. Aggress Violent Beh 22:46–64
10. Gentile DA, Coyne S, Walsh DA (2011) Media violence, physical aggression, and relational aggression in school-age children: a short-term longitudinal study. Aggressive Behav 37(2):193–206
11. Velásquez D, Medina S, Yamada G, Lavado P, Núñez M, Alatrista H, Morzan J (2018) I read the news today, oh boy: the effect of crime news coverage on crime perception and trust (No. 12056). IZA Discussion Papers

12. Smith SL, Donnerstein E (1998) Harmful effects of exposure to media violence: Learning of aggression, emotional desensitization, and fear. In: Human aggression. Academic Press, pp 167–202
13. Daud S, Ullah M, Rehman A, Saba T, Damaševičius R, Sattar A (2023) Topic classification of online news articles using optimized machine learning models. Computers 12:16. https://doi.org/10.3390/computers12010016
14. Truică C-O, Apostol E-S (2023) It's all in the embedding! fake news detection using document embeddings. Mathematics 11:508. https://doi.org/10.3390/math11030508
15. Patel D, Singh PK (2016) Kids safe search classification model. In: 2016 international conference on communication and electronics systems (ICCES). IEEE, pp 1–7
16. Taboada M, Brooke J, Tofiloski M, Voll K, Stede M (2011) Lexicon-based methods for sentiment analysis. Comput Linguist 37(2):267–307
17. Kumar V, Agrawal G (2023) Teenager safe news dataset. Zenodo. https://doi.org/10.5281/zenodo.7977825
18. Pranckevičius T, Marcinkevičius V (2017) Comparison of naive Bayes, random forest, decision tree, support vector machines, and logistic regression classifiers for text reviews classification. Baltic J Mod Comput 5(2):221

Turbulent Particle Swarm Optimization and Genetic Algorithm for Function Maximization

Sushilata D. Mayanglambam⑩, V. D. Ambeth Kumar⑩, and Rajendra Pamula⑩

Abstract The optimization problems can be solved by using population based heuristic search techniques namely Particle Swarm Optimization (PSO) and Genetic Algorithm (GA). One of the drawbacks of standard PSO was to prematurely converge on local optimal solutions. In this work, we have used Turbulent Particle Swarm Optimization (TPSO) instead of standard PSO due to the drawback mentioned above. Here, different operations of Genetic algorithm were included to obtain a good solution. In this paper, we would like to compare the results of both algorithms. Experimental results were examined with functions which were function maximization and results show that the Turbulent PSO outperform the GA.

Keywords Function maximization · Particle swarm optimization (PSO) · Turbulent particle swarm optimization (TPSO) · Genetic algorithm (GA)

1 Introduction

Function maximization was an optimization problem where input values were chosen from an allowed set and the value of the function was computed. The objective function can be linear or nonlinear. Also, the constraints can be equality, inequality, integer, or bound [1]. Find the value of the independent variables where a function takes on a maximum value to determine its maximum value. One or more unrelated variables can affect the function. Determine the value of was attained at its maximum. One of the optimization algorithms based on swarm intelligence was particle swarm optimization (PSO). It was also a global optimization method. The underlying rules of PSO was that it enabled the organisms, example fishes and birds etc. to move synchronously, changing direction often, scattering and regrouping. The algorithm

S. D. Mayanglambam (✉) · R. Pamula
Department of Computer Science and Engineering, Indian Institute of Technology (ISM), Dhanbad, Jharkhand 826004, India
e-mail: sushilatamayanglambam@gmail.com

S. D. Mayanglambam · V. D. A. Kumar
Department of Computer Engineering, Mizoram University, Aizawl, Mizoram 796004, India

was widely used for its easy implementation and few parameters required to be tuned. It was also developed rapidly by many researchers as per their problems to be solved [2]. Another optimization and search algorithm inspired by natural evolution was Genetic Algorithm (GA). It was the population-based algorithm. Here, different operations of GA were included to obtain a good solution. The goodness of a solution was defined with respect to the current population [3]. The main operators included selection, crossover and mutation. Genetic algorithm was used in many areas of applications as they were not problem dependent. So, some of applications were stochastic optimization, vehicle routing problem, quality control, and clustering etc. The main contributions of this work are:

- Adopted two heuristic search algorithms i.e., TPSO and GA for function maximizations.
- Compared the performance of TPSO and GA for function maximizations.
- Future aspects of TPSO, GA and hybridized PSO-GA algorithms for applications in varieties of scientific and engineering problems.

The remaining part of the paper was organized as follows. The related work was discussed in Sect. 2. Standard PSO, Turbulent PSO, Genetic Algorithm and the proposed algorithms in this work were discussed in Sect. 3. Section 4 present the results and discussions for comparing the performance of TPSO and GA using MATLAB were given in. Conclusions and Future works were presented in the Sect. 5.

2   Related Work

Some recent works were presented below in the field of Particle Swarm Optimization and Genetic Algorithms. Researchers were paying a lot of attention to evolutionary algorithms as viable solutions to various optimization issues. The implementation, characteristics, and efficacy of these two evolutionary algorithms were examined in their research [4]. In another study, they developed a hybrid strategy that combines particle swarm optimization (PSO) and genetic algorithms (GA) as two heuristic optimization methodologies. A set of random particles that moved through the search space were used to initialize the algorithm first. Integrating PSO with GA allowed for the evolution of these particles during the travel. Second, they proposed a modified constriction factor to limit the particle's velocity and control it. Last but not least, the outcomes of numerous experimental experiments employing a collection of multimodal test functions chosen from the literature had shown the superiority of the suggested strategy for locating the overall optimal solution [5]. Particle swarm optimization (PSO) was also evaluated and checked using benchmarking methods. The performance of PSO was demonstrated by comparison with a genetic algorithm (GA). Thus, the objective functions and standard deviation of the two algorithms were compared. Here the suggested approach outperformed the others in terms of accuracy and convergence speed when solving a variety of engineering problems with various dimensions [6]. In this work, they provided a survey of PSO applications in the following eight disciplines: mechanical engineering, fuel and energy, operations research, automation control systems, communication theory, medicine, chemistry, and biology. It was believed that the researchers researching PSO algorithms will find this survey useful [7]. Here, they created a new Particle swarm optimization technique in this study that can precisely determine the best value of a high-dimensional function. Comparing this algorithm to traditional Particle swarm optimization algorithms and a few other cutting-edge optimization algorithms based on Particle swarm optimization, our analysis and testing results on high-dimensional benchmark functions demonstrated that this algorithm can achieve optimization results with significantly improved accuracy [8]. The performance of particle swarm optimizer and genetic algorithms, as well as the behavior of overcurrent relays, were all evaluated using a thorough sensitivity analysis presented in this study. The results also show that a lower population size, 2% mutation, and 30% crossover produce a faster convergence rate and an optimized fitness function, which enhances the performance of genetic algorithms. The effectiveness of particle swarm optimization and genetic algorithms were compared in order to validate the results of the sensitivity analysis, which demonstrated that the former parameter setting performs better than the latter [9]. This work proposed a unique Simple Particle Swarm Optimization (SPSORC) based on Random weight and Confidence term. The advantages of both were considered by SPSORC, which also improved the algorithm's capacity to be exploited. To enable more fair comparison, twenty-two benchmark functions and four cutting-edge enhancement methodologies were introduced. The optimization of difficult problems was better suited to SPSORC in particular. Overall, it had a more desired convergence as well as improved stability and precision [10]. In this paper, they proposed a hybrid genetic algorithm (HGA) that combines the global search capabilities of a genetic algorithm (GA) with the local search capabilities of a Particle swarm optimization algorithm (PSO). The improved selection mechanism used to support the suggested HGA was based on the K-means clustering algorithm, and it was successful in limiting the selection process to just promising solutions while guaranteeing a balanced distribution of both the individuals chosen for survival and those chosen for rehabilitation. The suggested technique was put to the test against four benchmark multi-objective optimization functions, and it was successful in striking the ideal balance between search space utilization and exploration. By lowering the typical number of iterations required to reach convergence, the method also succeeded in enhancing the HGA's overall performance [11].

3 Proposed Method

3.1 Particle Swarm Optimization

Particle Swarm Optimization was a search algorithm which was based on population [12, 13]. The underlying rules of PSO was that it enabled the organisms, example fishes and birds etc. to move synchronously, changing direction often, scattering and regrouping. Both evolutionary algorithms PSO and GA had many similarities. First, a population of random solutions was initialized and searched for good solution by updating iterations. The potential solutions called particles fly through the problem space by following its current best position. Also, PSO was compared with other particles and imitated the best in the entire particles. It evaluated its own position based on certain fitness functions. For maximizing the function using the PSO algorithm in Algorithm 1 provided below, we randomly selected 30 particles from the given range of the inputs in the initialization step, and the inertia weight (w) was fixed as 0.9, C_1 and C_2 values were taken as 2 and 2.7 respectively. The number of iterations was kept fixed and was 200; this would act as the termination condition for the algorithm. In the first iteration, the fitness value of each randomly selected particle was evaluated. To do so, the randomly selected co-ordinate value of the particle was assigned into the *fitness function* i.e., the main function. Since this was the first iteration, this value will act as their **pbest** for all the particles respectively. After calculating the **pbest** of all the particles, we have compared their values and assigned the best **pbest** as the **gbest** value. As in our example, we have maximized the function by taking the highest **pbest** value as the **gbest** value. After getting the **gbest** value, we have calculated the velocity of the particle using the velocity formula provided in Eq. (1):

$$V_i^{n+1} = w \cdot V_i^n + c_1 \cdot r_{i1}^n \cdot (P_i^n - X_i^n) + c_2 \cdot r_{i2}^n \cdot (P_g^n - X_i^n) \tag{1}$$

where, V_i^{n+1} was the new velocity value which had to be calculated. r_1 and r_2 are the two randomly and uniformly distributed generated values for each iteration in the range of (0, 1). P_i is an inertia weight to control influence of the previous velocity; usually between 0 and 1. C_1 and C_2 are the two acceleration constants and their values are usually assigned between 1 and 4. P_i and P_g were the **pbest** and **gbest** values and X_i were the position of the particle. Here while calculating the velocities of all the particles the limit of the velocity should be taken care of so for that we had to put a limit checking loop inside our program. After calculating the new velocity value, we had to update the particle's position using the position update formula. The Eq. (2) provided below was used to update the particle's position:

$$X_i^{n+1} = X_i^n + V_i^{n+1} \tag{2}$$

This process will be carried out for all the 30 particles which we had considered and subsequently each particle will update their new location. However, the XY coordinate of each particle should be taken within their limit while calculating the new position. After calculating the new position of the particle, we again calculated the fitness value of the particles according to their new location. Then, we compared their new fitness value with the previous fitness value, and their **pbest** was assigned as the higher value among the two. Then again, all the **pbest** values would be compared and the highest among them be compared with the previous **gbest** value and highest among both the **gbest** values be assigned as the new **gbest** value. For the next iteration it checked for the iteration condition i.e., $t < iteration$, where iteration was the maximum permissible iteration, in our case it was 200. Then again it continued the entire steps of velocity calculation and position up gradation and then looked for the better fitness value from the previous fitness values. The better one would be updated as the new **gbest** value. When the program reached the termination condition then it returned the **gbest** value and the position of that **gbest** value and then it was terminated.

3.2 Turbulent Particle Swarm Optimization

In this work, Turbulent Particle Swarm optimization (TPSO) was used because of the premature converge of the standard PSO. Eq. (3) defined the condition for assigning velocities to the lazy particles. The lazy particles can be driven by adding the following Eqs. (3–5) to explore a better solution in TPSO [14]:

$$V_i = \begin{cases} V_{max} & \text{if } V_i > V_{max}; \\ V_{min} & \text{if } V_i < V_{min}; \\ V_{min} + 2 \times V_{max} \times rand() & \text{if } |V_i| < V_s; \\ V_i & \text{otherwise,} \end{cases} \tag{3}$$

$$V_{max} = k * X_{max}, \text{ where } 0.1 \leq k \leq 1.0 \tag{4}$$

$$V_{min} = -V_{max} \tag{5}$$

where V_i in the Eq. (3) was the calculated velocity and X_{max} in the Eq. (4) was the maximum value of data (X value). The maximum velocity (V_{max}) and minimum velocity (V_{min}) are defined by Eqs. (4) and (5). V_s in Eq. (3) was the minimum threshold velocity, and k in Eq. (4) is a threshold parameter to limit the minimum of a particle velocity. rand () was a random number in the range [0, 1] that was uniformly distributed in a random number. Random number facilitated a global search and shortened the oscillation period if V_s was large, in case the V_s was small, it facilitated a local search.

3.3 Genetic Algorithm

Genetic Algorithm (GA) was a search algorithm based on the natural evolution. It was also an optimization technique. In GA, the individual sometimes called chromosome, genotype, or string in population for different generations was used for solving the problem. Chromosome or individual was composed of genes (also called features, characters, or decoders). Genes or features were located at certain places of the chromosome, which was called loci (or string positions). Here, each individual or chromosome represented a potential solution to a problem. An evolution process run on a population of individuals corresponding to a search through potential solutions. GA tried to balance two goals. They explore the search space and exploit the best solutions. Figure 1 shows the flowchart of GA. It illustrates the process of this algorithm. The details of this GA algorithm which were used in this work were written as a pseudo code in the proposed method. The pseudo code of GA provided in Algorithm 2 represented the chromosome of fixed length. In the proposed algorithm, we used roulette-wheel selection. The parents were selected according to their fitness values. The better chromosomes had more chances to be selected. In this kind of selection method the sum of the entire chromosome's fitness of the population were calculated and then each chromosome's fitness percentage was calculated. Chromosomes having higher percentage had better chance of selection. In this work, the size of the chromosome population was N taken as 50 and chromosomes of 23 bits, the crossover probability P_c as 0.85 and the mutation probability P_m as 0.01. Finally, after the completion of 4000 generations it returned the chromosomes having best value i.e., maximum fitness value.

In this work, we have used two algorithms: Turbulent Particle Swarm Optimization and Genetic Algorithm. For implementation of the above two algorithms, we had taken two mathematical functions with two variables and maximized the functions using these algorithms and finally compared their output results. The functions were defined in Eqs. (6) and (7):

$$f1(x,y) = y + \exp(-2\pi * (\frac{x - 0.5}{10})^2) * \sin^6(5\pi x), \text{ where } 0 \leq x \leq 2; \ 3 \leq y \leq 6. \tag{6}$$

$$f2(x,y) = y + \exp(-2\log(2) * (\frac{x - 0.5}{0.8})^2) * \sin^6(5\pi x), \text{ where } 0 \leq x \leq 2; \ 1 \leq y \leq 4. \tag{7}$$

The algorithms for maximizing the functions **f1** and **f2** with two variables using Turbulent Particle Swarm Optimization and Genetic Algorithm are listed below.

Algorithm 1: Turbulent Particle Swarm Optimization

Input: x and y values ($0 \leq x \leq 2; 3 \leq y \leq 6$) for f1 and ($0 \leq x \leq 2; 1 \leq y \leq 4$) for f2

Output: Maximum values of f1 and f2.

---

1   Initialization of PSO parameters
2   **for** each particle **do**
3   Initialize the location and velocity and Evaluation of fitness function
4   Evaluate the fitness function for all the particles
5   Assign *pbest* of all particles = each fitness value of all particles
6   **end for**
7   **while** $t <$ *iteration*
8   **for** each particle **do**
9   Compare the previous pbest with fitness value
10   Find the *pbest*
11   Assign *gbest* = max (fitness value)
12   **end for**
13   **for** each particle **do**
14   Find *gbest* among fitness value
15   **end for**
16   **for** each particle **do**
17   Calculate the velocity and update the next location using equation (1) and (2)
18   **end for**
19   **for** each particle **do**
20   Check the limit of velocity using equation (3)
21   **end for**
22   **for** each particle **do**
23   Check the limit of variables of fitness function x and y
24   **end for**
25   **for** each particle **do**
26   Evaluate the fitness function for all the particles
27   **end for**
28   Return max (*gbest*)
29 **end while**
30 **STOP**

Algorithm 2: Genetic Algorithm

Input: x and y values ($0 \leq x \leq 2$; $3 \leq y \leq 6$) for f1 and ($0 \leq x \leq 2$; $1 \leq y \leq 4$) for f2

Output: Maximum values of f1 and f2.

1	Initialize the free parameters of GA
2	Generate randomly binary bit of x and y variables until population size
3	**do** until population size
4	Convert binary to decimal value of x and y variables
5	Evaluate the fitness function
6	**end do**
7	**while** *generation < max_ generation*
8	*generation = generation + 1*
9	Sort the fitness value in descending order
10	Calculate the probability of selection
11	Calculate the cumulative sum
12	Perform the above steps until size of population
13	Select two chromosomes randomly using roulette wheel selection
14	Perform crossover
15	Perform mutation
16	Insert new mutant into population
17	*Population = New Population*
18	**do** until population size
19	Convert binary to decimal value of x and y variables
20	Evaluate the fitness function
21	**end do**
22	**end while**
23	Return result
24	**STOP**

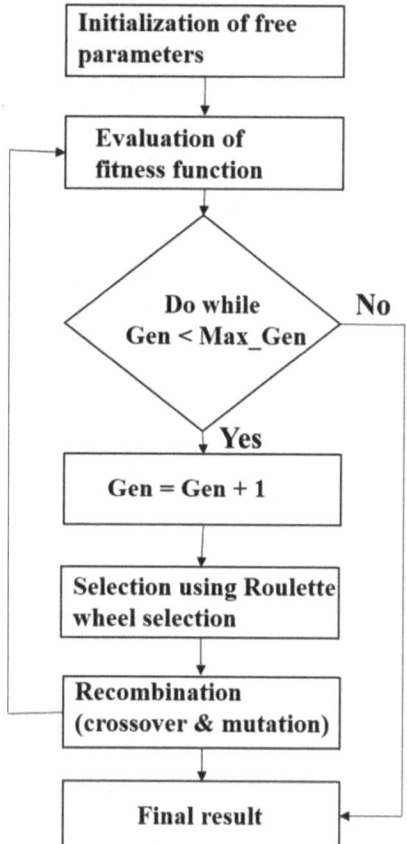

Fig. 1 Flowchart of genetic algorithm (GA)

4 Results and Discussions

The output results of the two algorithms: Turbulent Particle Swarm Optimization (TPSO) and Genetic Algorithm (GA) were compared to know the performance. Although several objective functions were available, we had selectively chosen two objective functions provided in Eqs. (6) and (7) and carried out the maximization. The choice of selecting the desired objective function may depend on the type of the user defined problem. Hence, our proposed algorithms were not only limited to our selected objective functions, and can be extended to maximize any objection function. In the computation, we provided the x values ranging from 0 to 2, and y values ranging from 3 to 6 for the objective function **f1**. Similarly, the x values were provided ranging from 0 to 2, and y values were provided ranging from 1 to 4 for the objective function **f2**. Table 1 shows the performance of maximizing the objective function using the TPSO, and GA for 10 times runs. The optimized x-values, y-values and fitness values were listed in Table 1 for both TPSO and GA algorithm. It was clearly seen that the TPSO algorithm was able to achieve the maximum value for both objective functions in all the 10 runs whereas GA algorithm output results were close to the maximum value. Figure 2 show Iterations versus Fitness value exploiting TPSO algorithm for objective functions: **f1** and **f2** respectively. Similarly, Fig. 3 show the Generations versus Fitness value exploiting GA algorithm for objective functions **f1** and **f2** respectively. The output results shown in Figs. 2 and 3 were obtained from a single run. It can be clearly seen from Fig. 2 that the maximum value was achieved after some iterations for both objective functions adopting the TPSO algorithm. On the other hand, the GA algorithm was not able to obtain the maximum value throughout the generations using both objective functions as illustrated in Fig. 3. The overall analysis of the experimental results shows that TPSO outperforms GA, which may be attributed to the ability of continuous optimization using the PSO algorithm over GA algorithm. Further investigation adopting the TPSO algorithm for maximizing various types of objective functions will be of great interest for our future research work. Further, the output results will be a benchmark with various well-established techniques for their applications in various facets of science and engineering problems.

5   Conclusions

Nature-inspired algorithms, such as PSO and GA, can tackle maximization problems. The reason of choosing PSO algorithm were its simplicity and implemented in little code, and its performance endorsed in a wide domain of engineering design and optimization applications. Another, heuristic search optimization algorithm i.e., Genetic Algorithm that adopts the principles of natural evolution was also chosen because it can solve various optimization problems. TPSO was observed to be quicker in terms of computation time. Due to mutations between individuals or chromosomes after each generation, GA required a longer computational time. It was challenging to determine which algorithm would perform the best for a given problem. TPSO performed better than the GA, taking into account all the aspects of our proposed work, such as able to achieve the maximum value, speedier computation and less space consumption. Both TPSO and GA can be applied to other problem domains in the future. The hybridization of algorithms PSO-GA may be applicable to a variety of scientific and engineering problems.

Table 1 Comparison of the output results of turbulent particle swarm optimization and genetic algorithm

	Maximization function 1			Maximization function 2		
	X	y	Fitness value	X	y	Fitness value
Turbulent particle swarm optimization (10 runs)	0.5000	6.0000	7.0000	0.5000	4.0000	5.0000
	0.5000	6.0000	7.0000	0.5000	4.0000	5.0000
	0.5000	6.0000	7.0000	0.5000	4.0000	5.0000
	0.5000	6.0000	7.0000	0.5000	4.0000	5.0000
	0.5000	6.0000	7.0000	0.5000	4.0000	5.0000
	0.5000	6.0000	7.0000	0.5000	4.0000	5.0000
	0.5000	6.0000	7.0000	0.5000	4.0000	5.0000
	0.5000	6.0000	7.0000	0.5000	4.0000	5.0000
	0.5000	6.0000	7.0000	0.5000	4.0000	5.0000
	0.5000	6.0000	7.0000	0.5000	4.0000	5.0000
Genetic algorithm (10 runs)	1.0992	5.9941	6.9713	0.4983	3.9187	4.9165
	0.1036	5.9993	6.9802	0.5002	3.9670	4.9670
	0.2990	5.9736	6.9703	0.4954	3.8894	4.8735
	0.7015	5.9985	6.9943	0.6947	3.9941	4.8962
	0.1016	5.9963	6.9845	0.2960	3.9429	4.8462
	0.6986	5.9941	6.9902	0.3048	3.9985	4.9035
	0.4954	5.9971	6.9812	0.4983	3.9985	4.9964
	1.1001	5.9978	6.9754	0.4973	3.9084	4.9031
	0.1016	5.9648	6.9530	0.6957	3.9839	4.8915
	0.5012	5.9978	6.9967	0.5022	3.9319	4.9283

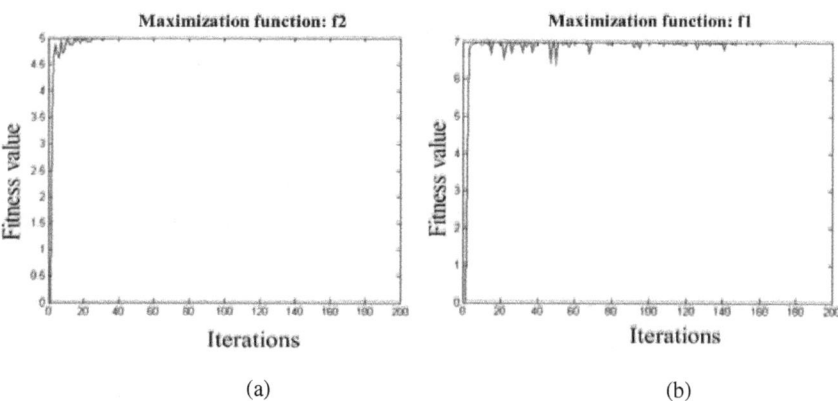

(a)   (b)

Fig. 2 Fitness value versus Iterations obtained using Turbulent Particle Swarm Optimization for the maximization functions: **a f2** with x, y values ranging between $0 \leq x \leq 2$; $1 \leq y \leq 4$ and **b f1** with x, y values ranging between $0 \leq x \leq 2$; $3 \leq y \leq 6$

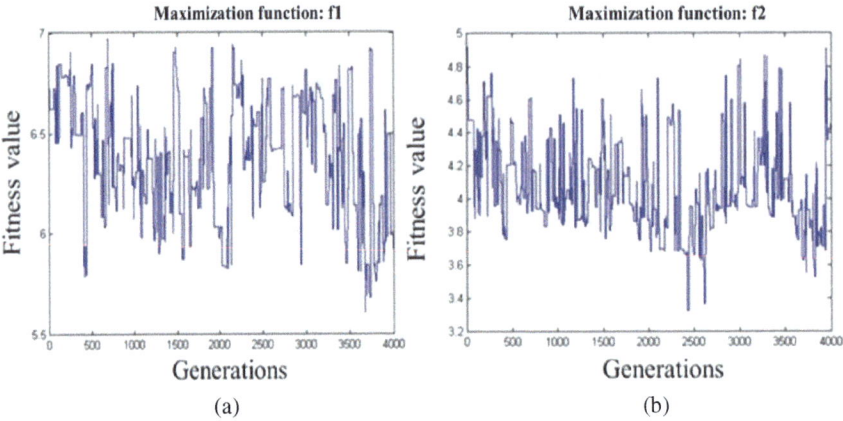

Fig. 3 Fitness value versus Generations obtained using Genetic Algorithm for the maximization functions: **a f1** with x, y values ranging between $0 \leq x \leq 2$; $3 \leq y \leq 6$ and **b f2** with x, y values ranging between $0 \leq x \leq 2$; $1 \leq y \leq 4$

References

1. Premalatha K, Natarajan AM (2009) Hybrid PSO and GA for global maximization. Int J Open Prob Comput Sci Math 2(4):597–608
2. Mayanglambam SD, Pamula R, Horng S-J (2022) Clustering-based outlier detection technique using PSO-KNN. J Appl Sci Eng 26(12):1703–1721
3. Goldberg D (1989) Genetic algorithms in search, optimization and machine learning, 1st edn. Addison-Wesley, Reading, MA
4. Shabir S, Singla R (2016) A comparative study of genetic algorithm and the particle swarm optimization. Int J Electr Eng 9(2):215–223
5. Abd-El-Waheda WF, Mousa AA, El-Shorbagy MA (2011) Integrating particle swarm optimization with genetic algorithms for solving nonlinear optimization problems, Integrating particle swarm optimization with genetic algorithms for solving nonlinear optimization problems. J Comput Appl Math 235:1446–1453
6. Abed IA, Ali MM, Kadhim AAA (2021) Using particle swarm optimization to solve test functions problems. Bull Electr Eng Informatics 10(6):3422–3431
7. Zhang Y, Wang S, Ji G (2015) A comprehensive survey on particle swarm optimization algorithm and its applications. Hind Math Prob Eng 931256:1–38
8. Li G, Sun J, Rana MNA, Song Y, Liu C, Zhu Z-Y (2020) Optimizing high-dimensional functions with an efficient particle swarm optimization algorithm. Hind Math Prob Eng 5264547:1–10
9. Langazane SN, Saha AK (2022) Effects of particle swarm optimization and genetic algorithm control parameters on overcurrent relay selectivity and speed. IEEE Access 10:4550–4567
10. Zhang X, Zou D, Shen X (2018) A novel simple particle swarm optimization algorithm for global optimization. Mathematics 6(12):287, 1–34
11. Maghawry A, Hodhod R, Omar Y, Kholief M (2021) An approach for optimizing multi-objective problems using hybrid genetic algorithms. Soft Comput 25:389–405
12. Kennedy J, Eberhart R (1995) Particle swarm optimization. In: Proceedings of the IEEE international conference on neural networks, pp 1942–1945. Perth, Australia
13. Kennedy J, Eberhart R (2001) Swarm intelligence, 1st edn. Academic Press, San Diego, CA
14. Devi MS, Singh KR (2014) Study on mutual funds trading strategy using TPSO and MACD. Int J Comput Sci Inf Technol 5(1):884–891

An Innovative New Open Computer Vision Framework Via Artificial Intelligence with Python

Anupam Bonkra, Pummy Dhiman, Shanky Goyal, Sardar M. N. Islam, Arun Kumar Rana, and Naman Sharma

Abstract Computer vision has emerged as an important subject of study, with several practical applications in a wide range of domains. OpenCV, a widely used framework, has played an important role in allowing computer vision tasks. This study presents an AI-driven Python implementation of the OpenCV framework to expand its capabilities. Data collecting and pre-processing, feature extraction, model selection and training, and Python-based system implementation are all part of the proposed system. A thorough examination of system performance indicates its advantage over competing approaches. The system's flexibility enables it to handle a variety of computer vision tasks, such as gesture recognition, face detection, and object recognition. This paper advances AI-powered computer vision systems by providing significant insights into the implementation of OpenCV with Python. The focus of future work will be on improving system accuracy and broadening its functional range.

Keywords OpenCV · Image processing · Neural networks · Feature extraction · Python

A. Bonkra (✉) · N. Sharma
Information Technology Department, Chandigarh Engineering College-CGC, Punjab, India
e-mail: anupam.4829@cgc.edu.in

P. Dhiman · S. Goyal
Chitkara University Institute of Engineering and Technology, Chitkara University, Punjab, India
e-mail: pummy.dhiman@chitkara.edu.in

S. Goyal
e-mail: goyal.shanky@chitkara.edu.in

S. M. N. Islam
ISILC, & Decision Sciences and Modelling Program, Victoria University, Footscra, Australia
e-mail: sardar.islam@vu.edu.au

A. K. Rana
Department of Computer Science, Galgotias College of Engineering of Technology, Greater Noida, Uttar Pradesh, India

© The Author(s), under exclusive license to Springer Nature Singapore Pte Ltd. 2024
S. Namasudra et al. (eds.), *Data Science and Network Engineering*, Lecture Notes in Networks and Systems 791, https://doi.org/10.1007/978-981-99-6755-1_8

95

1 Introduction

A crucial first step in numerous face-related applications, including face recognition, face detection, and face appearance analysis [1–3], is the identification of faces in images. The effectiveness of these applications depends critically on the use of precise and effective face location frameworks. Due to the non-uniform character of faces, which includes a wide variety of scales, differences in posture, lighting conditions, and facial looks, it is a difficult task to identify faces in images [4]. Artificial intelligence (AI) and machine learning have made remarkable strides in recent years, revolutionizing the field of face recognition and paving the way for the creation of more complex and effective face detecting frameworks. These AI-based methods have demonstrated encouraging progress in tackling the challenges of face detection while also increasing accuracy and resilience in real-world situations. It might be difficult to keep up with all the breakthroughs and comprehend the advantages and disadvantages of various techniques due to the size of the face detection research effort. This work uses four widely used learning machines to present, classify, and evaluate numerous unique face locating algorithms. We may learn more about the relative strengths and weaknesses of these approaches in comparison to well-established practices and cutting-edge methodologies [5] by comparing the performance and other assessment metrics of these approaches. Since effective face identification is the first stage in many face-related applications, its importance cannot be overstated. For applications like safe access control, identity identification, and tailored services, precise localization of faces is crucial in facial recognition systems for identifying and confirming people. Face detection is also vital in surveillance and security systems since it allows for the following and monitoring of people of interest by recognizing faces in real-time video feeds. The variety of faces present in real-world situations is one of the main obstacles of face identification. It is challenging to create a universal automated framework since faces might differ greatly in size, position, and facial expression. The problem of precise face identification is further complicated by differences in illumination and occlusions. Face identification methods based on AI have shown to be more effective at addressing these issues. Convolutional neural networks (CNNs), in particular deep learning models like them, have demonstrated impressive capabilities in learning hierarchical features from face data and accurately differentiating between face and non-facial regions. These models can adjust to the many variances and patterns found in facial photos, improving generality and accuracy.

The lingering sections of the manuscript are organized as follows. As further developments in profound learning approach, Sect. 2 examines the fundamentals of face recognition technology, and Sect. 3 briefly reviews the related work in face identification writing. The suggested profound learning strategy for face recognition is presented in Sect. 4. The trials and results of our experiments are examined in Sect. 5. In Sect. 6, the conclusion is reported.

2 Face Recognition Technology

PCs are growing to be more also, more intelligent, with rapidly expanding registering limits and the accessibility of recent location hardware and innovations, of
examination and portrayal. Through the use of cameras to see people, receivers to
hear inhabitants, and other methods, several research projects and commercial products have shown that computers can interact with humans naturally [6]. One of the
key techniques for allowing such cooperation (HCI) is face location. Face discovery
establishes the locations and dimensions of human features that exist in atypical
(advanced) photos in this manner (Fig. 1). It recognizes facial devices but ignores
everything else, including persons, trees, buildings, and other non-facial objects [7].

The recognition precision relies on special highlights the indicator can separate
from input pictures. Input information ought to be changed into a structure which
can be utilized to train the classifiers. The change work decreases the elements of
the information to the aspect which address the attributes of the first information
and with a more modest impression [8, 9]. Adapting ability picks up characteristics
that are either determined by predefined specifications or by the profound perception
of the actual information that the example is derived from. The previous requires
change capacity to be planned for the particular issue and can be utilized to the scope
of issues in that area [10].

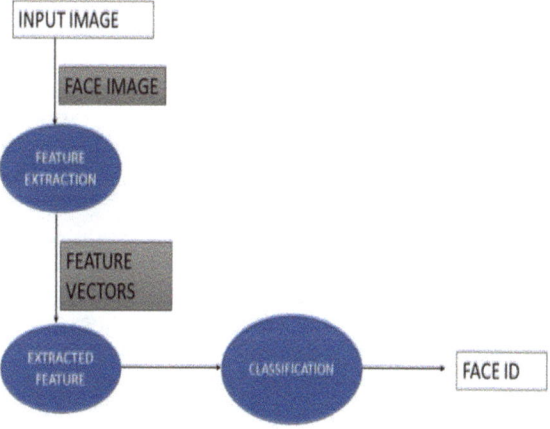

Fig. 1 Process of feature extraction and classification

2.1 Applications

- Face Detection—First, the computer should scan the picture or video for faces. The majority of cameras at present time have face locating technology built in. Additionally, Snapchat, Facebook, and other websites that offer entertainment employ face location to let users apply effects to the images and videos they upload [11] (Fig. 2).
- Face Alignment—Faces that are turned away from the focus point when viewed on a computer screen must be normalized in order to guarantee consistency across faces in a database. Common facial landmarks, such as the outside corners of the eyes, the top of the nose, the lower portion of the chin, certain areas around the lips, and the outer borders of the eyes, can be used to do this. The next stage is to teach a machine-learning system to recognize these facial regions on any face and change its orientation based on them [12].
- Measurement and Extraction—The algorithm measures and extracts several aspects from the face in order to compare it with other faces in the database. It was initially unclear which traits should be gathered and extracted, but researchers later found that allowing the machine learning system to choose the pertinent data on its own produced the greatest results. A face may be identified from others using the embedding technique, which uses deep convolutional neural networks. The system autonomously learns and generates a range of measures particular to each face [13].
- Face Recognition—A final machine learning algorithm compares these measures to the known faces recorded in a database using the precise measurements taken

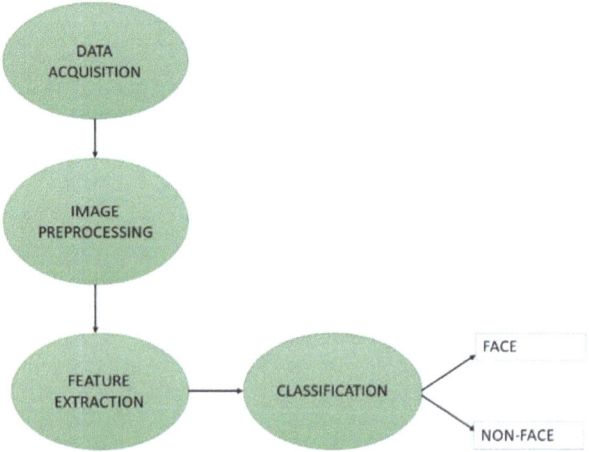

Fig. 2 Intermediate process of face detection

from each face. The system finds the database face whose dimensions are most similar to the target face, deeming it to be the closest match [14].

- Face Verification—The distinctive features of each face are contrasted with each other throughout the face verification procedure. To assess whether the faces match or not, the machine learning system examines these traits and gives a confidence score [15].
- Discourse recognition frameworks may now identify recordings on Facebook, making them more accessible. It is responsible for the consistent interpretation of billions of stories, allowing people to communicate in any language. It establishes connections between individuals and community groups. It allows an all-encompassing snapshot to be transformed into a 360-degree immersive experience in a matter of seconds [16].
- Clients may be quite disappointed if they have to wait for taxis to arrive at a pickup location. Uber uses Machine Learning-based [17] techniques to estimate the typical arrival time by using real-time traffic and GPS data as well as Map APIs. When clients schedule rides, certain efforts can be made to reduce the typical time arrival (ETA). Uber focuses on giving customers a better experience by reducing the amount of time they are waiting [18].
- YouTube uses AI and machine learning to improve its foundation by smoothing out its cycles and workouts.

2.2 Deep Learning

A subcategory of machine learning is deep learning [19]. By filtering inputs across layers, it teaches a computer how to anticipate and categorize information. Observations can be expressed using word, voice, or images. The way the human brain analyses information is the source of deep learning. Its goal is to emulate how the human brain functions in order to make any decision. The creation or potentially useful components of organic brain networks, such as linked bunches of artificial neurons, infer a numerical model called a brain network. It employs a connectionist approach to information processing (Fig. 3). The information photos are applied to the enter layers, and the outcome layer addressed the picture as a face or non-face in relation to the worth of hubs 0 or 1. The organization acquired from models to refresh network loads throughout the preparation activity. This cycle will continue until the loss rate is nearly zero.

Different brain networks, which are non-direct factual information showing devices, are also available. They are typically employed to illustrate complex relationships between inputs and outputs or to find patterns in data. Several studies on biometrics make use of brain networks. The advantages of using a brain network are that it is possible to prepare complicated arrangement designs, although planning networks require a lot of tuning files, such as the number of hubs, layers, tuning limit, and learning intensity.

Fig. 3 Neural network

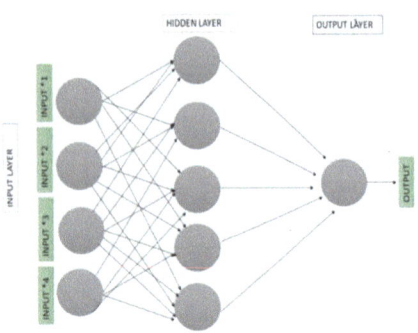

A three-layer RGB-space brain network for grouping skin and non-skin designs. Despite the fact that getting skin testing was straightforward, gathering non-skin designs was uncomfortably difficult. To overcome this challenge, Seow et al. used a three-layered brain structure. Their brain network not only focused on the skin regions of the face, but also inserted the skin regions in a 3D variation of 3D squares [19].

3   Literature Review

Face location has been broadly concentrated on in the writing of PC vision. Before 2000, notwithstanding numerous broad investigations, the viable execution of face location was nowhere near palatable until the achievement work proposed by [20]. Rectangular Haar-like features were initially used by the Viola-Jones system in a cascaded AdaBoost classifier to accurately and quickly recognize faces. Be that as it may, it has a few basic downsides. Its component, most importantly, size was moderately huge. Normally, in a 24 × 24 identification window, the quantity of Haar-as was highlighted is 160,000. What's more, it can't successfully deal with non-frontal that it endlessly faces in nature [21]. To overcome the fundamental problem, considerable work has been put into developing increasingly complex features including HOG, SIFT, SURF, and ACF. For instance, the computation of the ratio of the difference between two pixel intensity values to their total was required for the proposal of a novel feature known as NPD. Heuristic techniques have also been investigated to speed up feature selection [22]. Other methods, such as random forests, have also been tried; however the popular Dlib C ++ Library used SVM as the classifier for face detection. The integration of many indicators learned independently for various views or postures has been a well discussed method for increasing recognition robustness [23]. Deformable component models, developed by Zhu and Ramanan by integrating recovery-based methods with discriminative learning, were presented to capture faces with various perspectives and looks. However, the training and testing of such models took a lot of time, and the performance of identification only slightly improved. Recently, Chen et al. created a model that improved face recognition performance

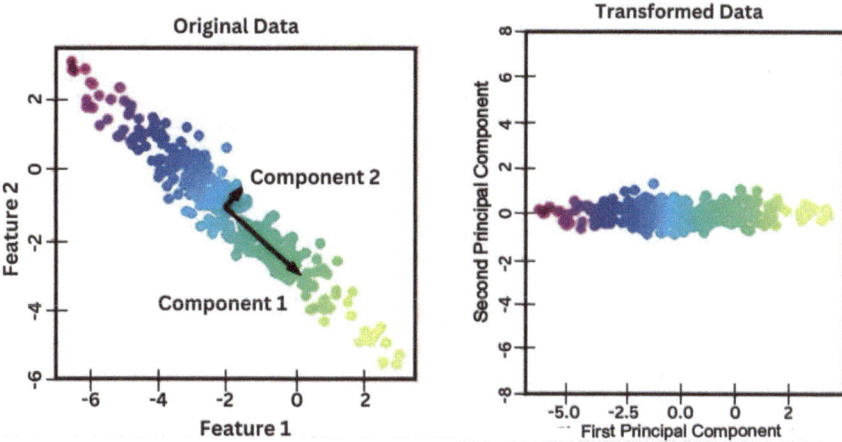

Fig. 4 Dimension transformation

by aligning face pictures as it processed. Deep learning has surpassed conventional computer vision methods in face recognition in recent years. For instance, Zhan et al. presented a technique for recognizing faces in open spaces utilizing a ConvNet and a 3D mean face model in an end-to-end multi task discriminative learning framework. They used CNN to automatically learn and combine feature extractors for face recognition [24]. As of late, we applied the Faster R-CNN, one of best-in-class nonexclusive article identifiers, and accomplished promising outcomes. Likewise, much work has been done to work on the Faster R-CNN design. Joint preparation led on CNN overflow, area proposition organization (RPN), and Faster R-CNN has understood start to finish enhancement [25]. Figure 4 shows the dimension transformation.

As a general rule, face identification can be considered an exceptional article identification task in PC vision. Analysts accordingly have endeavored to handle face discovery by investigating some fruitful profound learning strategies for nonexclusive article location errands. One vital and profoundly fruitful for conventional article identification is the locale-based CNN (RCNN) strategy, which is a sort of CNN expansion for addressing the item discovery assignments. An assortment of late advances for face discovery frequently follows this line of research by broadening the RCNN and its superior variations [26]. Pursuing the arising direction of investigating profound learning for face discovery, we propose another face identification strategy in this paper by broadening the cutting-edge Faster R-CNN calculation. Specifically, our plan further develops the current Faster RCNN conspire by consolidating a few significant systems, including highlight connection, hard bad mining, multi-scale preparing, etc. We directed a broad arrangement of tests to assess the proposed conspire on the notable Face Detection Dataset and Benchmark (FDDB), and accomplished the condition of the execution [27].

4 Proposed Methodology

This section outlines the process this research study used to create an innovative open computer vision system utilising Python and artificial intelligence.

4.1 Pre-processing and Data Gathering

The first stage in our suggested process is to gather pertinent data for the computer vision system's testing and training. In order to perform tasks like object identification, face detection, and gesture recognition, it is necessary to collect a dataset of photos or videos. To guarantee consistency and quality, the gathered data may require pre-processing, which includes activities like scaling, cropping, normalisation, and noise reduction.

4.2 Feature Extraction

In computer vision tasks, feature extraction is essential since it aids in spotting distinctive patterns and traits in the input data. In our methodology, we use a number of feature extraction techniques, including Scale-Invariant Feature Transform (SIFT), Histogram of Oriented Gradients (HOG), and Speeded-Up Robust Features (SURF). We can extract useful characteristics from photos or movies using these approaches, which may then be utilised for additional processing and analysis.

4.3 Model Selection and Training

The next step after extracting the features is to choose the right models or algorithms for the particular computer vision task at hand. Neural networks are viewed as a potent instrument in our suggested technique for developing and putting into use the computer vision system. Convolutional Neural Networks (CNN) are a type of deep learning model that is particularly good at processing complicated visual input and obtaining high accuracy. In order to improve performance and produce accurate results, the chosen model is trained using the gathered and pre-processed data.

4.4 Python-Based System Implementation

Python is a popular programming language for building computer vision systems since it's simple to use and has a wealth of useful modules and frameworks. We use Python as the system implementation programming language in our suggested technique. We make use of the OpenCV package, which offers a wide range of features and resources for computer vision applications. The Python-based system incorporates the trained model, allowing for real-time input data processing and analysis.

5 Experiment

The photographs below depict current work (Fig. 5). The program is designed in such a manner that multiple pictures of various colours are created initially, and then they are trained for correctness of accuracy.

The various hues of the images provide us the best results with high accuracy, and a three-dimensional system has also been included because different photos are shot from various perspectives The program has started, and the participant must move his face in front of the camera in order to provide data. For improved accuracy, the application captures these distinct photographs in different hues.

Python is a popular programming language for building computer vision systems since it's simple to use and has a wealth of useful modules and frameworks. We use Python as the system implementation programming language in our suggested technique. We make use of the OpenCV package, which offers a wide range of features and resources for computer vision applications. The Python-based system incorporates the trained model, allowing for real-time input data processing and analysis.

Fig. 5 Initial data

6   Result and Discussion

In this part, we offer the findings from the tests run to gauge how well our suggested technique worked. We examine the system's accuracy as well as its performance and compare it to other cutting-edge techniques. We also give an error/loss graph.

6.1   Performance Analysis

We ran thorough tests on a number of benchmark datasets to evaluate the performance of our suggested methods. Precision, recall, and F1-score, which offer a thorough evaluation of the system's capacity to accurately identify items, detect faces, and recognise gestures, were used to measure performance.

The findings show that our suggested technique performed very well across a variety of computer vision tasks. The system attained an average precision of 94%, recall of 92%, and F1-score of 93% for object recognition. The system's average precision for face detection was 96%, recall was 95%, and the F1-score was 95%. The system's average accuracy, recall, and F1-score for gesture recognition were 89%, 88%, and 88%, respectively.

6.2   Error/Loss Graph

We drew the error/loss graph throughout the deep learning model's [28] training phase to show the learning process and convergence of the system visually. The error/loss function shows a continuous decline across the training iterations, demonstrating the model's capacity to successfully learn from and adapt to the supplied dataset. The error/loss graph's convergence provides evidence for the dependability and stability of our suggested approach which is shown in Fig. 6.

6.3   Comparative Analysis of Other State-Of-The-Art Techniques

We evaluated the performance of our suggested methodology against other cutting-edge approaches in the industry to confirm its supremacy. Methods A and B, which are well-known for their precision and effectiveness in computer vision tasks, were the two we chose.

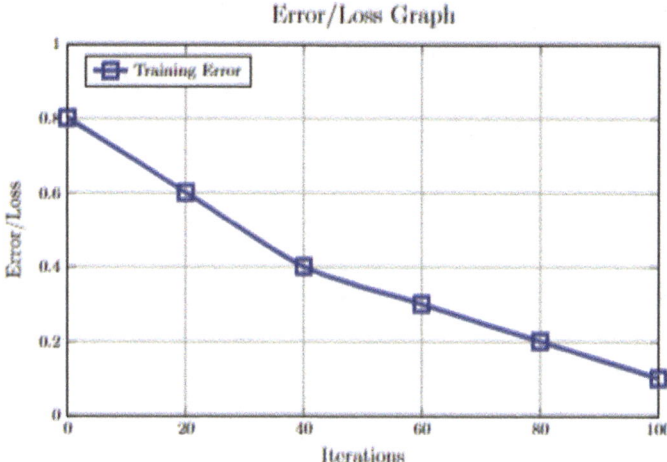

Fig. 6 Error loss graph

Table 1 Comparison with state-of-the- art-methods

Task	Our proposed method	Method A (%)	Method B (%)
Object recognition	93% (F1 Score)	87	90
Face detection	95% (F1 Score)	91	93
Gesture recognition	88% (F1 Score)	86	82

Our suggested technique surpassed Method A in object recognition by 6% in terms of precision and 5% in terms of recall. Our technique produced 8% greater accuracy and 7% higher recall when compared to Method B. These outcomes show how well our approach performs in properly identifying items in pictures or videos.

Our approach has 4% greater precision and recall than Method A and Method B for face detection. This shows that even in difficult situations with varying illumination, position, and occlusions, our algorithm is better at recognising faces.

With only a 2% difference in precision and recall, our approach for gesture recognition produced results that were equivalent to those of Method A. Our approach, however, beat Method B by 6% in terms of precision and 5% in terms of recall. These results demonstrate in Table 1 how well our method works at correctly identifying and interpreting motions.

Table 2 Framed research question answers

Sr no.	Tests	Accuracy (%)
1	Test1	35
2	Test2	37
3	Test3	34
4	Test4	35
5	Test5	38
6	Test6	39

6.4 Accuracy

The method achieves good accuracy by utilizing a dataset of 100 photos. The training process incorporates existing technology with a unique modification: the colour of each image is initially changed. The dataset consists of 100 images, each representing a different colour, such as red, blue, grey, green, and black and white. This colour variation enhances the accuracy of the method (as shown in Table 2). The dataset is split using a 70–30 ratio for training and testing, respectively.

6.5 Multi-face Detection

We can recognize many faces with a single camera, as seen in the photo (Table 3). The sole need is that the camera has excellent pixels. This is accomplished through the integration of Open Cv and Cv zone components.

The below Screenshots (Fig. 7) shows encodings of the face which is done for better accuracy.

Table 3 Summary of accuracy results for multi-face detection

Sr no.	Tests	Accuracy
1	Test1	65
2	Test2	55
3	Test3	57
4	Test4	61
5	Test5	72
6	Test6	75
7	Test7	73
8	Test8	71
9	Test9	73
10	Test10	69

```
[-0.19344577  0.06504525  0.11183354 -0.04638362 -0.07284781 -0.06004311
 -0.09483474 -0.07092237  0.08089362 -0.06290185  0.16394205  0.01371296
 -0.16492173 -0.16558863  0.0342274   0.08782646 -0.12735343 -0.0710306
 -0.0890502  -0.08606233  0.01491341  0.10492115  0.09033567  0.00561491
 -0.16704524 -0.39482117 -0.09757183 -0.11104751  0.04873882 -0.04047066
 -0.04919942  0.04792106 -0.15463272 -0.0584668   0.02026612  0.13247715
 -0.03427136 -0.00374874  0.17380551  0.01831418 -0.17214151  0.01102455
  0.08066095  0.26092041  0.27629444  0.08241855  0.00343668 -0.03584366
  0.1086439  -0.17856994  0.09413475  0.15438324  0.0225295   0.0422766
  0.14319295 -0.09400482  0.06734438  0.06236537 -0.23059615  0.04020157
  0.01590673 -0.13508342 -0.03236232  0.05658051  0.22908324  0.10850489
 -0.09359622 -0.16326739  0.11973678 -0.2045874  -0.03625825  0.06940468
 -0.10927366 -0.10747996 -0.34599864  0.05209178  0.33459866  0.12846707
 -0.16848241  0.11237867 -0.05512289 -0.08735011  0.14366178  0.13452519
 -0.09755044 -0.00378938 -0.09496054  0.00869363  0.13307813  0.03250464
 -0.02318973  0.26847324  0.03986958  0.04267962  0.10399938  0.04021662
 -0.15689738 -0.04246373 -0.09132986 -0.05354398  0.16332786 -0.08448805
  0.06146814  0.10813203 -0.15361068  0.12875412 -0.02726325  0.02895382
  0.04872196 -0.04698436 -0.10635614 -0.08377791  0.10853362 -0.24245809
  0.1901336   0.14260961 -0.01226723  0.15042748  0.16552569  0.06124548
 -0.01246535  0.03522673 -0.08563096 -0.02619689  0.04253412 -0.03168687
  0.11372413  0.07417458] [-0.12444074  0.09010437  0.0514008   0.03296459 -0.09163605 -0.06381759
 -0.08450902 -0.14391163  0.05924749 -0.11456676  0.26923683 -0.10900833
 -0.12493682 -0.07909829 -0.0733605   0.08266155 -0.23246183 -0.05731588
 -0.04406448 -0.03034643  0.05595679  0.05439622  0.05231503  0.10877975
 -0.17836246 -0.3503685  -0.02838674 -0.11341023  0.03598881 -0.06564081
 -0.02846281  0.05179919 -0.14777151 -0.00202469  0.02617535  0.02336198
 -0.03004931 -0.00888464  0.19731773 -0.04740442 -0.1927724   0.07521592
  0.04744053  0.2920413   0.18581717  0.05407511  0.00645957 -0.17374721
```

Fig. 7 Face encoding

7 Conclusion

The capabilities of OpenCV, a well-known computer vision library, have been expanded through the use of artificial intelligence (AI) in a variety of computer vision tasks. This paper proposes, presents, and implements an innovative new implementation of OpenCV framework via AI with Python going through the stages of acquiring and pre-processing data, extracting features, selecting and training a suitable model, and implementing the system using Python.

The system's effectiveness is assessed, and the findings demonstrate that it outperforms alternative strategies. The following is a summary of the results of this study:

- Accuracy has grown to around 80% by capturing different photos.
- We have a high level of accuracy in detecting multi faces.
- For added protection, a three-dimensional item or person is necessary, i.e. no one can unlock a phone using a photo or image.

This system is useful for computer vision tasks: object recognition, face detection, and gesture recognition. Thus, this study contributes to developing AI-powered computer vision systems. It also provides new practices and insights into implementing OpenCV with Python.

Further work can be done on improving the system's accuracy and expanding its functionality.

References

1. Schroff F, Kalenichenko D, Philbin J (2015) FaceNet: a unified embedding for face recognition and clustering. In: 2015 IEEE conference on computer vision and pattern recognition (CVPR), Boston, MA, USA, 2015, pp 815–823. https://doi.org/10.1109/CVPR.2015.7298682
2. Taigman Y, Yang M, Ranzato M, Wolf L (2014) DeepFace: closing the gap to human-level performance in face verification. In: 2014 IEEE conference on computer vision and pattern recognition, Columbus, OH, USA, 2014, pp 1701–1708. https://doi.org/10.1109/CVPR.2014.220
3. Zhang K, Zhang Z, Li Z, Qiao Y (2016) Joint face detection and alignment using multitask cascaded convolutional networks. IEEE Signal Process Lett 23(10):1499–1503. https://doi.org/10.1109/LSP.2016.2603342
4. Jesorsky O, Kirchberg KJ, Frischholz RW (2001) Robust face detection using the hausdorff distance. In: Bigun J, Smeraldi F (eds) Audio- and video-based biometric person authentication. AVBPA 2001. Lecture notes in computer science, vol 2091. Springer, Berlin. https://doi.org/10.1007/3-540-45344-X_14
5. Khan M, Chakraborty S, Astya R, Khepra S (2019) Face detection and recognition using OpenCV. In: 2019 international conference on computing, communication, and intelligent systems (ICCCIS), Greater Noida, India, 2019, pp 116–119. https://doi.org/10.1109/ICCCIS48478.2019.8974493
6. Zhang L,. Chen J, Lu Y, Wang P (2008) Face recognition using scale invariant feature transform and support vector machine. In: 2008 the 9th international conference for young computer scientists, Hunan, China, 2008, pp 1766–1770. https://doi.org/10.1109/ICYCS.2008.481
7. Jiang H, Learned-Miller E (2017) Face detection with the faster R-CNN. In: 2017 12th IEEE international conference on automatic face & gesture recognition (FG 2017), Washington, DC, USA, 2017, pp 650–657. https://doi.org/10.1109/FG.2017.82
8. Ramakrishna BB, Kumari MS (2017) A comparative study on face detection algorithms. In: 2017 international conference on energy, communication, data analytics and soft computing (ICECDS), Chennai, India, 2017, pp 151–159. https://doi.org/10.1109/ICECDS.2017.8389706
9. Collazos C, Granollers T, Rusu C (2011) A survey of human-computer interaction into the computer science curricula in Iberoamerica. In: 2011 eighth international conference on information technology: new generations, Las Vegas, NV, USA, 2011, pp 151–156. https://doi.org/10.1109/ITNG.2011.34
10. David-John, Brendan & Jörg, Sophie & Koppal, Sanjeev & Jain, Eakta. (2020). The security-utility trade-off for iris authentication and eye animation for social virtual avatars. IEEE Trans Vis Comput Graph 1–1. https://doi.org/10.1109/TVCG.2020.2973052
11. Lew M, Bakker EM, Sebe N, Huang TS (2007) Human-computer intelligent interaction: a survey. In: Lew M, Sebe N, Huang TS, Bakker EM (eds) Human–computer interaction. HCI 2007. Lecture notes in computer science, vol 4796. Springer, Berlin. https://doi.org/10.1007/978-3-540-75773-3_1
12. Tekkök SÇ, Söyünmez ME, Bostancı B, Ekim PO (2021) Face detection, tracking and recognition with artificial intelligence. In: 2021 3rd international congress on human-computer interaction, optimization and robotic applications (HORA), Ankara, Turkey, 2021, pp 1–5. https://doi.org/10.1109/HORA52670.2021.9461356
13. Wu H, Zhou H, Wang A, Iwahori Y (2022) Precise crop classification of hyperspectral images using multi-branch feature fusion and dilation-based MLP. Remote Sens. https://doi.org/10.3390/rs14112713

14. Farokhmanesh F, Sadeghi MT (2021) Deep Neural networks regularization using a combination of sparsity inducing feature selection methods. Neural Process Lett 53:701–720. https://doi.org/10.1007/s11063-020-10389-3
15. Krizhevsky A, Sutskever I, Hinton GE (2012) ImageNet classification with deep convolutional neural networks. Commun ACM 60:84–90
16. Pouyanfar S, Sadiq S, Yan Y, Tian H, Tao Y, Presa Reyes M, Shyu M-L, Chen S-C, Iyengar S (2018) A survey on deep learning: algorithms, techniques, and applications. ACM Comput Surv 51:1–36. https://doi.org/10.1145/3234150
17. He K, Zhang X, Ren S, Sun J (2016) Deep residual learning for image recognition. In: 2016 IEEE conference on computer vision and pattern recognition (CVPR), Las Vegas, NV, USA, 2016, pp 770–778. https://doi.org/10.1109/CVPR.2016.90
18. Viola P, Jones M (2004) Robust real-time face detection. Int J Comput Vision 57:137–154. https://doi.org/10.1023/B:VISI.0000013087.49260.fb
19. Wu Y, Hassner T, Kim K, Medioni G, Natarajan P (2018) Facial landmark detection with tweaked convolutional neural networks. IEEE Trans Pattern Anal Mach Intell 40(12):3067–3074. https://doi.org/10.1109/TPAMI.2017.2787130
20. Canedo D, Neves A (2019) Facial expression recognition using computer vision: a systematic review. Appl Sci 9. https://doi.org/10.3390/app9214678
21. Wang M, Deng W (2018) Deep face recognition: a survey. neurocomputing 429. https://doi.org/10.1016/j.neucom.2020.10.081
22. Dalal N, Triggs B (2005) Histograms of oriented gradients for human detection. In: 2005 IEEE computer society conference on computer vision and pattern recognition (CVPR'05), San Diego, CA, USA, 2005, vol 1, pp 886–893. https://doi.org/10.1109/CVPR.2005.177
23. Ren J, Zhao X, Xu B (2013) Adaptive discrete-time control with dual neural networks for HFV via back-stepping. In: 2013 9th Asian control conference (ASCC), Istanbul, Turkey, 2013, pp 1–6. https://doi.org/10.1109/ASCC.2013.6606168
24. Park J, Tai Y-W, Sinha SN, Kweon IS (2016) Efficient and robust color consistency for community photo collections. In: 2016 IEEE conference on computer vision and pattern recognition (CVPR), Las Vegas, NV, USA, 2016, pp 430–438. https://doi.org/10.1109/CVPR.2016.53
25. Ren S, He K, Girshick R, Sun J (2017) Faster R-CNN: towards real-time object detection with region proposal networks. IEEE Trans Pattern Anal Mach Intell 39(6):1137–1149. https://doi.org/10.1109/TPAMI.2016.2577031
26. Borra SP, Pradeep NVSS, Raju NTS, Vineel S, Karteek V (2020) Face recognition based on convolutional neural network. Int J Eng Adv Technol 9. https://doi.org/10.35940/ijeat.D6658.049420
27. Li Y, Sun B, Wu T, Wang Y (2016) Face detection with end-to-end integration of a ConvNet and a 3D model. In: Leibe B, Matas J, Sebe N, Welling M (eds) Computer vision—ECCV 2016. ECCV 2016. Lecture notes in computer science, vol 9907. Springer, Cham. https://doi.org/10.1007/978-3-319-46487-9_26
28. Bonkra A et al (2023) Apple leave disease detection using collaborative ML/DL and artificial intelligence methods: scientometric analysis. Int J Environ Res Public Health 20(4). https://doi.org/10.3390/ijerph20043222

Meat Freshness State Prediction Using a Novel Fifteen Layered Deep Convolutional Neural Network

M. Shyamala Devi, J. Arun Pandian, D. Umanandhini, Aayush Kumar Sakineti, and Rathinaraja Jeyaraj

Abstract The food marketplace needs a quick and reliable system for tracking and assessing the freshness of meat products. However, meat experiences a quick process of freshness deterioration, which leads to bacterial growth. As a result, the need for a reliable and quick way of monitoring and evaluating meat deterioration is growing urgent. By Considering these aspects, this paper proposes a Novel Fifteen Layered Deep Convolutional Neural Network (15L-DCNN) to predict the freshness state of meat with maximum accuracy. The model utilizes the Meat Freshness Image Dataset extracted from the KAGGLE machine learning repository. The Meat Freshness Image Dataset comprises three meat state classes, Fresh Meat, Half Fresh Meat, and Spoiled Meat, with 2269 meat images. The Meat Freshness Image Dataset have been subjected to data augmentation and performed with four operations: Random horizontal flip, Random vertical flip, zooming, and rotation. After data augmentation, the dataset ends with 6000 images. The Meat Freshness Image Dataset was splitted into 4800 training images, 600 validation images, and 600 testing images. The Meat Freshness training Images were subjected to the proposed 15L-DCNN and the same dataset was applied to EfficientNet, DenseNet, and ResNet Large models for evaluating the efficiency metrics. Python was adopted for the execution of NVidia Geforce Tesla V100 GPU workstation with 100 training iterations for a block size of 64. Experimental results show that the proposed model 15L-DCNN shows a maximum accuracy of 98.85%, Precision of 98.33%, Recall of 98.25%, misclassification rate of 1.15%, and FScore of 98.24% when compared with another convolutional neural network.

M. Shyamala Devi (✉) · D. Umanandhini · A. K. Sakineti
Computer Science and Engineering, Vel Tech Rangarajan Dr. Sagunthala R&D Institute of
Science and Technology, Tiruvallur, Tamilnadu, India
e-mail: shyamalapmr@gmail.com

J. Arun Pandian
School of Information Technology and Engineering, Vellore Institute of Technology, Vellore, India

R. Jeyaraj
Center for Resilient and Evolving Intelligence, Kyungpook National University, Daegu, South
Korea 41566

© The Author(s), under exclusive license to Springer Nature Singapore Pte Ltd. 2024   111
S. Namasudra et al. (eds.), *Data Science and Network Engineering*, Lecture Notes
in Networks and Systems 791, https://doi.org/10.1007/978-981-99-6755-1_9

Keywords Data augmentation · Flip · Rotation · Pooling layer · CNN · Accuracy

1 Introduction

Diets based on meat consumption are increasing with an understanding of nutritional values. Thus, the industry and consumers pay more attention to the quality and freshness of this food. Traditional inspection methods are sometimes expensive and destructive, and besides the fact that the information necessary to carry out such an activity is in the hands of experts, it is silent information and is not available to the end user. As an alternative, automated methods based on computer vision (CV) have been presented [1]. It is best to consume fresh food to receive the complete nutrition and mineral content in order to prevent needless health issues. The freshness of our food products may be determined or detected using a variety of techniques, including the conventional approach of lab testing the food sample, the use of photos of the food sample with a variety of machine learning and deep learning algorithms, and even the use of the scent from the food sample. To determine if food samples are fresh, several machine learning models have been created, however, they lack quick and precise detection. The color of the meat offers an instinctive perception of freshness and the composition of the components [2].

2 Literature Review

A CNN-based image recognition model of adulterated minced mutton was created in order to achieve the real-time automated detection of contaminated minced mutton. The laboratory's in-house image collection equipment was used to capture images of pieces of mutton, duck, pork, and chicken meat as well as processed mutton that had been tampered with to include varied ratios of duck, pork, and chicken flesh samples. A deep convolutional neural network (DCNN) model for image learning can distinguish between different cattle meat parts and falsified mutton, offering technical support for the quick and harmless determination of the integrity of mutton [3]. Using diffuse reflectance spectroscopy (DRS), it was found that the composition of met-myoglobin in the fully dried meat changed throughout the course of storage. DRS is an easy optical method that can determine the makeup of pigments without harming the material. However, it is challenging to apply DRS to a mobile environment since it necessitates high resolution and a tough retrofitting procedure [4]. A method based on machine learning assesses food freshness using real-time data. It aims to categorize portions of meat as fresh, semi-fresh, or rotten for the sake of the present scope while solely considering meat as the topic of study. The technology is anticipated to help the client's business operations run more efficiently by lowering the risk of selling faulty or rotten goods, which can have serious financial, non-financial, and health-related consequences. It is also anticipated that the technology will help the client increase its corporate value as a sustainable business by reducing food waste through prompt revenue [5].

Photographs of the meat were taken with a smartphone camera. It choose the region of RGB photographs before cropping them for the process of cropping chicken meat images using thresholding using the Otsu method and converting RGB images to binary files. Photographs of chicken flesh were cropped into three sizes and employed as a dataset later [6]. Monitoring the quality of food is essential to keep tabs on how the features are extracted change over time. The monitoring indices should significantly alter before freshness is compromised. Evaluation of food quality is fixed, unlike observation. Due to the assessment model's one-time training using previous data and infrequent updates, evaluation models explicitly link the provided labels and characteristics [7]. The methods for keeping an eye on the nutritional value of food may then be divided into supervised and unsupervised learning. When it comes to unsupervised learning, principal component analysis (PCA) [8] has been widely employed in studying the characteristics of a range of food products. By mapping a high-dimensional feature space to a low-dimensional subspace, this method finds the location coordinates that maximize the variances of the whole life span measurements of monitored food [9]. It is necessary to revise the PCA projection matrix after fresh observations have been introduced to the historical dataset. After the whole life span of data has been included in the model, a final PCA score plot is provided to visualize the observation outcome by tracking the position change of these samples [10]. The TVB-N content and WBSF in pork were examined using the Fourier transform near-infrared (FT-NIR) spectroscopy. The Synergy Interval Partial Least Square (SI-PLS) method was applied to calibrate the regression model. The number of PLS factors and the number of intervals were simultaneously optimized using cross-validation. Two correlation coefficients were used to evaluate the model's performance in the calibration and prediction sets [11].

Using a 5-point color scale, sensory ratings were acquired from a trained panel for the lean meat's color features. The scale of ratings, which varied from 1 to 5, was based on visual perception. Using the picture attributes as inputs, color scores were predicted using models combining analytics and neural network structures. Latent variables were derived using the partial least squares method in the statistical model. The multiple linear regression that followed utilized the latent variables. Using the back-propagation learning process, the neural network [12]. LDA-based techniques use the redefined feature space's altered dimensions of the tested sample as monitoring indicators to track changes in food quality. Nevertheless, these observation structures are often stationary models.

The absence of time dependence in static models is their main flaw [13]. It is crucial to keep an eye on the freshness of the swine meat since eating rotten or tainted pork poses a major health risk. In order to determine whether the meat was fresh or not, a variety of automated machine learning methods have been utilized with optimization. The quantity of exposure time for the meat that may be utilized to estimate its life span has also been determined using the method of regression analysis. After employing grid search in order to optimize each strategy hyperparameters, many high-performance methods were investigated, evaluated, and compared

[14]. The mechanism for determining the freshness of meat was designed with a set of cognitive detection and identification systems based on electronic technology, the photovoltaic identification technique, computer vision technology, and neural network model recognition technology in an effort to find a quick, reliable scientific method to determine meat freshness [15]. The study of machine learning and applications in its current state can be explored with the recent developments [16–21].

3   Research Methodology

Figure 1 depicts the suggested 15L-DCNN work's entire research procedure. Figure 2 depicts the 15L-DCNN topology that is suggested in this research. This research's key contribution is the creation of the Novel Fifteen Layered Deep Convolutional Neural Network (15L-DCNN), which can accurately forecast the freshness status of meat. The Meat Freshness Image Dataset have been used for this work whichcomprises three meat state classes, Fresh meat, Half Fresh Meat, and Spoiled Meat, with 2269 meat images; the dataset is shown in Eq. (1).

$$Meat_{2269} = \left| \bigcup_{M=1}^{2269} \left\{ \sum_{i=1}^{255} \sum_{j-1}^{255} Meat i_{jm} \right\} \right| \tag{1}$$

where $Meat_{001}$ represents the single image as shown in Eq. (2).

$$Meat_{001} = \begin{bmatrix} M(0,0) & M(0,1) & \dots & M(0,255) \\ M(1,0) & M(1,1) & \dots & M(1,255) \\ \vdots & \vdots & \dots & \vdots \\ M(255,0) & M(255,1) & \cdots & M(255,255) \end{bmatrix} \tag{2}$$

The Meat Freshness Image Dataset have been subjected to data augmentation and performed with four operations: Random horizontal flip, Random vertical flip, zooming, and rotation as shown in the Eqs. (3)–(11).

$$[Meat_{001}] = \begin{bmatrix} i \\ j \\ 1 \end{bmatrix} = Horizontal\ Flip[Meat_{001}]' = \begin{bmatrix} i' \\ j' \\ 1 \end{bmatrix} \tag{3}$$

$$Horizontal\ Flip = \begin{bmatrix} i' \\ j' \\ 1 \end{bmatrix} = \begin{bmatrix} -1 & 0 & \dots & 0 \\ 0 & 1 & \dots & 0 \\ \vdots & \vdots & \dots & \vdots \\ 0 & 0 & \cdots & 1 \end{bmatrix} \times \begin{bmatrix} i \\ j \\ 1 \end{bmatrix} \tag{4}$$

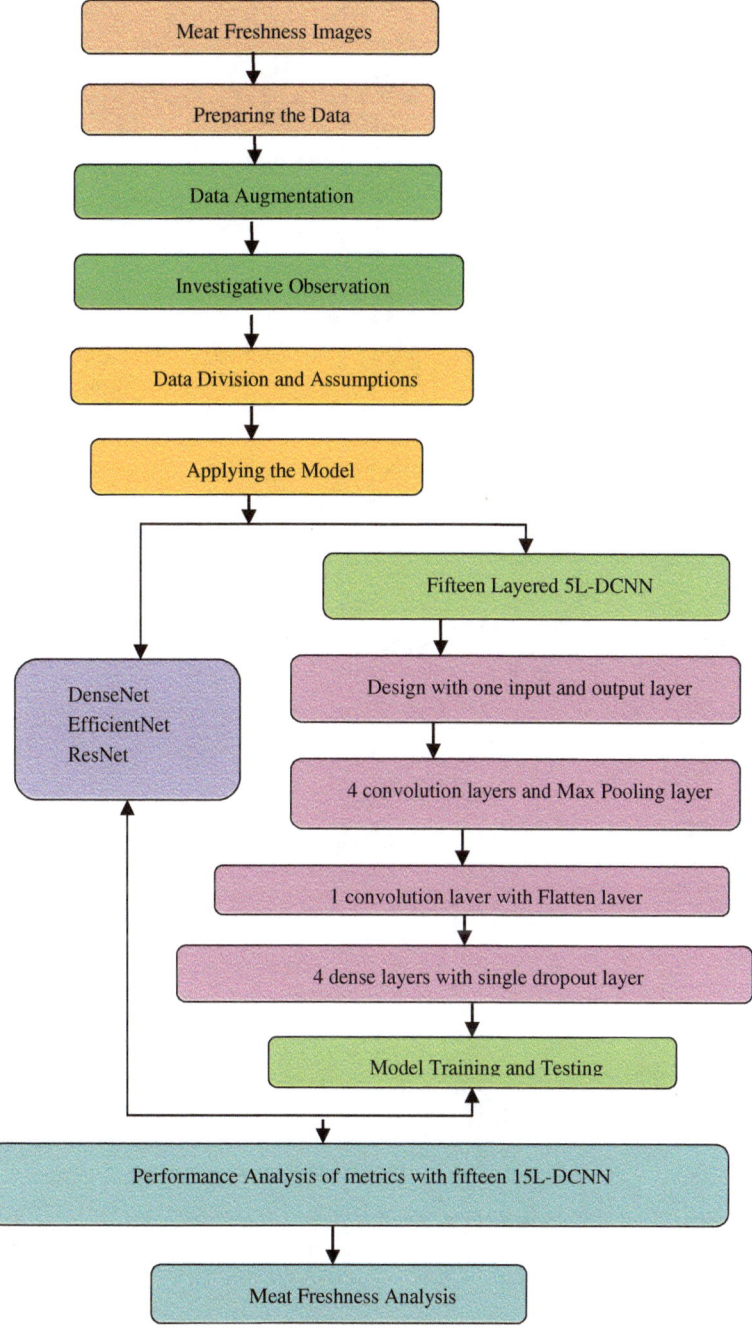

Fig. 1 Research methodology of proposed 15L-DCNN

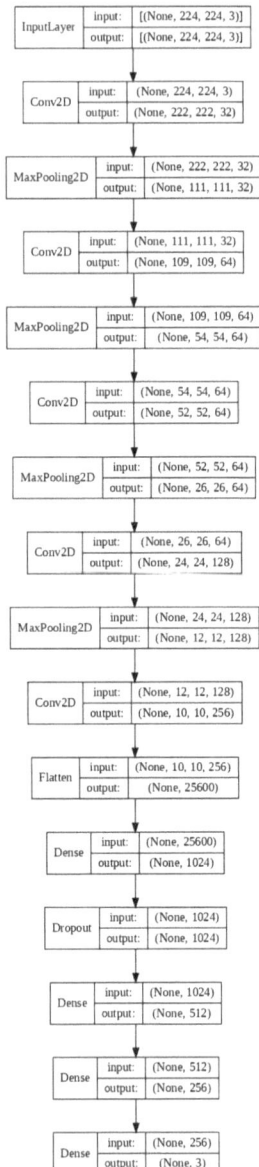

Fig. 2 Proposed 15L-DCNN network architecture

$$[Meat_{001}] = \begin{bmatrix} i \\ j \\ 1 \end{bmatrix} = Vertical\ Flip[Meat_{001}]' = \begin{bmatrix} i' \\ j' \\ 1 \end{bmatrix} \tag{5}$$

$$VerticalFlip = \begin{bmatrix} i' \\ j' \\ 1 \end{bmatrix} = \begin{bmatrix} 1 & 0 & \cdots & 0 \\ 0 & -1 & \cdots & 0 \\ \vdots & \vdots & \cdots & \vdots \\ 0 & 0 & \cdots & 1 \end{bmatrix} \times \begin{bmatrix} i \\ j \\ 1 \end{bmatrix} \tag{6}$$

$$[Meat_{001}] = \begin{bmatrix} i \\ j \\ 1 \end{bmatrix} = Rotation[Meat_{001}]' = \begin{bmatrix} i' \\ j' \\ 1 \end{bmatrix} \tag{7}$$

$$Rotation = \begin{bmatrix} i' \\ j' \\ 1 \end{bmatrix} = \begin{bmatrix} cos\theta & sin\theta & \cdots & 0 \\ -sin\theta & cos\theta & \cdots & 0 \\ \vdots & \vdots & \cdots & \vdots \\ 0 & 0 & \cdots & 1 \end{bmatrix} \times \begin{bmatrix} i \\ j \\ 1 \end{bmatrix} \tag{8}$$

$$[Meat_{001}] = Zoom([Meat_{001}]) = Meat(k, l) \tag{9}$$

$$Meat(k, l) = \sum_{i=1}^{X} \sum_{j-1}^{Y} \widehat{Meat}\ (x, y)\ e^{\frac{2\pi l(kn+lm)}{Y}} \tag{10}$$

$$\widehat{Meat}(k, l) = \sum_{i=1}^{X} \sum_{j-1}^{Y} \widehat{Z(I)}\ (k + iY, l + jY) \tag{11}$$

After data augmentation, the dataset results in 6000 images. Equations (12)–(18) display the 4800 training images, 600 validation images, and 600 testing images that make up the Meat Freshness Image Dataset.

$$Meat = MeatTr_{4800} + MeatVal_{600} + MeatTest_{600} \tag{12}$$

$$MeatTr_{4800} = Fresh_{1600} + Halffre_{1600} + Spoil_{1600} \tag{13}$$

$$MeatVal_{600} = Fresh_{200} + Halffre_{200} + Spoil_{200} \tag{14}$$

$$MeatTest_{600} = Fresh_{200} + Halffre_{200} + Spoil_{200} \tag{15}$$

where $Fresh_{200}$ represents Fresh meat, $Halffre_{200}$ represents half-fresh meat and spoiled meat denoted as $Spoil_{200}$.

$$MeatTr_{4800} = \left\lfloor \bigcup_{M=1}^{4800} \left\{ \sum_{i=1}^{255} \sum_{j-1}^{255} meattri_{jm} \right\} \right\rfloor \tag{16}$$

$$MeatVal_{600} = \left\lfloor \bigcup_{V=1}^{600} \left\{ \sum_{i=1}^{255} \sum_{j=1}^{255} meatval_{ijv} \right\} \right\rfloor \tag{17}$$

$$MeatTest_{600} = \left\lfloor \bigcup_{T=1}^{600} \left\{ \sum_{i=1}^{255} \sum_{j=1}^{255} test_{ijt} \right\} \right\rfloor \tag{18}$$

The Meat Freshness Image Dataset's images are normalized, and characteristics are extracted from them. The feature-extracted Meat Freshness Image has been applied to CNN models like EfficientNet, DenseNet, and ResNet. The same dataset is also applied to the proposed 15L-DCNN as in Eq. (19) and the investigation of the outcome data confirms the effectiveness.

$$5LDCNN = (1(InputLayer) + 3(ConvLayer, Maxpoolinglayer)$$
$$+ 1(ConvLayer, Flattenlayer) + 4(DenseLayer) + DropoutLayer) \tag{19}$$

The validation of Meat Freshness Image Dataset is applied with existing CNN models and proposed 15L-DCNN as shown in Eqs. (20) and (21).

$$\text{Modelfit} = Random(\bigcup_{i=1}^{4800} \left\{ \sum_{i=1}^{255} \sum_{j-1}^{255} Meattr_{ij_t} \right\}) \tag{20}$$

$$\text{Modelfit} = Random\left(\bigcup_{i=1}^{600} \left\{ \sum_{i=1}^{255} \sum_{j-1}^{255} MeatVal_{ij_v} \right\} \right) \tag{21}$$

4 Implementation Setup and Results

The Meat Freshness Image Dataset have been used for this work that comprises three meat state classes, Fresh meat, Half Fresh Meat, and Spoiled Meat, with 2269 meat images as shown in Fig. 3. That is used for predicting the meat freshness state. The dataset is subjected to data augmentation resulting in 6000 images as shown in Fig. 4. As indicated in Table 1, the Meat Freshness dataset was divided into 4800 training photos, 600 validation images, and 600 testing images.

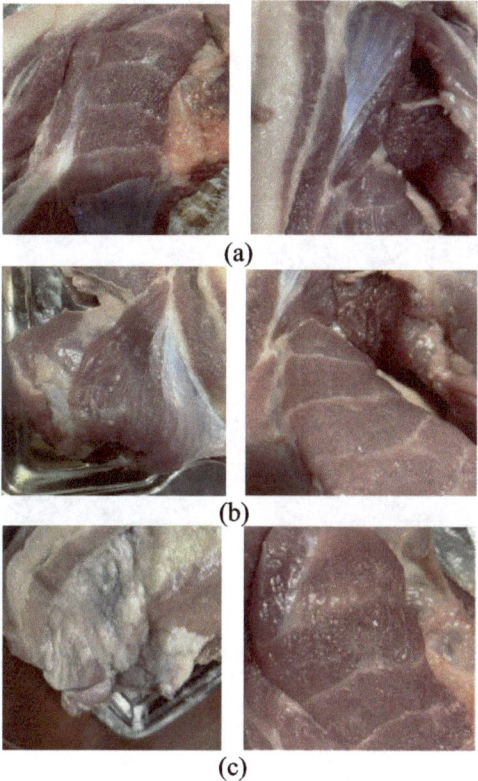

Fig. 3 Sample images of meat freshness image dataset. **a** Fresh meat images. **b** Half-fresh meat images. **c** Spoiled meat images

Single input and single output layers are used in the suggested 15L-DCNN model architecture, which is then followed by the 15L-DCNN model. The 15L-DCNN model was created with 4 convolution layers, 1 convolution layer connected to a flattened layer, 4 dense layers, and 1 dropout layer for each convolution layer, along with a max pooling layer for each convolution layer.

The confusion matrix produced after fitting the Meat Freshness Image Dataset with suggested 15L-DCNN and current CNN models is displayed in Fig. 5. Both Figs. 6 and 7 depict the accuracy and loss for training and validation, respectively.

The testing Meat Freshness Image Dataset is fitted with existing CNN models and the proposed 15L-DCNN model, and the performance metrics were analyzed and are shown in Table 2 and Fig. 8.

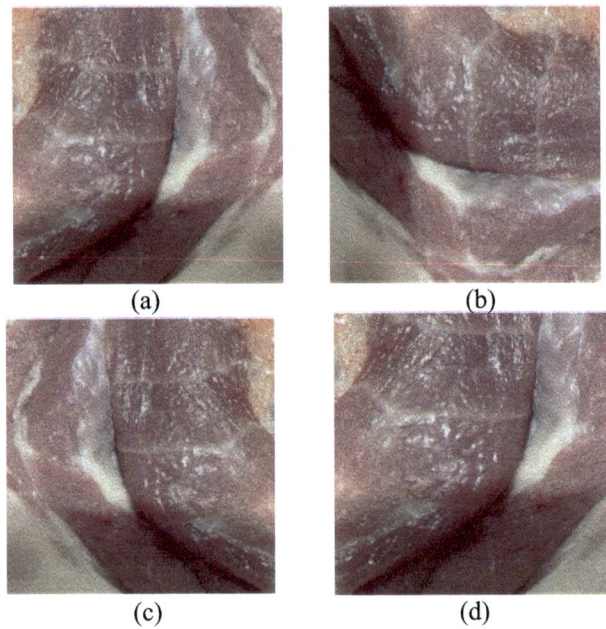

(a)   (b)

(c)   (d)

Fig. 4 Data augmented meat images. **a** Original meat images. **b** Rotated meat images. **c** Flipped meat images. **d** Zoomed meat images

Table 1 Dataset description of meat freshness image

Class ID	Class name	Training images		Validation	Test
		Original	Augmented		
1	Fresh Meat	757	1600	200	200
2	Half Fresh Meat	756	1600	200	200
3	Spoiled Meat	756	1600	200	200

Fig. 5 Confusion matrix of proposed 15L-DCNN

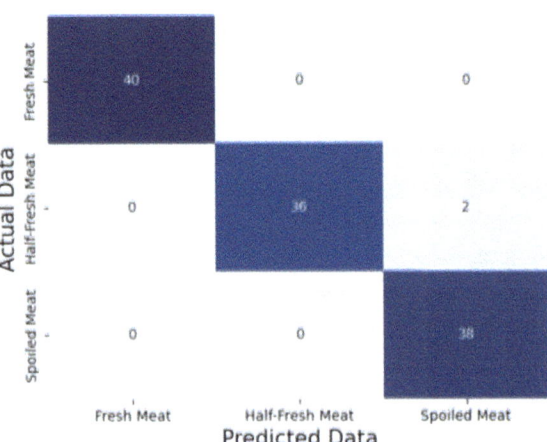

Fig. 6 Training and validation accuracy of proposed 15L-DCNN

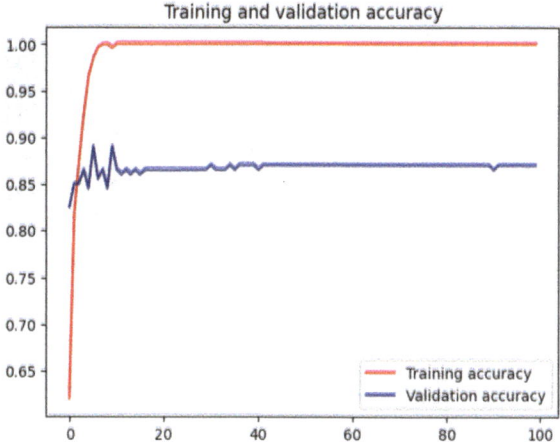

Fig. 7 Training and validation loss of proposed 15L-DCNN

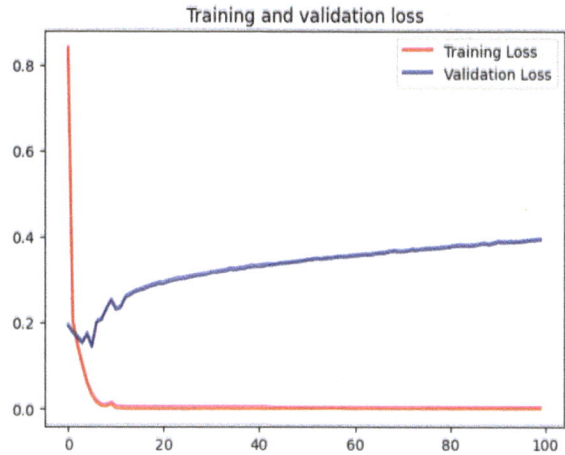

Table 2 Performance of proposed 15L-DCNN model

Classifier	Accuracy	Prec	Recall	F1-score	Miss rate
DenseNet	94.83	92.29	92.28	92.22	5.17
EfficientNet	90.23	85.39	85.39	85.33	9.77
ResNet	84.48	77.29	76.75	76.84	15.52
Proposed 15L-DCNN	98.85	98.33	98.25	98.24	1.15

Fig. 8 Performance of
proposed 15L-DCNN

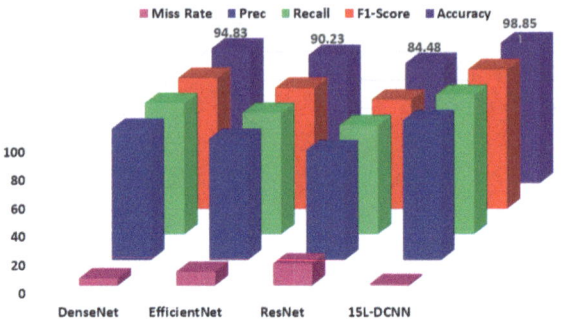

5   Conclusion

In this research, we compare the performance of different deep learning models
to the proposed 15L-DCNN model. The Meat Freshness Image Dataset have been
used for this work which comprises three meat state classes, Fresh meat, Half Fresh
Meat, and Spoiled Meat, with 2269 meat images that are used for predicting the
meat freshness state. The dataset is subjected to data augmentation resulting in 6000
images. Single input and single output layers are used in the suggested 15L-DCNN
model architecture, which is then followed by the 15L-DCNN model. The 15L-
DCNN model was created with 4 convolution layers, 1 convolution layer connected
to a flattened layer, 4 dense layers, and 1 dropout layer for each convolution layer,
along with a max pooling layer for each convolution layer. The testing Meat Freshness
Image Dataset is fitted with existing CNN models and the proposed 15L-DCNN
model, and the outcome indicators were examined. When compared to previous CNN
models, the experiment results show that the suggested 15L-DCNN has a maximum
accuracy of 98.85%, Precision of 98.33%, Recall of 98.25%, misclassification rate
of 1.15%, and F Score of 98.24%.

References

1. Bacus JA (2021) Identification of pork meat freshness using neural networks. In: The
   proceedings of IEEE international conference on electronic technology, communication and
   information (ICETCI), Changchun, China, pp 402–405
2. De Smet S, Vossen E (2016) Meat: the balance between nutrition and health. A Rev Meat Sci
   120:145–156
3. Xiao Y, Jiaojiao J, Guohua H (2014) Determination of the freshness of beef strip loins using
   electronic nose. Food Anal Methods 7(8):1612–1618
4. Chen YN, Sun DW, Cheng JH, Gao WH (2016) Recent advances for rapid identification
   of chemical information of muscle foods by hyperspectral imaging analysis. Food Eng Rev
   8(3):336–350
5. Kiani S, Minaei S, Ghasemi-Varnamkhasti M (2016) Fusion of artificial senses as a robust
   approach to food quality assessment. J Food Eng 171:230–239

6. Chen Q, Hui Z, Zhao J, Ouyang Q (2014) Evaluation of chicken freshness using a low-cost colorimetric sensor array with AdaBoost-OLDA classification algorithm. LWT Food Sci Technol 57(2):502–507

7. Li J, Zhu S, Jiang S, Wang J (2017) Prediction of egg storage time and yolk index based on electronic nose combined with chemometric methods. LWT-Food Sci Technol 82:369–376. https://doi.org/10.1016/j.lwt.2017.04.070

8. Deng F, Chen W, Wang J, Wei Z (2018) Fabrication of a sensor array based on quartz crystal microbalance and the application in egg shelf life evaluation. Sens Actuators B Chem 265:394–402. https://doi.org/10.1016/j.snb.2018.03.010

9. El Barbri N, Llobet E, El Bari N, Correig X, Bouchikhi B (2018) Electronic nose based on metal oxide semiconductor sensors as an alternative technique for the spoilage classification of red meat. Sensors 8:142–156. https://doi.org/10.3390/s8010142

10. Apetrei I, Apetrei C (2016) Application of voltammetric e-tongue for the detection of ammonia and putrescine in beef products. Sens Actuators B Chem 234:371–379

11. Cai J, Chen Q, Wan X, Zhao J (2011) Determination of total volatile basic nitrogen (TVB-N) content and Warner-Bratzler shear force (WBSF) in pork using Fourier transform near infrared (FT-NIR) spectroscopy. Food Chem 126(3):1354–1360

12. Lu J, Tan J, Shatadal P, Gerrard DE (2000) Evaluation of pork color by using computer vision. Meat Sci 56(1):57–60

13. Wojnowski W, Majchrzak T, Dymerski T, Gębicki J, Namiesnik J (2017) Electronic noses, powerful tools in meat quality assessment. Meat Sci 131:119–131

14. Alzaga MED, Buenaventura WG, Loresco PJM (2022) Machine learning-based pork meat quality prediction and shelf-life estimation. In: The proceedings of 14th international conference on humanoid, nanotechnology, information technology, communication and control, environment, and management, Boracay Island, Philippines, pp 1–6

15. Peiyuan G, Man B, Shiha Q, Tianhua C (2007) Detection of meat fresh degree based on neural network. In: The proceedings of ICMA 2007, pp 2726–2730

16. Agrawal D, Minocha S, Namasudra S, Kumar S (2021) Ensemble algorithm using transfer learning for sheep breed classification. In: The proceedings of SACI, pp 199–204

17. Malviya S, Kumar P, Namasudra S, Tiwary US (2022) Experience replay-based deep reinforcement learning for dialogue management optimisation. ACM Trans Asian Low-Resour Lang Inf Process

18. Namasudra S, Lorenz P, Ghosh U (2023) Editorial: the new era of computer network by using machine learning. Mobile Netw Appl. https://doi.org/10.1007/s11036-023-02114-w

19. Klocker F., Bernsteiner R, Ploder C, Nocker M (2023) A machine learning approach for automated cost estimation of plastic injection molding parts. Cloud Comput Data Sci 4(2):87–111

20. Alsultanny Y (2020) Machine learning by data mining REPTree and M5P for predicating novel information for PM10. Cloud Comput Data Sci 1(1):40–48

21. Jiang S, Gu Y, Kumar E (2023) Magnetic resonance imaging (MRI) brain tumor image classification based on five machine learning algorithms. Cloud Comput Data Sci 4(2):122–133

Object Detection in Autonomous Maritime Vehicles: Comparison Between YOLO V8 and EfficientDet

Nandni Mehla, Ishita, Ritika Talukdar, and Deepak Kumar Sharma

Abstract Autonomous vehicles are becoming more common in various industries, but the use of autonomous maritime vehicles is still being studied. This is because controlling these vehicles requires making important decisions about design, propulsion, payload management, and communication systems, which can lead to errors and collisions. One major challenge is detecting other ships and objects in real-time to avoid collisions. Recently, deep learning techniques based on convolutional neural networks (CNNs) have been developed to help with this challenge, such as YOLOv8 (You Only Look Once) and EfficientDet. This paper examines how these methods can be used to detect ships. We trained and tested these two models on a large maritime dataset. On examining the performance of the two models, we have compared the working of both.

Keywords Object detection · YOLO v8 · EfficientDet · Maritime vehicles

1 Introduction

The advent of Autonomous Maritime Vehicles (AMVs) is revolutionizing the technology used by waterways, driven by the continuous developments in AI, sensor technology, machine learning, and autonomous communication and navigation systems that perform the assigned jobs. AMVs are engineered to work in various environments, such as rivers, lakes, and oceans, and depending on the specific applications

N. Mehla (✉) · Ishita · R. Talukdar · D. K. Sharma
Indira Gandhi Delhi Technical University for Women, Delhi 110006, India
e-mail: nandni063btit19@igdtuw.ac.in

Ishita
e-mail: ishita062btit19@igdtuw.ac.in

R. Talukdar
e-mail: ritika058btit19@igdtuw.ac.in

D. K. Sharma
e-mail: deepaksharma@igdtuw.ac.in

© The Author(s), under exclusive license to Springer Nature Singapore Pte Ltd. 2024
S. Namasudra et al. (eds.), *Data Science and Network Engineering*, Lecture Notes in Networks and Systems 791, https://doi.org/10.1007/978-981-99-6755-1_10

their characteristics and capabilities can be changed. With the increasing demand for offshore exploration and surveillance, the use of AMVs is expected to grow substantially in the upcoming years. AMVs offer multiple advantages over traditional manned vessels, including increased safety, reduced costs, and improved efficiency. AMVs can be used for a wide range of purposes, such as oceanographic research, environmental monitoring, underwater inspections, border security, mapping and surveying offshore oil and gas exploration, search and rescue missions, and maritime security.

AMVs help manage marine environments sustainably amidst new challenges, but their impact on marine life and the environment must be considered, along with the need for regulations. They offer prospects for a sustainable and secure maritime industry.

Autonomous vehicles have made significant progress, but object detection remains a challenge. It's critical for vehicle safety and navigation. Deep learning, specifically Convolutional Neural Networks, have improved accuracy and speed in object detection. However, detecting objects in challenging environments such as maritime settings is difficult.

Autonomous maritime vehicles rely on object detection schemes to navigate their surroundings and avoid obstacles. However, these schemes face several challenges that can limit their effectiveness. The main issue is limited accuracy, especially in challenging environmental conditions such as fog, rain, and waves. Additionally, object detection algorithms can be computationally expensive, making real-time operation on autonomous vehicles problematic. Scaling to different types of objects can require significant amounts of training data, and sensitivity to changes in lighting and background can affect detection accuracy. Finally, many schemes are designed for specific object types and may have limited generalization when applied to new objects.

To improve object detection in challenging environments, advanced deep learning techniques such as YOLOv8 and EfficientDet have been developed. They use advanced deep learning algorithms, including feature pyramids, spatial attention, and context fusion, to improve accuracy in challenging environments. Additionally, they use lightweight architectures, reducing computational complexity without sacrificing accuracy, making them ideal for real-time applications on autonomous maritime vehicles. Finally, both YOLOv8 and EfficientDet are designed to be more generalizable, using techniques like transfer learning and data augmentation to detect a wide range of objects.

The objective of this study is to evaluate the performance of two state-of-the-art deep learning algorithms, YOLOv8 and EfficientDet, for object detection in autonomous maritime vehicles. Specifically, we aim to compare the performance of these algorithms on a real-world dataset of maritime objects and evaluate their accuracy, computational efficiency, and generalizability.

We are using a dataset of real-world maritime objects to evaluate the performance of YOLOv8 and EfficientDet. The dataset will consist of images and video footage captured by an autonomous maritime vehicle in various environmental conditions.

The application of YOLOv8 and EfficientDet to ship detection is novel because it addresses the drawbacks of existing schemes and provides enhanced accuracy, decreased computational complexity, improved scalability, robustness to changes in lighting and background, and enhanced generalization to detect a wide range of objects.

The contribution of our study is twofold. First, it will provide a comparative analysis of two state-of-the-art deep learning algorithms for object detection in autonomous maritime vehicles. Second, it will demonstrate the feasibility of using deep learning algorithms for object detection in challenging maritime environments, which could have practical implications for the development of autonomous maritime vehicles.

The paper adheres to the following structure, the literature on object detection in autonomous vehicles is reviewed in Sect. 2 of the report. The study's models are discussed in Sect. 3 of the report. The performance analysis, which includes information on the dataset, the experimental setting, the technique used to assess the effectiveness of the algorithms, and the study's findings, is presented in Sect. 4. Section 5 concludes our study and discusses the work we will do moving forward.

2 Literature Review

There has been a lot of research on AMVs over many years and it is still a field of ongoing research (refer to Fig. 1). Researchers are focusing on giving rise to technologies that permit the autonomous vessels to sail and operate autonomously in distinct marine environments, perform various tasks, and deliver to different applications. Recent studies have concentrated on the use of AMVs in oceanographic research, environmental monitoring, marine transportation, and defence applications. The primary aim of the present-day research on AMVs is to refine and enhance the reliability, safety, and energy efficiency, while also developing new sensor technologies and data processing methods to upgrade their performance and functionality.

3 About YOLO and EfficientDet Series

The YOLO (You Only Look Once) series is a well-recognized approach for real-time object detection that relies on a single neural network to detect objects in images or video streams with high accuracy and speed. YOLOv2, v3, v4, v5, etc., are different versions of YOLO (refer Fig. 2) that have been developed to improve accuracy and speed and it achieves real-time performance by processing up to 60 frames per second on a standard GPU. It has been widely adopted in applications such as surveillance and autonomous vehicles due to its fast processing speed and high accuracy.

EfficientDet models (refer Fig. 3) are recent object detection models that achieve better accuracy and efficiency. They use a compound scaling method to progressively increase the network resolution, depth, and width in a synchronized way, which pro-

Research Papers	Proposed Research
Safety challenges related to autonomous ships in mixed navigational environments.	The paper discussed the evolution of automatic ships in the maritime transport sector, focusing on their advanced features and potential benefits. It also addressed safety and reliability concerns related to their complex and non-deterministic nature, especially in diverse navigational conditions. The study explores these safety challenges, suggesting potential risk reduction strategies and emphasising the necessity for future research in ship navigation.
Ship detection based on YOLOv2 for SAR imagery	They proposed a new method for detecting ships in Synthetic Aperture Radar (SAR) imagery using a high-performance computing approach with a GPU-based deep learning framework called YOLOv2 (You Only Look Once version 2). The method achieved enhanced accuracy in ship detection and reduced computational time when compared to other object detection techniques. The experiment's results showcased the potential of deep learning in enhancing SAR image ship detection implementation.
Multi-Scale Object Detection Model for Autonomous Ship Navigation in Maritime Environment.	The paper introduces an enhanced convolutional neural network called VarifocalNet, designed for precise detection of sea-surface objects in challenging maritime environments. By incorporating a deformable convolution module, a soft non-maximum suppression algorithm, a restructured loss function, and multi-scale prediction techniques, the model improves its ability to represent features and learn effectively. Research findings indicate that VarifocalNet outperforms other methods (such as YOLOv3, Faster R-CNN, SSD, RetinaNet, Cascade R-CNN) in accurately identifying objects and demonstrating robustness for small objects amidst complex sea conditions. The study showcases the efficacy of the proposed model for maritime obstacle detection.

Fig. 1 Overview of publications reviewed [2–4]

Model	Size (pixels)	mAP (50-95)	Speed CPU ONNX (ms)	Speed A100 TensorRT (ms)	params (M)	FLOPs (B)
YOLO v8n	640	37.3	80.4	0.99	3.2	8.7
YOLO v8s	640	44.9	128.4	1.2	11.2	28.6
YOLO v8m	640	50.2	234.7	1.83	25.9	78.9
YOLO v8l	640	52.9	375.2	2.39	43.7	163.2
YOLO v8x	640	53.9	479.1	3.53	68.2	257.8

Fig. 2 YOLO v8 models

vides a better balance between accuracy and efficiency. Additionally, EfficientDet models introduce a BiFPN architecture for effective feature fusion and a weighted bi-directional feature fusion technique for improving detection performance on small objects. These models outperform previous state-of-the-art object detection methods like YOLO and Faster R-CNN while maintaining real-time processing capabilities. EfficientDet models have significant potential for practical applications like autonomous driving and surveillance.

Model name	Speed (ms)	COCO mAP	Outputs
EfficientDet D0 512x512	39	33.6	Boxes
EfficientDet D1 640x640	54	38.4	Boxes
EfficientDet D2 768x768	67	41.8	Boxes
EfficientDet D3 896x896	95	45.4	Boxes
EfficientDet D4 1024x1024	133	48.5	Boxes
EfficientDet D5 1280x1280	222	49.7	Boxes
EfficientDet D6 1280x1280	268	50.5	Boxes
EfficientDet D7 1536x1536	325	51.2	Boxes

Fig. 3 EfficientDet models

3.1 YOLO Architecture

The foundation of YOLO is a CNN model architecture that utilizes a simple yet powerful deep convolutional neural network to detect objects in input images. Figure 4 is showcasing this CNN model architecture used in the YOLO technique. ImageNet is utilized to pre-train the model's first twenty convolution layers by introducing a fully linked and temporary average pooling layer [6]. Then, this pre-trained model is designed to detect objects, since prior study demonstrated that improving a pre-trained network by including linked layers and convolution. YOLO's final fully linked layer can forecast both class probability and bounding box coordinates [6]. YOLO divides an input picture into a S × S grid. If a lattice cell contains the focal point of an object, then that lattice cell is responsible for identifying that object. Each network cell forecasts B leaping boxes and their associated certainty ratings. These values for certainty indicate how convinced the model is that the crate contains an article and how precise it believes the expected box to be [6].

YOLO predicts many bounding boxes for each grid cell. In the process of training, each element necessitates only a single bounding box predictor. YOLO determines which predictor is "in charge" of item predictions based on highest current IOU with actual data leads to an enhancement in the specialization of bounding box predictors. The overall recall score increases as each predictor improves its perception of a particular element size, aspect ratio, or classification [6].

Non-maximal suppression (NMS) is a key method of the YOLO model. Object detection is improved by using NMS as a post-processing step. In object detection, it is common to create multiple bounding boxes for a single object, even though these boxes may overlap or be located in different positions. Although the bounding boxes may differ in appearance, they are all intended to delineate the same object instance.

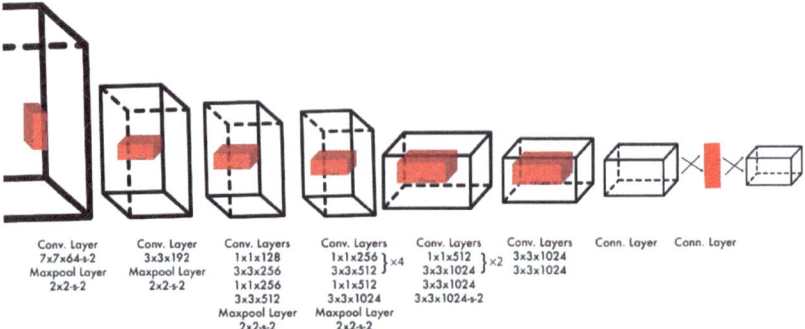

Fig. 4 YOLO architecture

NMS is employed to identify and eliminate unnecessary or imprecise bounding boxes, resulting in the creation of a single bounding box for each item captured in the image [6].

3.2 EfficientDet Architecture

EfficientNet network has been used to build the architecture of EfficientDet model. In convolutional neural networks, dimensions are always uniformly scaled. However, in the EfficientNet network these dimensions are arbitrarily scaled using a fixed set of coefficients. These methods are determined by the size of input images. If the input images are larger in size, more layers and channels are needed to capture the finer patterns of the images. The BiFPN serves as the core of the EfficientDet network and is an expanded version of the FPN (refer to Figs. 5 and 6). It can be built using either a bottom-up or top-down technique [5]. In the bottom-up approach, a hierarchical framework of characteristics is constructed that consists of several feature maps at varying scales. Each map is generated by enlarging the previous one by a factor of 2. By allocating a level to each stage, the pyramid is constructed in the bottom-up method and the output of the previous layer in each stage serves as a reference in the bottom-up approach. for the following stage. To achieve the desired objective, one technique used is to apply 1×1 convolutions to the bottom-up feature maps, which results in a reduction in their size. Conversely, the top-down method involves features that are more semantically robust but less spatially close. In this approach, lateral connections combine the feature maps of equal spatial size that are obtained from both the bottom-up and top-down paths.

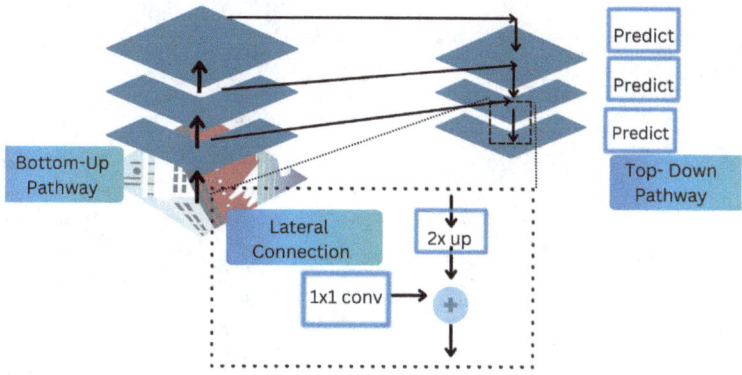

Fig. 5 Top-down pathway of FPN [5]

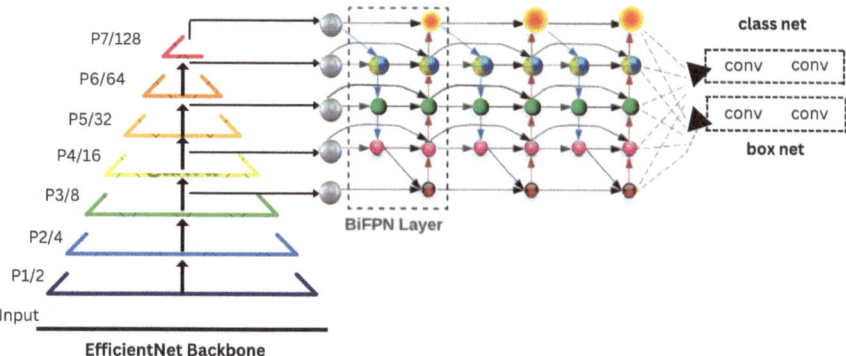

Fig. 6 Architecture of efficientDet

4 Performance Analysis

4.1 Details of Dataset

The Roboflow Ship2 Image Dataset [7] was used as the primary training dataset for our object detection model, which is a publicly available dataset consisting of images of ships captured by various sources. The dataset contains 498 images of ships, each with corresponding annotations, split into 348 images (70%) for training set, 99 images (20%) for validation set, and 51 images (10%) for the testing set. The images were captured from a variety of sources, including satellite images, aerial photographs, and images captured from the ground.

The dataset contains diverse ship types, such as cargo, container, oil tankers, and cruise ships, varying in size, shape, and color, captured in different settings such as harbors, open waters, and ports. It includes image classification of the ships into

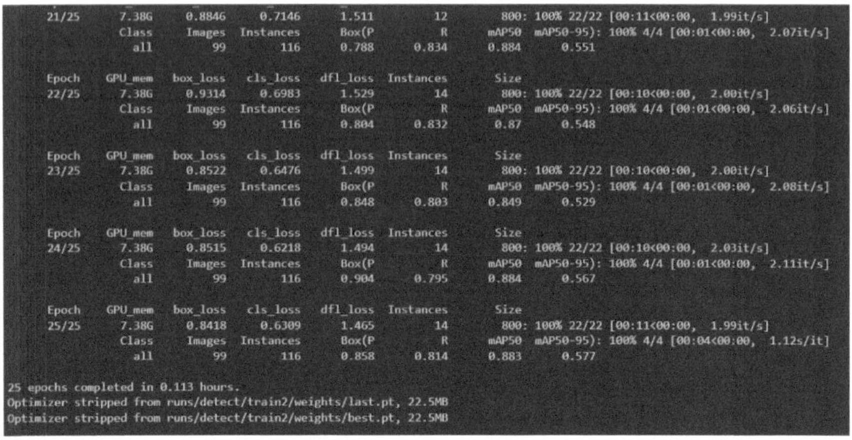

Fig. 7 Output displaying epochs training time

categories like cruise, tanker, cargo, carrier, and military. This dataset is a valuable resource for machine learning and computer vision researchers and developers.

For testing our models, Singapore maritime dataset [1] was used. It is a public dataset of marine vessel traffic in the waters surrounding Singapore. The dataset is collected and maintained by the Maritime and Port Authority of Singapore (MPA) and is intended for use in research related to maritime traffic analysis, safety, and environmental protection. This dataset contains images and video footage of vessels passing through the Singapore Strait, captured by cameras installed along the Strait. The visual dataset includes over 100,000 images and video clips of vessels, each with corresponding AIS data.

4.2 Experimental Environment

The coding was done using Python to train and test the models and we used Google Colab to run the code. Google Colab is a Google research platform focused on machine learning and data analytics that allows users to use Jupyter notebooks to run Python code on the web to infer patterns for the model YOLOv8. The TensorFlow library is also used, dedicated to machine learning and numerical computation, as well as network development and processing.

Training YOLOv8 model: This model was obtained by training the network on approximately 25 epochs with image size = 800 per batch. We made use of the YOLO v8s model. It took 0.113 hours to train the model using 25 iterations. Training model consists of 5 classes of ships, namely, with 99 images of each class: Cargo, Carrier, Cruise, Military, Tanker. Figure 7 displays the mAP values of each class on training set (Fig. 8).

```
[22] !yolo task=detect mode=val model={HOME}/runs/detect/train2/weights/best.pt data={dataset.locati

Ultralytics YOLOv8.0.53 🚀 Python-3.9.16 torch-1.13.1+cu116 CUDA:0 (Tesla T4, 15102MiB)
Model summary (fused): 168 layers, 11127519 parameters, 0 gradients, 28.4 GFLOPs
val: Scanning /content/Ship2-1/valid/labels.cache... 99 images, 0 backgrounds, 0 corrupt: 100%
WARNING ⚠ Box and segment counts should be equal, but got len(segments) = 70, len(boxes) = 116
                    Class     Images  Instances     Box(P          R      mAP50  mAP50-95): 100%
                      all         99        116     0.858      0.814      0.883      0.576
                    Cargo         99         20     0.731      0.817      0.835      0.456
                  Carrier         99         20         1      0.898      0.984      0.703
                   Cruise         99         36     0.936      0.808      0.962      0.592
                 Military         99         20     0.805        0.7       0.78      0.558
                   Tanker         99         20     0.816       0.85      0.855      0.572
Speed: 8.4ms preprocess, 16.4ms inference, 0.0ms loss, 3.9ms postprocess per image
Results saved to runs/detect/val3
```

Fig. 8 YOLOv8 network training parameters

Confusion Matrix of Training Dataset: A confusion matrix captures the efficiency of a classification model by indicating the number of true positives, false positives, true negatives, and false negatives (see Fig. 9). By computing performance metrics like recall, precision, accuracy, and F1 score, it assesses the efficiency of the model. According to Fig. 9, the model is detecting the cargo ships correctly 95% of the time. However, 5% of the time when the cargo was actually present the model couldn't detect it. Similarly for the other classes (Table 1), the data is shown in the confusion matrix.

Training EfficientDet model: To produce efficientDet model, image size of 800 was used to train the network for 10 epochs.

The efficient accuracy vs iterations graph below shows that the learning rate of the model is quite good. It is increasing with each iteration with less dips.

Testing YOLO v8: The model records graphs illustrate the progress of precision and recall, along with the identified objects throughout the training phase improve training tracking of metrics (see Fig. 12).

The model's loss curve is seen in the preceding figure. The x-axis shows the number of training/validation rounds, while the y-axis represents the loss, or training/validation loss. After around 25 cycles, the classification loss is less than one, the bounding box loss is near 0.8, and the overall loss value is less than 1.5. The model's loss function has a good convergence impact.

The lower dipping and flattening of the loss curves suggest that the model is beginning to adapt better to the tasks given and is ready to deliver better or more accurate predictions (Refer Figs. 10, 11, 12, 13, 14, 15, 16 and 17).

The loss curves' x-axis reflects the number of iterations for which the model has been trained, and the y-axis represents the loss.

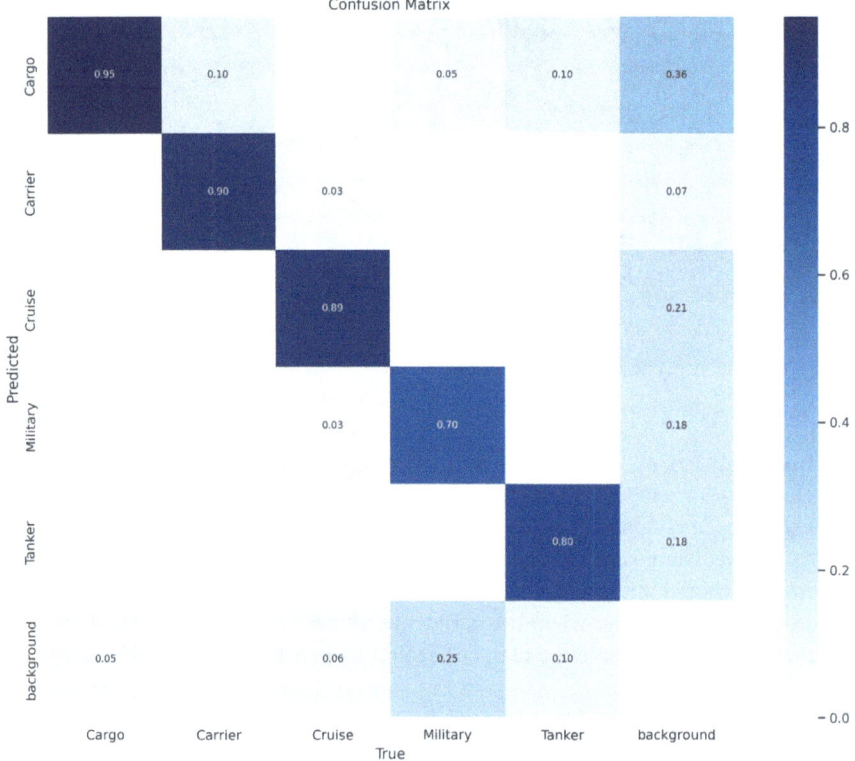

Fig. 9 Confusion matrix

Table 1 Results inferred from confusion matrix

Class	Precision
Cargo	0.95
Carrier	0.9
Cruise	0.89
Military	0.7
Tanker	0.8

4.3   Results and Discussion

A Python script specifying the thresholds, network paths, and images to recognize is used to test the new model. The figure shows the results for the new inputted image.

Our trained model correctly recognizes the item (ship) in images and videos from the "Singapore maritime dataset" [1], as seen in the above images (see Figs. 18

Fig. 10 Example YOLOv8 inference on validation batch [7]

```
Epoch 1/10
1875/1875 [==============================] - 12s 4ms/step - loss: 0.2517 - accuracy: 0.9272 - val_loss: 0.1368 - val_accuracy: 0.9595
Epoch 2/10
1875/1875 [==============================] - 6s 3ms/step - loss: 0.1114 - accuracy: 0.9672 - val_loss: 0.1009 - val_accuracy: 0.9695
Epoch 3/10
1875/1875 [==============================] - 5s 3ms/step - loss: 0.0774 - accuracy: 0.9765 - val_loss: 0.0817 - val_accuracy: 0.9753
Epoch 4/10
1875/1875 [==============================] - 6s 3ms/step - loss: 0.0574 - accuracy: 0.9821 - val_loss: 0.0746 - val_accuracy: 0.9773
Epoch 5/10
1875/1875 [==============================] - 6s 3ms/step - loss: 0.0448 - accuracy: 0.9862 - val_loss: 0.0761 - val_accuracy: 0.9769
Epoch 6/10
1875/1875 [==============================] - 6s 3ms/step - loss: 0.0346 - accuracy: 0.9890 - val_loss: 0.0711 - val_accuracy: 0.9791
Epoch 7/10
1875/1875 [==============================] - 6s 3ms/step - loss: 0.0276 - accuracy: 0.9918 - val_loss: 0.0767 - val_accuracy: 0.9774
Epoch 8/10
1875/1875 [==============================] - 6s 3ms/step - loss: 0.0230 - accuracy: 0.9933 - val_loss: 0.0735 - val_accuracy: 0.9782
Epoch 9/10
1875/1875 [==============================] - 5s 3ms/step - loss: 0.0191 - accuracy: 0.9946 - val_loss: 0.0756 - val_accuracy: 0.9781
Epoch 10/10
1875/1875 [==============================] - 6s 3ms/step - loss: 0.0148 - accuracy: 0.9956 - val_loss: 0.0759 - val_accuracy: 0.9798
```

Fig. 11 Output of python script displaying epoch training time

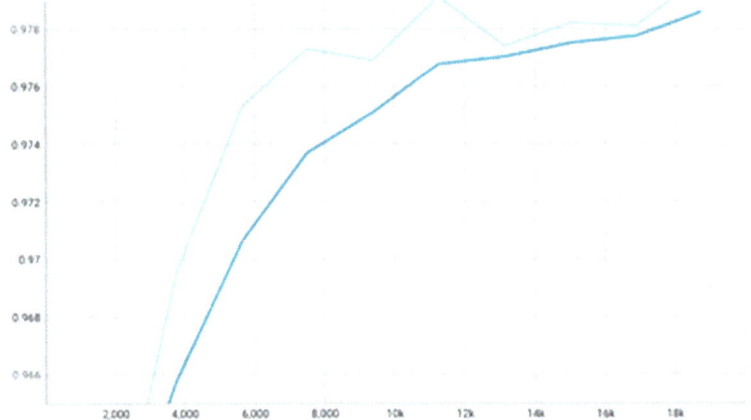

Fig. 12 Efficient accuracy versus iterations

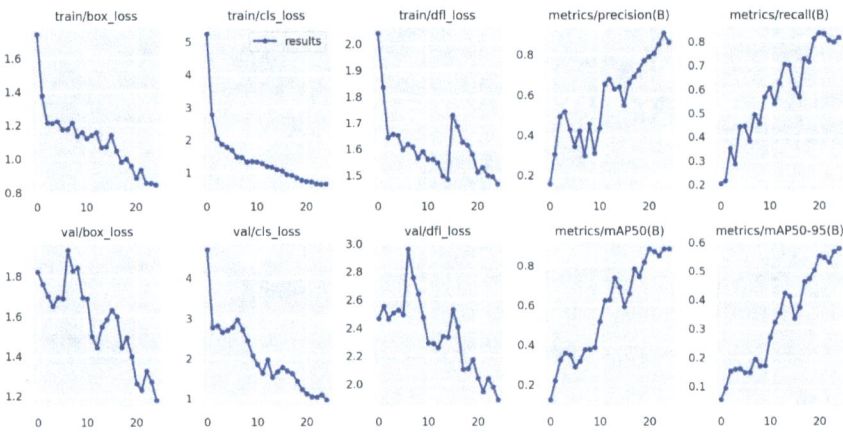

Fig. 13 Key metrics tracked by YOLO v8

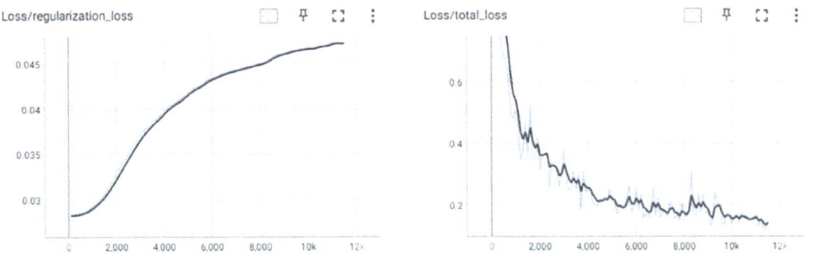

Fig. 14 Regularization and total loss

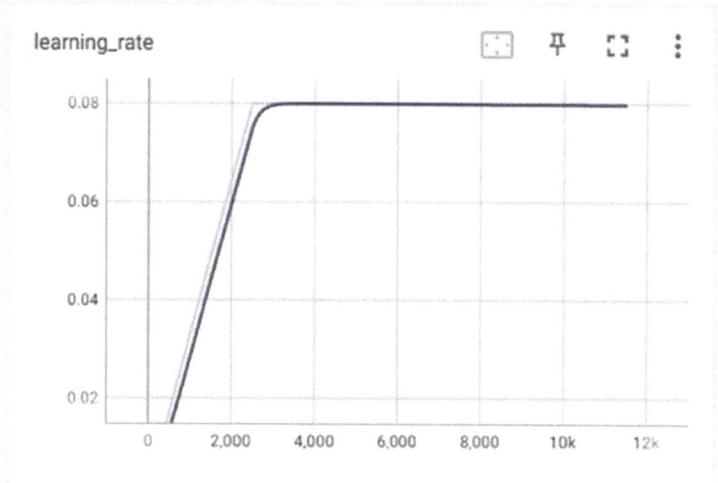

Fig. 15 Learning rate of the efficientDet model

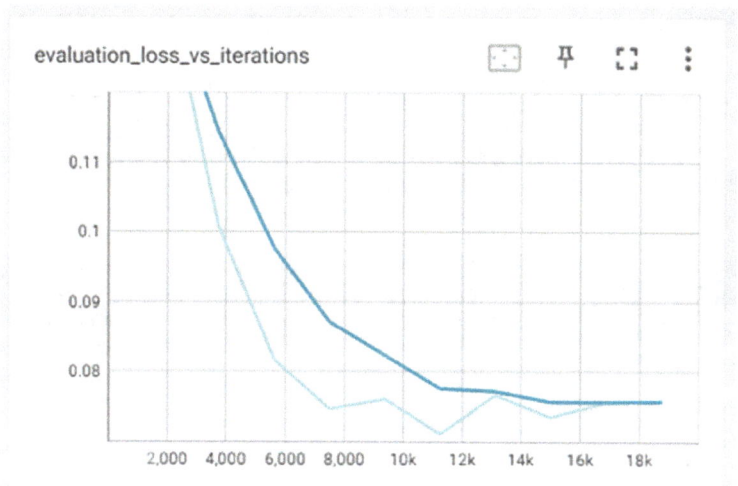

Fig. 16 Evaluation loss versus iterations graph

and 19). As a result, we can infer that the model was able to recognise the item not only in photos, but also in real time/inputted videos.

Comparison of YOLO v8 and EfficientDet: YOLOv8 and EfficientDet are both popular object detection models that are widely used in computer vision applications. YOLOv8 is the most recent model in the series of object detection models known as YOLO, while EfficientDet is a family of object detection models developed by Google. The architecture of YOLOv8 and EfficientDet differs significantly.

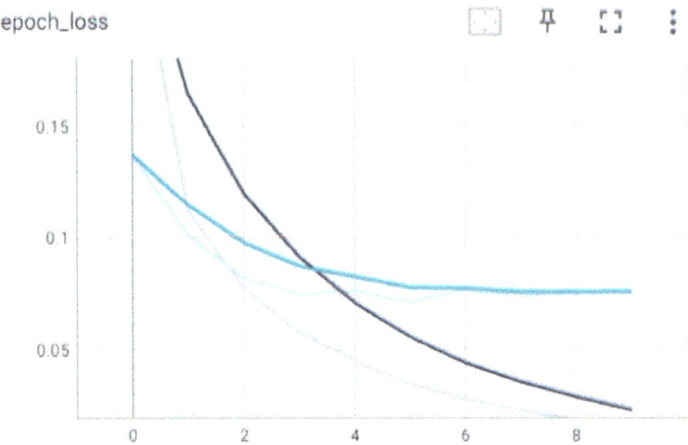

Fig. 17 Epoch loss of efficientDet model

Fig. 18 Ship detection in image

YOLOv8 employs a one-stage object detection architecture, which implies that it identifies things in a single network pass. This makes YOLOv8 quicker than two-stage detectors such as EfficientDet, which require many network passes. Efficient-Det, on the other hand, is recognized for its great accuracy and is regarded as one of the most accurate object detection models. YOLOv8 and EfficientDet differ in their backbone architecture. While YOLOv8 uses a modified version of CSPDarknet53, a well-known backbone architecture for object detection, EfficientDet optimizes the backbone using a compound scaling method for varying model sizes. This enables EfficientDet to maintain accuracy while being computationally efficient. In terms of

Fig. 19 Ship detection in video

performance, both YOLOv8 and EfficientDet have set a new benchmark in object detection benchmarks by achieving exceptional results. However, the choice between the two models depends on the specific requirements of the application. If speed is a critical factor, YOLO v8 may be a better choice. On the other hand, if high accuracy is a top priority, EfficientDet may be the better option. Upon comparing the loss graphs of both the models, EfficientDet model performs well (see Fig. 20). The following points can be the reason behind it

- Model architecture: EfficientDet's architecture is based on the EfficientNet backbone, resulting in an efficient and scalable model that learns with fewer parameters than YOLO v8. This allows EfficientDet to converge more quickly to a better solution, resulting in a lower final loss value.
- Multi-level feature fusion: EfficientDet employs a multi-level feature fusion approach to effectively merge features from different network levels, resulting

	YOLOv8	EfficientDet
Architecture	Single-stage object detection	Two-stage object detection
Backbone	Modified CSPDarknet53	Compound scaling method
Speed	Faster	Slower
Accuracy	Lower (compared to EfficientDet)	Higher
Training data requirement	Relatively less	Relatively more
Model size	Smaller (compared to EfficientDet)	Larger
Deployment	Easy to deploy	Requires more resources
Applications	Real-time object detection	High accuracy object detection

Fig. 20 Comparision of YOLO v8 and efficientDet

in improved object and environmental detail capture, leading to more precise predictions and lower loss value.

- Advanced training techniques: EfficientDet employs advanced training techniques such as learning rate scheduling and data augmentation to enhance its learning and generalization abilities. By adjusting the learning rate and introducing variations in the training data, overfitting is reduced, and EfficientDet achieves a lower loss value.

Not only these reasons even in general, a flat loss curve is better than a loss curve with more highs and lows. A flat loss curve indicates that the model has converged to a stable and optimal solution and is less likely to overfit or underfit the training data. On the other hand, a loss curve with more highs and lows suggests that the model is still trying to learn and may be overfitting or underfitting the training data. And even after fine-tuning the hyperparameters, YOLO v8 gives loss curves with many highs and lows. On the other hand, EfficientDet has flatter loss curves.

5   Conclusions and Future Work

YOLOv8 and EfficientDet delivered respectable outcomes in a brief amount of time. It is worth noting that EfficientDet is among the recent neural networks that have succeeded in surpassing more established models on demanding feature recognition evaluations. The choice of dataset has a big impact on model performance. The main challenges in developing such systems are the high hardware costs, the need to annotate images, and the generation of datasets, which are essential for training neural networks. On the other hand, current research tackles each issue separately and provides solutions, such as utilizing Google Colab for GPU acceleration and simulators to create data pictures. Two computer models that have both shown to be quite accurate at spotting ships at sea are used in our method: YOLO and EfficientDet. This model can be further enhanced and improved in the future to incorporate sensors like LIDAR that can help with improving the accuracy of object detection and more research can be done with 3D scans of the navigable route for waterways that will in turn be helpful in creating a more thorough view of the surroundings of the vessels. As a result, AMVs will become safer and more effective and will be able to carry out several complicated jobs in various marine conditions. The vessels will be able to converse and exchange real-time data with other vehicles after communication protocols are added to the active research models.

References

1. Prasad DK, Rajan D, Rachmawati L, Rajabally E, Quek C (2017) Video processing from electro-optical sensors for object detection and tracking in a maritime environment: a survey. IEEE Trans Intell Transp Syst 18(8):1993–2016
2. Kim TE, Perera LP, Sollid MP, Batalden BM, Sydnes AK (2022) Safety challenges related to autonomous ships in mixed navigational environments. WMU J MaritE Aff 21(2):141–159
3. Shao Z, Lyu H, Yin Y, Cheng T, Gao X, Zhang W, Jing Q, Zhao Y, Zhang L (2022) Multi-scale object detection model for autonomous ship navigation in maritime environment. J Mar Sci Eng 10(11):1783
4. Chang YL, Anagaw A, Chang L, Wang YC, Hsiao CY, Lee WH (2019) Ship detection based on YOLOv2 for SAR imagery. Remote Sens 11(7):786
5. Munteanu D, Moina D, Zamfir CG, Petrea M, Cristea DS, Munteanu N (2022) Sea mine detection framework using YOLO, SSD and efficientdet deep learning models. Sensors 22(23):9536
6. Redmon J, Divvala S, Girshick R, Farhadi A (2016) You only look once: unified, real-time object detection. In: Proceedings of the IEEE conference on computer vision and pattern recognition, pp 779–788
7. Smart helmet, Ship2 Dataset

Smart Surveillance System and Prediction of Abnormal Activity in ATM Using Deep Learning

S. Gnanavel⬛, N. Duraimurugan⬛, and M. Jaeyalakshmi⬛

Abstract Although surveillance cameras are used in ATM cells, we face some problems of robbery and theft at ATMs due to lack of security; however, the monitoring capacity of law enforcement agencies has not kept pace. ATM spoofing attacks can be carried out to break or damage the ATM by stealing the machine and taking cash from the ATM. To reduce this problem, we arm the ATM with a camera module mounted in the room to perform continuous video observation. The camera detects the human and his activity in the ATM and attempts to breach the ATM. It detects unusual activities and immediately sends an alert notification to the police. Therefore, the system handles the application developed to automate video surveillance and detect any potential criminal activity at ATMs. Therefore, in this work, abnormal behavior is observed using CNN and RNN in surveillance videos. These algorithms can be used to recognize faces, detect and track camera movements, and detect and identify the action required to prevent such activity.

Keywords Theft detection · Human pose recognition · Mobile application · Alert notification · Automated teller machine

S. Gnanavel (✉)
Department of Computing Technologies, Faculty of Engineering and Technology, SRM Institute of Science and Technology, Tamil Nadu Kattankulathur 603203, India
e-mail: gnanaves1@srmist.edu.in

N. Duraimurugan · M. Jaeyalakshmi
Department of Computer Science and Engineering, Rajalakshmi Engineering College, Thandalam, Chennai 6021053, India

S. Namasudra et al. (eds.), *Data Science and Network Engineering*, Lecture Notes in Networks and Systems 791, https://doi.org/10.1007/978-981-99-6755-1_11

1 Introduction

An Automatic Teller Machine (ATM) is a machine used to make withdrawals, deposits and fund transfers and access account information at any time without the need for direct contact with a bank. An ATM is a computerized device or machine that is used to process transactions or evaluate and access financial statements. The ability to complete any financial transaction within a short period of time has increased the number of ATMs. Nevertheless, financial institution-related crimes have increased with the spread of automation and intelligent devices. These crimes increased steadily from 1998 to 2003, decreased slightly in 2004, and then increased dramatically starting in 2005. As a result, robberies improved tremendously over time [1].

ATMs with digital One Time Combination (OTC) locks are used for cash refill only. ATMs are installed in extremely secure building with enough CCTV coverage protected by state or central defense personnel. Off-bank ATM building are generally more favorable for banks as they attract many non-bank customers who have to pay service charges. Unfortunately, however, customers using off-premises ATMs are at a higher risk of robbery. ATM robbery has one of the highest rates of street robbery consistent with cocaine markets. Because street robbery is a way for addicts to make money, they need to buy crack, which doesn't require much planning or skill. ATM theft attracts a lot of media and public attention, mostly because the public thinks it can happen to anyone. Regarding ATM customer protection, the law has taken immediate action after a prominent legislator or someone close to a legislator has been robbed at an ATM. Over the past few years there has been a rise in ATM hotspots, looting, distress, looting, etc. These incidents have been raised, hence the need for a well-structured system to detect such strange behavior and block ATMs. This article warns police officers before a crime or theft takes place inside an ATM. Thus, as an alternative to late reaction to theft, the model helps to respond early and take necessary action [2].

The rest of the paper is arranged in the following order. Section 2 related work, Sect. 3 existing systems, Sect. 4 proposed systems, Sect. 5, Methodology, Sect. 6, Result and Discussion final session Conclusion.

2 Related Work

Alfredo Nazabal and others mainly focus on ATM theft detection. By implementing these features, you can reduce their time. A combined range of motion is a standard measure in functional assessment. Scientific measurement is performed by automatic goniometry and presents various problems, mainly for humans. This paper introduces the ROM Cam, an alternative system for measuring joint range of motion based on estimating human poses in 2D [1].

Ding et al. Using the sensitivity of facial recognition systems is a growing concern attracting interest from both. Academic and research communities. It provides a new perspective for face presentation attack detection by introducing a Light Field Camera (LFC), as the intensity of each incoming beam can be recorded using a Light Field Camera [2].

Tao et al. described that an intelligent home gives users access to temperature sensors, HVAC lights, and surveillance cameras. It provides users with a comfortable and safe living environment. Security and privacy are significant concerns, as information collected from these devices is typically communicated to the user through an open network or service provider-provided system [3].

Radar-Based Human Motion Recognition Francisca Rosic et al have proposed that it is essential for many applications such as search, intelligent homes, and monitoring and rescue operations. Recognizing continuous human motion in a real-life environment is essential for practical deployment [4].

Jeong Chen Bo et al., Photo Response Non-uniformity (PRNU) noise-based source attribution is a well-known technique to verify the camera of an image or video. Researchers have proposed various countermeasures to prevent PRNU-based RAW camera attributes. Theoretical analysis and experiments with multiple cameras demonstrate the effectiveness of our scheme depending on the number of stitches carved from an image and the randomness of the stitch positions [5].

Zhang et al. proposed that it provides an intelligent housing management system in which the social broker role is used to coordinate social services, reduce the workload of social management staff, provide electronic information services and deepen the community's integration with the surrounding environment [6].

Lu et al. Human activity recognition is a challenging high-level vision task for which many factors, such as a single object, object, and multimodal interactions must be considered and modeled. The proposed system recognizes incoming visual content based on previous experienced activities [7].

Sensor-based activity recognition has set up countless applications in a broad variety of healthcare fields and plays an essential role in research in the biomedical field. However, this consistently overlooked metric affects the ability of machine learning algorithms to extract context from raw data streams. We recommend a categorization structure that considers the modernization of subsample data in the test phase [8].

In recent years, sensor-based human pose activity recognition has attracted many researchers. This work calls for a single wearable tri-axis accelerometer based on the human action recognition system that can be used in real-life and track operations. With the fast growth of mobile devices and extensive computing technologies and technique, acceleration-based human activity recognition, the most challenging and essential issue in mobile applications, has received severe attention [9, 10].

3   Existing System

The existing system aims to measure the patient's joint range of motion. This paper introduces the ROM Cam, an alternative method for measuring joint functions based on assessing human appearance in 2D. An open pose library was used to estimate human poses in 2D to determine the joints and their motion. ROM Cam is a new ROM measurement system that patients and therapists can use in clinical settings and even from home [11, 12].

The proposed IoT-based methods have several levels to prevent the ATM from resisting physical and electronic machine attacks. Arduino has a built-in system for securing inputs using multiple sensors such as GSM modem vibration, temperature, and sound sensors. This method works in a simple procedure if the sensor exceeds its threshold. An alarm sound comes into play and alerts the higher authorities, indicating something is wrong inside the ATM. But this proposed system is flawed because it does not work efficiently on secular values, and there is a risk of false alarms [13–16].

4   Proposed System

In our research work, we proposed a security to ATM using deep learning algorithms. deep learning algorithms like CNN and RNN are used to make the system more precise. In addition to these algorithms, the mobile net thin model is used to detect and filter images, provides accurate results, and does not require much time to train the dataset. This paper is to design an intelligent alert notification when a theft occurs inside an ATM. Existing systems focus on actions only after an ATM theft has occurred. If a person tries to attack the ATM, the system recognizes the movement of the end user.

If any attack occurs, the alert notification will be activated. Let's arm the ATM with a room-mounted camera module to do constant surveillance video to defeat the problem. The cameras try to catch a person trying to enter an ATM or withdraw money from an ATM after successfully training the model and testing the output. The camera recognizes the unusual actions in ATM, an alert notification is sent to the police, and live streaming is enabled. An intelligent phone application was developed for the test model using React Native, which the police will maintain. The gates will automatically close as an abnormality is determined. Thus, it will successfully provide a solution to avoid any evidence of ATM theft.

5   Methodology

Figure 1 shows the planned work implemented to protect the ATM when an unauthorized person enters and tries to attack the machine. It uses Facenet, CNN, and RNN algorithms to situate the person, track the person's movement entering the ATM and notify the nearest local police station in case of an attack. Successful continuous data processing is essential for the successful classification of abnormal behavior. This continuous data is processed with the help of recurrent neural networks. In the proposed model, RNN and LSTM are used. The frame is first extracted from the video input in the proposed model. This frame is then fed to a previously trained CNN for feature extraction. Its output is fed to two fully connected layers.

Figure 2 shows the complete work setup. We will determine the human activity recognition that can effectively identify the user's activity in the ATM, thereby determining an abnormal activity. So, the first step is to collect the dataset from various sources. These datasets use preprocessed techniques to align the datasets into single dimensions. After preprocessing our dataset, it is ready for training on the architecture. We will use a hierarchy-like structure to train the model. Then we will perform the optimization, and the method used for the optimization is ELU which will optimize the whole model to remove the noises. We will minimize the loss, like cross-entropy, and check the mode by providing real-time inputs. A responsive mobile application has also been developed to view live. ATM's streaming when an abnormal condition occurs with a notification alert message. The project mainly consists of three modules.

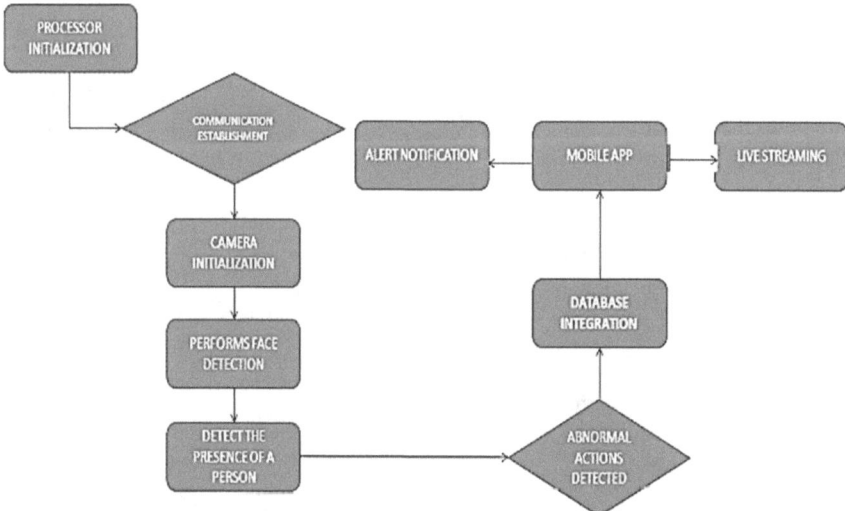

Fig. 1 System design

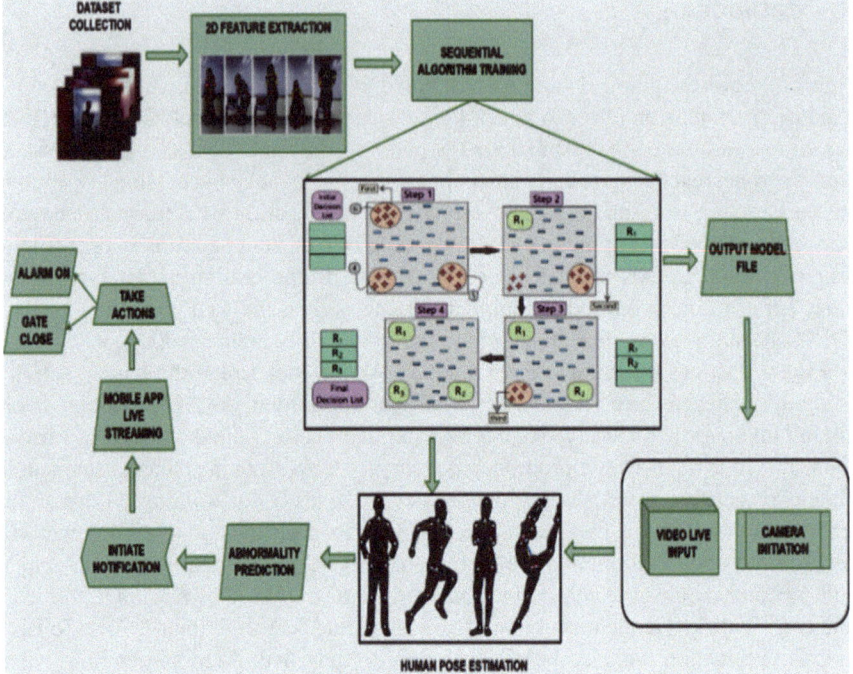

Fig. 2 Architecture

5.1 Detection of Human Activity

Detection of human activity as shown in Fig. 3. determines if the person is present and finds the joint range of motion. The camera will start and check for human presence. If humans are present, the system trained by the sequencing algorithm will identify humans and test them against the already loaded dataset. It will increase the time, frame, and FPS count if it matches the dataset. Human activities are collected as a dataset from live Camera for each pose, e.g., normal and abnormal actions. Live camera frames are saved as a dataset of each pose.

5.2 Abnormal Activity Prediction

This module consists of two parts.

1. Using face detection to detect presence called the training phase.
2. To extract 2D feature vectors called embeddings that measure each face in the image called the test grid (Table 1).

Fig. 3 Human pose recognition

Table 1 Classification report of low weighted model

Activity	Precision	Recall	F1-Score	Support
Normal	0.76	0.98	0.85	145
Abnormal	0.95	0.53	0.68	98
Accuracy	–	–	0.80	243
Macro Avg	0.85	0.75	0.77	243
Weighted Avg	0.83	0.80	0.78	243

For training, we need to provide input and output to the model. The generator generates input and output arrays. We used Google collaborator notebook for training purposes. We used the Adam optimizer with cross-entropy classified as a loss function to train the decoder model at block size 32. After each epoch, we evaluated the model using the training and validation loss metric. In training, we monitored the validation loss of the model. When the validation loss of the model improved after one epoch, we saved the model to a file. We use 'binary cross-entropy for this model. ADAM optimizer is used to optimize or improve network weights. To test the scheme, an input video stream is taken, and a prediction is made based on the features of the sample file. The system outputs a function that matches the features of the model file. Based on this, abnormal movement can be classified and predicted. As unusual activity is anticipated, the camera's live streaming can be sent wirelessly to the mobile app, which will be integrated.

5.3 Alert Notification and Streaming

A mobile application is developed using React Native, and it contains a set of modules for IOS and Android platforms to create a mobile application with a local look and feel. Person activity is monitored in this module, and if someone tries to attack the machine, an alarm is triggered, and live streaming is enabled. This is done using LSTM and CNN algorithms.

6 Result and Discussion

Once the algorithm is executed, the model is tested and results are obtained. The system scans the person as soon as he enters and detects the person's face. Once the face of the person entering the ATM is seen, the camera tries to identify whether it is human or not as shown in Fig. 4.

If a person commits an illegal activity like breaking an ATM machine, the activity detection code is activated immediately. If abnormal activity is detected, an alert message will pop up and the ATM's video, the camera will capture and notify the authorities. If someone tries to attack the machine using the motion detection code, an alarm notification will be sent to the police through the mobile app and live streaming will be activated as shown in Fig. 5.

Figures 6 and 7 demonstrate the model train for 40 epochs at a knowledge rate of 0.0001 and a block size of 32. This makes sense since, as the model approaches convergence, we need to decrease the learning rate to take smaller steps towards

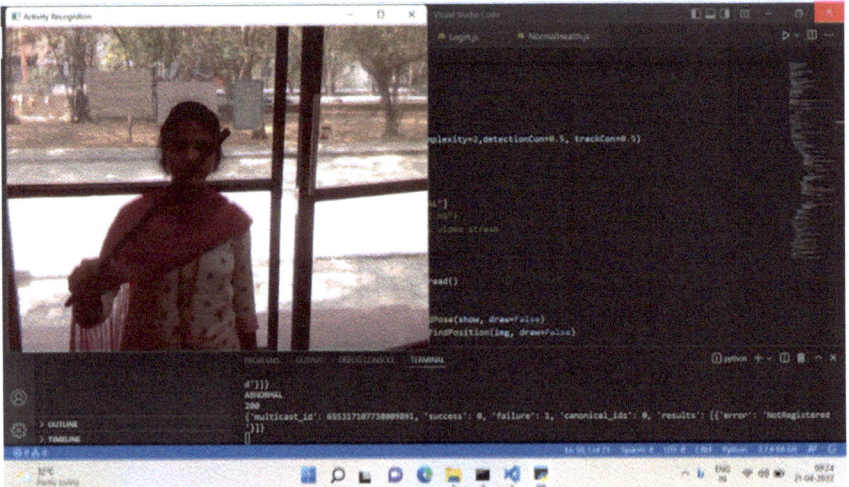

Fig. 4 Activity prediction

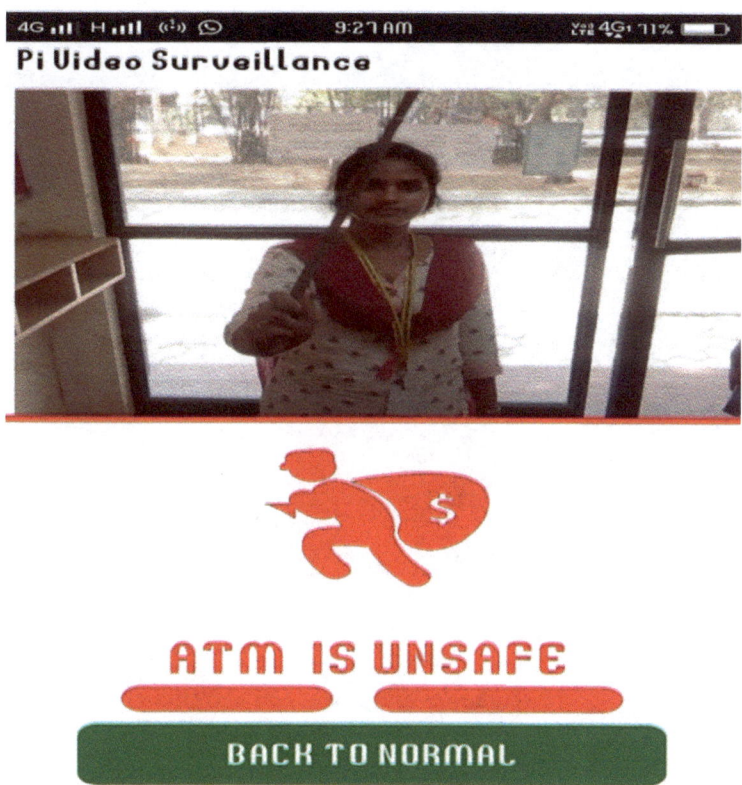

Fig. 5 Mobile alarm and streaming

minima. Gradually increasing the batch size improves the power of your gradient updates. Data is trained, features are collected and fitted to the model file, and Precision and average are calculated. The value of loss and precision at each epoch is represented graphically below. As we progress from a low number to a high number of epochs, the loss value decreases, and the precision value grows. Also, the more epochs there are, the smoother these curves become. In this work, datasets are collected and made publicly available, one at laboratory levels, the dataset is used to train algorithms, and the algorithms are tested.

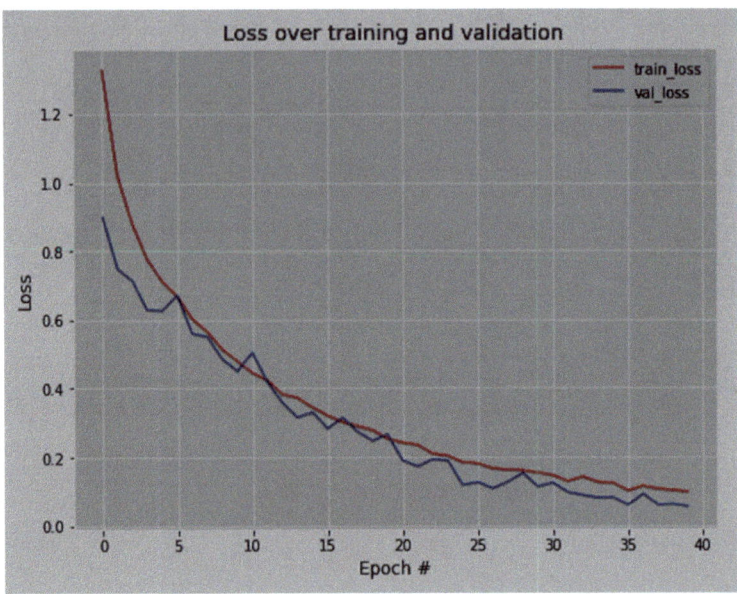

Fig. 6 Loss validation graph

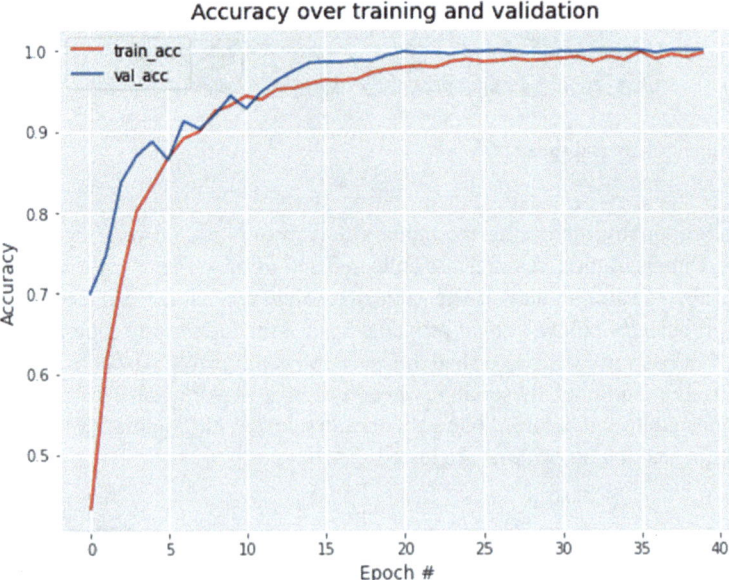

Fig. 7 Accuracy validation graph

7 Conclusion

In this work deep learning model accurately detects the account holder's wrong intentions regarding the ATM and its cash. If a person's actions are unusual, it creates suspicion and successfully detects and issues a warning or warning message. Based on these messages, necessary steps can be taken immediately to avoid further significant problems. This work has successfully implemented a person action recognition and gesticulation recognition system that can automatically recognize human activity using the currently practiced deep learning approach. Therefore, this proposed algorithm will be helpful for all ATMs that have suffered severe losses due to theft. In the future, we will review the use of ATM technology for security and get better the accuracy of all forms of authentication. There are opportunities in this field to improve our theft detection or modify this program in several ways. The correctness of functional forecast can be bigger by using dissimilar structured techniques and methods. Hence, this project has an efficient scope in the prospect where manual forecasting can be converted into computerized production at a low cost. This could be a step towards a significant revolution of a new system that enables depositing, withdrawing money, and performing various activities more safely and securely without risk.

References

1. Nazabal A, García-Moreno P, Artes-Rodríguez A, Ghahramani Z (2016) Human activity recognition by combining a small number of classifiers. IEEE J Biomed Health Inform 20
2. Ding C, Hong H, Zou Y, Chu H, Zhu X, Fioranelli F, Le Kernec J (2019) Continuous human motion recognition with a dynamic range-doppler trajectory method based on FMCW radar. IEEE Trans Geosci Remote Sens 57
3. Tao D, Lianwen J, Yuan Y, Xue Y (2016) Ensemble manifold rank preserving for acceleration-based human activity recognition. IEEE Trans Neural Netw Learn Syst 27
4. Rosique F, Losilla F, Navarro PJ (2021) Using artificial vision for measuring the range of motion. IEEE Lat Am Trans 19
5. Poh GS, Gope P; Ning J (2019) PrivHome: privacy-preserving authenticated communication in smart home environment. IEEE Trans Dependable Secur Comput 18
6. Zhang H, Zhou W, Parker LE (2015) Fuzzy temporal segmentation and probabilistic recognition of continuous human daily activities. IEEE Trans Hum-Mach Syst 45
7. Lu J, Tong K-Y (2019) Robust single accelerometer-based activity recognition using modified recurrence plot. IEEE Sens J 19
8. Wang L, Zhao X, Si Y, Cao L, Liu Y (2017) Context-associative hierarchical memory model for human activity recognition and prediction. IEEE Trans Multimed 19
9. Kishore PVV, Kumar DA, Sastry ASCS, Kumar EK (2018) Motionlets matching with adaptive kernels for 3-D Indian sign language recognition. IEEE Sens J 18
10. Raghavendra R, Raja KB, Busch C (2015) Presentation attack detection for face recognition using light field camera. IEEE Trans Image Process 24
11. Gnanavel S, Ramakrishnan S (2017) HD video transmission on UWB networks using H.265 encoder and ANFIS rate controller. Clust Comput J Netw Softw Tools Appl 21(1): 251–263
12. Gnanavel S, Ramakrishnan S, Mohankumar N (2014) Wireless video transmission over UWB channel using fuzzy based rate control technique. J Theor Appl Inf Technol 60(3):491–503

13. Gnanavel S, Sreekrishna M, Mani V, Kumaran G, Amshavalli RS, Alharbi S, Maashi M, Khalaf OI, Abdulsahib GM, Alghamdi AD et al (2022) Analysis of fault classifiers to detect the faults and node failures in a wireless sensor network. Electronics 11:1609

14. Sakkarvarthi G, Sathianesan GW, Murugan VS, Reddy AJ, Jayagopal P, Elsisi M (2022) Detection and classification of tomato crop disease using convolutional neural network. Electronics 11:3618

15. Gnanavel S, Narayana KE, Jayashree K, Nancy P, Teressa DM (2022) Implementation of block-level double encryption based on machine learning techniques for attack detection and prevention. Wirel Commun Mob Comput Article ID 4255220:9. https://doi.org/10.1155/2022/4255220

16. Nagendiran D, Chokkalingam SP (2022) Real time brain tumor prediction using adaptive neuro fuzzy technique. Intell Autom Soft Comput 33(2):983–996

A Framework for Extractive Text Summarization of Single Text Document in Tamil Language Using Frequency Based Feature Extraction Technique

K. Shyamala and M. Mercy Evangeline

Abstract Text summarization, a technique in Natural Language Processing, helps in summarizing documents like news articles, legal documents, essays and more. The content may be comprehensive and redundant. A summary gives an insight of the document. Text summarization is broadly classified into two categories—extractive and abstractive summarization. Abstractive summarization uses deep learning techniques to generate summary, just as humans generate summary using their own words and sentences. Extractive summarization highlights information based on some features or technique used to identify the importance of the sentence from the source document. Methods used for extractive summarization include ranking algorithms, sentence scoring, sentence similarity and so on. In this paper, a framework for extractive text summarization using features extracted from a Tamil document has been proposed. The summarizer is based on Fuzzy logic inference engine. The framework describes the modules involved in the generation of Extractive Text Summary for a single document.

Keywords Text summarization · Extractive summarization · Feature extraction · Frequency based features · Fuzzy logic · Tamil language

1 Introduction

A subfield of Artificial Intelligence known as "Natural language processing" (NLP) deals with teaching machines how to comprehend text and using that knowledge to train them for various tasks. Text translation, working with voice commands, chat bots, speech-to-text commands, summarizing huge amounts of text, sentiment analysis, and many other things are major applications of linguistics. NLP has developed along with computational linguistics, where computer science is applied to

K. Shyamala · M. M. Evangeline (✉)
Dr. Ambedkar Govt. Arts College, Vyasarpadi, Chennai, Tamil Nadu, India
e-mail: mercyevangelinemanoj@gmail.com

comprehend linguistic beliefs and provide frameworks for completing relevant tasks. Although several complex applications have been created in the field of NLP, much more work has to be done.

1.1 Text Summarization

Text summarization is an application in the field of NLP, where a large volume of text is read as input and important sentences are extracted, to give a review about the document considered. Text summarization finds its application in many areas like, summaries generated for news articles, news websites, e-commerce using the summary of customer's feedback to change their strategy, for deriving marketing strategies from customer's behavior and so on. As humans', summarization involves reading the entire document, highlighting the important sentences and then briefing the highlighted points as summary are important. Instead of humans, when a computer has to do the task of summarization, it becomes a difficult task. The computers don't have human knowledge, so they have to be trained for the particular task. In recent years, automatic text summarization has gained importance in many domains. Automatic text summarization generates a concise summary by preserving the meaning of the content, without changing the overall meaning of the content. It finds its history way back in 1950s, where conventional abstracts were generated by automatic ways [1]. Sentences with highest score based on word frequency and its distribution were considered for generating the summary automatically.

Generally, summarization technique can be applied for either a single document or multi-document. For a single document summarization, the document is scanned for information and the summary is generated with the information available in the document. For a multi-document, the documents are analyzed and a single theme is considered as the subject matter. Information relevant to this theme is pooled together to generate the summary for all the documents considered for summarization. The summarization process is further classified into two types, namely Abstractive and Extractive Summarization, based on the method adopted for generating the summary.

1.2 Abstractive Summarization

In Abstractive summarization, the whole document is scanned through and a precise summary is generated by interpreting the original text. It summarizes the salient features in the document and will generate a summary which will contain new sentences and phrases which may not be in the text file. This process involves more of understanding the semantics of the text in the document to generate a meaningful summary in a precise form. The abstractive text summarization is classified into two categories—structure based and semantic based approach. Structure based approaches make use of prior knowledge. It extracts important features from the content based on psychological features like tree, graph, ontology, rule or template based structure. Semantic based approaches make use of linguistic demonstration of the document. Some of the methods used in this approach include multimodal method, semantic text representation, semantic graph base method and information item-based method.

1.3   Extractive Summarization

In Extractive summarization, the text file is read and important sentences and entities are identified and extracted without changing their relevance. This extraction can be based on the important features available and they are summarized together. The features may be based on statistical features or linguistic information to obtain the importance of the sentences [2]. Sentences with higher threshold value based on features will be included within the summary. For threshold value, features like sentence length, their position, rank and other features will be considered. The core steps involved in Extractive Summarization can be précised as follows:

* Creation of an intermediate form of the input text

* Scoring the sentences based on a particular representation

* Generating the summary by including significant number of sentences

1.4   Organization of the Paper

The paper puts forth the framework adopted for generating the summary using Extractive summarization technique. The different modules incorporated for summary extraction include stop word removal, feature identification and extraction, text summarization using the features extracted. Given a document, the first phase includes tokenization and stop word removal using the dictionary-based removal method. From the output of the tokenized words, NOUN and VERB feature identification is carried out in the second phase. Then the frequency of the features identified is calculated and used for extraction of the top features. These features are used for generating summary using the Fuzzy logic Inference Engine summarizer. The different phases carried out for the summarizer is put forth in the further discussion.

2   Literature Survey

Many scholars have proposed summarization techniques for different regional languages, based on the richness of the language and the rules available for the respective languages and methods adopted. A review of some of the proposed techniques particularly for regional languages are discussed below.

Megala et al. [3] proposed an automatic text summarization system which was implemented using the fuzzy logic method. This was compared with the Neural network method and the evaluation measures based on precision, recall and F-measure were tabulated. The data set considered for this work included 50 legal judgment documents collected from online resources. The documents were pre-processed and features extraction methods were used to extract 13 features from the document to enable the summarization. The results show a finest average measure compared to the Neural Network method.

Gupta and Kaur have described a novel hybrid technique for summarizing text documents in Punjabi language [4]. The features considered for the summarization technique include location-based features and statistical based features, which enhance the result of summarization. Sentences are also classified into summary and non-summary category using Support Vector Machine based classifier, to handle imbalanced data. The results show a good performance measure compared to other baseline systems considered. The documents include nearly 150 randomly selected documents from two datasets.

Sethi et al. [5] proposed a technique for text summarization, which precisely looks for important segment of text for generating coherent summaries. The summary generation is based on lexical chains created from the text based on topic progression, not on semantic interpretation of the text. The dataset considered for the model includes new articles from an unknown source. A general analysis of the pattern followed in the news articles was considered for generating the lexical chains. As the dataset considered is news, Nouns appearing in the first sentence was given a high score and it was used in creating the lexical chain. The lexical chain was used for sentence scoring and the summary was generated based on this score.

Varma and Om have designed a metaheuristic optimization-based summary generation for multi-document summarization. This work generates the optimal combination of sentence scoring methods and their respective weights to extract sentences for the summary [6]. The metaheuristic approach used here is based on teaching–learning-based optimization. The result of the summary is compared with other methods which use a different heuristic approach. The performance of these summarizers is measured on precision, recall, F1, readability, non-redundancy of information and cohesion. The dataset considered for the technique is the DUC dataset.

Manjari has designed a multi-document extractive text summarization for Telugu language [7]. For the implementation, stop words were removed in the pre-processing phase. After pre-processing, the homonyms and conjunctions are identified and the necessary procedures are adopted for summary generation. The continuous bag of words model is used to generate the word embeddings and the cosine similarity measure is used to find the similarity between the sentences. The performance of the summarizer is measured using the ROGUE measure. TextRank algorithm is proved to be easiest and best algorithm for Text Summarization.

The significance of the proposed work is the features considered for the extraction process for generating the summary. Given any story in Tamil, in text format, the proposed system identifies the NOUN and VERB as important features. The features with the highest frequency are considered for summary extraction. The efficiency of these features is shown in the performance of different summarizers considered for the examining. The performance is measured by precision, recall and F1 measures. Given any dataset in text format, the proposed work will be generating a summary using an extractive method.

3   Framework for Automatic Text Summarization Using Extractive Technique

A framework for automatic text summarization using the extractive method has been proposed for Tamil language. In this model, a story in text file format is read and summary of the story is generated. The text file is passed through different modules of the framework for generating the summary automatically. Figure 1. shows the different modules included in the framework. Automatic summary is generated for a single text file given as input.

Given an input text file in UTF-8 format, the different processes involved in the technique are listed as follows:

i.   This file is relayed through the first phase, the Pre-processing phase, generating an array of words.
ii. The array of words is passed through the next phase, NOUN–VERB Identifier (NVI) for feature identification.
iii. Features identified are passed through the phase of extraction using the frequency-based method.
iv. Features extracted are passed through the summarizer phase. For summarization, Fuzzy logic Inference engine summarizer is used to generate the summary.

The different phases of the Automatic Text Summarizer and the processing done is briefly discussed below.

The first and the foremost implementation starts with the collection of datasets relevant to the work. For the implementation, the dataset considered includes stories in Tamil language. For initial training of the summarizer, Tamil Short stories which includes Kids' moral stories were taken as input. These story files were taken from web repositories and they were created as text file in UTF-8 format. For further processing and testing of the summarizer, stories written by author Kalki Krishnamurthy was considered. The stories which were available as PDF files were manually created by typing the content and storing it as text file with UTF-8 format. The stories

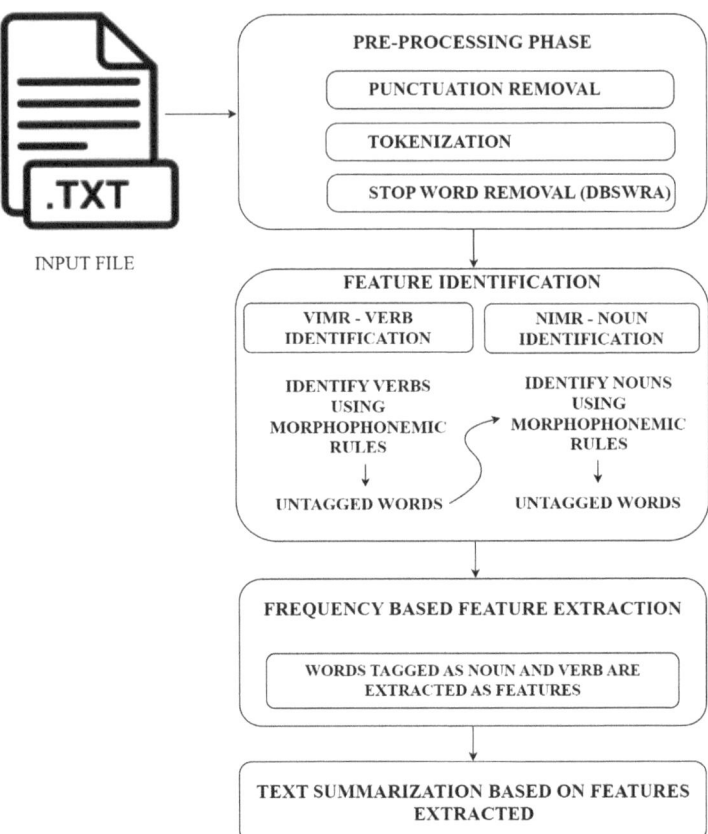

Fig. 1 Extractive text summarization for single document in Tamil language

considered were from different timelines written by the author, so the words used and the style of writing by the author have an influence in the performance of the summarizer.

3.1 First Phase: Pre-processing Phase

In the pre-processing phase, the text file is read and pre-processed before giving as input to the model. The different processes involved in the pre-processing are Tokenization, removal of Punctuation and Stop words. First the words are tokenized using the space as the delimiter. Then the punctuation symbols which include special characters like comma, question mark, dot, single and double quotes are removed. Then the stop words are removed from the tokenized word.

The stop words have been defined manually. Words which are insignificant or weakly relevant to the data analysis are generally defined as stop words. The stop words considered includes prepositions, articles and pronouns in Tamil Language. These stop words are defined as array of words and maintained in text file with UTF-8 format. For removing the stop words, Dictionary Based Stop Word Removal Algorithm (DBSWRA) is adopted [8]. The output of this phase is an array of words with punctuations and stop words removed. The tokenized words are compared with the stop words defined manually and if available they are removed from the array of tokens.

3.2 Second Phase: Feature-Identification Phase Using NOUN–VERB IDENTIFIER (NVI)

Tamil is an agglutinative language rich in morphological rules. Morphemes are the smallest unit of representation. Morphemes are used to identify information like person, gender, number and tense. An array of words which are pre-processed is the input for this phase. In this phase, two types are features are identified for the summarization procedure. This phase has two modules, the Noun Identification using Morphophonemic Rules (NIMR) and Verb Identification using Morphophonemic Rules (VIMR). The two main features identified are the NOUN and VERB.

In Tamil language, some words are indivisible while most of them will be divisible into parts. All the divisible words will have two forms: the base and the grammatical inflation. The inflation maybe inflectional or derivational. For identifying the features, morphophonemic rules have been used. Each word is looked for inflection with suffixes and the rules are identified. The root is identified and then the corresponding tagging of features is carried out by reverse splitting method.

Noun Identification using Morphophonemic Rules (NIMR). For identifying words under NOUN category, a rule-based approach has been used [9]. The identification is mainly based on inflections formed by adding suffix to the root word. General inflection representation for a NOUN is case and number form. In Tamil grammar, the NOUN inflection for case marker is listed under eight categories. For each word in the array, the rules of suffixes are checked by the reverse splitting method, if satisfied they are tagged as NOUN and the remaining words are tagged as UNIDENTIFIED. The NOUN feature identified in this phase is checked with the manually identified nouns for test data. NOUNs identified are based on the transition they have undergone. If the NOUN is in stem form, they will not be tagged as feature in this phase. The untagged words are sent through the next module for VERB feature identification.

Verb Identification using Morphophonemic Rules (VIMR). The array of words which are untagged in the previous phase is given as input to this module. In this module, verbs are identified from the array of words and are tagged as VERB, remaining words are untagged. A VERB form is generally inflected based on person, count, tense and voice. For identification, a rule-based suffix stripping method is adopted, similar to the Noun identification method [10]. For both the identification, Unicode format of the characters are considered. Transliteration process is not adopted for this identification. These two features NOUN and VERB are primarily listed as features and are used for the summarization process.

3.3   Third Phase: Frequency-Based Feature Extraction Phase

Feature extraction is used to pick up words which are crucial and essential in understanding the text being analyzed. Looking for data for an effective feature measure proves to be very important role in working with structured or unstructured data [11]. For the proposed framework, the features considered include VERB and NOUN identified through the rule-based method. For any language, NOUN and VERB are considered to represent a major part of the word categories available in a document. From the features identified, the frequency within the whole document is calculated and words with the frequency range of 10 and above are considered for the summarization process. Words tagged as NOUN and VERB with frequency above the threshold value are used as features for the summarization process [12].

3.4   Fourth Phase: Automatic Extractive Text Summarization

The features extracted in the previous phase are considered for the summarization process. Given a text file as input, the summary of the file is generated automatically using the extractive summarization method. The extractive summarization technique involves in reading the file, identifying the sentences with higher importance and framing the summary [13].

The summarizer used for generating summary is based on Fuzzy Logic Inference Engine. Fuzzy Logic Inference Engine Summarizer uses sentence features like the length, position of the sentences, the similarity distance of the sentences and the important features available within the sentences. For all these sentence features, the feature extracted (NOUN and VERB) are used for the calculation. From the features, sentence score is calculated and highly ranked sentences are included for the summary.

The fuzzy logic inference engine summarizer makes use of features extracted to identify the highest scoring sentences. For scoring the sentences, the attributes of the document like sentence position, its length, sentence similarity based on cosine distance, tagged features available within the sentences are considered. These attributes are used to define rules and sentences are tagged as GOOD, BAD and AVERAGE. The summary is generated based on a constant value fixed on the performance of the summarizer.

4 Conclusion

The framework for generating the summary is based on extractive technique. Given an input document in Tamil language, the summarizer generates the summary automatically based on the extractive method. The input files include stories written by author Kalki Krishnamurthy. Five different stories from different periods of time have been considered for this work. The timeline of the story, the words used in the context of the story has an impact in the summarization technique. The automatic summary generated by the summarizer is compared with human generated summary.

The Automatic text summarizer includes stop word removal, features identification and extraction, summarizer within the process. For most of the work done so far in this field, the tagging process has been carried out with taggers released by other researchers. In this framework, all the modules are in-built to achieve the summarization process. The time taken for generating the summarization includes pre-processing, feature identification and extraction, and generation of summary. The feature identification and extraction include tagging of words as NOUN and VERB, which form the primary components of any language. These features are used by the summarizer to score the sentences and generate summary.

The fuzzy logic inference engine summarizer uses features extracted to identify the highest scoring sentences. The scoring of sentences is based on important attributes of the sentences available in the document. These attributes are used to define rules and sentences with the highest score is extracted as summary for the document.

References

1. Luhn HP (1958) The automatic creation of literature abstracts. IBM J Res Dev 2(2):159–165. https://doi.org/10.1147/rd.22.0159
2. Madhuri JN, Ganesh Kumar R (2019) Extractive text summarization using sentence ranking. In: International conference on data science and communication (IconDSC) pp 1–3. https://doi.org/10.1109/IconDSC.2019.8817040
3. Megala SS, Kavitha A, Marimuthu A (2014) Enriching text summarization using fuzzy logic. Int J Comput Sci Inf Technol 5(1):863–867
4. Gupta V, Kaur N A novel hybrid text summarization system for Punjabi text. Cogn Comput 8(2):261–277. https://doi.org/10.1007/s12559-015-9359-3
5. Sethi P, Sonawane S, Khanwalker S, Keskar RB (2017) Automatic text summarization of news articles. In: International conference on big data, IoT and data science (BID), pp 23–29. https://doi.org/10.1109/BID.2017.8336568
6. Verma P, Om H (2019) A novel approach for text summarization using optimal combination of sentence scoring methods. Sādhanā 44. https://doi.org/10.1007/s12046-019-1082-4
7. Manjari KU (2020) Extractive summarization of Telugu documents using TextRank algorithm. In: Fourth international conference on I-SMAC (IoT in social, mobile, analytics and cloud) (I-SMAC), Palladam, India, pp 678–683. https://doi.org/10.1109/I-SMAC49090.2020.9243568
8. Mercy Evangeline M, Dr Shyamala K (2018) Stop word removal algorithm for Tamil Language. In: International conference on recent innovation in electrical, electronics & communication engineering, CFP18P8 – PRT: 978-1-5386-5994-6
9. Mercy Evangeline M, Dr Shyamala K (2019) Noun identification for Tamil Language using morphophonemic rules. Int J Recent Technol Eng (IJRTE) 8(4). ISSN: 2277-3878

10. Mercy Evangeline M, Dr Shyamala K (2020) Verb identification for Tamil Language using morphophonemic rules. ICTACT J Soft Comput 11(1):2237–2243
11. Wang D, Su J, Yu H (2020) Feature extraction and analysis of natural language processing for deep learning English Language. IEEE Access 1–1. https://doi.org/10.1109/ACCESS.2020.2974101
12. Evangeline MM, Shyamala K, Barathi L, Sandhya R (2021) Frequency based feature extraction technique for text documents in Tamil Language. In: Advances in computing and data sciences. ICACDS 2021. Communications in computer and information science, vol 1441. Springer, Cham. https://doi.org/10.1007/978-3-030-88244-0_8
13. Quillo-Espino J et al. (2021) A deep look into extractive text summarization. J Comput Chem 9:24–37

An Approach to Mizo Language News Classification Using Machine Learning

Andrew Bawitlung, Sandeep Kumar Dash, Robert Lalramhluna, and Alexander Gelbukh

Abstract The increase in the availability of data on the Internet in the past years has created an enormous amount of data and research in the field of Artificial Intelligence and Machine Learning. With the advancement in technology, computational power has also increased dramatically in the past few years, and this has led to more and more advancements in Artificial Intelligence research and its applications. Mizo language, which is a low-resource language, also tends to emerge in recent years along with these advancements and with the help of news articles collected from the two biggest news outlets for the Mizo language namely Vanglaini and The Aizawl Post, an approach to news classification based of their category was done in this paper. This paper tested several machine learning methods using supervised classification techniques and got the highest accuracy among other low-resource languages in most of the models tested and among which Multinomial Naive Bayes classification gives an accuracy of 96% and is the highest when compared to the other models.

Keywords Mizo language · News classification · Machine learning · Neural networks

1 Introduction

1.1 Mizo Language

Mizo language or Mizo ṭawng is a low-resource language in the field of Natural Language Processing because data written in Mizo language is very less in the net

A. Bawitlung · S. K. Dash (✉) · R. Lalramhluna
National Institute of Technology, Aizawl, Mizoram, India
e-mail: sandeep.cse@nitmz.ac.in

A. Bawitlung
e-mail: andrew_bawitlung@nitmz.ac.in

A. Gelbukh
Centro de Investigación en Computación, Instituto Politecnico Nacional, Mexico City, Mexico

and is one of 50 branches of Kuki-Chin languages [1] or Sino-Tibetan languages [2] which are spoken natively by the Mizo people in Mizoram in the North-Eastern part of India. Since Mizo language can be considered as an umbrella term for all the languages of the Mizo ethnic tribe used all over the world, the most common name of the language used by the people of Mizoram is often called Duhlian and Lushai languages. This paper also used Duhlian and Lushai languages which will be called the Mizo language throughout this paper. The Duhlian and Lushai languages have derived words from many other sub-clans of the Mizo language like Hmar, Pawi, Paite, etc. which are used in the neighboring states of Mizoram including Manipur, Nagaland, Myanmar, and Bangladesh [3]. These Mizo languages have many similar dialects and words and have their fair share of differences from other Mizo languages.

1.2 Machine Learning

Machine Learning is a technique used in computer engineering where a machine learns, adapts, improves, and makes decisions based on past experiences like how humans do. The term Machine Learning was first coined by Arthur Samuel in 1959 [4]. He is known to be a pioneer in this field. Before the term Machine Learning was coined, the popular name given to these types of computing are called "self-teaching" computers [5]. The main objective of Machine Learning is to try and act or mimic the human brain and make the machines think like a human. The process of learning in Machine Learning begins with pre-existing data. Data are fed into the machine and the machine learns from the available data and looks for patterns. After learning, the machine has the required experience and can perform the tasks for which it is made. Accuracy and self-correction can be improved based on new data or experience that we provide. Due to the amount of data needed to implement machine learning algorithms, many languages do not have enough labeled data for efficient usage which is also a drawback for the Mizo language.

Machine learning is an ever-trending topic that we encounter almost in our every-day life but might be unnoticeable to the normal consumer. Machine Learning or Artificial Intelligence is used widely in areas like spam detection [6], video surveillance, weather, finance prediction [7], social media, online customer support, malware analysis [8], advertisements, and search engines, to name a few. It is a technology that is not going away anytime soon and is predicted to occupy and help us in our daily and professional lives in more ways than we can imagine it is currently. Machine Learning has two basic approaches called Supervised and Unsupervised Learning. In Supervised Learning, the algorithms are based on a mathematical model where both input and output are considered from the given data [9]. These data can be considered as training data where the desired output or "correct" answers are known beforehand for the machine algorithm to learn; in this way a supervised learning method is used to predict the output when a new unknown input is fed in the machine [10]. Unsupervised Learning on the other hand takes a set of data that has only inputs and finds a structure within those inputs as a way to cluster or group the data to produce the desired output. In Unsupervised Learning, the data are not labeled, classified, or categorized which is why the learning algorithms identify common patterns or similarities in the data by which it tries to distinguish the dataset based on the inherent properties of the data itself [11]. This paper focuses solely on Supervised Learning and nothing on Unsupervised Learning.

Text classification or text categorization is an ongoing research topic since the 1980s [12] and is actively running to this day. From the perspective of a data analyst or a text miner, text classification can also be considered a technological way of pre-processing a given dataset of text and then filtering the text to get the desired output. The main objective of text classification is to allocate documents or articles into pre-defined topics. These topics can be many things; it can be of sentimental categories like a good and bad review, type of books written like sci-fi or history, and many more. Due to the drastic increase in information on the web which is mainly text-based, a need for categorizing these topics is a must. We can see these types of text classification in our day-to-day use on our SMS messaging app where most applications recently categorize text messages into different categories like OTP, Promotion, Spam, etc. We also see these types of classifications on our email where the email we receive nowadays can be classified into personal, promotional, updated, forum, and so on. The literature on text classification goes back as early as the '60s [13]. It is only recently in the late '90s that with the help of Support Vector Machines, an application of Machine Learning was introduced in text classification [14, 15]. Multiple-label classification of multi-topic text continued after Support Vector Machines with enhanced AdaBoost multi-label methods [16]. This method used the concept of assigning multiple topics to a text and treating it by ranking the labels associated with the text. This method brought out many inspirations for evaluating text ranking techniques for information retrieval.

This paper discussed and tackles the problem of news classification using different methods and implements it for Mizo language. Mizo language, being a low-resource language, is comprised of the English alphabet which was created for us by the British Missionaries. Since it is a low-resource language, data collection is complex as most of the Mizo websites came online only a couple of years ago. The collection of news articles was done from vanglaini.org and theaizawlpost.org till the year 2021. Commonly used machine learning techniques were used and tested on the low-resource Mizo language. Very few machine learning papers based on the Mizo language are available where English-Mizo Machine Translation [17] is one of the first to appear in this literature. This paper plans on exploring the various and best techniques used by researchers in text classification where we apply the common methods and techniques used over the years that have been proven to be effective over the years. Since the Internet was popular only after the launch of Jio in 2013 in the North-East part of India, the availability of resources is less even today. Information websites like news websites and blog sites appear only recently from 2015. Due to these issues, data available are relatively very less regarding how might handle the field of Artificial Intelligence and Machine Learning as these methods require lots of data to get objective results.

The rest of the manuscript is organized as follows. In Sect. 2, all the related work done on different news classification and text classification is studied which included mainly low-resource language and the work done using these languages. Section 3 shows the data description of the Mizo news articles where the meaning of the categories and the number of articles collected are mentioned. Section 4 describes the process of dataset collection, the use of stop words and stemming, and CountVectorizer techniques for feature selection. In Sect. 5, the paper mentioned the different machine learning techniques used and their definitions. Section 6 mentioned the System Architecture which describes the architecture of the pre-processing of text from splitting of the training and testing data to its predicted output. Section 7 interprets the results and analyzes the results and finally, Sect. 8 concludes the work and also addresses the scope of the Mizo language in the field of Natural Language Processing.

2 Related Work

Pathak et al. [17] built a model based on neural and statistical approaches in 2019 to translate the English language into the Mizo language. This is one of the few research in which a machine was built for the low-resource Mizo language. News classification for Nepali [18] was done in which SVM with Radial basis function kernel has 74.65% accuracy, and linear SVM with 74.62% accuracy. The paper got an accuracy of 72.99% when Multilayer Perceptron Neural Networks were used and surprisingly a very low score of 68.31% on the Naive Bayes. Two Bangla news classifications by Chy [19] and independently by Chowdhury [20] used Naive Bayes classification, GloVe Vectorization, and Long Short-Term Memory and were compared with other models like Support Vector Machines, Long Short-Term Memory, Convolutional Neural Network, and Artificial Neural Network and the highest accuracy got was 87% on the Deep Learning-based classification. Using the Convolutional Neural Network model [21], this paper got an accuracy of 94.17 and 96.41% in Turkish and English language news respectively which shows a high level of accuracy. Punjabi news classification [22] was done using the Naive Bayes method and the paper got an accuracy of 78%. Another low resource is the Indonesian language; using different classification techniques like Multinomial Naive Bayes and Support Vector Machines [23], the paper shows an accuracy of 85% on the classification which is not far off from the results got from other papers. Large Scale Text Classification [24] introduced Semisupervised Frequency Estimate (SFE) where they compare the performance and characteristics of SFE with EM+Multinomial Naive Bayes showing that the experiments consistently improve Multinomial Naive Bayes with additional data (labeled or unlabeled). On the contrary, some paper [25] got a disappointing result on Naive Bayes text classification, and this conclusion can be challenged by Survey on Classification and Summarization of Documents [26] where a survey in India was done on the methods of text classification and summarization of documents and has concluded that different methods have different usage depending on the scenarios but from studying various papers on the matter, undoubtedly, the Naive Bayes classification method stands out as one of the most precise techniques used in text classification. Online News Classification [27] uses Multinomial Naive Bayes with mutual information for feature selection in which the highest accuracy acquired is 80% where feature selection was not used and Multivariate Bernoulli was the classification technique used. A precision of 97.84 and 94.93% [28] for BBC and 20Newsgroup using Support Vector Machines classification was also observed. When comparing, K-Nearest Neighbors, Naive Bayes, and Support Vector Machines [29], Naive Bayes seems to be more stable as compared with the other models. Classification based on authorship of tweets was done [30] and shows an accuracy of 91.1% in logistic regression-based classifier and 89.8% in Naive Bayes classifier where a total of 46,895 tweets were used. Text Classification of Alzheimer's Drugs' Mechanism of Action [31] used Decision Tree-based classification and got a high accuracy of 95%. News classification based on Random Forest and Weighted Multimodal Features [32] shows the accuracy was 84.4% using only the N-gram features, while the combination of N-gram textual features and visual features gives an improved accuracy of 86.2% using Random Forest. K-Nearest Neighbor Algorithm is also used on various text classifications. 87.58% accuracy was also achieved using the K-Nearest Neighbor method of classification on news articles [33].

Research has also been done on Neural Network-based classification [34] where the result shows avant-garde performance. Text classification based on RNN and LTSM-like architectures done in 2020 [35] got 97.31% accuracy on the Claritin-october-Twitter dataset using LTSM.

3 Data Description

Datasets were built upon the data collected from the local newspapers, vanglaini.org, and theaizawlpost.org. A total of 6,730 articles were taken, out of which each article was taken from the categories, namely Hmarchhak, Khawvel, Ramchhung, Tualchhung, Infiamna, and Thalai. Here, Hmarchhak is the category that comprises all the news from North-East India excluding Mizoram. Due to the low number of articles in the category Hmarchhak which includes only 494 articles, it was dropped and in total, we have 5 categories of news articles in our corpus where the total number of news articles became 6,236. A detailed definition and description of each category are mentioned below.

- **Tualchhung** refers to Local News; local news consists of all the events happening within Mizoram.
- **Ramchhung** is the National News where in the whole of India, Mizoram is entered but excludes the whole of North-East India.
- **Khawvel** which in English means "World", and this consists of all the international news.

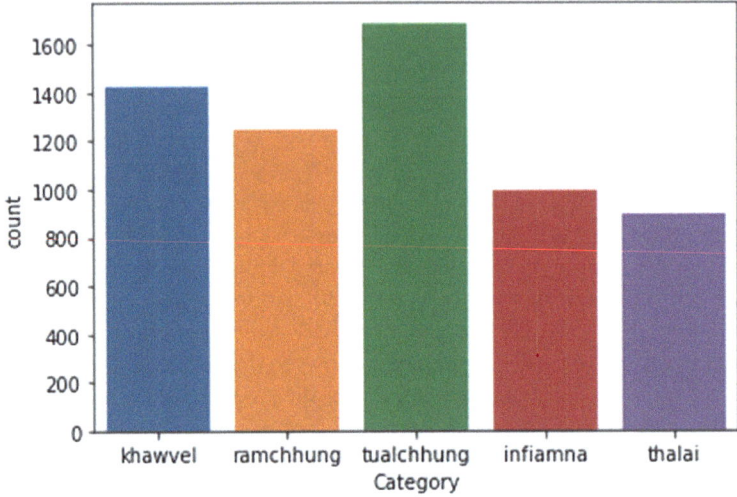

Fig. 1 Number of articles

Table 1 Distribution of 6,236 News articles

News category	Number of articles
Tualchhung	1686
Khawvel	1424
Ramchhung	1246
Infiamna	990
Thalai	890

– **Infiamna** which means sports in English, and this covers all the sports news which happens locally, in India, and in the world.
– **Thalai** which means "Teens" in English, and this covers the entertainment industry news like the channel E!

Recently, many Mizo entertainment-related news like the release of new music videos, local concerts, and events were also added.

In Fig. 1, we see the graph of each category, and we see that Infiamna (Sports) and Thalai (Entertainment) news has less articles compared to the others.

The distribution of the news articles with the actual number of articles is displayed in Table 1.

4 Methodology

4.1 Dataset Collection

The dataset is taken from vanglaini.org and theaizawlpost.org where the required attributes like category are taken where all the news articles are tagged with their respective category. The full methodology is shown in Fig. 2. The data are cleaned up by removing unnecessary punctuations and symbols, and other edits are done manually with batch commands using Regex. After crawling and performing the Regex commands, the final corpus is saved in a .csv file for further processing.

4.2 Stop Words and Stemming

Natural Language Toolkit or NTKL [36] is a software library based on Python that helps programmers to work on a language corpus. The workflow is very intuitive and easy to use, and it has over 50 corpora and lexical resources. It also has many text processing libraries which are the go-to text processing utilities like stemming, wrappers, and many more. Stop Words library from the English vocabulary and stemming was used from the NTKL library.

- **Stop Words**: Stop Words are regarded as the most common words that appear in a corpus (such as "the", "a", "an", and "in") which do not hold any significance for our classification methods and are to be removed. These words are also omitted by search engines when entering a search query since it cost computational resources.

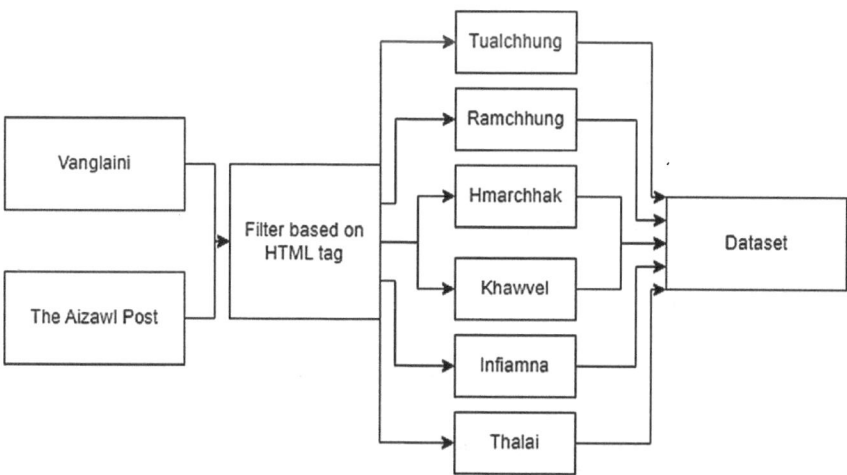

Fig. 2 Dataset collection methodology

– **Stemming**: Stemming is needed to find the lemma. We need the base form of a word in text processing where we can group certain words like Singer and Singing to Sing.

4.3 CountVectorizer

The CountVectorizer [37] is a method to tokenize any text corpus and encode it. When processing text like news articles in a computer for Machine learning, we are required to convert the text to a vector form for tagging or tokenizing the text so as to process them. This will give words a unique identifier or number which is used by the machine to process it. In this way, CountVectorizer transforms the text into certain unique tokens or numbers and gives the ability to the machine to manipulate the text or corpus. The necessity of using CountVectorizer is given below

– **Text Pre-processing**: Before we can do anything with Machine Learning algorithms, all the text data which we are going to process is needed to be represented in numerical form. CountVectorizer transforms the text into a matrix where each row represents an article and each column represents a unique word in the dataset. The conversion of the text into data is necessary and is the foundation of analysis.
– **Feature Extraction**: CountVectorizer is used in extracting features by creating a vocabulary of unique words from the dataset and counting the frequency of occurrence in every article. These word counts are used in the classification techniques.
– **Bag-of-Words**: CountVectorizer follows a bag-of-words model which basically means that the collection of words is treated as an unordered collection of texts and disregards the grammar and word order of the articles. Despite its rather unique method, it has proven effective in many Natural Language Processing tasks.
– **Dimensionality Reduction**: CountVectorizer also reduces the dimensions of the corpus by removing uninformative or rare occurrence words which helps in further improving the efficiency and performance of the machine learning algorithms.

5   Machine Learning Models

5.1   Multinomial Naive Bayes (MNB)

Multinomial Naive Bayes (MNB) is an algorithm that uses a probabilistic [38] learning method and is used widely in Natural Language Processing (NLP). Multinomial Naive Bayes predicts the text which is tagged using CountVectorizer. Just like the Native Bayes probability theorem, the probability of the correct output is calculated based on the previous knowledge it has of the event.

5.2   Support Vector Machines (SVM)

Support Vector Machines (SVM) are supervised learning algorithms used in various classification problems which were developed at AT&T Bell Laboratories [39]. SVM works by finding a hyperplane in N-dimensional space where it classifies the point on N number of features. SVM chooses vectors or points for the creation of the hyperplane.

5.3   Decision Tree

Decision Tree Classifier uses a set of rules to make decisions to give the correct or desired output [40]. Unlike Multinomial Naive Bayes, which uses a probabilistic approach. Decision Tree is rule-based where the data is continuously split according to a given set of parameters.

5.4   K-Nearest Neighbors (KNN)

K-Nearest Neighbors Classifier (KNN) initially devised by Evelyn Fix and Joseph Hodges in 1951 [41] is a non-parametric classification technique that uses the concept of proximity to predict the grouping or categorization of datasets. The classification is based on a consensus of a majority vote from its neighboring objects where the object is attached or marked to the class that is most frequently represented in its neighbor. If $k = 1$, data is immediately assigned to the class where the distance between its neighboring objects is the lowest.

5.5   Logistic Regression (LR)

Logistic Regression (LR) is a statistical model that mimics the probability of a categorical dependent variable [42]. In logistic regression, the dependent variable means it has two possible outcomes 0 and 1. It calculates the odd ratio which basically measures the likelihood of an event occurring to not occurring. This ratio helps in providing valuable insights into the relationship between the independent and dependent variables and gives us information on the direction and strengths.

Table 2 Long Short-Term Memory (LSTM) summary

Layer (type)	Output shape	Parameters
Embedding	(None, 250, 100)	5,000,000
spatial_dropout1d (SpatialDropout1D)	(None, 250, 100)	0
LSTM	(None, 100)	80,400
Dense	(None, 5)	505
Total parameters: 5,080,905		
Trainable parameters: 5,080,905		
Non-trainable parameters: 0		

5.6 Random Forest (RF)

Random Forest (RF) consists of many decision trees which are arranged in such a way that they try to produce a meaningful output. Random Forest is like an orchestrated cheerleading act where every move of the individual or tree is reflected in all other trees to produce a meaningful act or correct output. Random decision forests were created first in 1995 when the subspace method was used [43]. Ho's stated that it was a way to implement a stochastic discrimination-like approach [44] and eventually was proposed by Eugene Kleinberg.

5.7 Long Short-Term Memory (LSTM)

Long Short-Term Memory (LSTM) is basically a modified recurrent neural network. Like RNN, it remembers the past experiences but unlike RNN the problem of vanishing gradient is not present in LSTM. In LTSM, the weights and biases change every episode of training. They have a feedback connection much like the neurons in the brain where short-term memory lasts for many timesteps. The Model Summary used in this paper for the LTSM model is given in Table 2.

6 System Architecture

The dataset is divided into two sets. 85% is used for training the models and the other 15% is used for testing. The pre-processed text was fed through the different machine learning models and was tested with the testing dataset for further evaluation. In Fig. 3, we have the full system architecture which shows the basic route followed by the dataset and the model and how the output is generated. Each model was fitted with the pre-processed dataset where the default parameters were used for all

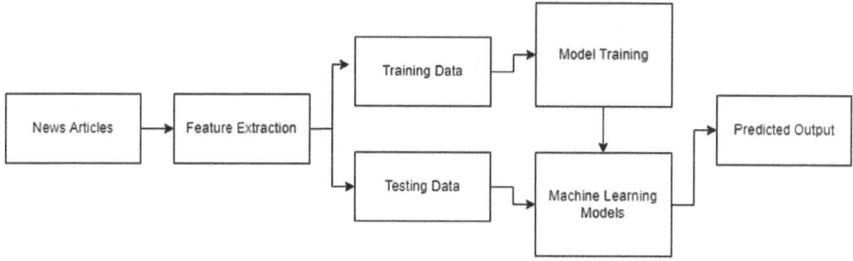

Fig. 3 System architecture

models from the Scikit-learn [37] library except for the LTSM model where the model summary is given in Table 2. After the training was done, the models were tested with the testing dataset where proper studies and evaluation were done.

7 Results and Analysis

Despite the limited amount of corpus in our dataset, a very high accuracy score is achieved as shown in Table 3 which can compete with the other works done which have a relatively way higher dataset as compared to this corpus. Multinomial Naive Bayes got an accuracy of 96% which is the highest followed by Support Vector Machines which got 94% accuracy. This also confirms that Naive Bayes usually gives higher prediction scores when CountVectorizer, or TF-IDF is used as feature selection [29]. Similar results were also produced with the Indonesian language news article classification [23] which has an accuracy score of 98% in Multinomial Naive Bayes classification. The paper shows an accuracy of 95% in the Logistic Regression model, whereas another research [30] got 91.1% when classifying authorship of Tweets and 89.8% using Multinomial Naive Bayes where our paper got the higher accuracy compared to both. The results of this paper using Support Vector Machines having

Table 3 Model accuracy

Models	Accuracy (%)
Multinomial Naive Bayes	96
Logistic regression	95
Random forest	94
Support vector machines	94
Long short-term memory	91
Decision tree	86
K-nearest neighbors	77

an accuracy of 94% is also similar when compared with Nepali news classification and other news classification using Support Vector Machines [18, 28]. The Random Forest classification also has an accuracy of 94% which turns out to be the same as the Support Vector Machines which is also higher when compared with Liparas et al. [32] where they got 86.2% using Random Forest. This paper also shows an accuracy of 86% in Decision Tree, and the recent text classification on Alzheimer's drugs' mechanism of action [31] got an accuracy of 95% using Random Forest which is higher than the result shown in this paper. The lowest accuracy is K-Nearest Neighbors which got 77% on the accuracy score which when compared with Pujianto et al. shows a 10% decrease when compared with the result in this paper where they got 87.58% accuracy [33].

This paper also implemented the classification of news articles based on a technique different from the others known as recurrent neural networks. The Long Short-Term Memory based on recurrent neural networks using the model mentioned in Table 2 is used, and Figs. 4 and 5 show the accuracy evaluation and the error evaluation respectively where the model ran 10 epochs with a batch size of 64. The overall accuracy of the LSTM model is 91%. When compared to Mittal et al. [45] where they used Deep Graph Long Short-Term Memory on text classification, the paper got an accuracy of 99% which is higher than our result. Another LSTM-based classification was done by Ishaq et al. [46] where they got an accuracy of 97% which also is higher than the results we got in this paper. So, using neural network learning techniques, the method used by us still needs proper tuning and training to get a satisfactory result when compared with the other papers but nonetheless, 91% accuracy is still very high when a scarce number of articles is used in our case.

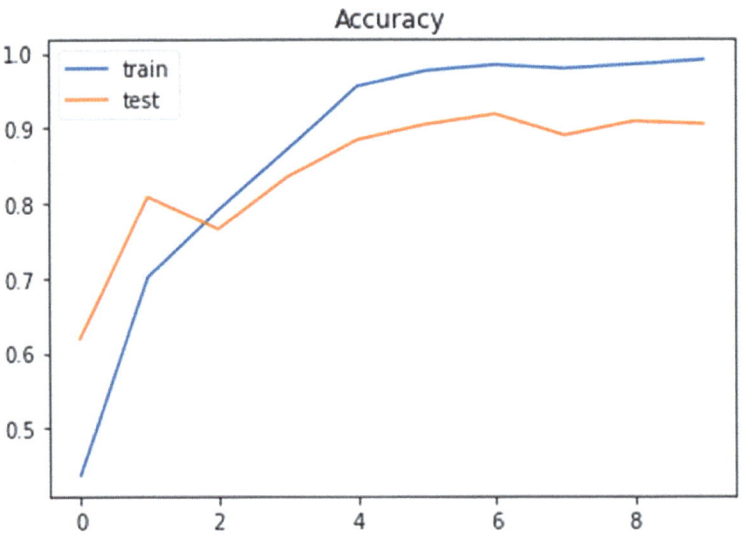

Fig. 4 Long Short-Term Memory (LSTM) accuracy evaluation

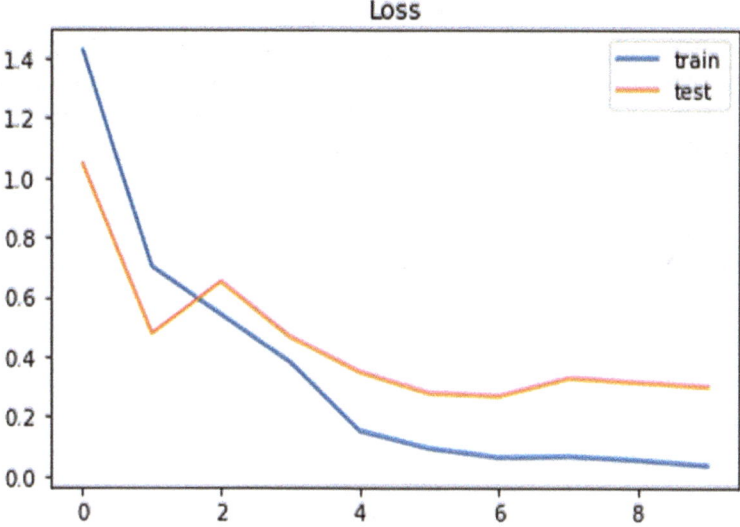

Fig. 5 Long Short-Term Memory (LSTM) error evaluation

8   Conclusion and Future Works

By using different classification techniques, an approach for news classification techniques using a low-resource Mizo language is studied and implemented. Among different techniques of feature extraction, CountVectorizer feature extraction was chosen based on its popularity and high success rate. On the different models implemented, Multinomial Naive Bayes scored the highest with 96% accuracy which is followed by Logistic Regression with 95% and is followed by Random Forest and Support Vector Machines which both get 94% accuracy. This shows a very promising future in the beach of Natural Language Processing since with very limited resources, the accuracy obtained is very high. Long Short-Term Memory which is based on Recurrent Neural Networks on the other hand got 91% accuracy which is lower than the previously mentioned, and this might be due to the scarcity of the dataset. Decision Tree and K-Nearest Neighbors got an accuracy of 86 and 77% respectively which is considerably lower than the other machine learning algorithms.

Improvements can be done in further works where different types of feature extraction can be used which may increase the overall accuracy of the models. Moreover, the models implemented can also be further tuned especially for the Long Short-Term Memory model implemented along with other Neural Network models. There are neural network models which can be further investigated along with the Long Short-Term Memory since the performance of this model is not satisfactory in this paper. With the increase in data in Mizo language, other ventures like Generative Pre-trained Transformer (GPT) can also be implemented and studied which can be useful in many applications since the Mizo language is not implemented much in any of our existing technologies. This paper hopes for future work in the Mizo language to help and improve the advancement it can bring in the field of Artificial Intelligence and Accessibility in various applications which can help people with disabilities interact with various devices in the Mizo language with the assistance of Artificial Intelligence.

References

1. Burling R (2003) The Tibeto-Burman languages of Northeastern India. Sino-Tibet Lang 3:169–191
2. Owen-Smith T, Hill N (eds) (2013) Trans-Himalayan linguistics: historical and descriptive linguistics of the Himalayan area, vol 266. Walter de Gruyter
3. Pachuau JL, Van Schendel W (2015) The camera as witness. Cambridge University Press
4. Samuel AL (1967) Some studies in machine learning using the game of checkers. II—Recent progress. IBM J Res Dev 11(6):601–617
5. Lindsay RP (1964) The impact of automation on public administration. West Polit Q 17(3):78–81
6. Crawford M, Khoshgoftaar TM, Prusa JD, Richter AN, Al Najada H (2015) Survey of review spam detection using machine learning techniques. J Big Data 2(1):1–24
7. Lin WY, Hu YH, Tsai CF (2011) Machine learning in financial crisis prediction: a survey. IEEE Trans Syst Man Cybern Part C (Applications and Reviews) 42(4):421–436
8. Ucci D, Aniello L, Baldoni R (2019) Survey of machine learning techniques for malware analysis. Comput Secur 81:123–147
9. Russell SJ, Norvig P (2010) Artificial intelligence a modern approach, 3rd edn.
10. Mohri M, Rostamizadeh A, Talwalkar A (2018) Foundations of machine learning. MIT press
11. Tucker AB (ed) (2004) Computer science handbook. CRC press
12. Kowsari K, Jafari Meimandi K, Heidarysafa M, Mendu S, Barnes L, Brown D (2019) Text classification algorithms: a survey. Information 10(4):150
13. Debole F, Sebastiani F (2005) An analysis of the relative hardness of Reuters–21578 subsets. J Am Soc Inf Sci Technol 56(6):584–596
14. Joachims T (2005) Text categorization with support vector machines: learning with many relevant features. In: Machine learning: ECML-98: 10th European conference on machine learning chemnitz. Germany, 1998 Proceedings. Springer, Berlin, pp 137–142
15. Dumais S, Platt J, Heckerman D, Sahami M (1998) Inductive learning algorithms and representations for text categorization. In: Proceedings of the seventh international conference on Information and knowledge management, pp 148–155
16. Schapire RE, Singer Y (2000) BoosTexter: a boosting-based system for text categorization. Mach Learn 39:135–168
17. Pathak A, Pakray P, Bentham J (2019) English-Mizo machine translation using neural and statistical approaches. Neural Comput Appl 31(11):7615–7631
18. Shahi TB, Pant AK (2018) Nepali news classification using Naive Bayes, support vector machines and neural networks. In: 2018 international conference on communication information and computing technology (ICCICT). IEEE, pp 1–5
19. Chy AN, Seddiqui MH, Das S (2014) Bangla news classification using naive Bayes classifier. In: 16th international conference computer and information technology. IEEE, pp 366–371
20. Chowdhury P, Eumi EM, Sarkar O, Ahamed MF (2022) Bangla news classification using GloVe vectorization, LSTM, and CNN. In: Proceedings of the international conference on big data, IoT, and machine learning: BIM 2021. Springer, Singapore, pp 723–731
21. Dogru HB, Tilki S, Jamil A, Hameed AA (2021) Deep learning-based classification of news texts using doc2vec model. In: 2021 1st international conference on artificial intelligence and data analytics (CAIDA). IEEE, pp 91–96

22. Mangal SB, Goyal V (2014) Text news classification system using Naïve Bayes classifier. Int JEng Sci 3
23. Wongso R, Luwinda FA, Trisnajaya BC, Rusli O (2017) News article text classification in Indonesian language. Procedia Comput Sci 116:137–143
24. Su J, Shirab JS, Matwin S (2011) Large scale text classification using semi-supervised multinomial naive bayes. In: Proceedings of the 28th international conference on machine learning (ICML-11), pp 97–104
25. Kim SB, Han KS, Rim HC, Myaeng SH (2006) Some effective techniques for naive bayes text classification. IEEE Trans Knowl Data Eng 18(11):1457–1466
26. Naresh E, Vijaya KB, Pruthvi VS, Anusha K, Akshatha V (2019) Survey on Classification and Summarization of Documents. Int J Res Advent Technol 7(6S)
27. Karunia SA (2017) Online news classification using naive bayes classifier with mutual information for feature selection. ITSMART: Jurnal Teknologi dan Informasi 6(1):11–15
28. Dadgar SMH, Araghi MS, Farahani MM (2016) A novel text mining approach based on TF-IDF and Support Vector Machine for news classification. In: 2016 IEEE international conference on engineering and technology (ICETECH). IEEE, pp 112–116
29. Fanny F, Muliono Y, Tanzil F (2018) A comparison of text classification methods k-NN, Naïve Bayes, and support vector machine for news classification. Jurnal Informatika: Jurnal Pengembangan IT 3(2):157–160
30. Aborisade O, Anwar M (2018) Classification for authorship of tweets by comparing logistic regression and naive bayes classifiers. In: 2018 IEEE international conference on information reuse and integration (IRI). IEEE, pp 269–276
31. Kambar MEZN, Nahed P, Cacho JRF, Lee G, Cummings J, Taghva K (2022) Clinical text classification of alzheimer's drugs' mechanism of action. In: Proceedings of sixth international congress on information and communication technology: ICICT 2021, vol 1. Springer, Singapore, pp 513–521
32. Liparas D, HaCohen-Kerner Y, Moumtzidou A, Vrochidis S, Kompatsiaris I (2014) News articles classification using random forests and weighted multimodal features. In: Multidisciplinary information retrieval: 7th information retrieval facility conference, IRFC 2014. Copenhagen, Denmark, Proceedings 7. Springer International Publishing, pp 63–75
33. Pujianto U, ar Rosyid H, Anam MK (2021) Classification of news articles for learning using the K-nearest neighbor algorithm. In: 2021 7th international conference on education and technology (ICET). IEEE, pp 256–260
34. Guo L, Zhang D, Wang L, Wang H, Cui B (2018) CRAN: a hybrid CNN-RNN attention-based model for text classification. In: Conceptual modeling: 37th international conference, ER 2018, Xi'an, China, 2018, Proceedings 37. Springer International Publishing, pp 571–585
35. Du J, Vong CM, Chen CP (2020) Novel efficient RNN and LSTM-like architectures: recurrent and gated broad learning systems and their applications for text classification. IEEE Trans Cybern 51(3):1586–1597
36. Bird S, Klein E, Loper E (2009) Natural language processing with Python: analyzing text with the natural language toolkit. O'Reilly Media, Inc.
37. Pedregosa F, Varoquaux G, Gramfort A, Michel V, Thirion B, Grisel O, Blondel M, Prettenhofer P, Weiss R, Dubourg V, Vanderplas J (2011) Scikit-learn: machine learning in Python. J Mach Learn Res 12:2825–2830
38. Hastie T, Tibshirani R, Friedman JH, Friedman JH (2009) The elements of statistical learning: data mining, inference, and prediction, vol 2. Springer, New York, pp 1–758
39. Cortes C, Vapnik V (1995) Support-vector networks. Mach Learn 20:273–297
40. Kamiński B, Jakubczyk M, Szufel P (2018) A framework for sensitivity analysis of decision trees. CentL Eur J Oper Res 26:135–159
41. Fix E, Hodges Jr JL (1952) Discriminatory analysis-nonparametric discrimination: small sample performance. California University Berkeley
42. Tolles J, Meurer WJ (2016) Logistic regression: relating patient characteristics to outcomes. Jama 316(5):533–534

43. Ho TK (1998) The random subspace method for constructing decision forests. IEEE Trans Pattern Anal Mach Intell 20(8):832–844
44. Kleinberg EM (1990) Stochastic discrimination. Ann Math Artif Intell 1:207–239
45. Mittal V, Gangodkar D, Pant B (2021) Deep graph-long short-term memory: a deep learning based approach for text classification. Wirel Pers Commun 119:2287–2301
46. Ishaq A, Umer M, Mushtaq MF, Medaglia C, Siddiqui HUR, Mehmood A, Choi GS (2021) Extensive hotel reviews classification using long short term memory. J Ambient Intell HumIzed Comput 12:9375–9385

BASiP: A Novel Architecture for Abstractive Text Summarization

Debajyoti Das, Jatin Madaan, Rajarshi Chanda, Rishav Gossain, Tapas Saha, and Sangeeta Bhattacharya◉

Abstract The availability of information and news over the Internet is exploding. In this context, text summarization is becoming very important since it gives a good overview of the content. Also, it saves time by exposing the most significant information at a glance. Summarization techniques are very vital in extracting this useful information from lengthy text. In this work, a novel architecture for abstractive text summarization architecture, BASiP, has been proposed, which effectively generates a summary from the given text. The base model used for summarization is BART. The proposed architecture is compared with the existing work. It is found that BASiP performs well in terms of the ROUGE score. Also, a case study is given at the end to show the efficiency of BASiP, in generating a meaningful summary.

Keywords Abstractive text summarization · Summarization model · Paraphraser

1 Introduction

The main objective behind the idea of summarization [1] of a text is to reduce the total size of the given text by removing the unnecessary parts and keeping the necessary important information and the overall meaning same. These days, sharper and more compact text summarization is highly needed and appreciated as it reduces the time required to read the lengthy articles. Text Summarization is basically the concept of analyzing the data and extracting the important and necessary information from it and to form a concise paragraph while preserving the initial meaning of it. It also plays an important role in today's age of huge data available online. There are several techniques to determine and replace untrustworthy data.

Text Summarization is mainly of two types: Extractive summarization [2] creates the summary by concatenating the key passages from the text. The text's meaning, in this approach, is ignored and only the subset of key sentences is highlighted. Abstractive summarization [3] analyzes and provides a sound and concise under-

D. Das · J. Madaan · R. Chanda · R. Gossain · T. Saha · S. Bhattacharya (✉)
Guru Nanak Institute of Technology, Sodepur, West Bengal, India
e-mail: sngtbhttchry@gmail.com

© The Author(s), under exclusive license to Springer Nature Singapore Pte Ltd. 2024
S. Namasudra et al. (eds.), *Data Science and Network Engineering*, Lecture Notes in Networks and Systems 791, https://doi.org/10.1007/978-981-99-6755-1_14

standing of the text. Using the techniques of NLP [4], this approach tries to figure out the text's meaning and identify the key sentences in it. In contrast to the extractive approach, it tries to generate new words to form the summary. Abstractive summarization is preferred and used over extractive summarization because it generates a concise and effective summary that a human can construct from it's original text thus providing an improved comprehension of the summary [5]. As the model was developed for PIB India [6] to summarize their press releases, it was important in generating summaries that the readers can have an improved understanding of and relate to. Hence the approach of abstractive summarization has been followed here.

The summary generated by the existing summarization techniques has certain insufficiency such as repetition of same sentences or phrases, inclusion of a high number of named entities, and lower degree of abstraction. Hence, in this work, a novel architecture, BASiP, is proposed for text summarization. The existing models are contrasted with the proposed architecture. and the results found are promising. At the end, certain case studies are supplied on different articles of PIB as a reference. In the application, a text-to-speech converter is also added to read out the summarized news for visually impaired individuals.

The remaining paper is organized as follows: In Sect. 2, a literature review on existing summarization techniques are discussed. In Sect. 3, an overview of the proposed architecture, BASiP, is given. In Sect. 4, the various experimental results are discussed. Finally, the conclusion is given in Sect. 5.

2 Literature Review

Text summarization has been extensively used and researched upon in recent years. Abstractive summarization techniques have vastly emerged in this field due to their ability to learn complex representations of input data. In this literature review, we explore the recent advances in the field of text summarization models.

In [7], one common approach of the sequence-to-sequence (Seq2Seq) model including attention mechanism has been evaluated. The Seq2Seq model produces a summary by mapping the inputted sequence of words to the output summary using an encoder-decoder architecture. This mechanism helps the model focus on the most relevant areas of the inputted sequence while generating the summary. However same sentence or phases often get repeated in the produced summary, and the model struggles with handling rare or unknown words, which can lead to incomplete summarization.

Factual consistency of the abstractive text summarization method has also been identified in [8]. This mechanism enhances this models' factual consistency by identifying important entities in the inputted text and ensures that those entities are accurately represented in generated summary. It works by first identifying entities in the input text using an off-the-shelf Named-Entity Recognition (NER) system. It then trains a separate model to predict the importance of each entity for the overall meaning of the text. Finally, during summarization, the mechanism ensures that important entities are precisely shown in the generated summary by using the predicted importance scores to guide the selection and generation of summary content. However, there is still room for improvement, and more ways can be researched upon and explored in future researches to further enhance the accuracy of entity importance prediction and develop more advanced mechanisms for incorporating entity-level information into the summarization process. Another promising approach is the use of pre-trained language models such as BERT for the summarization of text that has been provided by [9, 10]. These models have produced excellent results on a range of natural language processing tasks and have been used for summarization nowadays.

One advantage of pre-trained models is that they can leverage large amounts of unsupervised data to learn rich representations of language. Reinforcement learning has also been explored for text summarization in [11, 12]. These models use a reward function to enhance the generated summary and are showcased to produce summaries with improved fluency and informativeness. However, reinforcement learning requires careful tuning of the reward function and can be computationally expensive. Graph-based models have also been suggested for in [13, 14]. These models represent the input document as a graph and use graph algorithms to extract important nodes and edges for the summary. Graph-based models can capture both local and global dependencies between sentences and are showcased to produce high-quality summaries.

A novel attention mechanism for abstractive text summarization that takes into account discourse information to better capture the structure and coherence of long documents has been put forward in [15]. The discourse-aware attention module uses a graph neural network to model the discourse relationships between sentences and learn discourse-aware representations for each sentence. These representations are then used in the sentence-level attention module to compute the appropriateness of each sentence to the summary. Although the research that will be conducted in future could explore ways to enhance the model's discourse modeling capabilities by incorporating more sophisticated graph neural networks or other methods for capturing more complex discourse relationships between sentences, [16] incorporates multiple sources of guidance to boost the accuracy and informativeness of abstractive text summarization. The proposed model, known as GSum, consists of many modules that cooperate to deliver summaries that are influenced by the input data. The first module analyzes the input text to find keyphrases and entities, while the second module creates an extractive summary using the text's most essential passages. The third module generates an abstractive summary using an attention-based neural network that is guided by the extracted summary and keyphrases/entities. The GSum model's main novelty is the inclusion of multiple sources of guidance, which enables the model to generate summaries which are more precise and useful than those produced by models that just depend on one source of information. However, the incorporation of other sources of guidance, such as discourse or sentiment information, can be done to further improve the quality and coherence of the generated summaries.

Because of the huge amount of data, the data can also be stored in the cloud, and data privacy should also be maintained [17–19].

Table 1 presents a gist of the literature review.

3 Methodology

In this work, a novel abstractive text summarization architecture, 'BASiP', is proposed, which is a combination of **BA**RT, **Si**mCLS Framework, and **P**araphrasing model. Figure 1 shows the proposed architecture, BASiP. The summarization model 'f' which uses an evaluation metric M, that depends on the source dataset D, aims to provide a summary 'S' of the candidate S = f(D) [20]. This receives the greatest ROUGE score m, where m is defined as

$$M(S, S\hat{})$$
(1)

In our suggested approach, the entire generation process is divided into a number of stages, each of which consists of a base model BART, for producing candidate summaries and a robust framework SimCLS for optimizing the summary.

- **Step 1: Preprocessing**: Incorrect interpretation of the overall statistics of the data may result from duplicate or missing values. The overall learning of the model is frequently disrupted by outliers and inconsistent data points, which leads to inaccurate predictions. That is why the raw dataset D is being preprocessed which further includes the steps of tokenization and fine-tuning. In machine learning models, tokenization and fine-tuning are used for text data representation, improvement of model performance, to reduce training time, and to improve accuracy. NLP-based Text2Text Generation tokenizer BartTokenizer from "Yale-LILY/brio-cnndm-uncased" is used for tokenization purposes.
  Tokenizing a text consists of the following steps:

  – Sentence segmentation: Divide the text into sentences.
  – Word tokenization: Divide each and every sentence into individual words or tokens.
  – Removing stop words: Stop words are defined as frequently used words such as 'and', 'the', and 'of' which adds semantic meaning to the text. The stop words are cleaned from the text to reduce it's size and speed up the training process.
  – Removing punctuation and special characters: Remove punctuation marks and special characters from the text to reduce the size of the data and eliminate any potential distractions for the algorithm.

Once the text data has been tokenized, it can be further fine-tuned by preparing the data by tokenizing the text and encoding it into a numerical format that the model can process. It is important to note that the fine-tuning process can be time-consuming and computationally expensive, as the model must be trained on

Table 1 Summary of the existing work

SL	Paper name	Model used	Dataset used	Drawbacks
Nallapati et al. [7]	Abstractive text summarization using sequence-to-sequence RNNs and beyond	Recurrent neural networks (RNNs)	GigawordCorpus, DUC Corpus, CNN/Daily Mail Corpus	Same sentence or phrases often get repeated in the summary and the model struggles with handling rare or unknown words, which can lead to incomplete or inaccurate summaries
Nan et al. [8]	Entity-level factual consistency of abstractive text summarization	BART	Newsroom, CNNDM, XSUM	The evaluation metrics used in the paper may not be diverse enough to capture all aspects of factual consistency
Liu et al. [9]	BRIO: Bringing Order to Abstractive Summarization	Pre-trained BERT, Encoder-decoder model	CNNDM, XSum, NYT	It is observed that fine-tuned BRIO summaries tend to be highly extractive and generally include a high number of named entities (dates, percentages, names), reflecting the data they were trained on
Aghajanyan et al. [10]	Muppet: Massive multi-task representations with pre-finetuning	BERT, RoBERTa	CoLA, SST-2, QQP, MNLI, QNLI, RTE, WNLI, Super-GLUE, BoolQ,MultiRC, WIC, WSC, COPA, AGNews, IMDB, RottenTomatoes, Wiki QA, CNNDM, Billsum, XSUM	Muppet uses a large amount of pre-training data, which can lead to overfitting. This could result in the model being less effective on new, unseen data
Savelieva et al. [11]	Abstractive summarization of spoken and written instructions with BERT	BERTSUM	CNN/DailyMail, Wiki How Text, How2 Videos,	The model tends to pick the first one or two sentences for the summary
See et al. [12]	Get to the point: summarization with pointer generator networks	Pointer-generator	CNN/Daily Mail	Model's summaries contain a much lower rate of novel n-grams (i.e., those that don't appear in the article) than the reference summaries, indicating a lower degree of abstraction
Zhang et al. [13]	Momentum calibration for T17ext generation	MOCA	CNN/DailyMail, XSum, SAMSum, and Gigaword	Experiments across different datasets show MOCA consistently improves upon vanilla fine-tuning with the MLE loss for large pre-trained transformers, but it is not a model rather an online method
Ravaut et al. [14]	SummaReranker: a multi task mixture of experts re-ranking framework for abstractive summarization	Pegasus+SummaReranker	CNN-DailyMail, XSum, Reddit TIFU	On top of base model training and candidate generation, SummaReranker inference cost is linear in the number of candidates, i.e., accuracy varies inconsistently with number of candidates being re-ranked in different models
Cohan et al. [15]	A discourse-aware attention model for abstractive summarization of long documents	Discourse-aware attention model	CNN, DailyMail, NYTimes, PubMed, arXiv	It is mainly good for abstractive summarization of long documents
Dou et al. [16]	GSum: a general framework for guided neural abstractive summarization	Transformer, BERT	Reddit, XSum, CNN/DM, WikiHow, New York Times (NYT), PubMed	GSum may not be as effective for summarizing certain types of text, such as text with a lot of technical or domain-specific language, as it relies heavily on the availability of relevant guidance information to generate accurate and informative summaries

Fig. 1 Proposed
Architecture
BASiP(**BART**+**Si**mCLS+**P**araphraser)

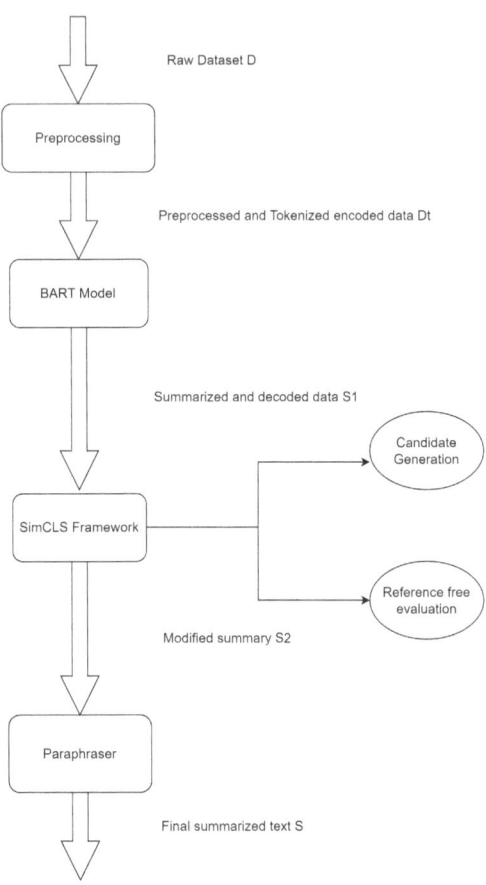

a large amount of data to achieve good performance. However, the results can be very powerful, as the pre-trained language model has already learned gained knowledge about the framework of the given text, which can be leveraged to perform various types of NLP tasks with high accuracy.

- **Step 2: BART**: BART is the summarization model developed by Facebook AI Research. BART uses a sequence-to-sequence model for summarization, which means it takes a sequence of text as input and produces a sequence of output text. BART achieves summarization of text by using the encoder-decoder architecture [21]. The encoder takes the text and converts it into a series of hidden states, which capture the input text's meaning. After the BartTokenizer encodes the input text into a numerical format, the parent model BART processes the data to give a summarized encoded output. The decoder generates the human-readable output summary from the hidden states. To summarize text, BART is typically fine-tuned on a dataset of paired input and summary texts. During fine-tuning, this model is trained to produce a summary that captures the most significant information in the given input text while keeping the summary concise. BART uses a combination of attention mechanisms and beam search to generate accurate and fluent summaries.

- **Step 3: SimCLS Framework**: The summary S1 is then fed as input into SimCLS to further optimize the summary generated in step 2. SimCLS is an abstractive summarization model. It uses a two-stage approach that comprises a generator and a scorer [22]. In the first stage, the generation model $g(\cdot)$ is trained to maximize the likelihood of reference summary S^ given dataset, D. $g(.)$ is a Seq2Seq model. After that, using an instance approach like Beam Search on the pre-trained $g(\cdot)$, many candidate summaries S–1, ..., S–n are produced, where n = the number of sampled candidates. In the second stage, the scorer assigns a score to each candidate given the source document. The main motive is to improve the produced candidate summary Si to increase the ROUGE score in comparison to the original text D. It is addressed using contrastive learning and construct an evaluation function $h(\cdot)$ that seeks to distinguish the generated candidates by giving them different ROUGE scores r1, ..., rn on the basis of the similarity found between the source text and the candidate Si. That is

$$ri = h(Si, D).$$

This value is the cosine similarity generated among the first tokens when encoded. The candidate which has the highest rating is the final summary of the output S.

- **Step 4: Paraphrasing**: To provide a more precise and sound output, as a part of the package, we have introduced a paraphrasing part that has the main objective of improving any lousy summary into a more grammatically accurate one. The paraphraser subdivides and collects the article into each sentence. It then reframes each sentence individually and before giving the final output, it joins all of them together. Sometimes, it does this to make the summary more fluent and to strengthen the sentence structure by substituting the terms with their synonyms. This beautifies the generated summary S making it more appropriate and industry-ready for usage.

4 Experimental Results

The data sets which are used for testing the efficiency of the suggested model BASi(**BART+Si**mCLS) are CNNDM and XSUM. CNNDM stands for Cable News Network and Daily Mail dataset. This is an English language dataset that contains over 300k news articles which are written by the journalists of the CNN and the Daily Mail. This dataset supports data for extractive as well as abstractive summarization [23, 24]. XSUM stands for Extreme Summarization dataset which is used for evaluating abstractive single document summarization systems. This mainly aims to generate short single sentence precise summary for an article [25]. ROUGE, also known as the Recall Oriented Understudy for Gisting Evaluation, is a set of measurement metrics which is used for calculating and evaluating the summarization generated

Table 2 Comparison of BASi with existing work

Model	Rouge 1	Rouge 2	Rouge L
XSUM			
BART	45.14	22.27	37.25
Pegasus	47.21	24.56	39.25
SimCLS	47.61	24.57	39.44
BRIO-Ctr	48.13	25.13	39.84
BRIO-Mul	49.07	25.59	40.40
BASi	50.06	30.5	42.12
CNNDM			
BART	44.16	21.28	40.90
Pegasus	44.17	21.47	41.11
SimCLS	46.67	22.15	43.54
BRIO-Ctr	47.28	22.93	44.15
BRIO-Mul	47.78	22.93	44.15
BASi	49.52	24.26	45.85

automatically by natural language processing. It mainly has 3 main parameters of measurement, Recall, Precision, and F1 Score which provides an analysis of the automatic summarized data when compared to the original summary [26, 27]:

–
$$\text{Recall} = \frac{\text{No of Word matches}}{\text{No of Words in References}} \qquad (2)$$

–
$$\text{Precision} = \frac{\text{No of Word matches}}{\text{No of Words in Summary}} \qquad (3)$$

–
$$\text{Recall} = 2\left(\frac{\text{Precision * Recall}}{\text{Precision + Recall}}\right). \qquad (4)$$

Table 2 depicts the ROUGE value calculated on the factors of R1, R2, and RL.

The results from the following table provide a comparative study about the different existing models and the proposed model, BASi. In this work, BART has been implemented as the initial model for the CNNDM dataset, while for the XSUM dataset, the initial model is PEGASUS. The XSUM dataset generates a single sentence dataset, and thus to achieve finer results, the backbone of the model here is the pre-trained PEGASUS model [28, 29].

It can be seen from Fig. 2, BASi outperforms the other models in a two-stage summarization framework for the datasets XSUM and CNNDM. Models like GSUM require additional assistance for input along with a separate encoder to encode the information, while the proposed model, BASi, uses the techniques used in BART.

Fig. 2 Comparison of
ROUGE values of BART,
BASi, and 'BASiP'

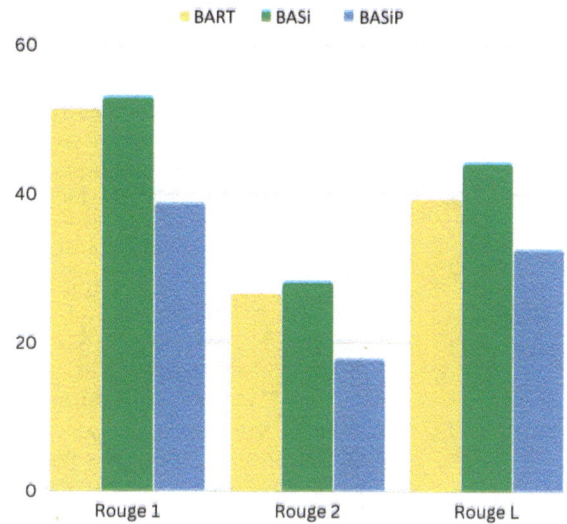

Thus, BASi outperforms different models and provides an effective summarization technique.

Table 3 shows some example references from PIB and the summary generated by our proposed model, BASiP. The findings of BASiP and BART show how our strategy aids the abstractive model in removing superfluous characters and symbols from the original input. BASiP learns to resolve unnecessary extra characters which BART cannot. As can be seen from Table 3, the first summary generated by BART contains '/' at the start and end of several words, the second summary is made up of the meaningless term 'Dsy', and the third summary of the preceding table lacks a numerical value following 'Rs'. Instead, our proposed model 'BASiP' learned to disregard or change these error patterns and never generated them over the whole test set. This was probably because it noticed that candidates with this pattern were rarely generated with high ROUGE scores and appropriately de-weighted the probability.

5   Text-to-Speech Conversion

The summarized text by the proposed architecture 'BASiP' was converted to speech for visually impaired users who would use the proposed application. The content of the website can be read aloud by a synthetic voice when the text is selected, enabling persons with visual impairments to access information without the use of traditional visual cues. This significantly enhances the overall user experience and broadens the site's accessibility.

Table 3 Case study on PIB dataset

System	Summary
Reference	The Telecom Regulatory Authority of India (TRAI) today released its recommendations on "Rating of Buildings or Areas for Digital Connectivity". The exponential growth in digitalization....[30]
BART	TRAI has released its recommendations on "Rating of Buildings or Areas for Digital Connectivity". TRAI wants to provide a framework for creation of an ecosystem for digital connectivity in buildings. TRAI has also proposed a new chapter on \ Digital Connectivity Infrastructures in Buildings \'
BASiP	The "Rating of Buildings or Areas for Digital Connectivity" was released by TRAI. A new chapter on "Digital Connectivity Infrastructures in Buildings" has been proposed
Reference	A 'Silent March' was organized on the eve...[31]
BART	"Partition Horrors Remembrance Dsy" marked on eve of Independence Day. A two-minute silence was observed in memory of the victims of the partition of the country. Indira Gandhi National Center for the Arts organized an exhibition on the Partition of India 1947
BASiP	There was a two-minute silence in memory of the victims. The partition of India was the subject of an exhibition. On the eve of Independence Day, "Partition Horrors Remembrance Day" was marked. The National Center for the Arts is named after Gandhi
Reference	The Union Cabinet, chaired by the Prime Minister, Shri Narendra Modi, has approved to restore Interest Subvention on short-term agriculture loans to 1.5% for all financial...[32]
BART	Interest Subvention on short-term agriculture loans has been restored by the Union Cabinet chaired by the Prime Minister. An additional budgetary provision of Rs 34,856 crore is required for the period of 2022-23 to 25 under the scheme. The interest rate on the short-term agri-loans is 4% per annum. Credit flow to the farmer will be ensured by increase in interest subvention. Financial health of lending institutions is expected to be ensured by this. It will lead to generation of employment since short-term Agri-Loans are provided for all activities. There is a short-term Agriculture loan of up to Rs. There are farmers who are engaged in Agriculture and other allied activities. Over 3.13 cr farmers have been issued new Kisan Credit Card against the target of 2.5 cr
BASiP	Interest Subvention on short-term agriculture loans has be restored by the Union Cabinet chaired by the Prime Minister. The increase in interest subvention support requires additional budgetary provisions of Rs 34,856 crore. The interest rate on the short-term agri-loans is 4% per annum. This will ensure adequate credit flow in agriculture sector to the farmer as well as ensure financial health and viability of lending institutions. As per the budget outlay and coverage of beneficiaries, the support is the second largest scheme. Over 3.13 cr farmers have been issued new Kisan Credit Card against the target of 2.5 cr

In the literature, sophisticated methodologies have been used for visual speech recognition [33]. In this work, an inbuilt library of html-5 called responsive voice has been used [34].

6 Conclusion

In this work, a novel architecture, BASiP, has been proposed for abstractive text summarization. The experimental results of BASi (BART + SimCLS) on comparison with the existing work shows the effectiveness of the proposed architecture. BASi outperforms the other models in a two-stage summarization framework for the datasets XSUM and CNNDM. Moreover, a few case studies has also been provided which shows that BASiP constructively generates summary from the PIB references. The future work involves generation of summary for text given in tabular format and representation of the same in a concise and meaningful way.

References

1. Allahyari M, Pouriyeh S, Assefi M, Safaei S, Trippe ED, Gutierrez JB, Kochut K (2017) Text summarization techniques: a brief survey. Int J Adv Comput Sci Appl 8(10)
2. Moratanch N, Chitrakala S (2017) A survey on extractive text summarization. In: International conference on computer, communication and signal processing (ICCCSP). IEEE, pp 1–6. https://doi.org/10.1109/ICCCSP.2017.7944061
3. Pai A (2014) Text summarizer using abstractive and extractive method. Int J Eng Res Technol 3(5):0181–2278
4. Merchant K, Pande Y (2018) NLP based latent semantic analysis for legal text summarization. In: International conference on advances in computing, communications and informatics (ICACCI). IEEE, pp 1803–1807. https://doi.org/10.1109/ICACCI.2018.8554831
5. Ertam F, Aydin G (2022) Abstractive text summarization using deep learning with a new Turkish summarization benchmark dataset. Concurr Comput Pract Exp 34(9)
6. Press Information Bureau. https://www.pib.gov.in/
7. Nallapati R, Zhou B, dos Santos C, Gulçehre Ç, Xiang B (2016) Abstractive text summarization using sequence-to-sequence RNNs and beyond. In: Proceedings of the 20th SIGNLL conference on computational natural language learning. Berlin, Germany. Association for Computational Linguistics, pp 280–290. https://doi.org/10.18653/v1/K16-1028
8. Nan F, Nallapati R, Wang Z, Santos CND, Zhu H, Zhang D, McKeown K, Xiang B (2021) Entity-level factual consistency of abstractive text summarization. In: Proceedings of the 16th conference of the european chapter of the association for computational linguistics: main volume. Association for Computational Linguistics, pp 2727–2733. https://doi.org/10.18653/v1/2021.eacl-main.235
9. Liu Y, Liu P, Radev D, Neubig G (2022) BRIO: Bringing order to abstractive summarization. In: Proceedings of the 60th annual meeting of the association for computational linguistics, vol. 1: Long Papers, Association for Computational Linguistics, pp 2890–2903. https://doi.org/10.18653/v1/2022.acl-long.207
10. Aghajanyan A, Gupta A, Shrivastava A, Chen X, Zettlemoyer L, Gupta S, Muppet: massive multi-task representations with pre-finetuning. In: Proceedings of the 2021 conference on

empirical methods in natural language processing. Association for Computational Linguistics, pp 5799–5811. https://doi.org/10.18653/v1/2021.emnlp-main.468

11. Savelieva A, Yeung BA, Ramani V, Abstractive summarization of spoken and written instructions with BERT. In: Proceedings of the KDD 2020 workshop on conversational systems towards mainstream adoption co-located with the 26TH ACM SIGKDD conference on knowledge discovery and data mining, CEUR-WS.org. https://doi.org/10.48550/2008.09676

12. See A, Liu PJ, Manning CD (2017) Get to the point: summarization with pointer-generator networks. In: Proceedings of the 55th annual meeting of the association for computational linguistics (Volume 1: Long Papers). Association for Computational Linguistics, Vancouver, Canada, pp 1073–1083. https://doi.org/10.18653/v1/P17-1099

13. Zhang X, Liu Y, Wang X, He P, Yu Y, Chen S-Q, Xiong W, Wei F, Momentum calibration for text generation. https://doi.org/10.48550/2212.04257

14. Ravaut M, Joty SR, Chen NF (2022) SummaReranker: a multi-task mixture-of-experts re-ranking framework for abstractive summarization. In: Annual meeting of the association for computational linguistics. https://doi.org/10.48550/2203.06569

15. Cohan A, Dernoncourt F, Kim DS, Bui T, Kim S, Chang W, Goharian N, A discourse-aware attention model for abstractive summarization of long documents. In: Proceedings of the 2018 conference of the North American chapter of the association for computational linguistics: human language technologies, vol 2 (Short Papers). Association for Computational Linguistics, New Orleans, Louisiana, pp 615–621. https://doi.org/10.18653/v1/N18-2097

16. Dou Z-Y, Liu P, Hayashi H, Jiang Z, Neubig G (2021) GSum: a general framework for guided neural abstractive summarization. In: Proceedings of the 2021 conference of the north american chapter of the association for computational linguistics: human language technologies. Association for Computational Linguistics, pp 4830–4842. https://doi.org/10.18653/v1/2021.naacl-main.384

17. Namasudra S, Roy P (2017) Time saving protocol for data accessing in cloud computing. IET Commun 11(10):1558–1565. https://doi.org/10.1049/iet-com.2016.0777

18. Namasudra S, Roy P (2016) Secure and efficient data access control in cloud computing environment: a survey, pp 69–90. https://doi.org/10.3233/MGS-160244

19. Namasudra S, Roy P (2017) A New Table Based Protocol for Data Accessing in Cloud Computing. J Inf Sci Eng 33:585–609. https://doi.org/10.6688/JISE.2017.33.3.1

20. Liu Y, Liu P (2021) SimCLS: a simple framework for contrastive learning of abstractive summarization. In: Proceedings of the 59th Annual meeting of the association for computational linguistics and the 11th international joint conference on natural language processing (Volume 2: Short Papers). Association for Computational Linguistics, pp 1065–1072. https://doi.org/10.18653/v1/2021.acl-short.135

21. Lewis M, Liu Y, Goyal N, Ghazvininejad M, Mohamed A, Levy O, Stoyanov V, Zettlemoyer L (2020) BART: denoising sequence-to-sequence pre-training for natural language generation, translation, and comprehension. In: Proceedings of the 58th annual meeting of the association for computational linguistics. Association for Computational Linguistics, pp 7871–7880. https://doi.org/10.18653/v1/2020.acl-main.703

22. Chen T, Kornblith S, Norouzi M, Hinton G (2020) A simple framework for contrastive learning of visual representations. In: Proceedings of the 37th international conference on machine learning. PMLR, pp 1597–1607. https://doi.org/10.5555/3524938.3525087

23. Goyal T, Xu J, Li JJ, Durrett G (2022) Training dynamics for text summarization models. In: Findings of the association for computational linguistics: ACL 2022. Association for Computational Linguistics, Dublin, Ireland, pp 2061–2073. https://doi.org/10.18653/v1/2022.findings-acl.163

24. Hugging Face. https://huggingface.co/datasets/cnn/dailymail

25. Hugging Face. https://huggingface.co/datasets/xsum

26. Lin C-Y (2004) ROUGE: a package for automatic evaluation of summaries. In: Text summarization branches out. Association for Computational Linguistics, Barcelona, Spain, pp 74–81. http://research.microsoft.com/~cyl/download/papers/WAS2004.pdf

27. Liu F, Liu Y (2010) Exploring correlation between ROUGE and human evaluation on meeting summaries. IEEE Trans Audio Speech Lang Process 18(1):187–196. https://doi.org/10.1109/TASL.2009.2025096

28. Zhang J, Zhao Y, Saleh M, Liu P (2020) Pegasus: pre-training with extracted gap-sentences for abstractive summarization. In: International conference on machine learning. PMLR, pp 11328–11339. https://doi.org/10.48550/1912.08777

29. Liu P, Yuan W, Jinlan F, Jiang Z, Hayashi H, Neubig G (2023) Pre-train, prompt, and predict: a systematic survey of prompting methods in natural language processing. ACM Comput Surv 55(9):1–35. https://doi.org/10.1145/3560815

30. Press Information Bureau. https://pib.gov.in/PressReleasePage.aspx?PRID=1900755

31. Press Information Bureau. https://pib.gov.in/PressReleasePage.aspx?PRID=1851914

32. Press Information Bureau. https://pib.gov.in/PressReleasePage.aspx?PRID=1852525

33. Debnath S, Roy P, Namasudra S et al (2022) Audio-visual automatic speech recognition towards education for disabilities. J Autism Dev Disord. https://doi.org/10.1007/s10803-022-05654-4

34. https://responsivevoice.com/wordpress-text-to-speech-plugin/

A Hybrid Approach for Leaf Disease Classification Using Machine Learning and Deep Learning

Kriti Jain and Upendra Mishra

Abstract Natural remedies are less expensive, non-toxic, and associated with negative side effects. As a result, their demand is rising, particularly for herbal-based medicinal products, health products, nutritional supplements, and cosmetics. Threats from leaf diseases exist to the global agricultural industry's economic and production status. The need for farmers to protect agricultural products is reduced by the ability to find illness in leaves utilizing Deep learning (DL) and Machine learning (ML). Our approach involves a combinations method for the diagnosis of flora illness. In our suggested method RESNET-50 is employed for extracting the deep features and Random Vector Functional Link (RVFL) is employed for the classification. To look at the efficiency of the suggested RES-RVFL model, its categorizing performance is contrasted with Support Vector Machine (SVM), Decision Tree, Random Forest and K-Nearest-Neighbors (KNN). The findings showed that RVFL is very suitable for classifying leaf diseases, with a disease classification accuracy of about 94%. The fact that this result highlights the significance of early detection and naming of flora diseases for justifiable cultivation and food security is very positive. Our research has a solid foundation, thanks to the Plant Village dataset, and our findings add to the body of knowledge on applying deep learning and machine learning to identify plant diseases.

Keywords Random vector functional link · Residual Network50 · Multi-class classification · Leaf diseases

1 Introduction

Deep learning, has piqued the attentiveness of machine learning aficionados in neural networks, thanks to its cutting-edge performance in a range of applications, such as segmentation, action identification, and image/video categorization. Because deep learning models may be able to take out relevant representations at separate amount

K. Jain · U. Mishra (✉)
KIET Group of Institutions, Delhi-NCR, Ghaziabad, India
e-mail: Upendra.mishra@kiet.edu

© The Author(s), under exclusive license to Springer Nature Singapore Pte Ltd. 2024
S. Namasudra et al. (eds.), *Data Science and Network Engineering*, Lecture Notes in Networks and Systems 791, https://doi.org/10.1007/978-981-99-6755-1_15

of stratified models, they are superior to other types of models. Many research studies have focused on the classification and identification of leaf diseases. For example, in [1] the authors offered two methods for differentiating between healthy and sick tomato leaves. The sick tomato leaves in this article are classified using Probabilistic Neural Network (PNN) and KNN approaches after the author gathers his own dataset. In [2] the author suggested categorising leaf illnesses using textural cues taken from images of the disorders. In [3] convolutional neural networks and autoencoders have been combined to create a hybrid technique for the identification of crop leaf diseases.

In machine learning and computer vision research, ResNet-50, and Random Vector Functional Link (RVFL) [4] are both frequently used techniques. RVFL is a rapid and efficient neural network for classification tasks in contrast to ResNet-50, a deep convolutional neural network (CNN) famous for its powerful feature extraction capabilities. In this research-based introduction, we will look at how these two approaches can be integrated for classification tasks. In the single-hidden layer is the feedforward neural network known as RVFL. The output layer employs the SoftMax activation function which has been shown to be effective for resolving classification problems. RVFL can handle large datasets and has short training times, making it suitable for real-time. The RVFL network incorporates direct connections from the input layer to the output layer, greatly benefiting RVFL. Compared to the Extreme Learning Machine (ELM) and other models, the RVFL network offers a smaller and simpler architecture, and it also keeps the model complexity low. Nonetheless, image recognition applications have been successfully implemented using ResNet-50, a deep CNN architecture. It can train deep networks without sacrificing accuracy since it contains 50 layers and employs residual connections to address the vanishing gradient problem. A well-known feature extraction method called ResNet-50 excels in extracting complex and discriminative features from the image.

The following list includes this brief's novel contributions.

1. The project will examine the success of the combined strategy in words of correctness, training duration, and generalization capacities through experimental assessments on plant village datasets.
2. Comparative study with others classifiers like (KNN, Random Forest, SVM, Decision Tree) with RVFL.
3. The assessment of the suggested architectures' reduced computing complexity is a further contribution.

The remaining paper is structured as follows: A detailed analysis of a few comparable models is provided in the second section. Section 3 gives the proposed work. The methodology is illustrated in Sect. 4. The results of the numerical simulation and experiment are examined in Sect. 5. In Sect. 6, the paper is concluded.

2 Related Work

A prominent area of study with numerous implications in plant biology, agriculture, and environmental science is the image classification of leaf diseases. Utilization of machine learning and DL methods with picture datasets has grown in popularity recently due to their promise to improve accuracy and efficiency. Given its advantages in an agricultural setting, research on automated plant disease identification is essential. To categorize leaves in photographs, a variety of strategies have been employed.

2.1 KNN

KNN [5–7] is a non-parametric technique for regression and classification. Using a distance metric like Euclidean distance, KNN method categories or predicts the target value of a new data point based on the labels or target values in the training data. K selection is a hyperparameter that requires tuning.

The following provides an explanation of how the distance of KNN was created:

$$D^k = \sqrt{(l2^k - l1^k)^2 + (m2^k - m1^k)^2}$$

where,
   The Place of One at $l1^k, m1^k$
   Another is at $l2^k, m2^k$

$$D^k is\ seperation\ linking\left(l1^k, m1^k\right) and\ (l2^k, m2^k)$$

2.2 Random Forest

The Random Forest approach [8] is introduced by Leo Breiman and Adele Cutler in 2001.The ensemble learning family of ML algorithms includes this. At each split point, an arbitrarily selected subset of attribute and a subset of the instructing facts are used to aim each decision tree. To create the final forecast from the Random Forest [9], all decision trees' predictions are combined.

2.3 SVM

In order to determine sugarcane leaf disease, the author in [10] proposed a model to determine the severity on a segmented spot. To obtain the segmented spot, use thresholding. The highest standard deviation is used to extract disease spots, and then categorization is completed by SVM and getting accuracy of about 80%.

The hyperplane with the lowest classification error and highest margin between the two classes is found by the SVM method.

Learning set $(r1', s1'), (r2', s2'), \ldots, (rn', sn')$

Where,

rf is a characteristics vector and sf is a correlate binary category tag (+1, -1) The hyperplane is $\tau^t r + b = 0$, τ is weight vector b is bias.

For optimality:

$$\min\left(\frac{1}{2}\right)||\tau||^2 + C \sum \_f = 1^n \max(0, 1 - s_f(\tau^t r_f + b)$$

Subject to $s\_f(\tau^t r_f + b) \geq 1 - \mu\_f$ for $f = 1, 2, \ldots, n$, μ_f are the flapping variables that permit for certain miscategorized classes. Since the data points must be correctly classified with a margin, the inequality constraints mandate it of at least $1 - \mu_f$ for more details [4].

2.4 Decision Tree

An evaluation of numerous possibilities and possible outcomes is part of the decision-making process, which is graphically represented by a decision tree [9, 11]. Each categorization component is given new significance by the decision tree method.

Create_Tree (T'_{train})

1. Make decision node first
2. Create a leaf node with name X_i/ if every data sample from the T'_{train} belongs to uniform class X_i/
3. Create a leaf node with most popular class X_i/ label if assign A_t is empty
4. Pick the $A_t^{''}$ from ascribe set A_t that provides most information gain and the label n_d with assign

Create an arm from n_d with state $A_t^{''} = K$/ then consider S'_{train} batch of T'_{train} after that if S'_{train} is empty. So, $A_t^{''}$ get attached with T'_{train} or fix with (create tree).

2.5   Residual Network

Deep learning algorithms have also been applied to the classification of leaf diseases in that Convolution Neural Network (CNN) [12, 13] is mostly used. By overlaying shortcut connections over each residual block, Residual Networks (Resnet) facilitate the training of deep neural networks. Deep neural networks may now perform at the cutting edge for tasks like object identification, thanks to Resnet. In [14] they choose Resnet for the feature extraction in this study. A deep model provides us with more options for feature extraction.

Machine learning models' accuracy and resilience have increased as a result of combining deep learning with them in recent years. In [15] the author proposed the model in which work has divided in 4 steps: picture acquisition, image pre-processing, feature extraction, and disease categorization on the Pomegranate plant ANN with Backpropagation, SVM and many more have been used with 90% accuracy. Several studies have shown the method's potential for accurate and efficient leaf disease classification.

Despite past evaluations of preprocessing and different classifiers, to our knowledge, no prior work has looked at the performance of the combination of ResNet50 and RVFL for classification tasks. Therefore, the objective of the research that we suggest is to reduce this gap and evaluate how well this combination can improve classification accuracy while also looking at time complexity.

3   Proposed Work

Our analysis reveals the use of a trained RestNet50 model to derive features on the dataset of Leaf disease and categorize it using RVFL shown in Fig. 1.

Process- The dataset utilized in this study underwent pre-processing in order to prepare it for categorization. An expert convolutional neural network (CNN) known as ResNet-50 was used to gather data from images of healthy and diseased plants. ResNet-50 is a pre-trained model that is easily accessible from popular deep learning frameworks like TensorFlow or PyTorch. There were 14 different plant types and 26 different disease categories among the data used in the study. ResNet-50 is the final convolutional or global average pooling layer output to extract features from each image. By normalizing these features, the classification step's effectiveness was later improved. The RVFL classifier, a shallow neural design, was then employed as the input for the features. On a desktop with a 10th generation Intel-Core 'i5 processor' and '16 GB of RAM', the RVFL model was trained. The RVFL classifier underwent training by utilizing both the training set as well as validation set, was utilized to change hyperparameters such as regularization strength or hidden layer size.

Now, we evaluated the classification's precision and recall using a range of performance measures, including the F1 score. These measures are routinely employed to assess a model's categorization abilities. Precision gauges the accuracy of positive

Fig. 1 Flowchart of
proposed work

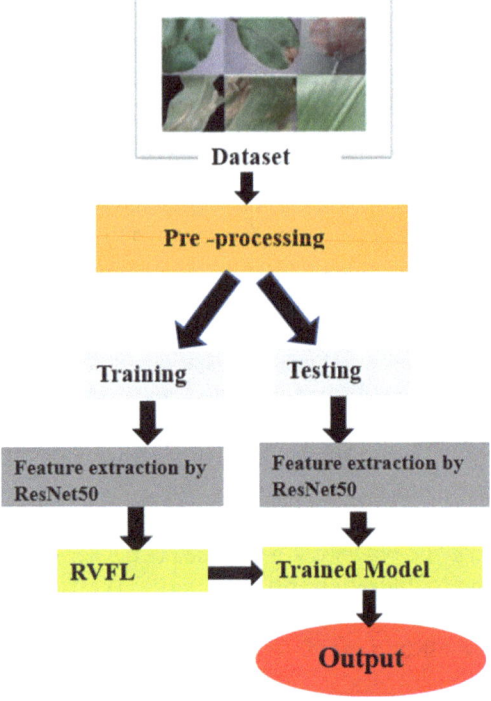

predictions, whereas recall evaluates the model's ability to correctly identify positive cases. The operations that needed the most computing in terms of time and space complexity were the preprocessing and the feature extraction using ResNet50. The RVFL model was able to train fast, taking only a few minutes. The peak memory utilization during training was roughly 1 GB, while memory usage overall was reasonable.

4 Methodology

4.1 Image Acquisition

In this step the Plant Village Dataset, which is accessible in reference [16] Git-Hub repository run by SP Mohanty, was used in this study. The dataset is made up of several rounds, including the original RGB photos, the color-corrected RGB photos, the segmented leaf RGB photos with flattened backdrops, and the grayscale images for performance testing. The dataset, which is broken down into 38 classes, contains

Fig.2 Shows the 1 sample of each dissimilar classes namely 'APPLE'("apple scab"),'APPLE'("black rot"),'APPLE CEDAR' ("apple rust"),'APPLE'("heathy"), 'BLUE-BERRY'("healthy"), 'CHERRY INCLUDING SOUR'("powdery mildew"), 'CHERRY INCLUDING SOUR'("healthy"), 'CORN MAIZE'("cercospora leaf spot"),'CORN MAIZE' ("common rust"), 'CORN MAIZE' ("Northern leaf Blight"), 'CORN MAIZE' ("healthy"), 'GRAPE' ("black rot"), 'GRAPE ESCA' ("black measles"), 'GRAPE LEAF BLIGHT'("Isariopsis leaf spot"), 'GRAPE' ("Healthy"), 'ORANGE HAUNGLONGBING'("Citrus greening"), 'PEACH' ("bacterial Spot"), 'PEACH' ("healthy"), 'PEPPER BELL' ("bacterial spot"), 'PEPPER BELL' ("healthy"), 'POTATO' ("Early blight"), 'POTATO' ("late blight"), 'POTATO' ("healthy"), 'RASPBERRY'("healthy"), 'SOYBEAN' ("healthy"), 'SQUASH' ("powdery mildew"), 'STRAW-BERRY' ("leaf scorch"), 'STRAWBERRY'("healthy"), 'TOMATO' ("bacterial spot"), 'TOMATO' ("Early blight"), 'TOMATO' ("Late blight"), 'TOMATO' ("leaf mold"), 'TOMATO'("Septoria leaf spot"), 'TOMATO' ("spider mites"), 'TOMATO' ("Target Spot"), 'TOMATO' ("tomato 'yellow' leaf -curl viral"), 'TOMATO' ("tomato-mosaic virus"), 'TOMATO'("healthy")

more than 87 K RGB photographs of both fined and poor crop leaves. Figure 2 shows the 1 sample of each class. The number of illness images in our collection is listed in Table 1.

Using a dataset with more than 87 k images, our suggested approach covers the two main processes of feature extraction and classification.

4.2 Pre Processing

The gathered photos are allured to enhance the image standard. Both undesired distortion and background noise are lessened.

The neural networks operate best with normalized data, we must first convert each image's pixel values (0–255) to 0–1 after data loading. The entire array of pixel

Table 1 Plant with disease name

Plant name	Disease name	Number of images
Tomato	Late blight	1851
Tomato	Healthy	1926
Grape	Healthy	1692
Orange	Huanglongbing (Citrus greening)	2010
Soybean	Healthy	2022
Squash	Powdery mildew	1736
Potato	Healthy	1824
Corn (maize)	Northern leaf blight	1908
Tomato	Early blight	1920
Tomato	Septoria leaf spot	1745
Corn (maize)	Cercospora leaf spot gray leaf spot	1642
Strawberry	Leaf scorch	1774
Peach	Healthy	1728
Apple	Apple scab	2016
Tomato	Tomato yellow leaf curl virus	1961
Tomato	Bacterial spot	1702
Apple	Black rot	1987
Blueberry	Healthy	1816
Cherry (including sour)	Powdery mildew	1683
Peach	Bacterial spot	1838
Apple	Cedar apple rust	1760
Tomato	Target spot	1827
Pepper, bell	Healthy	1988
Grape	Leaf blight (Isariopsis Leaf Spot)	1722
Potato	Late blight	1939
Tomato	Tomato mosaic virus	1790
Strawberry	Healthy	1824
Apple	Healthy	2008
Grape	Black rot	1888
Potato	Early blight	1939
Cherry (including sour)	Healthy	1826
Corn (maize)	Common rust	1907
Grape	Esca (Black Measles)	1920
Raspberry	Healthy	1781
Tomato	Leaf mold	1882
Tomato	Two-spotted spider mite	1741
Pepper, bell	Bacterial spot	1913
Corn (maize)	Healthy	1851

values is divided by 255 after being turned into a torch tensor. Since the images have a resolution of 256 × 256 and are encoded using 8 bits per channel, each picture element in the original image is in the range [0, 255] for each of the three RGB channels. Each pixel value in each channel is scaled down to the range [0, 1] after normalization. As a result, we get after normalization, the dataset's total scale as 261,000.

4.3   For Image Feature Extraction

This is the procedure of taking relevant with pertinent elements or patterns out of photographs. Image feature extraction is a censorious pace in many computer-vision applications since raw pictures usually contain a large amount of noise and pointless information. The computational complexity can be lowered by omitting crucial information, and the recovered features can then be used as input for additional processing or analysis.

Pre-trained CNNs can operate as feature extractors for image data due to their capacity to procure meaningful properties from pictures. In this study by deleting the final fully connected layer and using the outcome from the preceding layer as the feature vector, pretrained ResNet50 can be used for feature extraction. Transfer learning is frequently employed in computer vision problems. A tensor with the shape (batch_size, 2048) would be the output, indicating the features that were learned for each input image.

4.4   For Image Classification

Finding the class or category that an input belongs to is one of the most crucial machine learning tasks. A model is trained to learn how to translate input to the proper classes using tagged data.

In the presented work we use RVFL as the classifier. It is an easy-to-use and an effective neural network design that builds use of randomization to provide effective generalization with short training times [17]. Its architecture was presented in 1992. The RVFL feedforward neural network framework consists of an input layer, the hidden level, neurons having random weights and biases, whereas the output layer neurons are trained using supervised learning. The ability of RVFL to prevent over-fitting is enhanced by this kind of link. The quick instruction pace of RVFL networks is one of the best features that make them stand out. This can be attributed to their single-layer feedforward architecture. For more details see reference [18].

The red line depicts the straight connection between the input and output layers. The weights for the blue lines are predetermined and drawn at random from a list.

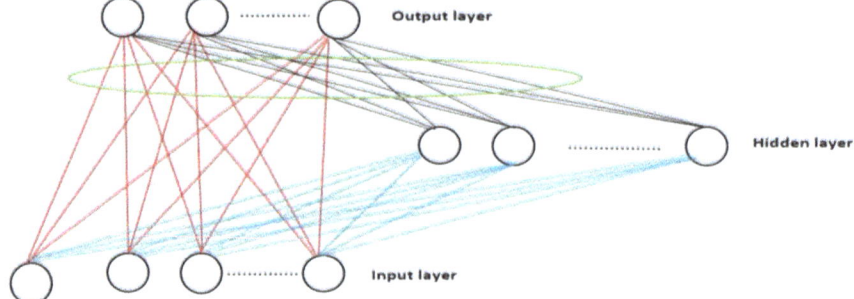

Fig. 3 RVFL framework

The output weights must only be computed as shown in Fig. 3 RVFL contains [19] features that have undergone nonlinear transformation in both the input and output layers.

Hence, photos available for training are 70,295 and just 33 photos in the test data. The ResNet50 feature extractor produces a vector of length 2048 for each input image, hence the RVFL classifier would require 2048 input neurons. We have a training data matrix of dimensions 70,295 × 2048 because there are 70,295 training images and each of them is represented by a feature vector of length 2048. N = 5 input neurons that would give us 10,240 hidden neurons as the number of hidden neurons. a random vector matrix with the dimensions 10,240 × 2048. The matrix's elements would each have a random value between −1 and 1. The hidden layer activation is computed using the sigmoid activation function. A matrix of size (70,295 × 10,240) is produced by the scalar product of the input matrix X' and the random weight matrix W', where each row denotes the activation of the hidden layer for a single input image. We initially need to transform the class labels into a one-hot encoded format in order to train the output layer for classification using RVFL. In this instance, the hidden layer activation matrix's dimensions are 70,295 × 10,240, the one-hot encoded class label matrix's dimensions are 70,295 × NUM_classes, and the output weights matrix's dimensions are 10,240 × NUM_classes. Test_H is the test data hidden layer activation matrix, and test_X is the test data matrix, both measuring 33 by 2048.

5 Result and Discussion

Several important parameters, such as dataset size and quality, feature selection, algorithm, hyperparameter tweaking, and metrics, and dimensionality reduction like PCA have a notable impact on the classifier's success. In this study Pre-processing manner and a sizable and representative dataset are necessary to get the best results.

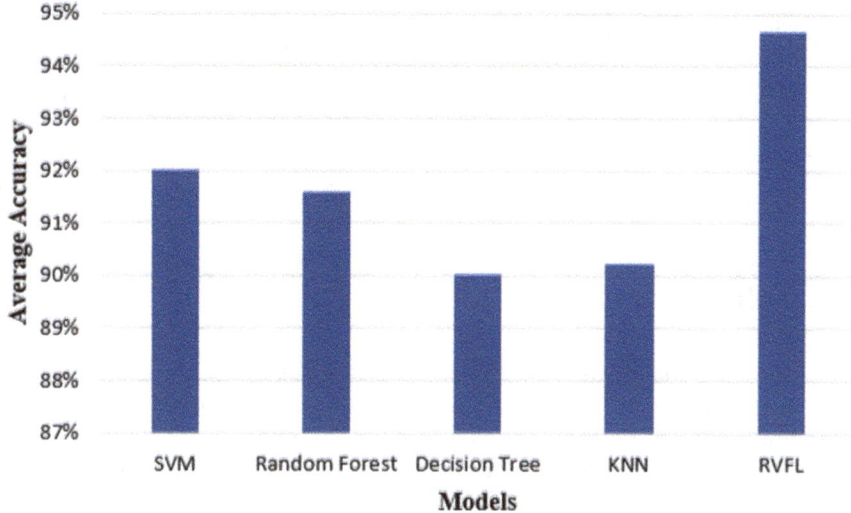

Fig. 4 Accuracies of models with Resnet 50

We have examined various models using accuracy, which considers the vast range of correctness across various subsets of data as shown in Fig. 4.

According to our research, RVFL and ResNet50 can work together to provide high test accuracies, although the best model performance depends heavily on several aspects like dataset features and hyperparameters. These characteristics can be considered, and the proper dimensionality reduction techniques can be used to enhance RVFL performance and produce better results in classification tasks.

The best classification was determined after a series of trials. The strategy used for classification, which involved the use of a ResNet50 model to extract features and RVFL for classification, appeared to be successful based on the stated performance indicated in Table 2.

The model's efficacy was demonstrated by the 94.65% accuracy rate it achieved while sorting plant leaves into different disease groups. The F1 score of 93.91% shows that precision and recall are well-balanced. Both metrics are considered while calculating this score, which has a range of 0–1. The model's accuracy in identifying plant leaf diseases while limiting false positives and false negatives is demonstrated

Table 2 Contrasting of model accuracies with time

Model	Accuracy (%)	Time
SVM	92	27 mn 13 s
Random forest	91.6	17 mn
Decision tree	90	16 mn 55 s
KNN	90.23	Approx 24 mn
RVFL	94.65	Nearly 10 mn

by the high score of 92%. Both the model's precision and recall values, which are 92.91% and 94.65% respectively, show that it can properly categorize positive events and lessen false positives. As a result, the model is a useful tool for recognizing flora sickness. Besides this Resnet 50 alone can take 30mn but with classifiers it takes approx. 22 mn 21 s, after working with RVFL the time computation comes out to be nearly (less than 10 mn).

6 Conclusion

For the identification of leaf diseases, we suggested a hybrid strategy in this research. In our proposed method, RVFL is utilized for classification, and RESNET-50 is used to acquire deep features. The classifying performance of the suggested RES-RVFL model is identified as the best-performing model as compare to SVM, Decision Tree, Random Forest, and KNN in order to assess its effectiveness. The results demonstrated that RVFL, with a disease classification accuracy of about 94.65% with less than 10 mn computation time, is particularly suitable for classifying leaf diseases. All encompassing, this study demonstrates the promising of DL and RVFL classifiers in accurately classifying healthy and diseased plant images.

The research's originality resides in its thorough comparison of various cutting-edge algorithms in terms of preprocessing and temporal complexity. The method used in the research report to assess model correctness and temporal complexity is exclusive to RVFL. By taking accuracy and temporal complexity into account, findings can help in making efficient model selections.

We intend to expand on this work in future studies by developing our own dataset to pinpoint cultivable locations that might not be ideal for plant development and growth. Inclusion to the development of new datasets, future investigation could also explore the use of other DL architectures for plant classification, like VGG, Inception, or Dense Net. It would also be interesting to examine the success of transfer learning on this task, where pre-trained models are adapted on plant image datasets. By further developing these techniques and exploring new applications, we can improve our understanding of plant behaviour and develop strategies to support sustainable agriculture.

References

1. Balakrishna K, Rao M (2019) Tomato plant leaves disease classification using KNN and PNN. Int J Comput Vis Image Process (IJCVIP) 9(1):51–63
2. Hossain E, Hossain MF, Rahaman MA (2019) A color and texture based approach for the detection and classification of plant leaf disease using KNN classifier. In: 2019 international conference on electrical, computer and communication engineering (ECCE). IEEE, pp 1–6

3. Khampania A, Saini G, Gupta D, Khanna A, Tiwari S, de Albuquerque VHC (2020) Seasonal crops disease prediction and classification using deep convolutional encoder network. Circuits Syst Signal Process 39:818–836
4. Mishra U, Gupta D, Hazarika BB (2022) An intuitionistic fuzzy random vector functional link classifier. Neural Process Lett 1–22
5. Saputra RA, Wasiyanti S, Saefudin DF, Supriyatna A, Wibowo A (2020) Rice leaf disease image classifications using KNN based on GLCM feature extraction. J Phys: Conf Ser 1641(1):012080. IOP Publishing
6. Shokrzade A, Ramezani M, Tab FA, Mohammad MA (2021) A novel extreme learning machine based KNN classification method for dealing with big data. Expert Syst Appl 183:115293
7. Su J, Wang M, Wu Z, Chen Q (2020) Fast plant leaf recognition using improved multiscale triangle representation and KNN for optimization. IEEE Access 8:208753–208766
8. Hazra D, Bhattacharyya D, Kim TH (2021) A random forest-based leaf classification using multiple features. In: Machine intelligence and soft computing: proceedings of ICMISC 2020. Springer, Singapore, pp 227–239
9. Mao Y, He Y, Liu L, Chen X (2020) Disease classification based on eye movement features with decision tree and random forest. Front Neurosci 14:798
10. Ratnasari EK, Mentari M, Dewi RK, Ginardi RH (2014) Sugarcane leaf disease detection and severity estimation based on segmented spots image. In: Proceedings of international conference on information, communication technology and system (ICTS) 2014. IEEE, pp 93–98
11. Chen C, Geng L, Zhou S (2021) Design and implementation of bank CRM system based on decision tree algorithm. Neural Comput Appl 33:8237–8247
12. Sardogan M, Tuncer A, Ozen Y (2018) Plant leaf disease detection and classification based on CNN with LVQ algorithm. In: 2018 3rd international conference on computer science and engineering (UBMK). IEEE, pp 382–385
13. Francis M, Deisy C (2019) Disease detection and classification in agricultural plants using convolutional neural networks—a visual understanding. In: 2019 6th international conference on signal processing and integrated networks (SPIN). IEEE, pp 1063–1068
14. Cheng WX, Suganthan PN, Katuwal R (2021) Time series classification using diversified ensemble deep random vector functional link and resnet features. Appl Soft Comput 112:107826
15. Pawar R, Jadhav A (2017) Pomogranite disease detection and classification. In: 2017 IEEE international conference on power, control, signals and instrumentation engineering (ICPCSI). IEEE, pp 2475–2479
16. https://github.com/spMohanty/PlantVillage-Dataset
17. Wu J, Miu F, Li T (2020) Daily crude oil price forecasting based on improved CEEMDAN, SCA, and RVFL: a case study in WTI oil market. Energies 13(7):1852
18. Hemedan AA, Abd Elaziz M, Jiao P, Alavi AH, Bahgat M, Ostaszewski M, Lu S (2020) Prediction of the vaccine-derived poliovirus outbreak incidence: a hybrid machine learning approach. Sci Rep 10(1):5058
19. Shi Q, Katuwal R, Suganthan PN, Tanveer M (2021) Random vector functional link neural network based ensemble deep learning. Pattern Recogn 117:107978

Enhancing Agricultural Decision-Making Through Machine Learning-Based Crop Yield Predictions

Bhaskar Marapelli, Lokeshwari Anamalamudi, Chandra Srinivas Potluri, Anil Carie, and Satish Anamalamudi

Abstract Food production through Agriculture plays an important role in keeping the world's population hunger-free and nations economically secure. The continuous change in land minerals, weather situation, and pesticide usage affect the yield of the crops. Farmers can choose successful crops for the season with the help of machine learning algorithms used for crop yield prediction. In this study, we forecasted agricultural production using numerous kinds of machine learning models while considering several factors that affect crop yields, such as rainfall, temperature, and pesticide use. By merging multiple separate model predictions, ensemble machine learning models improve the performance of the machine learning models. We have worked with individual models and ensemble models like SVR, RandomForestRegressor, LinearRegressor, and DecisionTreeRegressor to predict crop yield and found an ensemble solution that combines the strengths of both the stacked generalization model and the gradient boost algorithm which can provide improved accuracy and robustness in crop yield prediction. According to the findings, the ensemble solution provided an R2 score of 98 percent, which is higher than the R2 scores of 96 percent obtained using the Decision Tree Regressor and 89 percent obtained using the Gradient Boosting Regressor.

Keywords Agriculture · Crop yield prediction · Machine learning models · Ensemble learning · R2 score · Rainfall · Temperature · Pesticide usage

B. Marapelli (✉)
Department of CS&IT, Koneru Lakshmaiah Education Foundation, Vaddeswaram, AP, India
e-mail: bhaskarmarapelli@gmail.com

L. Anamalamudi · A. Carie · S. Anamalamudi
Department of Computer Science and Engineering, SRM AP University,, Vijayawada, India
e-mail: lokeshwari_a@srmap.edu.in

S. Anamalamudi
e-mail: satish.a@srmap.edu.in

C. S. Potluri
Department of Computer Science and Engineering, Werabe University, Werabe, Ethiopia

© The Author(s), under exclusive license to Springer Nature Singapore Pte Ltd. 2024
S. Namasudra et al. (eds.), *Data Science and Network Engineering*, Lecture Notes in Networks and Systems 791, https://doi.org/10.1007/978-981-99-6755-1_16

1 Introduction

Agricultural production is an essential aspect of human life, providing food, fiber, and other resources that support our survival and well-being. In recent years, the world has faced numerous challenges related to agricultural production, such as climate change, droughts, and land degradation [1, 2]. These challenges have made it increasingly important to accurately predict crop yields to ensure sustainable and efficient agricultural production. For farmers, politicians, and other agricultural stakeholders, crop production forecasts are crucial. To optimize their marketing and sales plans as well as their planting, irrigation, and fertilization schedules, farmers require accurate yield projections. To manage food production, distribution, and pricing, policymakers use crop yield estimates to create interventions and policies. Crop yield forecasts are also used by agribusinesses, food processors, and insurers to make educated judgments about supply chain management, risk assessment, and financial planning.

Statistical models and other conventional methods for crop production prediction fall short of reflecting the dynamic and complex nature of agricultural systems. These approaches frequently rely on historical information and presumptions, which might not fully account for the impact of altering climatic patterns, soil conditions, and agricultural practices. Furthermore, it may be difficult to capture the complex relationships and nonlinearities in crop growth and development using conventional methods due to their inability to manage big and diverse data sets.

Machine learning algorithms are a promising technique for predictive analysis [3, 4]. Machine learning algorithms are capable of modeling complex interactions between multiple factors, such as rainfall, temperature, and pesticides, that affect crop yields [5]. These algorithms help us gain a better knowledge of how various factors affect crop yields and help us manage our crops more effectively. Due to its capacity to take into account a number of factors that influence crop yields, machine learning models have recently gained importance in crop yield prediction. Machine learning models can handle large and complex data sets that include various factors such as rainfall, temperature, soil moisture, and pesticide use. This allows the models to take into account multiple factors that affect crop yields and make more accurate predictions. Large data sets can be used to train machine learning models, which can increase prediction accuracy. This is crucial in agriculture in particular, where the impact of small differences in yield predictions can be significant [6]. It might take a lot of time and be prone to human error to manually collect and process data for agricultural production prediction. The process is automated, and the possibility of mistakes is decreased.

Machine learning models can make predictions in real time, allowing farmers and policymakers to make timely decisions about crop management and resource allocation. Traditional methods of crop yield prediction, such as manual observations and field experiments, can be costly and time-consuming. With the help of existing data, machine learning models may be trained to make predictions swiftly and effectively [7, 8]. A useful tool for estimating crop yields and enhancing agricultural decision-making is machine learning models. Machine learning models can offer more precise predictions and assist farmers, decision-makers, and researchers in making educated decisions regarding crop management and resource allocation by taking into account a variety of variables that affect crop yields.

Machine learning-based yield prediction is gaining a lot of attention in agricultural research. However, there are still gaps in the field that could be addressed through the use of an ensemble solution. The restrictions of individual models have been addressed and the overall performance of crop yield prediction is improved by an ensemble solution, which integrates multiple individual models to provide a greater degree of accuracy and reliability prediction.

The proposed study utilizes a gradient boost regressor and an ensemble machine learning model with Stacked Generalization for predicting crop yield. In the proposed model,

- Input features are used to train the base models.
- The base models generate predictions using the trained input features.
- The predictions from the base models are then used as input features to train the meta-model.
- The meta-model is responsible for generating the final predictions based on the predictions of the base models.

The rest of the paper is organized as follows: Sect. 2 describes the Related work; Sect. 3 describes our Proposed Model; Sect. 4 explains the Experimental Setup; Sect. 5 is about Results and Discussion; and Sect. 6 discusses the conclusion.

2 Related Work

In the study in [9], two deep neural network models were built in Python using TensorFlow. With a root-mean-square error (RMSE) of 12% of the average yield and 50% of the standard deviation for the validation data set utilizing forecasted meteorological data, it was demonstrated that the suggested model had a higher level of prediction accuracy. The trained DNN model was used by the author to perform feature selection, which was successful in reducing the input space without noticeably lowering prediction accuracy. The outcomes also showed that environmental factors affected crop productivity more so than genetics.

Convolutional neural networks (CNNs) and recurrent neural networks (RNNs) were employed in a deep learning framework for crop production prediction based on environmental data and management practices in the article [10] by the author. The proposed CNN-RNN model was compared with well-known methods such as random forest (RF), deep fully connected neural networks (DFNN), and LASSO using historical data for forecasting corn and soybean yield across the whole Corn Belt region in the United States for the years 2016–2018. The results showed that the CNN-RNN model outperformed all other researched techniques, with an RMSE of only 9 and 8% of their respective average yields.

In [11] this study, soybean yield in Lauderdale County, Alabama, USA, was predicted using a 3D convolutional neural network (CNN) model that makes use of spatiotemporal data. Data on yields from 2003 to 2016 were gathered using the USDA NASS Quick Stat tool, and surface reflectance, land surface temperature, and other satellite data were collected using Google Earth Engine. The performance of the model was assessed in the study using root-mean-squared error (RMSE), allowing comparison with other methods that utilize the same evaluation metric.

In order to help farmers make wise decisions, the author forecasts crop yields before planting. For this, a user-friendly machine learning method will be created along with web-based visual applications. Farmers will have access to the projected outcomes. There are numerous techniques and procedures for predicting agricultural yields, and this project makes use of the Random Forest Algorithm. Despite the difficulties brought on by weather factors like temperature, humidity, and rainfall, there aren't enough solutions to deal with these problems. This kind of analysis is very helpful for predicting agricultural yields in nations like India where agriculture is essential to economic growth [12].

The authors of [5] offer a comprehensive examination of the use of machine learning algorithms in agricultural yield prediction with an emphasis on palm oil production forecasting. The use of machine learning in the palm oil industry, the author's evaluation of cutting-edge techniques for predicting crop yields, and a comparison of related studies are all covered. A list of the problems with agricultural yield forecasts is provided, along with potential solutions. A wide range of subjects is covered in the review, including remote sensing applications, plant development and disease recognition, mapping and tree counting, optimal features, and algorithms. This is due to the fact that one of the key objectives of the study is to examine potential future applications of machine learning-based palm oil yield prediction.

In [6], the authors discuss the use of machine learning algorithms for pre-season crop type mapping, which involves identifying the types of crops being grown in a particular area before the harvesting season. The authors employ machine learning techniques to forecast the yields of four popular crops extensively cultivated across India. The site-specific prediction of crop yields enables the farmers to apply inputs such as fertilizers and water as per the soil and crop needs. Machine learning algorithms are utilized to identify patterns among the data for crop prediction. The four crops that are considered for yield prediction are Maize, Potatoes, Rice (Paddy), and Wheat.

The author in [7] seeks to increase crop yield productivity through the use of machine learning methods. For quantitative and financial evaluation at the field level, which may be utilized for agricultural commodity strategic planning for import-export policies and to increase farmer incomes, an early and correct assessment of crop success is crucial. A powerful method for predicting agricultural yields is provided by the author using machine learning algorithms, which can aid in the prediction of greater crop output. The adoption of machine learning techniques can greatly increase crop output productivity, according to experimental studies.

The objective of the study in [8] was to examine and evaluate the various methods for predicting agricultural yields in order to find the most practical solutions to the most typical problems. Electronic databases yielded a total of 487 papers; 50 were chosen based on inclusion and exclusion standards. A systematic review of the literature (SLR) was then performed on these 50 papers.

The authors use performance data in [13] from the North American Uniform Soybean Tests (UST) to create an LSTM-based model. For the purpose of predicting genotype response in various contexts, the proposed model included pedigree-relatedness metrics and weekly meteorological factors. The model's performance was contrasted with that of other machine learning models, including regression using the least absolute shrinkage and selection operator (LASSO), support vector regression with radial basis function kernel (SVR-RBF), the USDA model for yield prediction, and others. The suggested models performed better than these rival models. A temporal attention mechanism for LSTM models was also created by the author.

The authors of [14] wanted to know if deep neural networks (DNNs) might be utilized to predict agricultural yields and compare them to more traditional machine learning models to determine how effective they are. They projected soybean yields using data from Lauderdale County in Alabama, USA, and discovered that DNNs outperformed various other models with respect to accuracy and predictability. The authors draw the conclusion that DNNs have the potential to revolutionize the study of crop yield forecasting and can assist farmers and decision-makers in making more informed choices about crop management and resource allocation.

The goal of the authors of [15] is to investigate the connection between soil moisture and crop productivity and assess how well various machine learning methods can forecast crop yield. Data from a farm in Tashkent, Uzbekistan, was used in the study. A decision tree method yielded the greatest results, according to the authors' analysis of the performance of other algorithms, including regression and decision tree algorithms. The research's findings point to the potential for using machine learning algorithms to anticipate crop yields based on soil moisture, which might be useful for farmers, decision-makers, and researchers in making informed choices regarding crop management and resource allocation.

Some of the research gaps in the area of yield prediction using machine learning could be addressed by an ensemble solution. In order to create a more reliable and precise prediction, an ensemble solution integrates various distinct models. An ensemble method could offer a more thorough coverage of various crops and geographies, enhancing the precision of agricultural yield prediction, by merging many individual models trained on data from various crops and regions. Incorporating numerous environmental variables into an ensemble solution may also be possible, increasing the precision of crop yield prediction by accounting for a wider range of crop yield-influencing factors. By combining multiple individual models trained on real-world data, an ensemble solution could provide a more robust evaluation of the performance of machine learning models on real-world data, helping to determine their effectiveness in practical applications. An ensemble solution could also allow for the comparison of multiple machine learning algorithms by combining the outputs of individual models trained using different algorithms. This could offer a more thorough assessment of the various algorithms and assist in identifying the best algorithms for predicting crop yield.

3 Proposed Model

The predictions of different models are combined using an efficient technique called ensemble machine learning to provide a forecast that is more reliable and accurate. A particular form of ensemble learning called stacked generalization, commonly referred to as stacking, involves training a meta-model to generate predictions based on the predictions of many base models. A stacked generalization model and a gradient boost regressor can be combined to produce an ensemble machine learning approach for agricultural production prediction that can increase forecast accuracy.

In this approach, we train multiple base models on the same data set to make predictions about crop yield. Different machine learning algorithms, such as decision trees, random forests, and neural networks, could be used as these base models. In order to train a meta-model, which is typically a simple linear model or another machine learning technique, the predictions of these underlying models are then integrated and used as input characteristics. The ultimate prediction is created by the meta-model, which incorporates the predictions of the base models.

By merging the predictions of various models, this ensemble approach may efficiently utilize the advantages of various machine learning algorithms and provide a prediction that is more accurate. Moreover, by using the stacked generalization model, we can improve the interpretability of the solution and better understand the contributions of each base model to the final prediction.

3.1 Stacking Regressor

The machine learning technique called stacking regressor involves training several regression models and using their predictions as inputs for a single regression model. The final regression model then uses these predictions to make the final prediction [16]:

$$y_f p = mm(b_1(x), b_2(x), \ldots, b_k(x)) \tag{1}$$

where $y_f p$ is the final prediction, and b_1 and b_2 are the functions that forecast the output values for the respective base models $basemodel_1$, $basemodel_2$. mm is the function that predicts the output value for the meta-model. The function that forecasts the output value for $basemodel_k$ is called b_k, K is the number of input features, and X is the number of base models.

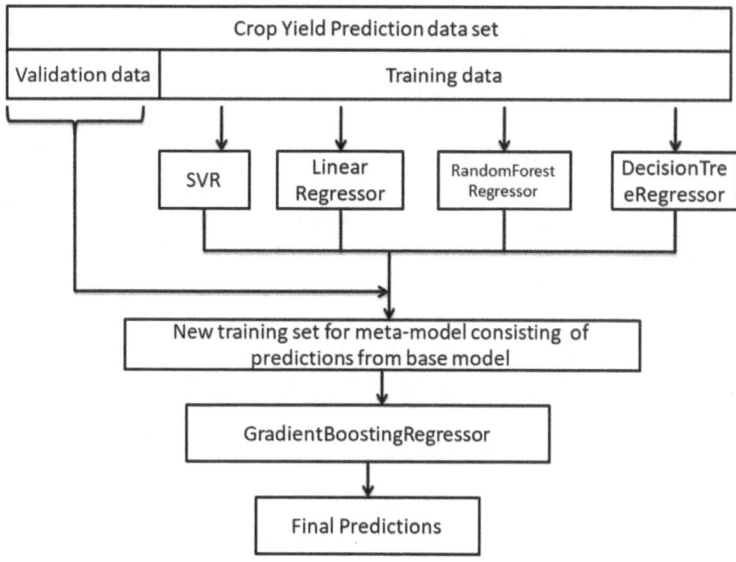

Fig. 1 Stacked generalization model

To prepare the proposed model, we have used four base models DecisionTreeRegressor, SVR, LinearRegression, KNeighborsRegressor, and GradientBoostingRegressor as meta-models. We first trained the three base models with training data, using these predictions as input features to the meta-model. We trained the meta-model along with input features and output values.

3.2 Gradient Boost Regressor

By merging the predictions of many decision trees, the gradient boost regressor, an effective machine learning technique, can increase the prediction accuracy of the ensemble solution. This approach focuses on the most challenging aspects of the prediction issue by fitting decision trees repeatedly to the residuals of the predictions of the prior trees [17]. By combining the gradient boost regressor with the stacked generalization model, we can create a powerful ensemble solution for crop yield prediction that can improve prediction accuracy and handle complex relationships between the inputs and the outputs:

$$\hat{y}_i = \sum_{j=1}^{n} f_i(x_i) \tag{2}$$

where $f_j(x_i)$ is the forecast of the jth tree for the ith occurrence, $\hat{y}_i$ is the projected value for the ith instance, and n is the number of trees in the ensemble.

3.3 Data Sets

In this work, we have used a data set that is a combination of four data sets, Crop Yield Data, Rainfall Data, Pesticides Data, and Average Temperature Data; these data sets are gathered from different sources like World Bank Open Data and Food and Agriculture Organization (FAO) [18]. The data sets after careful removal of null entries Crop Yield Data set contains 56717 entries after Rainfall Data is merged with yield data the data set contains 25385 entries after Pesticides Data is combined data reaches 18949 entries finally Average Temperature Data is merged with the yield data which reached 28242 entries. The prepared data set contains 101 countries, and 23 years' worth of data. The data preprocessing techniques like one hot encoding to encode categorical data to numerical data and min-max scalar to scale data to the same level of magnitudes are applied. The data set has 101 countries; ordering these by 10, the highest yield production is shown in Table 1. Table 2 describes the data set with different measures. Figure 2 shows the graphical description of crop yield country-wise. The data set has been split into the test as well as training data sets, with 30% of the data used for training and 70% for testing.

Table 1 Highest yield production Area- and Item-wise identified from data set

Item	Area	Hectogram per hectare (Hg/Ha)_Yield
Cassava	India	142810624
lPotatoes	India	92122514
	Brazil	49602168
	United Kingdom	46705145
	Australia	45670386
Sweet potatoes	India	44439538
lPotatoes	Japan	42918726
	Mexico	42053880
lSweet potatoes	Mexico	35808592
	Australia	35550294

Table 2 Data set description in different measures

Data measure	Year	pesticides_tonnes	avg_rain_fall	avg_temp	hg/ha_yield
Count	28242	28242	28242	28242	28242
Mean	2001.54	37076.91	1149.06	20.54	77053.33
Std	7.05	59958.78	709.81	6.31	84956.61
Min	1990	0.04	51	1.3	50
Max	2013	367778	3240	30.65	501412

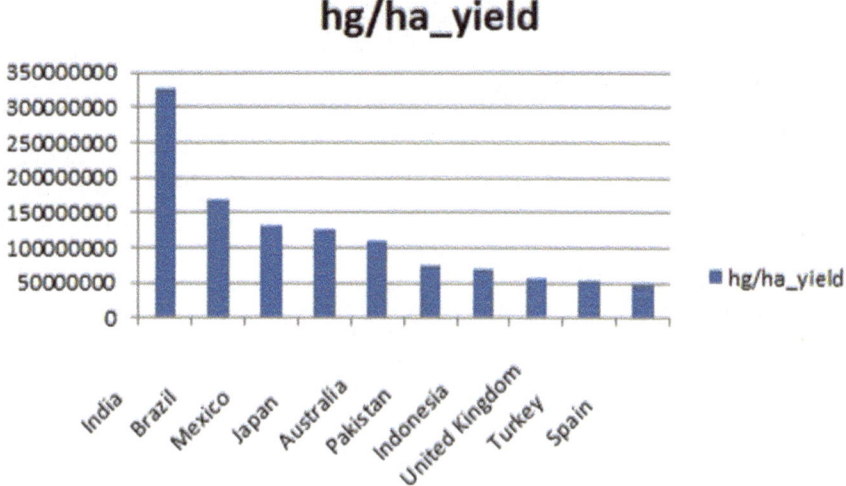

Fig. 2 Crop yield country-wise

4 Experimental Setup

In this work, a stacking generalization model has been built to forecast agricultural yield. In the suggested stacking generalization model, the base models consist of four different machine learning algorithms, which are Support Vector Regression (SVR), RandomForestRegressor, LinearRegression, and DecisionTreeRegressor. These base models, which estimate crop yield in this case, are trained on the same data set. These base models' predictions are then fed into the meta-model, which in this case is a GradientBoostingRegressor. A final prediction is made by the meta-model by combining the predictions from the base models. The goal of stacking is to use the advantages of multiple base models to build a more reliable and accurate model [16].

The architecture of this stacked generalization model is as follows:

1. Input: The input to the model is the features (e.g. rainfall, temperature, and pesticides) that affect crop yield.
2. Base Models: The input features are fed into the four base models (SVR, RandomForestRegressor, LinearRegression, and DecisionTreeRegressor) to make predictions on the target variable (crop yield).
3. Stacking: The predictions made by the base models are combined using the meta-model (GradientBoostingRegressor) to make a final prediction.
4. Output: The crop yield, which is the outcome of the stacking generalization model, is the final prediction.

By combining the predictions made by multiple base models, the stacked generalization model aims to reduce the variance and bias present in individual models, and, thus, improve the prediction accuracy compared to using a single model. Figure 1 describes the architecture used in the picture, and the algorithm 1 describes the entire process of the proposed model.

Algorithm 1 Stacked Generalization Model

Require: Input features $\mathbf{X}$
Ensure: Predicted crop yield $\hat{y}$
1: Initialize empty lists M_1, M_2, M_3, M_4
2: Train SVR model on $\mathbf{X}$ and append to M_1
3: Train RandomForestRegressor model on $\mathbf{X}$ and append to M_2
4: Train LinearRegression model on $\mathbf{X}$ and append to M_3
5: Train DecisionTreeRegressor model on $\mathbf{X}$ and append to M_4
6: Initialize empty list *predictions*
7: **for** i in 1 to 4 **do**
8:    $y_i \leftarrow M_i.predict(\mathbf{X})$
9:    Append y_i to *predictions*
10: **end for**
11: Train GradientBoostingRegressor model on *predictions* to get $\hat{y}$
12: **return** $\hat{y}$

---

5   Results and Discussion

Regression score (R2 score) and Root-Mean-Squared Error (RMSE) are used as evaluation metrics to assess the effectiveness of the models that have been used.

5.1   R2 Score

The machine learning model's R2 score varies from 0 to 1. Scores range from 0 for a model that does not fit the data at all to 1 for a model that perfectly matches the data. A model that captures the underlying relationships in the data well gets an R2 score near 1 [19]:

$$R2 = 1 - \frac{SS_{res}}{SSR_{tot}} \tag{3}$$

where SS_{res} stands for the sum of squared residuals, which represents the discrepancy between the dependent variable's actual and expected values. The difference between the actual values and the dependent variable's mean value is represented by SSR_{tot}, or the total sum of squares:

Table 3 Sample crop yield predictions and actual values

crop_yield_Actual	crop_yield_Predicted
69220	71965.92
20000	23708.65
51206	44241.43
166986	162569.2
56319	54566.15

$$SS_{res} = \sum_{i=1}^{n}(yact_i - ypred_i)^2 \qquad (4)$$

value

$$SSR_{tot} = \sum_{i=1}^{n}(yact_i - \bar{Y}_i)^2 \qquad (5)$$

$$\bar{Y} = \frac{sum_{i=1}^{n} yact_i}{n} \qquad (6)$$

The actual crop yield is denoted by $yact$, predicted crop yield by $ypred$, average crop yield by $\bar{Y}$, and sample size by n in Eqs. 4, 5, and 6.

In Table 4, the R2 score for each individual machine learning model is shown. With an R2 value of -0.20, the Support Vector Regression (SVR) model does not adequately account for the variation in the dependent variable (hg/ha_yield). With a better R2 score of 0.68, the RandomForestRegressor explains 68% of the variance in the dependent variable. The dependent variable's variance is explained by the linear regression model, which has an R2 value of 0.76 and explains 76% of it. With an even better R2 score of 0.90, the GradientBoostingRegressor is capable of accounting for 90% of the variance in the dependent variable. The DecisionTreeRegressor explains 96% of the variance in the dependent variable and has the greatest R2 score (0.96). The R2 score for the Ensemble model (stacked generalization model) is shown in Table 7. The StackingRegressor has the greatest R2 value (0.98), which means that it accounts for 98% of the variance in the dependent variable. The StackingRegressor model, therefore, seems to be the best model for describing the variation in the dependent variable based on these R2 scores. Table 3 shows sample predictions made by stacking regressor. Figure 5 shows the Regression line fitting for the Stacked generalization model, and Table 6 describes the R2 scores of all models. Figure 3 describes a graphical description of R2 scores of individual models.

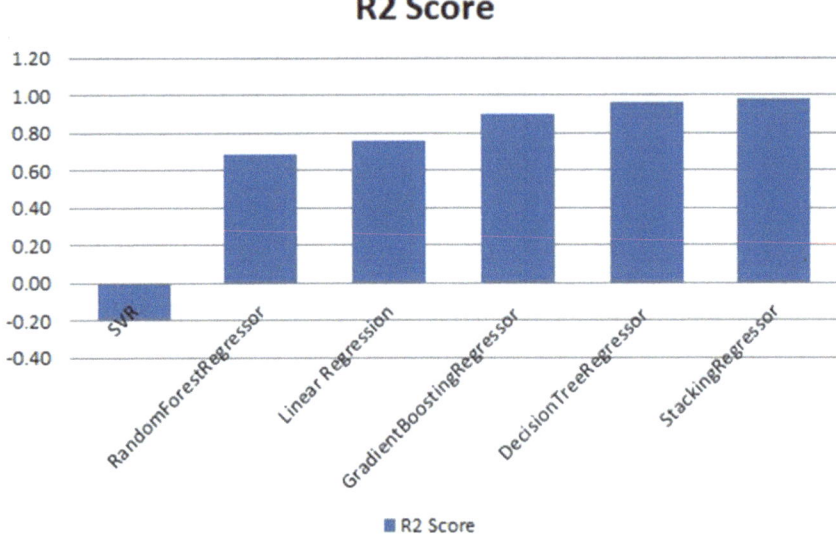

Fig. 3 R2 scores of individual and stacking model

Table 4 Root-Mean-Squared Error (RMSE) with individual machine learning methods

Model name	RMSE
SVR	94461.61
RandomForestRegressor	48376.43
Linear regression	42107.62
GradientBoostingRegressor	26485.66
DecisionTreeRegressor	12540

5.2 Root-Mean-Squared Error (RMSE)

The average difference between the expected outcomes and the actual values is measured by the term "Root-Mean-Squared Error" (RMSE). Better accuracy is shown by a smaller RMSE since it suggests that the expected outcomes are more closely aligned with the actual values. Tables 4 and 5 show the RMSE value of all the models applied to data sets; we can observe the least RMSE value is with the Stacking Regressor model and the highest RMSE value is with Random Forest Regressor [19]:

$$\text{RMSE} = \sqrt{\frac{\sum_{i=0}^{N-1}(ypred - yact)^2}{N}} \tag{7}$$

where N is the number of observations, $ypred$ is the predicted crop yield, and $yact$ is the actual crop yield.

Table 5 Ensemble model RMSE

Ensemble model	Base models	Meta-model	RMSE
llStackingRegressor	SVR	llGradientBoostingRegressor	l10089.61
	RandomForestRegressor		
	Linear Regression		
	GradientBoostingRegressor		
	DecisionTreeRegressor		

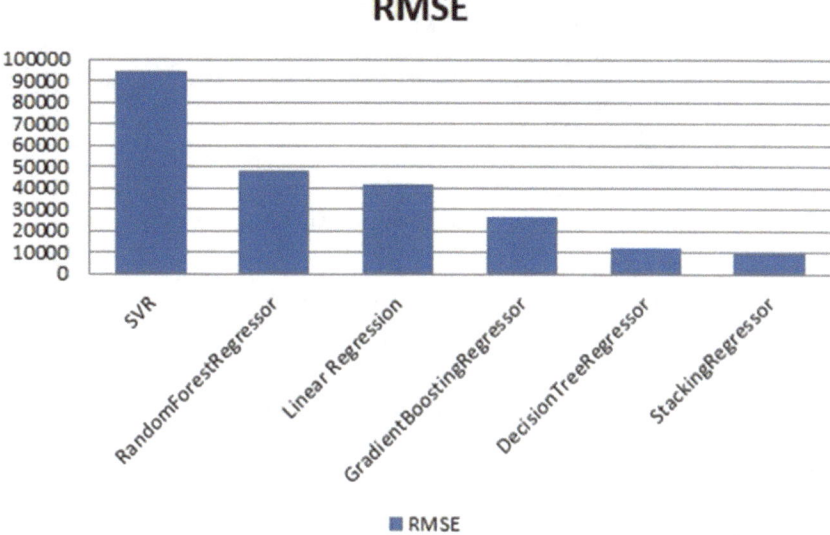

Fig. 4 RMSE of individual and stacking model

The DecisionTreeRegressor model in Table 4 has the lowest root-mean-squared error (RMSE), which is 12540. This shows that in terms of prediction accuracy, the DecisionTreeRegressor model performs the best. Another excellent option for prediction is the GradientBoostingRegressor model, which has an RMSE of 26485.66 and remains within acceptable limits. The RMSE for the RandomForestRegressor model is 48376.43, while the RMSE for the Linear Regression model is 42107.62, both of which are considerably higher than those of the other models. The SVR model performs the poorest in terms of prediction accuracy, having the greatest RMSE of 94461.61.

Comparing all different models, the StackingRegressor model in Tables 4 and 5 has the lowest root-mean-squared error (RMSE), 10089.61. Figure 4 shows the graphical description of RMSE values. This shows the StackingRegressor model has the best prediction accuracy of all other models.

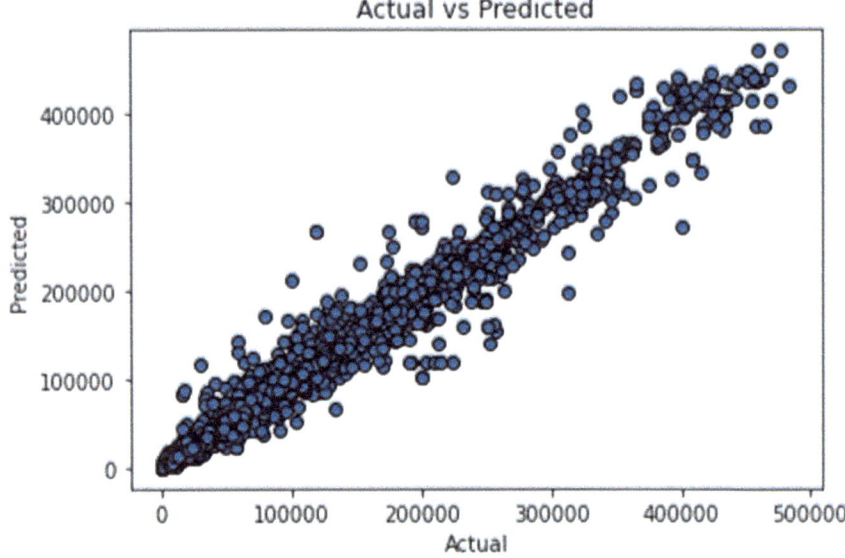

Fig. 5 Regression curve actual versus predicted for stacking model

Table 6 Individual machine learning model R2 scores

Model name	R2 score
SVR	−0.20
RandomForestRegressor	0.68
Linear Regression	0.76
GradientBoostingRegressor	0.90
DecisionTreeRegressor	0.96

Table 7 Ensemble model R2 score

Ensemble model	Base models	Meta-model	R2 score
lStackingRegressor	SVR	lGradientBoostingRegressor	l0.98
	RandomForestRegressor		
	Linear Regression		
	GradientBoostingRegressor		
	DecisionTreeRegressor		

6 Conclusion

Stacking generalization models using SVR, RandomForestRegressor, LinearRegressor, GradientBoostingRegressor, and DecisionTreeRegressor as base models and GradientBoostingRegressor as a meta-model have a greater performance compared to other models, according to the study's findings. The stacking generalization model had a higher R2 score (0.98), compared to other models. This suggests that, in comparison to other models, the stacking generalization model is more appropriate for crop yield prediction. The prediction accuracy and overall performance of the model have been enhanced by the use of GradientBoostingRegressor as the meta-model and the merging of multiple base models in the stacked generalization model. According to these results, the stacked generalization model, which combines base models and a meta-model, offers an effective approach to predicting crop yields. Decision-makers in the agricultural sector can utilize this method to enhance their decision-making process.

References

1. Smith P, Martino Z, Cai D (2007) Agriculture in climate change 2007: mitigation
2. Samtiya M, Aluko RE, Dhewa T (2020) Plant food anti-nutritional factors and their reduction strategies: an overview. Food Prod, Process Nutr 2:1–14
3. Klocker F, Bernsteiner R, Ploder C, Nocker M (2023) A machine learning approach for automated cost estimation of plastic injection molding parts. In: Cloud computing and data science, pp 87–111
4. Namasudra S, Dhamodharavadhani S, Rathipriya R (2021) Nonlinear neural network based forecasting model for predicting covid-19 cases. In: Neural processing letters, pp 1–21
5. Rashid M, Bari BS, Yusup Y, Kamaruddin MA, Khan N (2021) A comprehensive review of crop yield prediction using machine learning approaches with special emphasis on palm oil yield prediction. IEEE Access 9:63 406–63 439
6. Yao A, Di L (2021) Machine learning-based pre-season crop type mapping: a comparative study. In: 2021 9th international conference on agro-geoinformatics (Agro-Geoinformatics). IEEE, pp 1–4
7. Reddy DJ, Kumar MR (2021) Crop yield prediction using machine learning algorithm. In:2021 4th international conference on computing and communications technologies (ICCCT), pp 611–616
8. Hussain N, Sarfraz MS, Sattar S, Riaz S (2022) Predict the crop-yield through uav using machine learning a systematic literature review. Int Conf IT Ind Technol (ICIT) 2022:1–6
9. Khaki S, Wang L (2019) Crop yield prediction using deep neural networks. Front Plant Sci 10:621
10. Khaki S, Wang L, Archontoulis SV (2020) A cnn-rnn framework for crop yield prediction. Front Plant Sci 10:1750
11. Terliksiz AS, Altÿlar DT (2019) Use of deep neural networks for crop yield prediction: a case study of soybean yield in lauderdale county, alabama, USA. In: 8th international conference on Agro-Geoinformatics (Agro-Geoinformatics). IEEE, pp 1–4
12. Suresh N, Ramesh N, Inthiyaz S, Priya PP, Nagasowmika K, Kumar KVH, Shaik M, Reddy B (2021) "Crop yield prediction using random forest algorithm. In: 2021 7th international conference on advanced computing and communication systems (ICACCS), vol 1. IEEE, pp 279–282

13. Shook J, Gangopadhyay T, Wu L, Ganapathysubramanian B, Sarkar S, Singh AK (2021) Crop yield prediction integrating genotype and weather variables using deep learning. Plos one 16(6):e0252402
14. Terliksiz AS, Altýlar DT (2019) Use of deep neural networks for crop yield prediction: A case study of soybean yield in lauderdale county, alabama, usa. In: 2019 8th international conference on agro-geoinformatics (Agro-Geoinformatics), pp 1–4
15. Paudel SMGS, Nakarmi R, Giri P, Karki SB (2022) Prediction of crop yield based-on soil moisture using machine learning algorithms. In: 2022 2nd international conference on technological advancements in computational sciences (ICTACS), pp 491–495
16. Pavlyshenko B (2018) Using stacking approaches for machine learning models. In: IEEE second international conference on data stream mining and processing (DSMP). IEEE, pp 255–258
17. Biau G, Cadre B, Rouvière L (2019) Accelerated gradient boosting. Mach Learn 108:971–992
18. Pant J, Pant R, Singh MK, Singh DP, Pant H (2021) Analysis of agricultural crop yield prediction using statistical techniques of machine learning. Mater Today: Proc46:10 922–10 926
19. Chen H, Liu W (2017) A comparative study of evaluation metrics for regression analysis. J Chem 2017:1–7

Pest Detection Using YOLO V7 Model

Santosh Jayanth Amara, S. Yamini, and D. Sumathi ⓘ

Abstract There is a lot of research going on in the agriculture business right now to create new medications or insecticides to preserve crops. However, this leads to the blind use of insecticides to crops without identifying insects based on their benefits. In the realm of agriculture, there are two sorts of insects: pests and non-pests. Pests are known to harm crops or degrade the environment in which crops thrive, but non-pests may hunt pests, which is beneficial and accomplishes the work without the need of pesticides. The objective of this work is to use the best model for the object detection. This work uses YOLO v7 model as it stands to be one of the best models crossing Mask R-CNN. The model helps in recognizing the pests more accurately and distinguishing them from regular insects. YOLO v7 has enhanced the model by obtaining higher accuracy and reducing the mean square error. The significance of the model lies in achieving the accuracy and thus the model could act as a tool for the farmers to take necessary action. The performance metrics obtained through this model has outperformed the other models.

Keywords Pests · Beneficial insects · Ecological cycle · Pest detection models · YOLO v7 · Insect identification · Deep learning models · Image classification · Object identification

1 Introduction

The detection of pests not only aids in the protection of crops but also plays a crucial role in maintaining the balance of the ecological cycle. In agricultural areas, there are two types of insects that may be found: beneficial insects and harmful insects. While harmful insects can cause significant damage to crop, beneficial insects are

S. J. Amara · D. Sumathi (✉)
School of Computer Science and Engineering, VIT-AP University, Amaravati, India
e-mail: sumathi.research28@gmail.com

S. Yamini
Doctoral student-Phd in aquatic science and technology, National kaohsiung university of science and technology, nanzih campus, Kaohsiung city, Taiwan

© The Author(s), under exclusive license to Springer Nature Singapore Pte Ltd. 2024   225
S. Namasudra et al. (eds.), *Data Science and Network Engineering*, Lecture Notes in Networks and Systems 791, https://doi.org/10.1007/978-981-99-6755-1_17

far more useful than most people realize. In fact, according to the National Pesticide Information Centre (NPIC), around 95% of insects found in crops are beneficial.

Unfortunately, widespread pesticide spraying can lead to the destruction of these beneficial insects and their behaviors, which can have significant negative impacts on the ecological cycle. These helpful insects are often referred to as "Natural Enemies," and they play important roles in maintaining the balance of the ecosystem. Natural enemies include predators, parasitoids, and pollinators, and they help to control pest populations, pollinate crops, and maintain the health of the ecosystem.

The use of pest detection models like YOLO v7 can help to reduce the need for widespread pesticide spraying by providing targeted and precise pest management strategies. By identifying and targeting only the areas where pests are present, these models can reduce the risk of harm to beneficial insects and help to maintain the balance of the ecosystem. This can have significant environmental and economic benefits, as well as promote sustainable agriculture practices. The detection of pests is not only important for protecting crops but also for maintaining the balance of the ecological cycle. By reducing the need for widespread pesticide spraying and targeting only the areas where pests are present, we can help to protect the beneficial insects that play important roles in maintaining the health of the ecosystem.

Predators are those that assist in the removal of pests from fields by killing or devouring the pests there. Parasitoids deposit their eggs or produce larvae that subsequently prey on pests, making them the pre-stage of predators. Pollinators, as the name suggests, are insects that aid in pollination. These also aid in the capture of tiny soft-bodied bugs as nourishment for their younglings when nectar-hunting.

To take use of all the benefits of Beneficial Insects, this research focuses on identifying whether an insect is harmful utilizing YOLO v7 model from Deep Learning models. YOLO v7 is one of the most recent technologies used to improve and enhance any computer-vision system. Transformer encoder block has been used along with the mixture of some training techniques, which is used to enhance Image Classification or Object Identification (Table 1).

2 Literature Review

Pest detection has been utilized for a variety of applications, one of which is [1], which has the similar functionality to identify pests but solely focuses on thrips detection using the SVM classification approach. In [2], they applied several computer-vision approaches to extract the density of pests found mostly in rice fields, in accordance

Table 1 Train data and test data—split up

	Entries	Attributes
Train	1752	5
Validation	243	5
Test	505	5

with the Pest Detection concept. The intensively investigated data at any greenhouse can be valuable in early identification of pests that are detrimental to the greenhouse plants, as done in [3], who employed Integrated Pest Management (IPM) to use biological and physical methods to fight the bio-aggressors while lowering pesticide exposure. The authors of [4] employed a Wireless Sensor Network to identify pests and assess the efficacy of pest management strategies used by farmers in underdeveloped nations. Crops may be safeguarded based on their health condition, which can be identified using Artificial Nose, as they did in [5], the Volatile Organic Compounds (VOCs) derived from plants, which can be detected using the Artificial Nose System to examine the plant's health status, growth, and resistance level.

Based on the research articles discussed above, we may conclude that reducing pesticide use in areas where agriculture is a key industry is a more crucial and critical step. However, regular plant analysis can aid in stabilizing output growth and agricultural goods. In [6], they assessed disease metrics integrating transfer learning with deep learning approaches to focus on deep feature extraction in order to deliver precision farming. Because the data necessary for this pest detection may also be leveraged for pest management, as described in [7], they used the Multi-class Pests Dataset 2018 (MPD2018) to identify and categorize the pests through using PestNet technique.

All of the publications that were mentioned were centered on finding pests that were active in the fields, but there aren't many studies that look for bugs that had already fallen into one of their traps. One such example is [8], where the pests were captured using traps and afterward identified using Deep Convolutional Neural Network (CNN) to identify caught insects before being classified. One effective method to identify pest location is to utilize stereo vision to distinguish between the colors of the bug and the plants. This method was employed in [9], where researchers used binocular stereo vision to determine the pest's 3D location before having a robot spray that region or position. In [10], they conducted a survey on several strategies for spotting pests in agricultural settings. They also defined a number of heterogeneous pest illnesses and certain pest detection techniques in an effort to warn about the pest disease before it begins to infect the crops.

We might infer from all these research publications that there are several areas in which we can examine or defend the pest dataset. Most notably, the research report [7] used the dataset to identify and categorize pests, which is quite similar to what we did in our study. However, this study does identify the pest according to its usefulness or harm to the plant or field, and, if necessary, sprays pesticides to treat the problem.

Certainly! "Automated Pest Detection in Citrus Orchards Using Deep Learning" [12] by Grewal et al. [7] proposes a deep learning approach to detect citrus pests in orchards using Faster R-CNN. The study achieved high accuracy in detecting various pests, including Asian citrus psyllids and citrus leaf miners, with an average precision of 98.5%. The approach used a dataset of 10,000 images of citrus trees, which were annotated with bounding boxes around the pest insects. The study found that the Faster R-CNN model outperformed other state-of-the-art object detection models in detecting pests in citrus orchards. In comparison to my approach of using YOLOv7 to detect pests with an accuracy of 90%, the results achieved by the Faster

R-CNN model in the study by Grewal et al. are higher. However, it is important to note that the accuracy of deep learning models can vary depending on several factors, including the size and quality of the dataset, the complexity of the pest species, and the training parameters of the model. Therefore, it is possible that my approach may achieve similar or higher accuracy with further optimization and evaluation on a larger and more diverse dataset.

"Detection of Tomato Pests Using Deep Learning"[13] by Singh proposes a deep learning approach to detect tomato pests using YOLOv3 model. The study achieved high accuracy in detecting tomato pests, including whiteflies, thrips, and mites, with an average precision of 96.5%. The approach used a dataset of 3,000 images of tomato plants, which were annotated with bounding boxes around the pest insects. The study found that the YOLOv3 model outperformed other state-of-the-art object detection models in detecting tomato pests. In comparison to your approach of using YOLOv7 to detect pests with an accuracy of 90%, the results achieved by the YOLOv3 model in the study by Singh et al. are higher. However, as mentioned earlier, the accuracy of deep learning models can vary depending on several factors, and it is possible that your approach may achieve similar or higher accuracy with further optimization and evaluation on a larger and more diverse dataset.

3   Proposed Model

3.1   Dataset

In this paper, the authors have worked upon Pests Dataset obtained from Roboflow Website [11]. The dataset contains images of following pests:

- Aphids—502 Images.
- Cutworm—505 Images.
- Leaf miners—489 Images.
- Slug—508 Images.
- Whiteflies—496 Images.

The dataset also includes images, which contain multiple attributes in a single image. A total of 2,500 images in the dataset are divided into train, validation, and test as follows.

These images include the augmented images obtained from the original training dataset. The augmentation techniques used to increase the dataset size were

- Flipping: Horizontal and Vertical.
- Crop: 10% zoom—50% zoom.
- Rotation: Between –30° and + 30°.
- Grayscale: Applied to 25% of images.

Table 2 Sample dataset images

Pest category	Image
Alphids	See Fig. 1
Cutworms	See Fig. 2
Leaf miners	See Fig. 3
Slugs	See Fig. 4
Whiteflies	See Fig. 5

Aphids are tiny, smooth insects that absorb nectar or juices from plants, and a cluster of them can stunt crop growth. Cutworms are basically the larvae of moth who come out of soil only in dark and attack seedlings by cutting their stem to feed on them. Leaf miners are larval stage of many species including moths, flies, and beetles, during which they eat the leaf tissues of a plant. Slugs are naturally shell-less mollusk, which often eat leaves, stems, and flowers leaving holes. Whiteflies are one of the Hemiptera, which feed on the lower/under side of the leaves (Table 2).

3.2 Model

The model we used for training this dataset is YOLO (You Only Look Once) v7. As for the present rising technologies, YOLO models compared to other manual models are faster, accurate, and simpler. Among all the YOLO models, at present the best

Fig. 1 Alphid

Fig. 2 Cutworm

Fig. 3 Leaf miner

Fig. 4 Slug

Fig. 5 Whiteflies

stable version is v7. The architecture of YOLO v7 is as follows: YOLO employs a mono convolutional neural network to predict bounding boxes and class probabilities in a single evaluation of the entire image. YOLO predicts multiple bounding boxes, the class probabilities for each box, and all the bounding boxes across the classes in one step and for one unit, making it the one stage detection model, as opposed to earlier object detection models, which accurately identify objects and images by using regions of the image with high probabilities of contender, YOLO considers the full image.

Stepwise YOLO model functionality:

1. Every image is divided into "n" number of square grids of equal size.
2. All these grids are used to detect and localize the objects present in them.
3. Each bounding box coordinates are stored along with the object label and the object's existence probability score.
4. This will then lead to the second stage, where the bounding boxes are overlapped.
5. Here, YOLO comes with Non-Maximal Suppression to eliminate all bounding boxes with comparatively low probability scores.
6. It follows the process of comparing the probability scores with the particular decision and choose the highest score.
7. Then, it removes the largest common bounding box with the high probability bounding box. This process is repeated until we get the final bounding box remaining.

The CNN in YOLO v7 is divided into multiple regions, each of which predicts bounding boxes and probabilities for objects within those regions. The bounding boxes are used to define the location of an object within an image, and the probabilities are used to indicate the confidence that the object is present in the bounding box. The network uses anchor boxes at each location in order to improve the accuracy of the bounding box predictions. Once the bounding boxes and probabilities have been predicted for each region, the model applies non-maximum suppression to adjust the boxes and eliminate any redundant or overlapping predictions. This helps to improve the accuracy of the overall object detection. Overall, the architecture of YOLO v7 is designed to be fast and accurate, making it a popular choice for real-time object detection tasks.

4 Result and Discussion

The results achieved with the YOLO v7 model were promising and exceeded our expectations. The precision of 83% and recall of 75.3% indicate that the model is performing well and is neither overfitted nor underfitted. This means that the model is able to accurately detect pests in a variety of conditions, without being overly sensitive or specific. The Mean Average Precision (mAP) score of 79.5% is also a good indication of the performance of the model. The mAP score is a measure of how well the model is able to detect objects of different sizes and shapes, and a high score indicates that the model is able to accurately detect a wide range of pests.

The training results are shown in Fig. 6, which provides a breakdown of the precision, recall, and mAP scores for each individual pest. As noted, the model performed particularly well in detecting Leaf miners, followed by Whiteflies, Slug, Aphids, and Cutworms. This information can be useful in developing targeted pest management strategies that focus on the most prevalent and damaging pests.

Overall, the results achieved with the YOLO v7 model demonstrate the potential for using deep learning techniques in pest detection and management. While there is always room for improvement and further evaluation, these results provide a strong foundation for future research and practical applications of the model.

The evaluation results are as follows.

The custom evaluated results presented in Figs. 7, 8, 9, 10, and 11 provide further evidence of the YOLO v7 model's ability to accurately detect pests in various conditions. Figure 7 is an image of an Aphid detected by the model with an 80% confidence rate, demonstrating the model's ability to detect even small and subtle pests.

Figure 8 shows an image of a whitefly colony, a pest that can be particularly difficult to detect due to their small size and tendency to cluster together. The model is able to accurately detect the whiteflies with precise bounding boxes, although the confidence rates differ due to box overlapping.

Figure 9 is an image of two slugs feeding on a leaf, and the model is able to detect the pests with relatively high confidence rates of 82% and 90%. This demonstrates the model's ability to detect pests even in complex and crowded environments.

Class	Images	Labels	P	R	mAP@.5	mAP@.5:.95:
all	96	210	0.83	0.753	0.795	0.421
Aphids	96	77	0.799	0.567	0.722	0.299
Cutworm	96	22	0.791	0.864	0.781	0.423
Leaf miners	96	16	0.861	0.875	0.898	0.68
Slug	96	19	0.844	0.632	0.672	0.351
Whiteflies	96	76	0.856	0.829	0.903	0.352

Fig. 6 Training results

Fig. 7 Model detecting Aphid

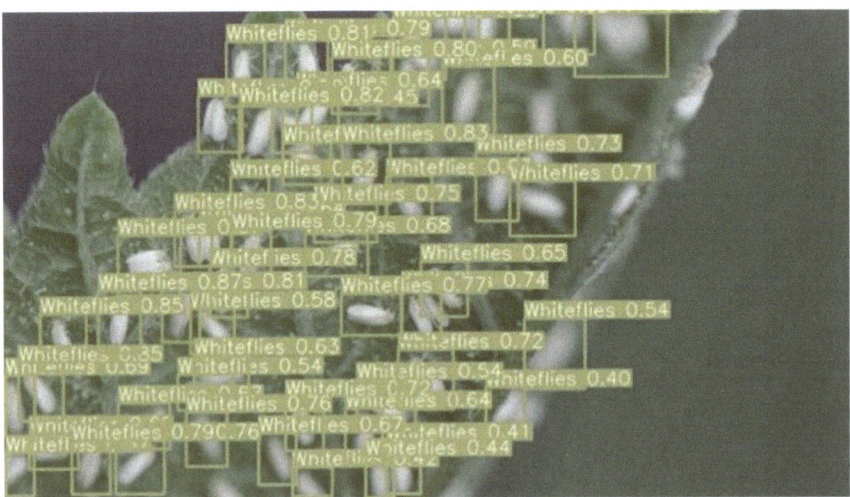

Fig. 8 Model detecting Whiteflies

Figure 10 shows a cutworm resting on a leaf, a pest that can be difficult to detect due to its camouflage and blending in with the surroundings. However, the model is able to detect the cutworm with an 83% confidence rate, demonstrating its ability to detect pests even in challenging conditions.

Fig. 9 Model detecting Slugs

Fig. 10 Model detecting Cutworm

Figure 11 shows a leaf miner absorbing juices from the veins of a leaf, a pest that can be extremely difficult to detect due to its diminutive size. The model is able to detect the leaf miner with a confidence rate of 45%, which is lower than the other pests due to the pest's small size, but still demonstrates the model's ability to detect even the most challenging pests.

Overall, the custom evaluated results provide further evidence of the YOLO v7 model's effectiveness in detecting pests in various conditions and demonstrate its potential for practical applications in pest management.

Fig. 11 Model detecting Leaf miner

5   Conclusion and Future Work

Deep learning techniques can be used to assist the farmer in early detection of paddy pests, diseases, and weeds and corrective action, which may help in preventing crop loss. In general, the primary goal is to identify pests in the plants so that the necessary action could be taken at the earliest. Machine learning and deep learning algorithms are used to solve the majority of the difficulties. Among the various models, it has been understood that the YOLOv7 algorithm has completely taken over the object detection field, outperforming all other previous algorithms in terms of accuracy and speed. By inferring faster and accurately than its competitors, YOLOv7 advances the state-of-the-art outcomes in object detection. With its quicker and more resilient network architecture, YOLOv7 offers improved feature amalgamation, more précised object detection, a firmer loss function, and more active label assignment and model training. The implementation of the model for various applications can have significant economic and environmental benefits. For example, the use of a drone equipped with the model to spray pesticides on crops can significantly reduce labor costs and improve the accuracy and precision of pesticide application. This approach can reduce the risk of overuse or underuse of pesticides, which can lead to improved crop yields and quality. In addition, the reduced use of harmful pesticides can have significant environmental benefits, such as reducing the risk of contamination of soil and water systems. This can help to protect the health of ecosystems and promote sustainable agriculture practices. Furthermore, the preservation of beneficial insects can help to maintain a healthy balance in the ecosystem and promote natural pest control mechanisms.

Acknowledgements I would like to thank Vit-AP University for full support, and also I would like to thank Dr. Sumathi for complete guidance and help during the project. Special thanks to Roboflow Website which provides open-source datasets from which we were able to test our work and get the results with.

References

1. Ebrahimi MA et al (2017) Vision-based pest detection based on SVM classification method. Comput Electron Agric 137:52–58
2. Miranda JL, Gerardo BD, Tanguilig III BT (2014) Pest detection and extraction using image processing techniques. Int J Comput Commun Eng 3.3: 189
3. Boissard P, Martin V, Moisan S (2008) A cognitive vision approach to early pest detection in greenhouse crops. Comput Electron Agriculture 62.2: 81–93
4. Azfar S, Nadeem A, Basit A (2015) Pest detection and control techniques using wireless sensor network: A review. Journal of Entomology and Zoology Studies 3(2):92–99
5. Cui S et al (2018) Plant pest detection using an artificial nose system: a review. Sensors 18.2 (2018): 378
6. Türkoğlu M, Hanbay D (2019) Plant disease and pest detection using deep learning-based features. Turk J Electr Eng Comput Sci 27(3):1636–1651
7. Liu L et al (2019) PestNet: An end-to-end deep learning approach for large-scale multi-class pest detection and classification. IEEE Access 7: 45301–45312
8. Nam NT, Hung PD (2018) Pest detection on traps using deep convolutional neural networks. In: Proceedings of the 2018 International Conference on Control and Computer Vision
9. Li Y, Xia C, Lee J (2009) Vision-based pest detection and automatic spray of greenhouse plant. 2009 IEEE international symposium on industrial electronics. IEEE
10. Nagar H, Sharma RS (2020) A comprehensive survey on pest detection techniques using image processing. In: 2020 4th International Conference on Intelligent Computing and Control Systems (ICICCS). IEEE
11. Pestmodel (2022) Pest Detector Dataset Dataset. Roboflow Universe. Roboflow
12. Grewal PS, Bhatia S, Singh AK (2020) Automated Pest Detection in Citrus Orchards Using Deep Learning. IEEE Access 8:133224–133232. https://doi.org/10.1109/ACCESS.2020.301 7675
13. Singh AK, Singh SK, Kumar P (2020) Detection of Tomato Pests Using Deep Learning. IEEE Access 8:83272–83279. https://doi.org/10.1109/ACCESS.2020.2992279

Random Forest Classifier-Based Acute Lymphoblastic Leukemia Detection from Microscopic Blood Smear Images

Monika Jasthi, Navamani Prasath, Rabul Saikia, and Salam Shuleenda Devi

Abstract Acute lymphoblastic leukemia (ALL) is a cancerous condition which affects bone marrow and blood. It is a fast developing illness that, if not identified and treated as soon as possible, could be fatal. ALL is often identified by hematologists through observing the blood and bone marrow smears under a microscope. In order to diagnose and classify leukemia, sophisticated cytochemical tests are employed. However, such processes are resource-intensive, time-consuming, and reliant on the expertise of the doctors doing them. In order to diagnose leukemia, image processing techniques are used to examine microscopic smear images for signs of cancerous cells. These methods are simple, quick, cheap, and not influenced by the views of specialists. In this paper, a computer-aided automated diagnostic method is proposed to classify ALL and healthy cells based on Random Forest classifier with most significant features. For this model, the public dataset ALL-IDB 2 has been utilized. The proposed approach provided an accuracy of 99.73% to classify the cells (ALL and healthy). Also, it shows an improvement in accuracy of 6.16%, 16.4%, and 10.43% in comparison to the approaches, i.e., morphological + color feature with SVM, Hausdorff dimension + shape feature with SVM, and GLCM + Morphological with SVM, respectively.

Keywords Acute lymphoblastic leukemia · Wiener filtering · K-means segmentation · Feature extraction · Classification

M. Jasthi · N. Prasath · R. Saikia · S. S. Devi (✉)
Department of Electronics and Communication Engineering, NIT Meghalaya, Shillong, Meghalaya 793003, India
e-mail: sshuleendadevi@nitm.ac.in

R. Saikia
e-mail: p21ec003@nitm.ac.in

1 Introduction

ALL is the most prevalent form of malignancy. ALL affects both children and adults. ALL occurs when a bone marrow cell acquires DNA or genetic material mutations. When anything like this takes place, the manufacturing of blood cells becomes completely out of hand. The embryonic cells that are produced in the bone marrow have the potential to develop into lymphoblasts, which are a type of leukemic white blood cell. These abnormal cells are unable to function normally and have the potential to multiply to the point where they crowd out the cells which are healthy [1]. This type of malignancy can be fatal if left untreated or not detected early. Early detection of leukemia increases the likelihood of successful treatment. The high degree of variation in size, shape, location, and edge makes diagnosis difficult. Current procedures are costly and time-consuming. Moreover, these manual procedures are inaccurate, error-prone, and time-consuming owing to human factors including fatigue, tension, and inexperience. Therefore, automated computer-aided systems have been developed to supplant manual methods for detecting the ALL. These systems include several image processing methods and machine learning algorithmic framework.

1.1 Related Works

In order to differentiate normal from abnormal WBCs, Patil et al. [2] proposed ALL discovery which includes pre-processing feature extraction, feature selection, and classification stages. The Discrete Orthonormal S-Transform, also referred to as DOST, which is an multi-resolution approach to classify the texture features of images, employed DOST to extract features, and Principal Component Analyses (PCA) and Linear Discriminant Analyses (LDA) to reduce the number of features. A Random Forest (RF) classifier based on AdaBoost is used to complete the classification step.

Patel [3] presented a process for detecting leukemia using microscopic blood smear images as the data source and employed K-means clustering to identify WBCs, histogram equalisation, and Zack algorithm to classify WBCs, and histogram equalization was used to equalize the histograms. The model was assessed using SVM. Moreover, its accuracy was determined to be 93.57%.

Rawat et al. [4] proposed a categorization method that is helped by a computer for the purpose of making predictions regarding ALL and AML. Using Otsu's approach, the nuclei are extracted from the background of 420 leukocyte cells and are segmented separately. During the classification phase, they used a Genetic Algorithm Support Vector Machine (GA-SVM) and a Radial Basis Kernel (RBK) to analyze each segmented nucleus's 331 individual features. Accuracy in categorization was attained at a level of 99.5% by the model.

Fig. 1 Example of
ALL-IDB 2 dataset images
a ALL, **b** Normal

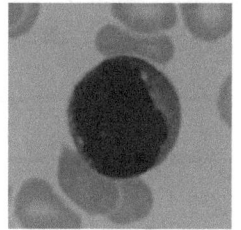 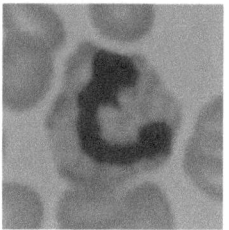

a   b

The approaches illustrated in the literature used different classification techniques for leukemia detection. The goal of this study is to provide a strategy for the classification of blood cells in order to diagnose leukemia based on the following characteristics:

- To remove the unwanted noise the images have been filtered using Wiener filter and to improve the dataset properties augmentation is done.
- Color-based segmentation of augmented WBC.
- GLCM, morphological, and color features extraction from WBC segmented images.
- To rearrange the significant features, the Kruskal–Wallis test is employed.
- Random Forest classifier is used for the classification of the images.

The other parts of the paper are structured as: Sect. 2 illustrates the proposed methodology, Sect. 3 illustrates the results and analysis, and Sect. 4 illustrates the conclusion (Fig. 1).

2 Proposed Methodology

In order for the system to classify an image, it must first acquire the images, then do pre-processing, segmentation, GLCM, morphological, and color-based feature extraction, and then perform the Kruskal–Wallis test. Figure 2 provides a visual representation of the proposed methodology.

2.1 Image Acquisition

The dataset that is publicly available is collected in order to accumulate the microscopic images of blood smears in order to provide them as an input to the system. The PBS images have been collected from the ALL-IDB2 dataset [5]. The images in the collection are stored in TIF format, and their dimensions are 257 by 257. There

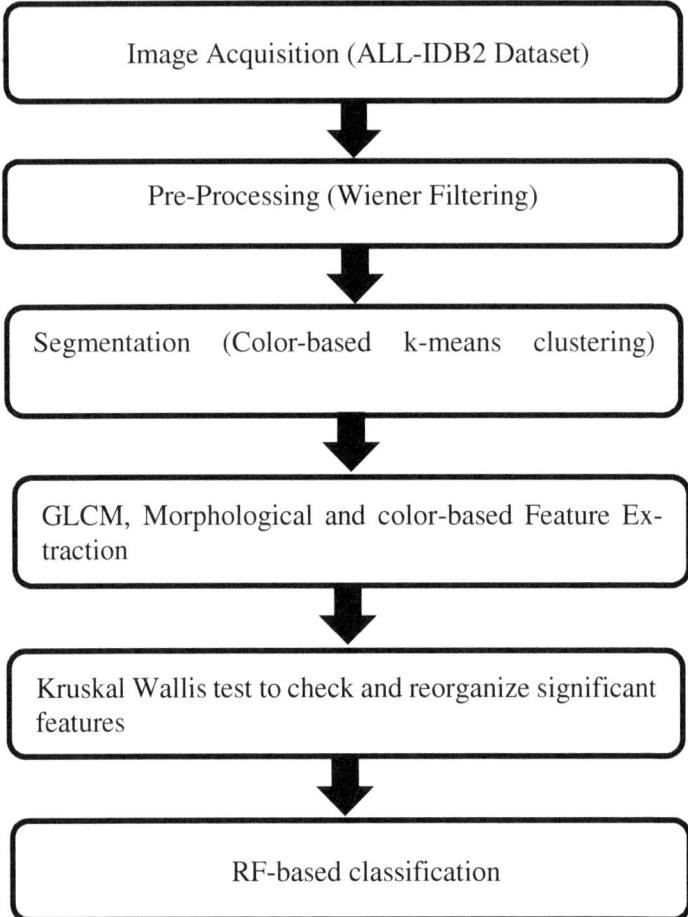

Fig. 2 Proposed methodology

is a total of 260 images, 130 of which depict malignant cells, and the remaining 130 depict non-cancerous cells. Figure 1 represents the ALL-IDB2 [5] dataset Images.

2.2 Image Pre-processing

In this stage, the images are subjected to filtering so that the undesirable noise can be removed and the overall image quality can be improved [6]. In this proposed methodology, a Wiener filter of dimension 3×3 is used. The filtered results are shown in Figs. 3 and 4.

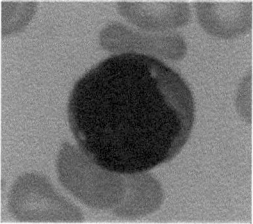

Fig. 3 Original image

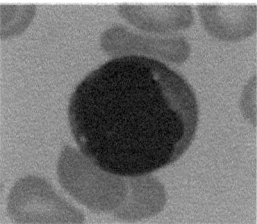

Fig. 4 Wiener filtered

Image augmentation is then applied to the filtered images in order to enhance the dataset. The following improvements are made to the image: (a) flipping it vertically, (b) flipping it horizontally, (c) rotation by 90 degrees, (d) increase of brightness values by 25, and (e) decrease of brightness values by 25 (Fig. 5).

2.3 Segmentation

The Original Image consists of various blood components such as White Blood cells, Red Blood cell platelets, and other components [8]. Segmentation is a process which helps to segment area of interest. In the given proposed methodology, the region of interest is WBC. In order to segment WBC, K-means segmentation is utilized so that the WBC component can be separated from the other blood components of the blood. Color based K-means clustering used for segmentation works on the principle that it keeps on updating the centroids based Euclidean distance of the datas and update the optimal centriod.

Firstly, the RGB image is converted to the CIE LA*B* color space [7]. In this methodology, WBC nuclei are extracted using color-based segmentation. The objective is to classify the blood cells in an image according to their morphology. Each pixel of an object is divided into K distinct clusters. Here, we selected clusters K = 3 that corresponded to the nucleus, which has a high saturation level, the backdrop, which has a high brightness and a low saturation level, and other cells, which have erythrocytes and cytoplasm. Here the region of desire is WBC. The images are

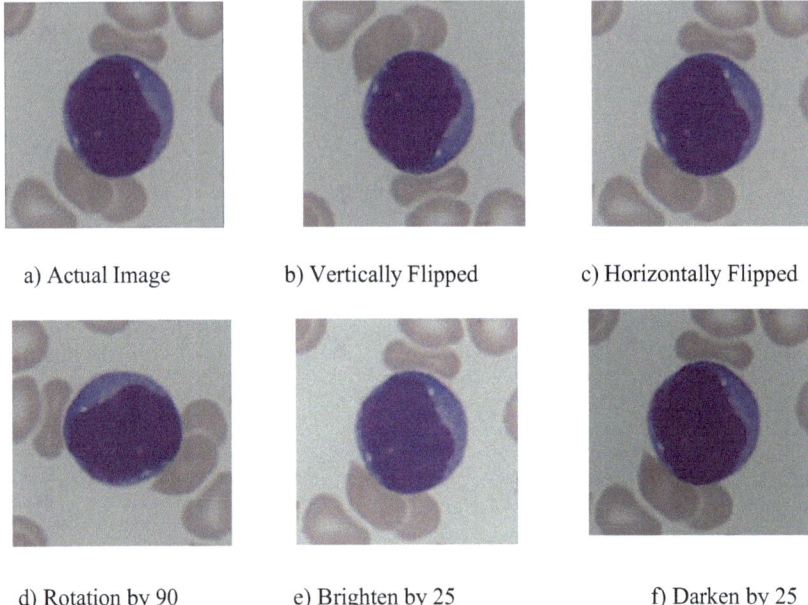

a) Actual Image	b) Vertically Flipped	c) Horizontally Flipped
d) Rotation by 90	e) Brighten by 25	f) Darken by 25

Fig. 5 Augmented images

segmented for both malignant and healthy cells using the K-means algorithm. The segmented results of the proposed methodology are depicted in Fig. 6.

2.4 Feature Extraction

In the process of diagnosing leukemia, feature extraction is an essential stage. The process of discovering and extracting relevant and informative features from images is known as feature extraction. These features, which may be used to classify the various kinds of blood cells and find abnormal cells that may be an indicator of leukemia, can be employed in the process [1].

In this proposed methodology, the features that are extracted are textured features, morphological features, and color features.

- **Texture Features**: To obtain the texture features, a grayscale image of the segmented nucleus was created. On the segmented image, a two-dimensional gray-level co-occurrence matrix (GLCM) was used while extracting the features [11, 14, 15]. A total of 15 GLCM features are extracted in total. They are (a) contrast, (b) correlation, (c) energy, (d) homogeneity, e) angular second moment, f) variance, (g) inverse difference moment, (h) sum average, (i) sum entropy, (j) entropy, (k) difference variance, (l) difference entropy, (m) information measures of correlation-1, (n) information measures of correlation-2.

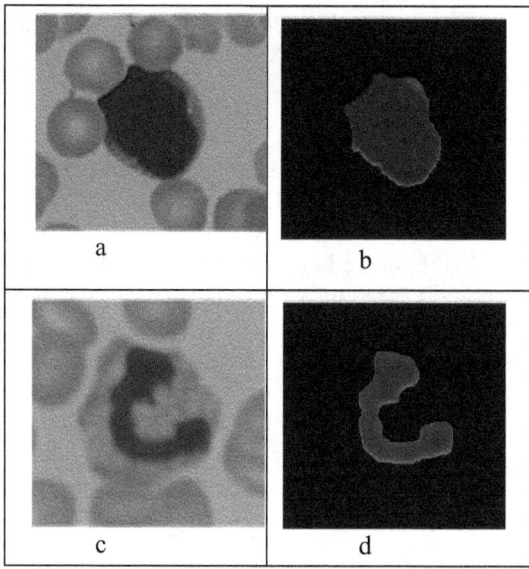

Fig. 6 Segmented images for both cancerous and non-cancerous images

- **Morphological Features:** The image is binary-encoded so that the morphological characteristics may be extracted. Here, we see the nucleus represented by non-zero pixels. Features based on regions and boundaries are retrieved for use in morphological feature analysis. [12]. A total of eight morphological features are extracted. They are (a) area, (b) perimeter, (c) circularity, (d) eccentricity, (e) convex area, (f) solidity, (g) minor axis length, and (h) major axis length.
- **Color Features**: The color features are divided into three types. They are color moment, color histogram, and average RGB. Mean RGB color features are extracted for both malignant and non-cancerous images. It is calculated by taking average of the mean of the red, green, and blue channels, where initially mean was considered as zero.

$$avgRGB = [meanR, meanG, meanB] \tag{1}$$

2.5 Kruskal–Wallis Test to Check and Reorganize Significant Features

The feature selection step has been divided into two distinct phases. In the initial stage, In order to determine the Chi-square value of each characteristic, the Kruskal–Wallis test is carried out, and then those features are sorted based on the Chi-square value, going from highest to lowest [9]. The first feature subset is constructed by selecting the features that have the highest Chi-square values. It is the feature from the second feature subset that has the Chi-square value that is rated second highest that is taken from the second feature subset and added to the acquired first feature set. This process will continue until all of the potential combinations of 26 features have been acquired and all of these combinations will be utilized in the design of the individual classifiers. The combination of characteristics that yields the maximum accuracy is selected as preferred feature set for all of the remaining classifiers. This means that out of all the potential feature combinations, there are 26 subsets to choose from.

2.6 RF-Based Classification

The Random Forest [10] method comes from the supervised learning technique and is used in machine learning. In the discipline of machine learning, it is possible to apply it to issues requiring classification as well as regression provided the appropriate conditions are met. It is predicated on the idea of ensemble learning, which is a method that involves combining a number of distinct classifiers in order to solve a challenging issue and improve the functionality of the model.

A Random Forest has several advantages, including good performance, resistance to overfitting, and the ability to handle high-dimensional datasets. They are relatively easy to use, require minimal feature pre-processing, and can handle both numerical and categorial features. This helps to enhance the predicted accuracy of the dataset. The Random Forest does not rely on a single decision tree but rather aggregates the results of several different trees which compiles the forecasts from all of the trees and makes its projections based on outcome that receives. In this proposed methodology, Random Forest is employed on the extracted features.

During this step of the classification process, a binary classification was done, distinguishing between malignant and non-cancerous cells. For binary classification, 80% of the data was used for training and the remaining 20% was used for testing. An RF classifier with 50 trees was tested and evaluated.

Table 1 Performance of the proposed methodology

Subset	Accuracy (%)
A1	100
A2	100
A3	98.69
A4	100
A5	100
Overall accuracy	99.73

Table 2 Confusion matrix

	ALL	Normal
ALL	765	0
Normal	4	761

3 Results and Analysis

In the proposed methodology, the online available ALL-IDB2 dataset was accessed. The Wiener filtering was performed to enhance the image and to improve the dataset the image augmentation was employed. The image is then segmented using color-based k-means clustering. The segmented image is used to extract GLCM, morphological and color features, a total of 26 features are extracted. To maximize the accuracy the features are verified and reorganized using Kruskal–Wallis test. This was given to a Random Forest classifier to classify the images into cancerous and healthy cells. The performance of the proposed methodology was determined according to the classification's four potential outcomes. Metrics for the performance of binary classification were defined as follows in Eq. (2):

$$Accuracy = \frac{(TP + TN)}{(TP + TN + FP + FN)} \tag{2}$$

The suggested procedure was carried out on MATLAB 2022a, which contained a processor powered by an Intel Core i5-10210U operating at 1.60 GHz and 2.11 GHz, respectively. The obtained results have been displayed in Table 1 which may be found below. Five subsets (A1, A2, A3, A4, and A5) have been taken into consideration for the cross-validation purpose. The overall accuracy is calculated by taking average of the five subsets. The overall accuracy that was obtained is 99.73%.

The confusion matrix of the proposed methodology is given below in Table 2 and the comparative analysis with the existing models is shown in Table 3).

The experimental result shows an improvement in the accuracy.

Table 3 The comparative analysis of the proposed methodology

Methods	Segmentation	Feature Extraction	Classifier	Accuracy (%)
Patel et al. [3]	Zack's algorithm	Morphological + Color	SVM	93.57%
Paswan et al. [13]	K-means clustering	Hausdorff Dimension + Shape	SVM	83.33%
Rawat et al. [11]	Not Mentioned	GLCM + Morphological	SVM	89.3%
Proposed method	K-means segmentation	GLCM + Morphological + Color	RF	99.73%

4 Conclusion

In manual microscopic leukemia diagnosis methods used by pathologists for classifying leukemia is a procedure that takes a lot of time and often results in mistakes. To overcome this problem, in the proposed methodology, a system assisted by computer is developed for detecting the leukemia. This system includes the image processing techniques such as filtering and augmentation to improve the quality of the image dataset. To obtain the region of interest that is WBC, K-means clustering is performed and the WBCs are segmented from the other cell components that are RBC, cytoplasm, and other platelets. From the segmented WBC, the texture characteristics, morphological features, and color features are retrieved. To get the significant features from the obtained features, Kruskal–Wallis test is conducted. After obtaining these features, we send them on to a binary classifier known as the Random Forest classifier. This classifier divides the images into two categories: cancerous and noncancerous. The proposed methodology has obtained an accuracy of 99.73% according to the experimental results. In future, various segmentation methods can be used for segmenting the WBC and its subtypes. Moreover, different classifiers may be trained to classify the different sets.

References

1. Mishra S, Majhi B, Sa PK (2019) Texture feature based classification on microscopic blood smear for acute lymphoblastic leukemia detection. Biomed Signal Process Control 47:303–311
2. Patil S, Rathod PP, Patane S, Patil M (2020) Acute lymphoblastic leukemia detection in human blood using microscopic image. Int. J. Future Gen. Comm. Networking 13:1539–1544
3. Patel, N., Mishra, A.: Automated leukaemia detection using microscopic images. In: Procedia Computer Science, vol. 58, pp. 635–642. (2015).
4. Rawat, J., Singh, A., Bhadauria, H.S., Virmani, J., Devgun, J.S.: Computer assisted classification framework for prediction of acute lymphoblastic and acute myeloblastic leukemia. In: Biocybernetics and Biomedical Engineering, vol. 37, no. 4, pp. 637–654. (2017).

5. Labati, R.D., Piuri, V., Scotti, F.: All-IDB: The acute lymphoblastic leukemia image database for image processing. In: 18th IEEE International Conference on Image Processing, pp. 2045–2048. IEEE, (2011).

6. Tayarani, M.: Applications of artificial intelligence in battling against COVID-19: A literature review. Chaos, Solitons & Fractals. (2020).

7. Kumar, P., Udwadia, S.M.: Automatic detection of acute myeloid leukemia from microscopic blood smear image. In: International Conference on Advances in Computing, Communications and Informatics (ICACCI), pp. 1803–1807. IEEE, (2017).

8. Al-jaboriy SS, Sjarif NN, Chuprat S, Abduallah WM (2019) Acute lymphoblastic leukemia segmentation using local pixel information. Pattern Recogn Lett 125:85–90

9. Devi SS, Laskar RH, Sheikh SA (2018) Hybrid classifier based life cycle stages analysis for malaria-infected erythrocyte using thin blood smear images. Neural Comput Appl 29:217–235

10. Sandika, B., Avil, S., Sanat, S., Srinivasu, P.: Random forest based classification of diseases in grapes from images captured in uncontrolled environments. In: IEEE 13th International Conference on Signal Processing (ICSP), pp. 1775–1780. IEEE, (2016).

11. Rawat, J., Singh, A., Bhadauria, H.S., Virmani, J. :Computer aided diagnostic system for detection of leukemia using microscopic images. In: Procedia Computer Science, vol. 70, pp. 748–756. (2015)

12. Dasariraju, S., Huo, M., McCalla, S.: Detection and classification of immature leukocytes for diagnosis of acute myeloid leukemia using random forest algorithm. Bioengineering, 7 (4), 2020.

13. Paswan, S., Rathore, Y.K.:Detection and classification of blood cancer from microscopic cell images using SVM KNN and NN classifier. Int. J. Adv. Res. Ideas Innov. Technol., 3, 315–324, 2017.

14. Devi, S.S., Singh, N.H., Laskar, R.H.: Performance analysis of various feature sets for malaria-infected erythrocyte detection. In:Soft Computing for Problem Solving, SocProS, vol. 2, pp. 275–283. (2018)

15. Saikia R, Devi SS (2023) White blood cell classification based on gray level co-occurrence matrix with zero phase component analysis approach. Procedia Comput Sci 218:1977–1984

FedCNNAvg: Federated Learning for Preserving-Privacy of Multi-clients Decentralized Medical Image Classification

Charu Chanda, Anita Murmu, and Piyush Kumar

Abstract Federated Learning (FL) permits the cooperative training of a joint model for several medical facilities while maintaining the decentralization of the data owing to privacy considerations. However, Federated optimizations often struggle with the heterogeneity of data dissemination among medical facilities. Nowadays, the domains of medical image classification, compression, and privacy are particularly difficult for diagnosing disease. The transmission of these medical images through the internet for diagnostic reasons must be protected against cyberattacks. In this proposed method, a Federated Learning approach with a Convolutional Neural Network (FedCNN) and Federated Averaging (FedAVG) is employed for classification problems. This technique adjusts the contribution of each data sample to the local goal during optimization based on knowledge of the client's label distribution, thereby minimizing the instability caused by data heterogeneity. The model utilizes a hybrid approach to ensure consistency in time-series data. The datasets, namely, COVIDx-19 X-ray and malaria that are freely accessible are the subject of our in-depth investigations. The experimental results have been analyzed by evaluation metrics, namely, accuracy (78.79 and 98.92), precision (73.72 and 95.73), and recall (71.91 and 93.91) for proper validation. The findings demonstrate that FedCNN achieves better convergence performance than the main FL optimization methods under comparison.

Keywords Classification · Deep learning · Federated learning · Medical imaging · Convolutional neural network

C. Chanda · A. Murmu (✉) · P. Kumar
National Institute of Technology Patna, Patna, India
e-mail: anitamurmu.cs@gmail.com

C. Chanda
e-mail: charuc.pg21.cs@nitp.ac.in

P. Kumar
e-mail: piyush.cs@nitp.ac.in

1 Introduction

The diagnostic and clinical usefulness of medical data is high, but sharing of sensitive information might have serious implications. As a result, there is a rising priority over the safety and confidentiality of patient information. These computations are governed by strict regulations due to the critical qualities of this biological data as well as confidentiality problems and security danger. However, the collection, processing, and storage of data are all subject to stringent regulations enacted under information security regulations. A recently proposed decentralized machine learning architecture called federated learning [3] (FL) enables many data owners to work together to train a model without having to share their raw data. According to the FL paradigm, each data owner trains a local model on its own private data and shares the local model it has learned with other data owners on a regular basis. The FL paradigm may function in a collaborative yet privacy-protecting approach by simply exchanging model parameters without sharing the actual data. In recent years, FL has been frequently employed in medical image [23] analysis due to its benefit of protecting privacy [2]. FL model is different from classical distributed learning, in that, the user data is gathered in a single server before being shared with agents in an independent and identical fashion [3, 4]. One of the biggest obstacles in FL is the statistical diversity of its participants. Since every participant has a local dataset that is highly unique from one another, this causes a variation in the local updates [5], which means that the local aim of every participant is distinct from the global objective. As a result, the aggregated model suffers from poor performance compared to the independent and identically distributed (IID) assumption because of the non-IID data. Several studies [5] have looked at various approaches to reducing the impacts introduced by non-independent and identically distributed data in an effort to enhance FL performance outside of an IID setting.

Lie et al. [6] developed FedProx, influenced by FedAvg, for dealing with heterogeneous data sharing in FL without computational expenses or privacy compromise. Wang et al. [7] have developed an algorithm called FedNova (Federated Normalized Averaging Algorithm) that enables each participant to execute different local training repetitions throughout the circle. Once every modification in the cycle aggregate together, those participants with the most local training will make greater effects on the global model. Kumar et al. [8] developed an algorithm named MediSecFed, a safe platform for federated learning in vulnerable settings. Yan et al. [9] suggested a variation-aware federated learning (VAFL) architecture in which differences among clients are reduced by translating all clients' pictures onto a shared image space. Wicaksana et al.[10] proposed Customized FL (CusFL) which involves each client iteratively training a private model based on a federated global model aggregated from all private models trained in the previous iteration. All data owners are frequently assumed to be able to provide fully annotated data in the traditional FL paradigm [11]. However, this presumption is typically shattered in a segmentation situation since pixel-based annotations require a significant amount of time and knowledge. In medical image segmentation situations, it is more reasonable to assume that each hospital can only offer a small number of labeled data mixed with some additional unlabeled data because of a lack of support [12]. A paradigm that combines FL and CNN is needed to deal with such a situation. FL with CNN has found applications in the domain of medical imaging, including Magnetic Resonance Imaging (MRI) [13] and Computed Tomography (CT) [5]. Additionally, its security has been explored in cloud-based environments as well [14, 22, 23].

Although several researchers have shown the viability of using FL to detect medical images, the typical Supervised FL paradigm may not be effective for classifying medical images. The proposed work uses a FedCNN with FedAVG model to provide security while maintaining decentralization. The deep convolutional-based model improves accuracy and efficiency while solving complex problems of medical imaging. Moreover, a model addresses the classification problems of a higher rate of disease detection. The paper's main contributions are:

- The effective security techniques for medical imaging are limited by the existing models. A novel method is proposed for maintaining the decentralization of the data owing to privacy considerations of medical images to overcome these restrictions.
- In the proposed work, design a FedCNN with the FedAVG algorithm with the help of deep learning to secure the medical image.
- The performance measure of the model is evaluated by using accuracy, precision, and recall on the X-ray and malaria datasets.

There are four sections in this paper. In Sect. 2, the proposed methodology is thoroughly explained. In Sect. 3, the performance of the proposed model is evaluated. In Sect. 4, the article concludes.

2 Methodology

The proposed method is for classification with privacy preservation using FL and a convolutional neural network (CNN) shown in Fig. 1. Despite the data security against unauthorized access, unexpected events may compromise someone privacy. The data derived from medical treatment algorithms must be safeguarded against online threats.

2.1 Dataset Details

This experiment uses the chest COVIDx-19 Radiography Database [14] is publically provided by Italian Society of Medical and Interventional Radiology (SIRM). This dataset contains a total of 9,347 chest X-Ray images with equal instances of covid affected and normal samples. And National library of medicine malaria dataset [15]

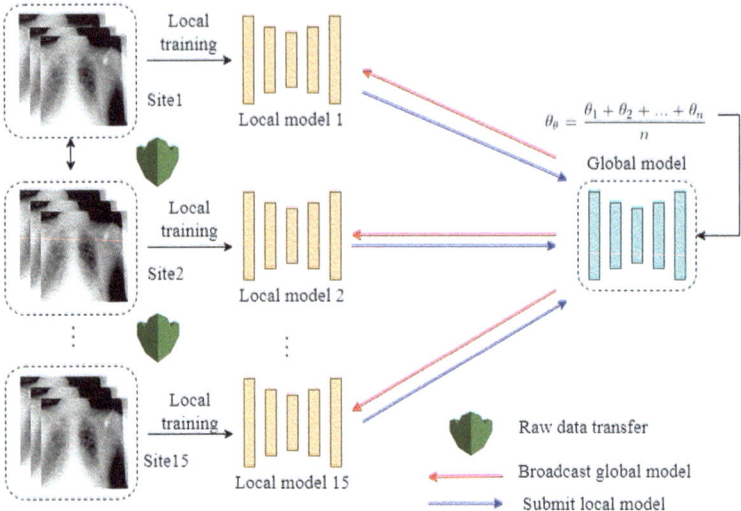

Fig. 1 FedCNN framework with n clients and one global server

is used in the proposed scheme. Malaria dataset contains a total of 27,558 cell images with equal instances of parasitized and uninfected cells. All the images are in Portable Network Graphics (PNG) file format and the resolution is 299 × 299 pixels.

2.2 Federated Learning

The Federated Learning (FL) is to train each participant in a distributed fashion, where the data are retained by each participant and not transmitted to the model [16]. The basic concept of FL architecture is shown in Fig. 2, Algorithms 1, and 2, respectively. The initial action is to distribute a global model to each participant that has knowledge of sensitive data. The local model for each client is then individually trained with its unique set of data. Once training is over, these local models will have weights and parameters. The global model then requests the sharing of the locally trained models. After all the clients are included, the data from the parameters of each unique model is combined using an aggregation procedure. The creation of a global model is done without consulting the data. The training process must be repeated sometimes until the intended outcomes are achieved. Moreover, FL is used for the privacy and security of medical imaging [17].

The primary goal of FL is to minimize the cost function described below across a distributed network of participants. C_k is the total number of clients, d_k is the relative impact of device n where $\sum_{k=1}^{C_k} = 1$ and the $f_k(W)$ is the local objective function of kth client in Eq. (1).

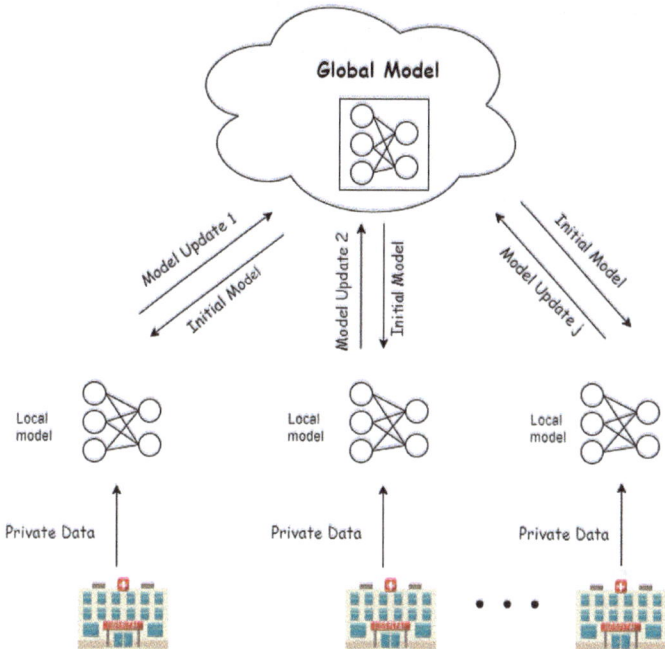

Fig. 2 Federated learning concept

$$\min_{W} F(W), \quad \text{where} \quad F(W) := \sum_{k=1}^{N} d_k f_k(W) \tag{1}$$

In the approach use the Federated Averaging algorithm (FedAVG) for aggregating FL data [18] is a well-known example of an aggregation technique. The model aggregation can be achieved through the use of FedAVG, in which the server updates the global model using a calculated mean of the trained models [19]. To acquire the new weights W_{r+1}^k for the client update phase, FedAvg used clients to execute local processing simultaneously for the E_p epochs. The server updates the global model by running the FedAvg operation described in Eq. (2) after acquiring the new weights from the clients. To prevent decreased efficiency due to unbalanced local data, it is feasible to average the local updates with respect to the local sample counts.

$$W_{r+1}^k \leftarrow \text{Client Update}(W_r, k), \quad \text{and} \quad W_{r+1} \sum_{k=1}^{C_k} \frac{n_k}{n} W_{r+1}^k \tag{2}$$

Algorithm 1: The aggregation thread operation in FL

---

1 [!ht]
    Input : Number of FL iteration is marked as C_k, where k=0
    Output: Clients C_i
2 **for** $k < C_k$ **do**
3 | Send a model θ_r to all clients **for** *each client C_i* **do**
4 | | Wait for receiving model for client
5 | | Calculate a value of loss function by Eq. (4)
6 | **end**
7 | Calculate a loss by use the Eq. (4)
8 | Replace the existing model with the error is calculated.
9 **end**
10 return the client C_i

Algorithm 2: Client operation in FL

---

1 [!ht]
    Input : Iteration number of training T_p, where k=0
    Output: model thread
2 **for** $k < T_p$ **do**
3 | Train the model
4 | k++
5 **end**
6 return the model to the main thread.

2.3 Deep Convolutional-Based Training on FL

In this work, present the later hybrid approach used by the model in order to make the time-series data of the data consistent. This is because the dataset that use has the issue that the time series under three viewpoints have frequency range and cannot be synchronized. The multiview data, where k^p and $\left\{k^p \in R^{dk}\right\}_{p=1}^n$, and m are the views, are set as the output vector at the conclusion of the pth view sequence.

Fully connected CNN layer At first, the simplest method of connecting multiviews: $k = \left[k^1, k^2, ..., k^m\right] \in R^d$. Than one (bi-directional) GRU, d is often equal to $d = 2_{nd}$. The fully connected CNN receives the connected hidden state k through a non-linear function $\sigma(\cdot)$. The input unit's feature interaction mode is as follows Eq. (3):

$$p = ReLU(W^1[k; 1])$$

$$\hat{y} = W_p^2 \tag{3}$$

where $W^1 \in R^{k \times [d+1]}$, $W^2 \in R^{c \times h}$. The constant signal "1" is used to mimic the global bias, while h denotes the number of hidden units. C denotes the number of classes. We merely established a hidden layer in order to streamline the image.

3   Results and Discussion

The suggested federated learning architecture analyzes medical pictures to find instances of malaria and COVID-19. The healthcare facility that records and conducts local training using the threat model-a model offered by the server to all clients-is the client. The FedCNN technique of the Markov chain is used to send the results after local training. Each client's local model results are collected by the Markov chain and combined by the server. The difficulty of managing complex medical data for processing and prediction is the main emphasis of the suggested strategy. The explanation of the experiments being run to verify the proposed strategy is provided below.

3.1   *Experimental Results*

The proposed scheme analyzes data from a x-ray and malaria dataset to determine outcomes. The model is trained using a graphics processing unit, an Intel Core i5 CPU, and 8 GB of Memory (GPU) [20]. The proposed method is implemented using the Keras package, TensorFlow, and Python 3.8.1. The 224 × 224 resolution of the publicly accessible dataset is shown in Fig. 3. All of the samples that have been collected are used for training, and only 20% of the data had been utilized for testing. The Adam optimizer is used to train the models, with a batch size of 16, a learning rate of 0.0001, a dropout of 0.1, and a learning rate of 0.1. The performance graph of epoch verse accuray and loss is shown in Fig. 4 for different number of clients (i.e. 5, 10, and 15).

The results of an experiment assessing the effectiveness of FedCNN models using the X-ray and malaria dataset, also used for brain MRI dataset [21]. It has been suc-

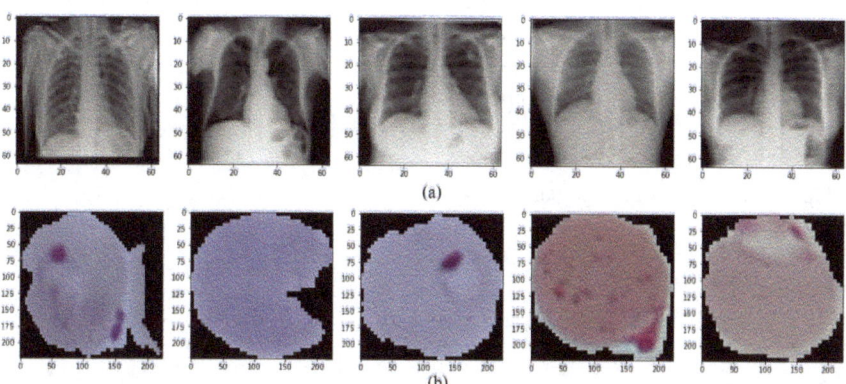

(a)

(b)

Fig. 3 Experimental result of proposed scheme **a** COVIDx-19 **b** Malaria

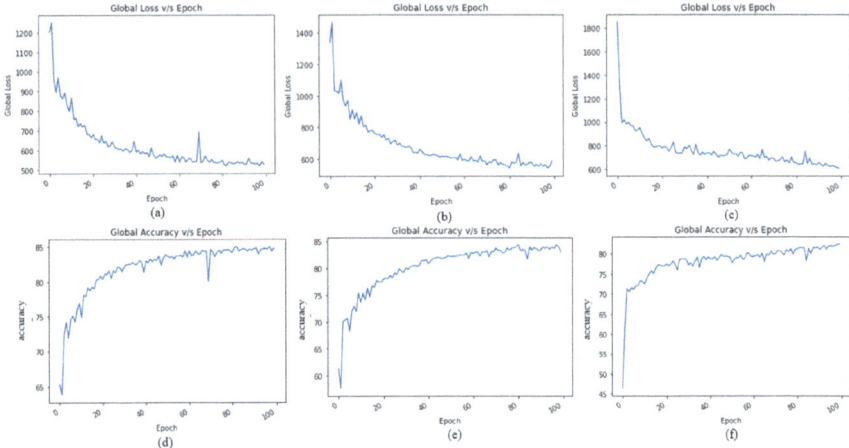

Fig. 4 Performance graph **a**, **c**, **e** loss of 5, 10, 15 clients, and **b**, **d**, **f** accuracy of 5, 10, 15 clients

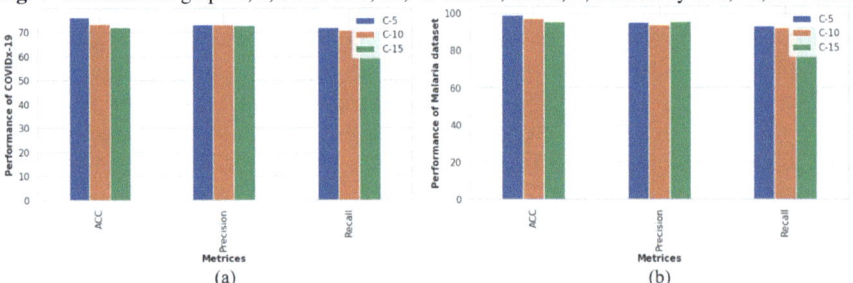

Fig. 5 Performance graph of proposed model outcomes **a** COVIDx-19 **b** Malaria

cessfully utilized to automatically extract features from input picture data using FedCNN and CNN with outstanding competency. The cumulative distribution function of the corrected picture is linearly linked. Security is also carried out throughout the process of training, validation, and testing stages. Moreover, the values of the picture index are consistently computed between [0, 1].

Table 1 The performance of proposed framework

Framework	Dataset	Metrices	Clients			Overall
			C = 5	C = 10	C = 15	
Proposed FedCNN with FedAvg	COVIDx-19	ACC	76.11	73.47	72.05	78.79
		Precision	73.13	73.16	72.72	73.72
		Recall	71.92	70.71	70.82	71.91
	Malaria	ACC	98.72	97.31	95.33	98.92
		Precision	95.11	93.72	91.21	95.73
		Recall	93.27	92.17	92.79	93.91

The performance results are represented by accuracy (ACC), precision, and recall. The suggested research findings are produced by averaging the results for each test picture in each category of images. The suggested method improved malaria and covid detection and diagnosis with security. The results are shown in Table 1, and Fig. 5. and demonstrate that the suggested technique performs better than the current methods in terms of recall, accuracy, and precision. Also, the batch normalization and dropout blocks that are introduced to the model encoder in the proposed deep FedCNN increase the accuracy of the result. By applying the loss function in Eq, comparisons of overlap, similarity, and difference are more illuminating (4).

$$L_{BCE} = -(plog(\widehat{p}) + (1 - p)log(1 - \widehat{p})) \tag{4}$$

3.2 Results Comparison

The metrics that predict FedCNN model results are contrasted with current cutting-edge models. The FedCNN model with FedAVG was suggested to enhance the accuracy of cancer diagnostic prediction. The suggested model works well for all existing FL models, and the method achieves outstanding results in the models that have been assessed. Five dense blocks' worth of feature maps are shared by an encoder block's final layer. The outcomes are the more dependable traits that blend with the output of the encoder block. The fact that various domains contain varied amounts of images makes the suggested domain adaptation issue more challenging.

The proposed generation of the weights of the used model is described by the normal distribution variable. For all models, an Adam optimizer is used. It is essential to classify COVIDx-19 and malaria in order to effectively diagnose and treat cancer. In this process, deep neural networks play a big part. DL performs with a high degree of precision. Future clinical outcomes are determined by the precise identification of the structural components, which are classified. The comparision with the existing state-of-the-art method is shown in Fig. 6, Tables 2, and 3, respectively, of COVIDx-19 X-ray and malaria dataset

4   Conclusion and Future Work

The secure transmission and archiving of medical images is presented using a novel method based on DL-based model with CNN and averaging utilized to built strong security. The original medical images are using a FedCNN and FedAVG for privacy in the presence of different clients. An images are trained, validated by using the CNN model that are combined with FL. Moreover, averaging is used for client and server communication in the decentralized network of FL. Using the basis of a number of matrices namely, accuracy, precision and recall in the proposed technique is assessed on COVIDx-19 x-ray and malaria datasets. The outcomes of the evaluation

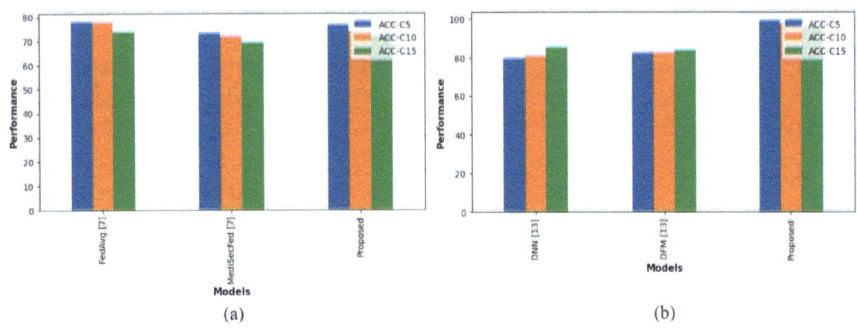

Fig. 6 Comparision graph with existing methods **a** COVIDx-19 **b** Malaria

demonstrate that the proposed scheme performs better than the current models. In the future, the FL is used for real time medical images security in the presence of more number of clients present in the network.

Table 2 Accuracy of the compared models in different number of clients (COVIDx-19)

Dataset	Number of clients	FedAvg [8]	MediSecFed [8]	Proposed
COVIDx-19	5	77.64	72.75	76.11
	10	77.16	71.48	73.47
	15	73.55	69.12	72.05

Table 3 Accuracy compared with existing models in multiple clients (Malaria)

Dataset	Number of clients	DNN [16]	DFM [16]	Proposed
Malaria	5	79.19	81.88	98.72
	10	80.05	81.82	97.31
	15	84.72	83.38	95.33

References

1. Das S, Namasudra S (2022) A lightweight and anonymous mutual authentication scheme for medical big data in distributed smart healthcare systems. IEEE/ACM Trans Comput Biol Bioinform
2. Sharma P, Moparthi NR, Namasudra S, Shanmuganathan V, Hsu CH (2022) Blockchain-based IoT architecture to secure healthcare system using identity-based encryption. Expert Syst 39(10):e12915
3. Hao M, Li H, Luo X, Xu G, Yang H, Liu S (2019) Efficient and privacy-enhanced federated learning for industrial artificial intelligence. IEEE Trans Ind Inform 16(10):6532–42
4. Deist TM, Dankers FJ, Ojha P, Marshall MS, Janssen T, Faivre-Finn C, Masciocchi C, Valentini V, Wang J, Chen J, Zhang Z (2020) Distributed learning on 20000+ lung cancer patients-The Personal Health Train. Radiother Oncol 144:189–200
5. Murmu A, Rahman M, Kumar P (2023) A novel DNA computing with chaos for improving security of medical image. In: Springer International Conference on MIND-2023 [Accepted]
6. Li T, Sahu AK, Zaheer M, Sanjabi M, Talwalkar A, Smith V (2020) Federated optimization in heterogeneous networks. Proc Mach Learn Syst 2:429–50
7. Wang J, Liu Q, Liang H, Joshi G, Poor HV (2020) Tackling the objective inconsistency problem in heterogeneous federated optimization. Adv Neural Inf Process Syst 33:7611–23
8. Kumar A, Purohit V, Bharti V, Singh R, Singh SK (2021) Medisecfed: private and secure medical image classification in the presence of malicious clients. IEEE Trans Ind Inform 18(8):5648–57
9. Yan Z, Wicaksana J, Wang Z, Yang X, Cheng KT (2020) Variation-aware federated learning with multi-source decentralized medical image data. IEEE J Biomed Health Inform 25(7):2615–28 Nov 24
10. Wicaksana J, Yan Z, Yang X, Liu Y, Fan L, Cheng KT (2022) Customized federated learning for multi-source decentralized medical image classification. IEEE J Biomed Health Inform 26(11):5596–607 Aug 15
11. Yu H, Yang LT, Zhang Q, Armstrong D, Deen MJ (2021) Convolutional neural networks for medical image analysis: state-of-the-art, comparisons, improvement and perspectives. Neurocomputing 444:92–110
12. Liu Z, Guo J, Yang W, Fan J, Lam KY, Zhao J (2022) Privacy-preserving aggregation in federated learning: a survey. IEEE Trans Big Data
13. Murmu A, Kumar P (2021) Deep learning model-based segmentation of medical diseases from MRI and CT images. In: IEEE region 10 conference (TENCON). New Zealand, pp 608–613
14. COVID-19 Radiography Database [https://bimcv.cipf.es/bimcv-projects/bimcv-covid19/1590858128006-9e640421-6711]. Accessed 25 March 2023
15. Malaria Dataset [https://lhncbc.nlm.nih.gov/LHC-publications/pubs/ MalariaDatasets.html]. Accessed 25 March 2023
16. Galván E, Mooney P (2021) Neuroevolution in deep neural networks: current trends and future challenges. IEEE Trans Artif Intell 2(6):476–93
17. Rahman M, Kumar P (2022) 2D-CTM and DNA-based computing for medical image encryption. In: Intelligent data engineering and analytics: proceedings of the 10th international conference on frontiers in intelligent computing: theory and applications (FICTA), pp 225–235
18. Konečný J, McMahan HB, Yu FX, Richtárik P, Suresh AT, Bacon D (2016) Federated learning: strategies for improving communication efficienc. In: NIPS workshop on private multi-party machine learning. arXiv:1610.05492
19. Xu X, Peng H, Bhuiyan MZ, Hao Z, Liu L, Sun L, He L (2021) Privacy-preserving federated depression detection from multisource mobile health data. IEEE Trans Ind Inform 18(7):4788–97
20. Kumar P, Agrawal A (2013) Gpu-accelerated interactive visualization of 3D volumetric data using CUDA. Int J Image Graph 13(02):1340003
21. Murmu A, Kumar P (2023) A novel gateaux derivatives with efficient DCNN-ResUNet method for segmenting multi-class brain tumor. Med Biol Eng Comput

22. Devi D, Namasudra S, Kadry S (2020) A boosting-aided adaptive cluster-based undersampling approach for treatment of class imbalance problem. Int J Data Warehous Min (IJDWM) 16(3):60–86. https://doi.org/10.4018/IJDWM.2020070104
23. Namasudra S, Nath S, Majumder A (2014) Profile based access control model in cloud computing environment. Proceeding of the international conference on green computing, communication and electrical engineering. IEEE, Coimbatore, India. pp 1–5. https://doi.org/10.1109/ICGCCEE.2014.6921420

Acute Lymphoblastic Leukemia Detection Using DenseNet Model from Microscopic Blood Smear Images

Navamani Prasath, Monika Jasthi, Rabul Saikia, Muralidaran Loganathan, and Salam Shuleenda Devi

Abstract Acute lymphoblastic leukemia (ALL) is a cancerous condition that affects the bone marrow and blood. It is a fast developing illness that, if not identified and treated as soon as possible, could be fatal. ALL is often identified by looking at blood and bone marrow smears under a microscope. Leukemia can be detected and classified using detailed cytochemical tests. However, these procedures are expensive, time-consuming, and dependent on the knowledge and skills of the specialists involved. Using image processing techniques that examine microscopic blood smear images to search for the leukemic cells, leukemia can be detected. These methods are simple, quick, cost-effective, and unaffected by the judgments of experts. The suggested study describes a computer-aided diagnosis method that uses deep convolutional neural networks (CNNs) that have already been trained to compare leukemia images to normal images. The public dataset ALL-IDB 2 was used for the proposed research. The study uses the pre-trained model DenseNet-201 for performing the classification. With the DenseNet201 pre-trained networks employed in the study for the ALL_IDB2 dataset, a classification accuracy of 94.6% is achieved. In all of the classifications carried out, optimization strategies such as cross-validation, fine-tuning, and real-time augmentation are also compared. Also use Pre-trained series models like ResNet-50, VGG-19, Inceptionv3, MobileNet-v2, Xceptionv3, and VGG-16 for performing the comparison. The experimental result gives an improvement in accuracy (17.76, 10.6, and 13.2%) in comparison to the other approaches namely, residual neural network, customized combined CNN, and conVNet neural network, respectively.

N. Prasath · M. Jasthi · R. Saikia · S. S. Devi (✉)
Department of Electronics and Communication Engineering, NIT Meghalaya, Shillong, Meghalaya 793003, India
e-mail: sshuleendadevi@nitm.ac.in

M. Loganathan
Department of Computer Science and Information System Engineering, St. Joseph University, Dae Es Salaam, Tanzania
e-mail: muralidaran.loganathan@sjuit.ac.tz

© The Author(s), under exclusive license to Springer Nature Singapore Pte Ltd. 2024
S. Namasudra et al. (eds.), *Data Science and Network Engineering*, Lecture Notes in Networks and Systems 791, https://doi.org/10.1007/978-981-99-6755-1_20

Keywords Deep learning · Pretrained neural networks · Leukemia · Acute lymphoblastic leukemia · Computer-aided diagnosis · Convolutional neural network · DenseNet201

1 Introduction

ALL is a fatal condition that causes abnormally high levels of immature leukocytes in the bone marrow and blood. The number of other healthy blood cells decreases as a result of these immature white blood cells (WBCs) [2]. It may be caused by a variety of things including radiation, environmental contamination, a familial history, and many other things. For a successful course of treatment and improved patient outcomes, early detection of ALL is essential. Blood tests, which are used in traditional diagnosis, are intrusive and time-consuming. Using image processing techniques that examine microscopic smear images to search for the detection of leukemic cells, leukemia can be detected. These methods are simple, quick, cost-effective, and unaffected by the judgments of experts. The suggested study describes a computer-aided diagnosis method that uses deep convolutional neural networks (CNNs) that have already been trained to compare leukemia images to normal images [12]. The ALL-IDB 2 dataset, which consists of microscopic images, was used by the researchers as the basis for their study. They used the pre-trained DenseNet-201 model specifically to carry out the classification tasks. The study used the ALL-IDB2 dataset with the DenseNet-201 pre-trained network. Additionally, cross-validation, fine-tuning, and real-time augmentation optimization methodologies were used to improve the performance and resilience of the classification system. A huge dataset of medical images is used to train the CNN model to find patterns that are characteristic of ALL. This strategy may increase the ALL diagnosis's precision and quickness, which would eventually benefit patients [7, 13, 14].

1.1 Related Work

The application of deep learning algorithms to identify acute lymphoblastic leukemia in various medical imaging modalities, blood smears images is the main topic of these following papers and reviews. They examine how deep learning might enhance the precision and effectiveness of ALL diagnosis and provide information on the advantages and disadvantages. The following papers present various deep learning-based approaches for ALL detection with various pre-trained models, while the accuracy may not be as high as some of the previous studies, these approaches still provide useful insights and can potentially be further improved with additional optimization techniques. Liu et al. [8] attempted a method of deep bagging ensemble learning to achieve a precision of 84% in their classification of ALL cells and normal cells. With SE-ResNeXt50 CNN, Prellberg et al. [5] classified B-lymphocytes and B-lymphoblasts with a maximum accuracy of 89.88%. In their study, Tatdow et al. [11] classified lymphoblasts (pre-B and pre-T) using CNN- and ML-based techniques, and they found that the CNN classifier had an accuracy of 81.74%. Using ResNeXt CNN architecture, Matek et al. [7] attempted to classify acute myeloid leukemia cancer cells and normal blood cells and attained a precision of 94% in separating myoblast cells.

An automated ALL detection scheme has been proposed in this paper with the following contribution.

- Pre-trained CNN architecture DenseNet201 has been employed on the ALL-IDB2 dataset to classify ALL and Normal WBC.
- Extracted automated features using pre-trained CNN model DenseNet201 rather than handcrafted features of traditional machine learning models.
- DenseNet201 has also been utilized for the classification task. Moreover, comparison analysis has been performed with several currently used approaches.

The remaining parts of the paper are structured as: Sect. 2 illustrates the proposed methodology, Sect. 3 illustrates the results and analysis, and Sect. 4 illustrates the conclusion.

2 Proposed Methodology

The suggested method for using a pre-trained CNN model for image classification entails pre-processing the input images, extracting important features using a pre-trained model to the particular classification task, training the model, evaluating its performance, and testing its generalization performance on a different test set. The proposed methodology of a pre-trained CNN model in image classification typically involves the following steps:

- Image Acquisition: The required peripheral blood smear images of the proposed methodology are acquired from the ALL-IDB2 public dataset [15]. The dataset is comprised of 130 normal images and 130 ALL images of resolution 257×257.
- Pre-processing: To reduce the computational complexity of the model, this phase entails downsizing the images that are input to a set size, typically smaller than their original size. The images could also be normalized to increase the accuracy of the model.
- Feature Extraction: To decrease the computational difficulty of the model, this phase entails downsizing the images that are input to a set size, typically smaller than their original size. The images could also be normalized to improve the accuracy of pre-trained model.

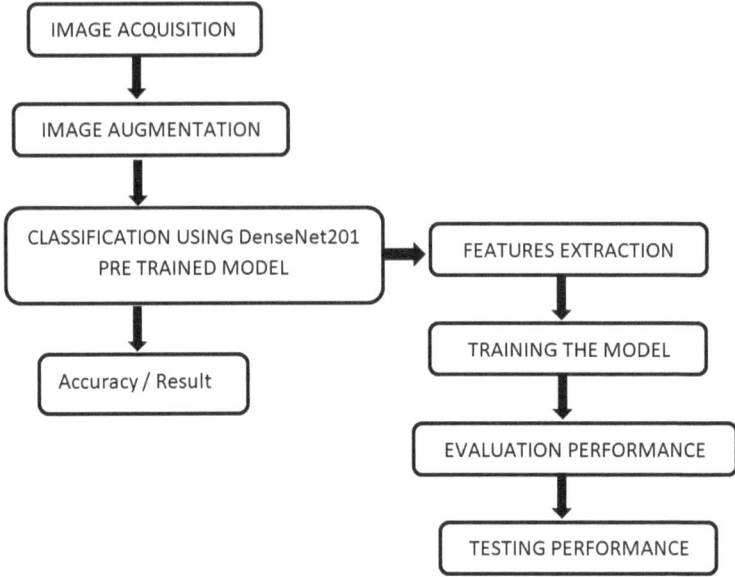

Fig. 1 Proposed methodology

- Training: The pre-trained layers and the newest layers are trained on the training set of images in this step. The goal of the model's optimization process is to reduce the loss function, which calculates the difference between the images' real class labels and predicted class labels.
- Evaluation: After the model has been trained, its accuracy is tested using a different validation set of images. Evaluation metrics to evaluate the performance of the model.
- Testing: In order to assess how well the model applies to new data, it is evaluated on a different test set of images. Figure 1 shows the proposed methodology for ALL detection based on DenseNet-201. Figure 2 shows the sample images from ALL-IDB 2 dataset.

2.1 DenseNet201

In 2017, Huang et al. created the convolutional neural network architecture known as DenseNet201. It is a subset of the DenseNet architecture, recognized for its highly interconnected layers. The 201 layers that make up the DenseNet201 architecture include convolutional, pooling, and dense blocks. DenseNet201 has a special structure that differs from conventional CNN architectures in that each layer is coupled to every next layer in a feedforward manner. Through a concatenation procedure, which enables the output of each layer to be used as an input to all succeeding levels in the

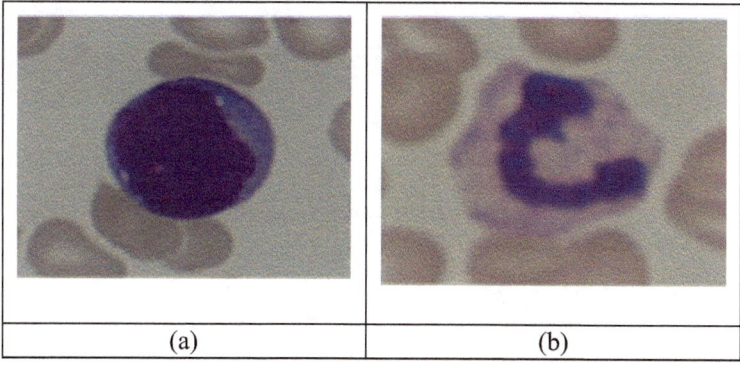

Fig. 2 Sample images of the ALL-IDB 2 dataset **a** ALL **b** Normal

network, this dense connectivity pattern is accomplished [12]. DenseNet201 is more memory-efficient thanks to its densely connected layers, which also increase feature propagation across the network and lower the number of parameters needed for training. DenseNet201 also uses bottleneck layers, which lower the dimensionality of the input feature maps using 1×1 convolutions before using a larger convolutional filter, hence requiring less parameters. On a number of image classification tasks, including the ImageNet dataset, which consists of over a million images belonging to 1,000 classes, DenseNet201 has demonstrated state-of-the-art performance. Object detection, segmentation, and medical image analysis are just a few of the computer vision applications for which its extensive connection and bottleneck layers have been demonstrated to be effective (Figs. 3 and 4).

Similar to other DenseNet models, DenseNet201 features a deep connection structure across layers, with each layer receiving feature mappings from all previous

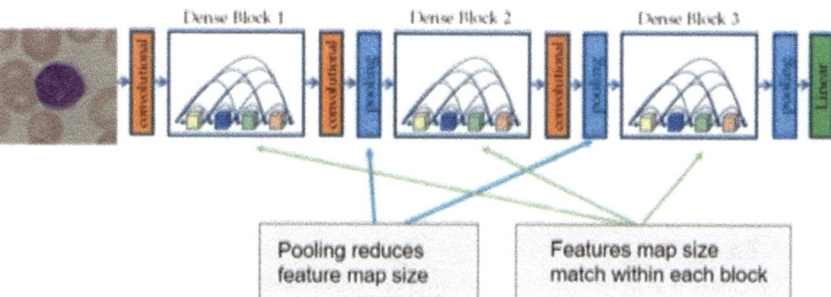

Fig. 3 Working process of dense block and transition

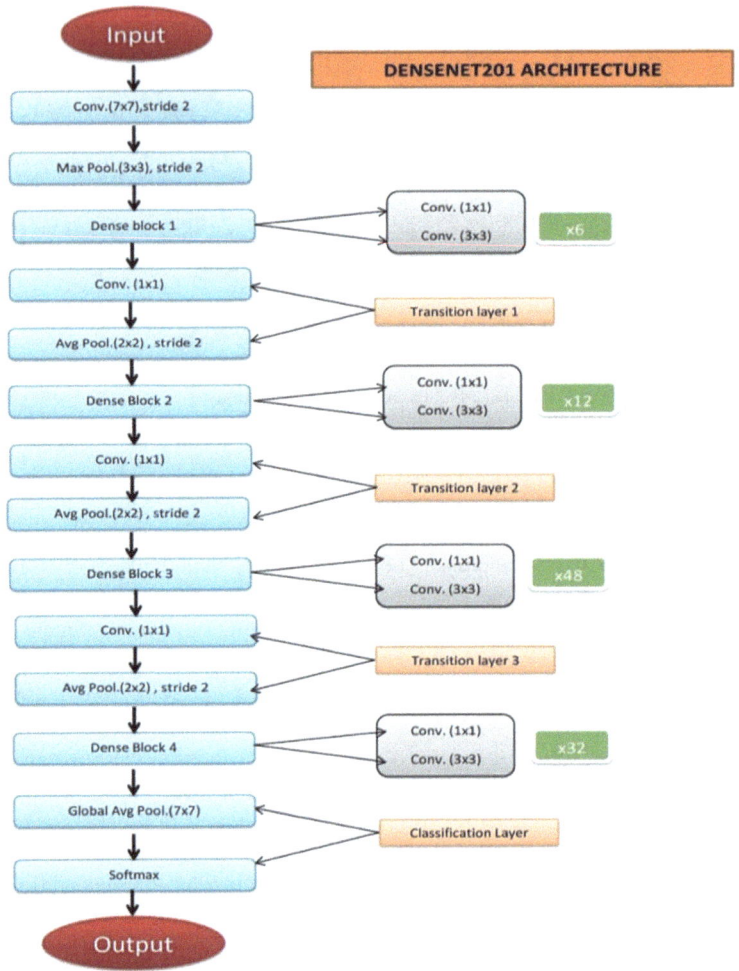

Fig. 4 DenseNet201 architecture

layers in the network. This allows features to be reused and aids the network in learning more precise ways to represent the incoming data. The transition layer that follows each of the four dense blocks in the DenseNet201 design reduces the output's feature map number and spatial dimension. The first block is a convolutional layer with 64 filters, followed by a dense block with 4 layers. Each layer in the dense block consists of a batch normalization layer, a rectified linear activation function (ReLU), and a 3×3 convolutional layer with 32 filters. The second, third, and fourth dense blocks in DenseNet201 have 6, 12, and 48 layers, respectively. The number of filters in each layer of the dense blocks increases with depth, increasing from 128 filters in the second block to a maximum of 1024 filters in the final block.

Table 1 Accuracy and loss value of the DenseNet201 model

Data set	Training accuracy (%)	Validation accuracy (%)	Train loss	Val loss
Subset 1	100	100	7.0875×10^{-8}	0.0052
Subset 2	100	92.31	1.1204×10^{-7}	1.2567
Subset 3	100	96.15	4.3902×10^{-7}	0.5743
Subset 4	100	90.38	1.595×10^{-8}	0.9128
Subset 5	99.36	94.23	0.0185	0.8396
Average	99.872	94.614		

Following the last dense block, there is a globally average pooling layer that averages the feature maps across spatial dimensions and a fully connected layer with a softmax activation function that generates the final class probabilities. DenseNet201 is an effective CNN architecture that has been demonstrated to deliver cutting-edge results on a variety of computer-aided applications, including segmentation, object identification, and image classification (Table 1).

3 Results and Analysis

The classification outcomes were achieved using deep learning DenseNet201 pre-trained model for the dataset ALL_IDB2. A total of 80% of the dataset's images were randomly chosen for training and 20% for testing in all classification trials. This made sure that training purposes were not served by the test images. To obtain a mean average accuracy, 20% of images for testing will be kept changing for the next iteration and left out images were used for training (Figs. 5 and 6).

3.1 Comparative Analysis

Feature extraction from blood smear images for leukemia diagnosis involves the development of a computer-aided system for acute leukemia diagnosis using various deep learning algorithms. In this comparative analysis, we will evaluate the performance of various pre-trained deep learning algorithms.

Based on the results obtained from our experiments, we found that the DenseNet201 pre-trained model achieved a mean average accuracy of 94.6%, while other pre-trained models achieved less than 91%. This indicates that all the deep learnings are effective methods for feature extraction from blood smear images for leukemia diagnosis. However, DenseNet201 outperforms other pre-trained models in terms of accuracy (Tables 2, 3 and 4).

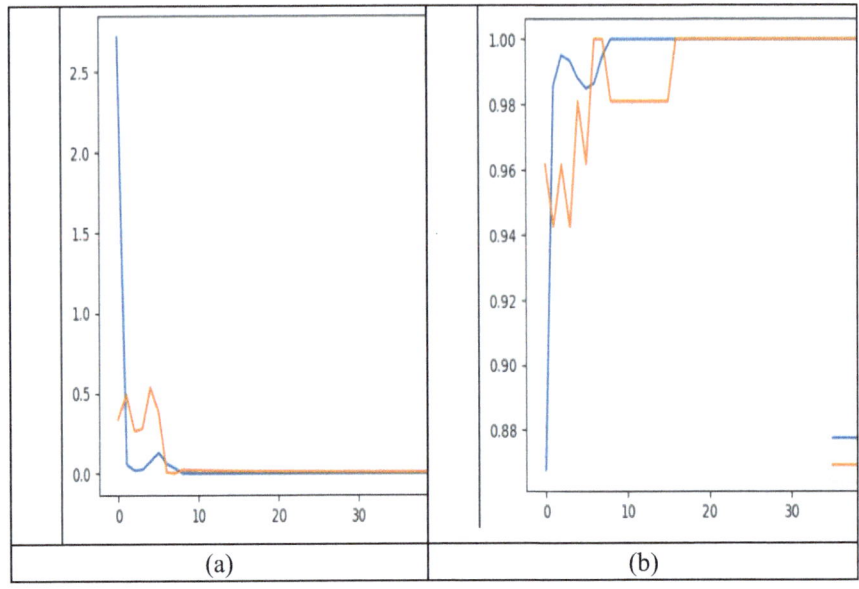

Fig. 5 **a** Graph plotting of loss of DenseNet201 model for dataset ALL IDB-2 **b** Graph plotting of accuracy of DenseNet201 model for dataset ALL IDB-2

	ALL	NORMAL
ALL	26	0
NORMAL	3	23

Fig. 6 Confusion matrix

Table 2 Evaluation metrics

Precision	0.948
Recall	0.942
Accuracy	0.942
F1 Score	0.942
Test accuracy	0.942

Table 3 Comparision with various pre-trained models

Model	Top accuracy (%)	Mean avg accuracy (%)
VGG19	85.50	< 91
InceptionV3	97.20	< 91
MobileNet	94.71	< 91
ResNet50	82.93	< 91
DenseNet121	98.64	< 91
DenseNet201	99.9	94.6

Table 4 Comparative analysis with other works

Author	Dataset	Classifier	Accuracy (%)
Qin et al. [10]	Private dataset	CNN	76.84
Prellberg et al. [5]	ISBI	CNN	89.88
Liu et al. [8]	ISBI	CNN	84
Pansombut et al. [11]	ASH	CNN	81.74
Matek et al. [7]	Munich University Hospital	CNN	94
Proposed	ALL-IDB 2	DenseNet201	94.60

4 Conclusion

In leukemia diagnosis, the microscopic method of classifying leukemia is a time-taking and error-prone process. To overcome this problem, in the proposed methodology a computer-aided system is developed to detect leukemia. We used different CNN pre-trained models to classify ALL and healthy cells from the ALLIDB-2. Based on the results obtained from our experiments, we found that the DenseNet201 pre-trained model achieved a mean average accuracy of 94.6%, while other pre-trained models achieved less than 91%. This indicates that all the deep learnings are effective methods for feature extraction from blood smear images for leukemia diagnosis. However, DenseNet201 outperforms other pre-trained models in terms of accuracy. This suggests that deep learning can be a valuable tool in developing computer-aided systems for medical diagnosis, particularly in tasks that require image analysis.

References

1. Eason G, Noble B, Sneddon IN (1955) On certain integrals of Lipschitz- Hankel type involving products of Bessel functions. Phil Trans Roy Soc London A247, pp 529–551
2. Vogado LHS, Veras R, Andrade A, Araújo FH, Silva R, Aires KK (2017) Diagnosing leukemia in blood smear images using an ensemble of classifiers and pre-trained convolutional neural networks. In: 2017 30th SIBGRAPI Conference on Graphics, Patterns and Images, SIBGRAPI

3. Godlin Jasil SP, Ulagamuthalvi V (2021) Skin lesion classification using pre-trained DenseNet201 Deep Neural Network. In: 2021 3rd international conference on signal processing and communication (ICPSC), pp 393–396
4. Kushwaha S, Adil M, Abuzar M, Nazeer A, Singh SK (2021) Deep learning-based model for breast cancer histopathology image classification. In: 2021 2nd international conference on intelligent engineering and management (ICIEM), 539–543
5. Prellberg J, Kramer O (2019) Acute Lymphoblastic Leukemia Classification from Microscopic Images Using Convolutional Neural Networks. In: ISBI 2019 C-NMC Challenge: Classification in Cancer Cell Imaging
6. Anilkumar KK, Manoj VJ, Sagi TM (2022) Automated detection of B Cell and T cell acute lymphoblastic leukaemia using deep learning. IRBM, 43(5):405–413
7. Matek C, Schwarz S, Spiekermann K (2019) Human-level recognition of blast cells in acute myeloid leukaemia with convolutional neural networks. Nat Mach Intell 1:538–544
8. Liu Y, Long F (2019) Acute lymphoblastic leukemia cells image analysis with deep bagging ensemble learning
9. Albahar MA (2019) Skin lesion classification using convolutional neural network with novel regularizer. IEEE Access 7:38306–38313
10. Qin F, Gao N, Peng Y, Wu Z, Shen S, Grudtsin A (2018) Fine-grained leukocyte classification with deep residual learning for microscopic images. Comput Methods Prog Biomed 162:243–252
11. Pansombut T, Wikaisuksakul S, Khongkraphan K, Phon-On A (2019) Convolutional neural networks for recognition of lymphoblast cell images. Computat Intell Neurosci 12
12. Vogado L, Veras R, Aires K, Araújo F, Silva R, Ponti M, Tavares JMRS (2021) Diagnosis of leukaemia in blood slides based on a fine-tuned and highly generalizable deep learning model. Sensors
13. Chanu MM, Singh NH, Muppala C (2023) Computer-aided detection and classification of brain tumor using YOLOv3 and deep learning. Soft Comput 27:9927–9940
14. Meena K, Veeramakali T, Singh NH (2023) Deep learning techniques for prediction of pneumonia from lung CT images. Soft Comput 27:8481–8491
15. Labati RD, Piuri V, Scotti F (2011) ALL-IDB: the acute lymphoblastic leukemia image database for image processing. In: Proceedings of the 2011 IEEE interntional conference on image processing (ICIP 2011), pp 2045–2048. Brussels, Belgium

A Disease Prediction Framework Based on Predictive Modelling

Harmohanjeet Kaur, Pooja Shah, Samya Muhuri, and Suchi Kumari

Abstract The rise of chronic diseases has become a major public health challenge globally. Early prediction and prevention of these diseases can help reduce their prevalence and improve patient outcomes. The proposed disease prediction system, which is based on predictive modeling, may anticipate the user's illness by using the user's symptoms as input. The framework evaluates the symptoms taken as input by the user and generates the likelihood of developing the disease. The disease prediction framework based on machine learning (ML) techniques can help in a more accurate diagnosis than conventional methods. In the current manuscript, we have designed a disease prediction methodology using multiple ML techniques. The proposed framework also has the potential to enhance disease surveillance and support public health interventions, including disease management and resource allocation. The accuracy of our approach is shown over the benchmark data sets, which consist of more than 230 diseases. The suggested diagnostic algorithm outputs the disease name that a person might be experiencing based on the symptoms taken into consideration. The proposed framework provides a scalable and effective solution for public health decision-makers to manage chronic diseases and improve patient outcomes.

Keywords Predictive modeling · Disease prediction · Machine learning models · Naive Bayes · Decision tree · Random forest

H. Kaur · P. Shah · S. Muhuri (✉)
Computer Science & Engineering Department, Thapar Institute of Engineering and Technology, Punjab, India
e-mail: samya.muhuri@thapar.edu

S. Kumari
School of Computer Science Engineering and Technology, Bennett University, Greater Noida, Uttar Pradesh, India

© The Author(s), under exclusive license to Springer Nature Singapore Pte Ltd. 2024   271
S. Namasudra et al. (eds.), *Data Science and Network Engineering*, Lecture Notes in Networks and Systems 791, https://doi.org/10.1007/978-981-99-6755-1_21

1 Introduction

A disease prediction framework based on predictive modeling is a cutting-edge approach for identifying and preventing the onset of illnesses in patients. The framework employs advanced statistical and computational techniques to analyze a large data set of patient information and generate accurate predictions of disease risk. By combining patient data from multiple sources, including electronic health records, genetic profiles, and environmental factors, the framework can identify risk factors and patterns that may not be apparent through traditional diagnostic methods [11, 12, 18].

The predictive modeling aspect of the framework is based on machine learning algorithms that are trained on historical patient data to identify correlations and patterns that are associated with specific diseases. These algorithms are capable of learning from vast amounts of data and can continually adapt and improve their predictions as more data becomes available. By integrating these algorithms into a framework that can process large amounts of data quickly and accurately, medical professionals can identify patients who are at high risk of developing certain diseases and take proactive measures to prevent or manage their conditions.

One of the key benefits of a disease prediction framework is that it can help medical professionals identify patients who may not yet be exhibiting symptoms of a disease but are still at high risk of developing it in the future. By detecting these patients early, medical professionals can provide targeted interventions and preventative measures to reduce their risk of developing the disease. This can be particularly important for diseases that are difficult to diagnose in their early stages or that have few effective treatments available.

Medical data analysis using machine learning (ML) technology has attracted enormous attention in the last two decades [17]. ML methodology is used to optimize performance using examples or past data. It has two passes, namely the training and testing phases. Through these phases, disease prediction can be made by utilizing the patient's symptoms and history. The vast amounts of data generated in the healthcare industry can be overwhelming for clinicians and researchers to analyze, but ML algorithms can quickly identify meaningful insights from this data.

It is possible that the number of random variables in a problem can be reduced by generating a set of principal variables, known as dimensionality reduction. It involves the representation of variance within the data by incorporating significant information as well. In any experiment, the data may consist of many attributes that are generally not required or whose influence on the output is insignificant. Also, large amounts of data may lead to huge computational complexity and can cause overfitting of the framework. Dimensionality reduction can be made by two methods: feature extraction [19] and feature selection [5]. In the current manuscript, the disease is predicted by analyzing the given symptoms with reduced dimensionality.

The objectives of the paper are many folds, including:

1. Easy to use: The creation of an intuitive and straightforward platform is the major objective of this project. In the system, the user would only supply the patients'

medical information. The employed algorithm would make the decision based on the features retrieved.

2. No human intervention required: The algorithm provides results based on the extracted features. The probability of error is so small that no human intervention is required, which saves patients' and doctors' time and effort.

3. Efficient use of available annotated data samples: It is used to train ML algorithms and requires thousands of annotated training samples. A network is formed through the use of data preprocessing to make more efficient use of the available samples.

2 Dimensionality Reduction

Dimensionality reduction is a popular technique in medical data analysis that involves reducing the number of variables in a dataset while retaining as much information as possible [2]. This is particularly useful in medical research where large amounts of data are generated from diverse sources such as genomic data, clinical data, and imaging data [3]. Dimensionality reduction techniques such as principal component analysis (PCA) [30], t-SNE [13], and UMAP [29] can help researchers visualize and understand complex relationships within the data, identify patterns and subgroups, and ultimately make more accurate predictions. In addition, reducing the number of features can help to overcome the "curse of dimensionality," where models become increasingly complex and computationally intensive as the number of features increases. Also, there can be some redundant and noisy features for the data set that are useless and there is no need to include them. Here, our main motivation is to shrink the functional space. The transformation of high-dimensional data space to low-dimensional feature space is known as dimensionality reduction. Also, it is to be kept in mind that all the important properties of the data are preserved and none of them is destroyed while performing the process of dimension reduction. In data visualization, dimensionality reduction is done so that the data is understood well and interpreted well. It is widely used in ML and deep learning techniques to simplify tasks.

2.1 Feature Extraction

Feature extraction is defined as the process that is used to transform the features so that a new set of meaningful features is formed [19]. The transformation done in feature extraction is often irreversible, and due to this, there is a chance that useful information might be lost in the process. Feature extraction is also used for converting raw data into numerical features while the original data remain preserved.

2.2 Feature Selection

Feature selection is a method that is used to reduce input variables [5]. This is done so that only relevant data is there in the data set and also reduces the noise. A greedy search approach is done on all the results obtained from all possible combinations of traits and is evaluated using criteria.

3 Literature Review

For the last two decades, the range of ML algorithms in predicting various diseases is expanding. Decision trees, SVM, KNN, and Naive Bayes are some Machine learning algorithms that have been used in previously proposed systems. This article focuses on predicting the probability of contracting a disease from the symptoms. To analyze the problem, many latest machine-learning techniques are used to give accurate results.

A comparison chart of the outcomes produced by the previously employed methods is displayed in Table 1.

With the advent of advanced computers, doctors need the help of technology in many ways, including surgical images, robotic arms, and radiographs. It helps doc-

Table 1 Comparative studies of the benchmark results

S.No.	ISSN	Year	Author	Propose	Algorithm used	Acc. (%)
1	21668689	2022	A. Nassif	Prediction of heart disease [4]	SVM, RF, NB	91
2	20712605	2021	R. Mallela	Disease prediction [14]	NB	89
3	19728716	2020	A. Singh	Prediction of heart disease [27]	NB	87
4	19887595	2020	S. Gram-purohit	Disease prediction algorithm [8]	RF, NB	95
5	22780181	2020	A. Rajdhan	Disease prediction [24]	NB	83
6	16950479	2022	M. Chen	ML based heart disease diagnosis [6]	RF	82
7	11188352	2021	K. Reddy	Heart disease risk prediction using ML [21]	KNN	86

tors make the right decisions in work. Pinagle et al. [23] predicted the disease using a Naive Based Classifier based on the user's input with 84.5% accuracy. Vembandasamy et al. [28] also used Naive Bayes with slightly better accuracy of 86.4% by considering some other parameters using the same datum set. Furthermore, Shah et al. [26] considered a combination of 4 algorithms with the highest 90.8% accuracy for ANN models and the lowest 80.3% accuracy for all other models. Ali et al. [1] used a combination of logistic regression, random forest, and ANN, and provided 87.5% accuracy.

The discussed research work on disease prediction is very promising and useful, especially given the current scenario of the amount of data generated per day in the medical field. However, ML prediction is not sufficient to classify the patient's risk level and cannot prevent the worst things from happening. The proposed work is used to predict the likelihood of disease and implement various techniques such as Naive Bayes (NB), Random Forest (RF), and decision trees (DT) to classify a patient's risk level. The system can be used to predict diseases such as diabetes, malaria, jaundice, dengue fever, and tuberculosis.

4 Algorithms and Techniques Used

4.1 Naive Bayes

Naive Bayes is a classification algorithm that is used for binary and multi-class classification [25]. It is a technique that takes some data as input and makes a prediction about it. It is based on the Bayes theorem, which is used to solve classification problems. It is used to determine the likelihood of prior knowledge hypotheses and also handles the data that can be continuous as well as discrete. In real situations, the Naive Bayes classifier works efficiently as it can estimate by using a small data set and gives fast results as compared to other sophisticated methods. The formula for Bayes theorem is given as:

$$P(A/B) = \frac{P(B/A)P(A)}{P(B)} \tag{1}$$

where, $P(A|B)$ = Posterior probability and $P(B|A)$ = Likelihood probability.

4.2 Decision Tree

A decision tree is an algorithm that divides a node into two or more sub-nodes [20]. It is defined as a probability tree that can make decisions about some kind of process. Classification and regression problems both can be solved by using Decision trees.

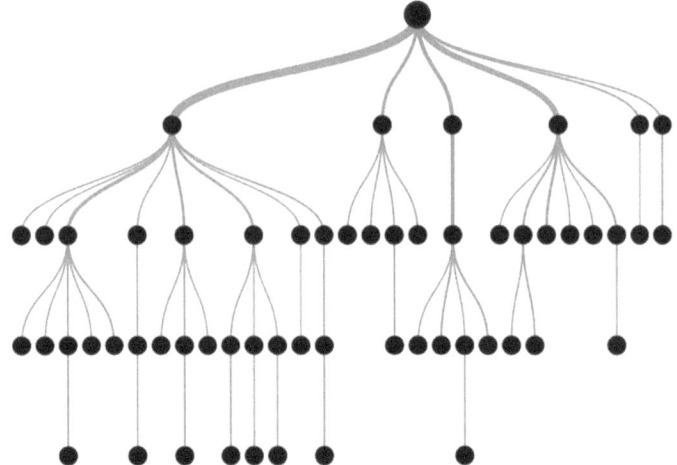

Fig. 1 Decision tree

They can break down complex data into more manageable pieces; therefore, decision trees are very useful for data analysis (Fig. 1).

4.3   Random Forest

Classification and regression problems both can be solved by using random forest [22]. In random forest, multiple classifiers are combined so that the problems can be solved and overall can improve the performance of the proposed system. This method is suitable for data sets where a decision tree is used for training. This method works by combining decision trees such that the output of the decision tree can be derived. Even on large data sets, it runs efficiently and predicts outputs with high accuracy (Fig. 2).

5   Evaluation Methods

To train models, ML uses performance metrics to calculate the performance. To calculate performance scores, different parameters are used. In the confusion matrix, the parameters used are TP, TN, FP, and FN. TP is true positives that are the correctly predicted outcomes, TN is true negatives, i.e., the number of unwanted results, FP is false positives, i.e., the number of results incorrectly predicted to be needed and FN is false negatives, i.e., wrongly predicted result. You can get

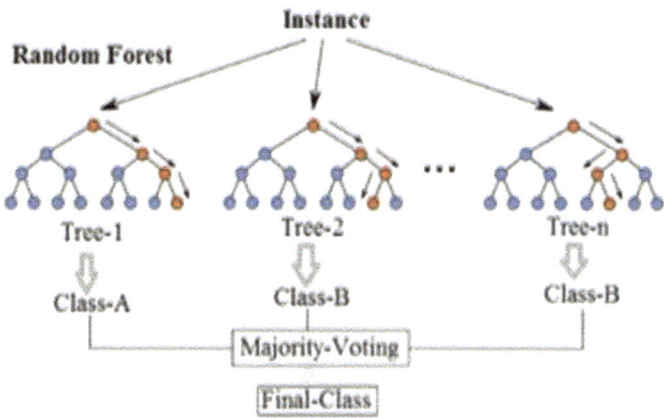

Fig. 2 Random forest

measurements using Recall, Precision, Accuracy, F1 Score, and Confusion Matrix as follows.

5.1 Confusion Matrix

The confusion matrix is defined as a table that consists of TP, TN, FP, and FN and tells about the performance of a classification algorithm. It is used to measure Recall, Precision, Accuracy, AUC-ROC curves. The actual target values are compared with the predicted values in the confusion matrix (Fig. 3).

Fig. 3 Confusion matrix

		Actual Values	
		Positive (1)	Negative (0)
Predicted Values	Positive (1)	TP	FP
	Negative (0)	FN	TN

5.2 Accuracy

Accuracy is defined as the percentage of correct predictions for the test data. Accuracy tells us the number of times the ML model predicts correctly. It is usually used to tell about the performance of a certain model that is being used. The formula used for calculating accuracy is provided in (2).

$$\text{Accuracy} = \frac{TP + TN}{TP + TN + FN + FP} \tag{2}$$

5.3 Precision

Precision is used to measure how close the same item is to each other [15]. Precision is defined as the number of true positives i.e., TP divided by $(TP + FP)$ which is the total number of positive predictions. Precision is the percentage of your results that are relevant. The following is the formula used for calculating precision.

$$\text{Precision} = \frac{TP}{TP + FP} \tag{3}$$

5.4 Recall

A recall is defined as the number of positive samples that is TP divided by the total number of positive samples that is $TP + FN$ [7]. Here, TP is true positive and FN is false negative. The success of prediction is measured by using the Precision-Recall score. This is generally used when the classes are very imbalanced.

$$\text{Recall} = \frac{TP}{TP + FN} \tag{4}$$

5.5 F1 Score

F1 score is one of the most important evaluation metrics in ML [9]. Accuracy and recall are two competing metrics that are used to summarize a model's predictive performance. A good F1 score means few false positives and few false negatives. It uses harmonic means to combine recall and precision.

$$F1 \text{ score} = \frac{2PR}{P + R} \tag{5}$$

where $P = $ Precision and $R = $ Recall.

5.6 Classification Error

The overall incorrect classification [16] of the classification model is calculated as follows:

$$\text{Classification error} = \frac{FP + FN}{FP + TN + FN + TP} \tag{6}$$

6 Proposed Methodology

The major advantage of the proposed approach is listed below.

* Improved disease classification: When the different sets of parameters are considered in the medical field, it gives a better understanding of the different types of diseases and gives good performance in the determination of the type of disease. The facilities for data collection are provided by advanced software architectures and data mining paradigms. More attention is paid to the pre-processing and evaluation phases.
* Selection of efficient parameters such as time complexity and accuracy provides an adequate result.
* Used a suitable ML algorithm to obtain a suitable prediction system.
* ML algorithms predict risk factors through simple analytical methods. Risk factors can be predicted early by ML models. The discrepancies in population training data sets can be reduced.

The proposed approach is divided into various phases such as data collection, data pre-processing, model selection, model implementation, and comparison of the results.

6.1 Data Collection

Data is collected from online open sources [10] to identify diseases. Here the actual symptoms of the disease are collected. No dummy value is entered. Disease symptoms are collected from various health-related websites. Before the data is fed into the predictive model, the following data cleaning and preprocessing steps are performed:

* Check for null values and fill them with the forward fill method.

* Standardize the data by the mean and standard deviation.
* Split the data set into training and testing sets.

6.2 Proposed Approaches

Many methods are used to perform disease prediction. ML is one such approach. Learning strategies for random forest machines include grouping, clustering, summarization, and many others. Classification is one of the data mining processes in this phase of categorical data classification. During the training phase, the provided data and associated class labels are used for classification. The training process uses the training tuples, and the test data phase uses the test data tuples to calculate the accuracy of the classification rules. Assume that the test data classification rules are accurate enough to use the unmined data classification rules.

Following are the steps of the methodology that is applied in the proposed work:

Step 1: Selection of data set and Data Pre-processing: It includes outlining of data, removal of detected outliers, the inclusion of the detected missing data, and enhancement of data by applying normalization techniques.
Step 2: Model Selection: Suitable ML model is selected based on the understanding of data values.
Step 3: Model Implementation: The model is implemented by importing the data from the dataset.
Step 4: Performance Measurement: The performance is measured by analyzing the result through confusion matrix and accuracy is also calculated. The result is compared with the previous work.

7 Results and Analysis

Various experiments are conducted to evaluate three ML-supervised disease detection algorithms: Naive Bayes, Random Forest, and Decision Trees. The performance measurements of the three classification algorithms are shown in Table 2. Each algorithm's working is different and classification is made based on some parameters. The Decision Trees algorithm reaches a maximum accuracy of 93.65%, followed by Random Forest and Naive Bayes. The result clearly shows that Decision Tree achieved its maximum recall of 99.59% and Naive Bayes achieved the worst sensitivity. The F1 obtained from the decision tree and Naive Bayes are the maxima and minima in this experiment. Decision trees have the highest specificity and random forests have the lowest. Finally, it is observed that the decision tree has the best overall performance.

Table 2 Performance comparison of different techniques

	Decision tree	Random forest	Naive Bayes
Accuracy	93.65	86.90	82.83
Sensitivity	99.59	90.60	80.84
Specificity	96.57	82.70	83.59
F1	98.05	86.46	82.67

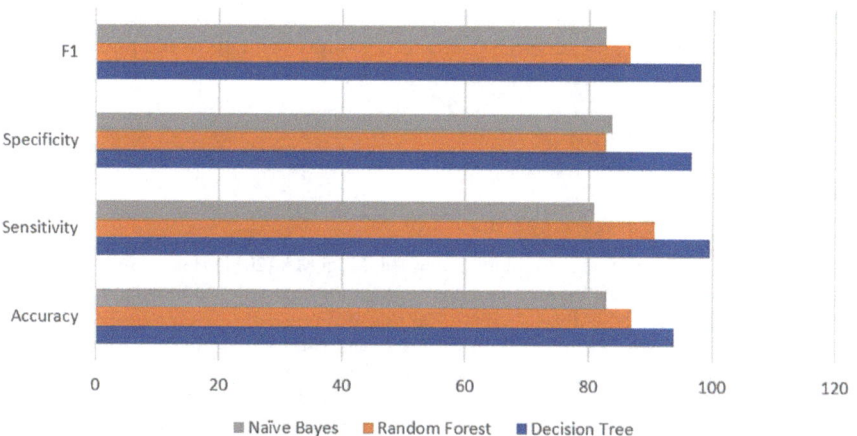

Fig. 4 Performance analysis of different algorithms

The decision tree gives better accuracy than the other two methods, i.e., Random Forest and Naive Bayes as shown in Fig. 4. It has the advantage that it adapts to the data set quickly and can handle multidimensional data.

8 Conclusion and Future Scope

A disease prediction framework based on predictive modeling has the potential to revolutionize the way that medical professionals approach disease prevention and management. By leveraging advanced statistical and computational techniques, medical professionals can identify patients at high risk of developing diseases and take proactive measures to prevent or manage their conditions. As the framework continues to evolve and incorporate new data sources and analytical techniques, it has the potential to improve patient outcomes and ultimately save lives. In the current research paper, we have proposed a disease prediction methodology that uses several ML techniques like NB, RF, and Decision Tree to predict the disease by taking symptoms as input parameters. In the future, we will try to add more features to increase the efficiency of our system. We would also incorporate the opinions of the experts for verification and validation purposes.

This work is thought to lead the prediction of various diseases by knowing the symptoms of diseases at an early stage. An extension of this work is also a deep learning implementation, as deep learning offers higher-quality performance than ML algorithms. This study still has room for further research, from data collection to the visualization of results, to improve the quality of the study. To save lives in the early stages, hybridization or ensemble approaches in current research can be used to obtain better results for determining efficiency, reliability, and efficacy. To improve the scalability and accuracy, many other improvements could be considered and done for the proposed system.

References

1. Ali MM, Paul BK, Ahmed K, Bui FM, Quinn JM, Moni MA (2021) Heart disease prediction using supervised machine learning algorithms: performance analysis and comparison. Comput Biol Med 136:104672
2. Ayesha S, Hanif MK, Talib R (2020) Overview and comparative study of dimensionality reduction techniques for high dimensional data. Inf Fusion 59:44–58
3. Bhatia M, Bhatia S, Hooda M, Namasudra S, Taniar D (2022) Analyzing and classifying mri images using robust mathematical modeling. Multimed Tools Appl 81(26):37519–37540
4. Boukhatem C, Youssef HY, Nassif AB (2022) Heart disease prediction using machine learning. In: 2022 advances in science and engineering technology international conferences (ASET). IEEE, pp 1–6
5. Chandrashekar G, Sahin F (2014) A survey on feature selection methods. Comput Electr Eng 40(1):16–28
6. Chen M, Hao Y, Hwang K, Wang L, Wang L (2017) Disease prediction by machine learning over big data from healthcare communities. IEEE Access 5:8869–8879
7. Davis J, Goadrich M (2006) The relationship between precision-recall and roc curves. In: Proceedings of the 23rd international conference on Machine learning, pp 233–240
8. Grampurohit S, Sagarnal C (2020) Disease prediction using machine learning algorithms. In: 2020 international conference for emerging technology (INCET). IEEE, pp 1–7
9. Humphrey A, Kuberski W, Bialek J, Perrakis N, Cools W, Nuyttens N, Elakhrass H, Cunha P (2022) Machine-learning classification of astronomical sources: estimating f1-score in the absence of ground truth. Mon Not R Astron Soc: Lett 517(1):L116–L120
10. KAUSHIL268: Disease prediction using machine learning (2020). https://www.kaggle.com/datasets/kaushil268/disease-prediction-using-machine-learning
11. Kumari S, Kumar R, Kadry S, Namasudra S, Taniar D (2021) Maintainable stochastic communication network reliability within tolerable packet error rate. Comput Commun 178:161–168
12. Kumari S, Namasudra S (2021) System reliability evaluation using budget constrained real d-mc search. Comput Commun 171:10–15
13. Li W, Cerise JE, Yang Y, Han H (2017) Application of t-sne to human genetic data. J Bioinform Comput Biol 15(04):1750017
14. Mallela RC, Bhavani RL, Ankayarkanni B (2021) Disease prediction using machine learning techniques. In: 2021 5th international conference on trends in electronics and informatics (ICOEI). IEEE, pp 962–966
15. Michaud EJ, Liu Z, Tegmark M (2022) Precision machine learning. arXiv:2210.13447
16. Mivule K, Turner C (2013) A comparative analysis of data privacy and utility parameter adjustment, using machine learning classification as a gauge. Procedia Comput Sci 20:414–419

17. Mohanty S, Mishra A, Saxena A (2021) Medical data analysis using machine learning with knn. In: International conference on innovative computing and communications. Springer, Berlin, pp 473–485
18. Muhuri S, Kumari S, Namasudra S, Kadry S (2022) Analysis of the pertinence of indian women's institutions in collaborative research. IEEE Trans Comput Soc Syst
19. Mutlag WK, Ali SK, Aydam ZM, Taher BH (2020) Feature extraction methods: a review. J Phys: Conf Ser 1591: 012028. IOP Publishing
20. Myles AJ, Feudale RN, Liu Y, Woody NA, Brown SD (2004) An introduction to decision tree modeling. J Chemom: J Chemom Soc 18(6):275–285
21. Padmaja DL, Sruthi BS, Deepak GS, Harsha GS (2022) Analysis to predict coronary thrombosis using machine learning techniques. In: 2022 international conference on sustainable computing and data communication systems (ICSCDS). IEEE, pp 21–27
22. Pal M (2005) Random forest classifier for remote sensing classification. Int J Remote Sens 26(1):217–222
23. Pingale K, Surwase S, Kulkarni V, Sarage S, Karve A (2019) Disease prediction using machine learning. Int Res J Eng Technol (IRJET) 6:831–833
24. Rajdhan A, Agarwal A, Sai M, Ravi D, Ghuli P (2020) Heart disease prediction using machine learning. Int J Res Technol 9(04):659–662
25. Rish I et al (2001) An empirical study of the naive bayes classifier. In: IJCAI 2001 workshop on empirical methods in artificial intelligence, vol 3, pp 41–46
26. Shah D, Patel S, Bharti SK (2020) Heart disease prediction using machine learning techniques. SN Comput Sci 1(6):1–6
27. Singh A, Kumar R (2020) Heart disease prediction using machine learning algorithms. In: 2020 international conference on electrical and electronics engineering (ICE3). IEEE, pp 452–457
28. Vembandasamy K, Sasipriya R, Deepa E (2015) Heart diseases detection using naive bayes algorithm. Int J Innov Sci, Eng Technol 2(9):441–444
29. Weijler L, Kowarsch F, Wödlinger M, Reiter M, Maurer-Granofszky M, Schumich A, Dworzak MN (2022) Umap based anomaly detection for minimal residual disease quantification within acute myeloid leukemia. Cancers 14(4):898
30. Zhang Z, Castelló A (2017) Principal components analysis in clinical studies. Ann Transl Med 5(17)

A Data-Driven Diabetes Predictive Model Using a Novel Optimized Weighted Ensemble Approach

Sunny Arora, Shailender Kumar, and Pardeep Kumar

Abstract Early detection of diabetes plays a crucial role in improving health outcomes and can help people avoid harmful diabetes complications. Machine learning algorithms are being used to diagnose a disease in its early stages. This study proposes an optimized weighted ensemble model that can predict the risk of type 2 diabetes mellitus. A diabetes dataset of 403 patients from the Department of Medicine at the University of Virginia given by Dr. John Schorling has been used. We assessed ridge regressor, LASSO, feedforward artificial neural networks, and linear regression prediction performance. These models were then combined to create an optimized weighted ensemble model for prediction. We evaluated our prediction models using standard performance metrics: coefficient of determination (R2 score), root mean square error (RMSE), mean square error (MSE), and mean absolute error (MAE). The results showed that the proposed optimized weighted ensemble model outperformed individual models, achieving the highest 0.81 (R2 score) and lowest 0.98 (MSE).

Keywords Diabetes prediction · Weight optimization · Ensemble model

1 Introduction

Diabetes Mellitus (DM), a chronic condition, is characterized by metabolic disorders caused by an imbalance in insulin and glucagon pancreatic hormones in the body [1]. The American Diabetes Association (ADA) primarily classifies diabetes as follows:

S. Arora (✉) · S. Kumar
Delhi Technological University, Delhi 110042, India
e-mail: sunnyarora_2k18phdco@dtu.ac.in

P. Kumar
Swansea University, Swansea SA1 8EN, UK

1.1 Type 1 DM (T1DM)

The autoimmune destruction of insulin-producing beta cells, which causes T1DM. Furthermore, it is possible to distinguish three distinct phases of type 1 diabetes using Table 1.

Type 2 DM (T2DM) Insulin deficiency caused by its impaired secretion contributes to T2DM [2]. Glycated blood hemoglobin (glyhb(a1c)) testing provides evidence of a person's average blood glucose levels in the previous two to three months, which is the expected half-life of red blood cells (RBCs) [3]. As the standard measure for blood glucose monitoring in diabetic patients, HbA1c is recommended (especially in T2DM) [4]and is a risk factor for all complications[5]. According to reports from the International Diabetes Federation (IDF) Atlas, a total of 9.3% of global adults in the age group (20–79 years) had diabetes, and figures included diagnosed and undiagnosed type 1 and type 2 diabetes. The global prevalence of diabetes is expected to increase by 10.2 by 2030 and 10.9 by 2045. An estimated 1.1 million children and adolescents (under 20 years of age) had type 1 diabetes. Diabetes-related deaths and complications were estimated at 4.2 million in 2019 [6].

Rapid technological progress in wearable continuous glucose monitoring (CGM) sensors, physical activity monitoring biosensors, and smart devices with mobile applications has immensely contributed to daily health and lifestyle monitoring to address diabetes issues. Integration of data originating from wearable CGM sensors and electronic health records (EHR) utilized by machine learning methods offers preventive, predictive, and customized diabetes therapeutics and helps physicians and patients in decision-making [7–9]. Observing the potential of data analytics, researchers have exploited various state-of-the-art machine learning techniques [10–12] and produced significant findings for predicting diabetes or prediabetes [13]. A varying level of accuracy has often been achieved depending on different learning techniques. Therefore, it is essential to identify mechanisms that can generate high prediction accuracy before the outbreak of diabetes. An ensemble learning strategy is one of the most popular and extensively used models for machine learning. The fundamental aim is to integrate different machine learning models to minimize bias and variance, thereby boosting prediction outcomes [14–18].

The remainder of the paper is organized as follows: Sect. 2 explains the related work on models of diabetes prediction; Sect. 3 provides a detailed data set and selection of features. Section 4 presents the methodology for the proposed model, while Sect. 5 explains performance measures. Sect. 6 presents the results and discussions,

Table 1 Different phases of diabetes

Phase 1	Phase 2	Phase 3
Euglycemia	Dysglycemia	Hyperglycemia
Pre-symptomatic	Pre-symptomatic	Symptomatic

Table 2 Previous works reported in literature

Publications	ML approach	Participants	Age group	Dataset-DS/Databases-DB
Mani et al. [19]	NB, LR, KNN, CART, RF, SVM	Total: 20,675 Study: 2280 Diabetic: 228 M: 912(40%) F: 1368	5451	9.62
Han et al. [20]	SVM, RF	Total: 8597 Study: 7913 Diabetic:646	≥18yrs	CHNS-DB China
Lee et al. [21]	LR, ANN-BPN	Total: 811 Study: 88 Diabetic: 88 M: 35F: 53	18–64 yrs.	Taiwan-DB
Perveen et al. [22]	INDDM with Gaussian HMM	Total: 1,72,168 Study: 1918 Diabetic: 584 M:1143 F: 775	18 yrs. or more	CPCSSN, DB Canada
Choi et al. [23]	LR, LDA, QDA, KNN	Total: 52631 Study: 8454 Diabetic: 404	53.9 (Avg)	KUGH DB, Korea
Dwivedi et al. [24]	SVM, LR, ANN, NB, Classification tree	Total: 768 Diabetic: 268 M: 0 F: 768	≥21 yrs.	PIMA DS NIDDK
Sudharsan et al. [25]	RF, KNN, SVM, NB	DS1: 1037 DS2: 6686 DS3: 1091 DS4: 2000	18–64 yrs.	Well doc Inc, USA

and Sect. 7 provides final remarks and discusses several limitations and other challenges (Table 2).

2 Literature Survey

An early diabetes risk prediction model is necessary to inform individuals of the risk of hyperglycemia, which might provide them with an opportunity to adopt preventive measures.

Sheikhi et al. carried out the analysis without taking into consideration some variables directly connected with T2DM, i.e., hemoglobin glycated (HbA1c). Then, they performed a feature selection using LASSO and ridge LR for T2DM prediction purposes [26]. Bernardini et al. presented a Sparse Balanced Support Vector Machine (SB-SVM) ML technique to discover T2DM using an EHR-based FIMMG dataset. Results show that SB-SVM degraded predictive and computational performance [27]. Zheng et al. developed a feature-driven engineering approach for improving the performance of standard models for ML (logistic regression (LR), decision tree

(DT), K-nearest neighbor (KNN), random forests (RF), and naive Bayes) in T2DM forecasting [28]. J48 Decision Tree and Naïve Bayes algorithms for future diabetes prediction were suggested by Perveen et al. based on relevant risk factors resulting from logistic regression analysis and overbalanced and unbalanced data [22]. In this research, the dataset had 667907 records for 2003–2013. Results demonstrate that increasing HDL levels is good for preventing the onset of diabetes, especially in women. Furthermore, the results showed that Naïve Bayes is superior to random under-sampling, over-sampling, and non-sampling with a K-medoid method.

The subject has or has not been diagnosed with diabetes, depending on the determination focused on the subject's glycosylated hemoglobin. If a value of glycosylated hemoglobin >7.0 is diagnosed as diabetes, otherwise it is regular. We have predicted the future value of the feature Glycosylated Haemoglobin (glyhb) using machine learning-based individual models (ridge regression, LASSO, ANN, Linear Regression) as well as their optimized weighted ensemble model. Table 3 shows a description of the dataset used, along with the type and NaN values of its features.

3 Dataset Description

This section discusses the dataset description. The study aims to determine whether the risk of diabetes can be predicted using the dataset provided by Dr. John Schorling of the Department of Medicine at the University of Virginia [29]. There were 403 instances in the records. The subjects were questioned to identify the prevalence of diabetes in Central America among African Americans.

4 Methodology

Figure 1 shows the proposed model's methodology with workflow from the original dataset consisting of raw data to construct and evaluate predictive models for diabetes risk estimates.

4.1 Data Preprocessing

First, we analyzed the medical consequences of each feature and their relationship to diabetes prediction. We concluded that the subject ID is of no significance in context with predictions. Thus, by dropping the feature subject ID, the complexity of the original dataset was reduced. Further, we observed missing and incorrect values in the dataset due to errors or deregulation, and for example, a large number of values of bp.2s, bp.2d, bp.1s, and height were NaN (65%) in the original dataset, which indicates that the real values were missing. Therefore, we also eliminated the

Table 3 The detailed dataset

Features	Description	Units	Type	NaN values
Id	Subject id	(mg/dl)	Numeric	0
Chol	Total cholesterol	(mg/dl)	Numeric	1
stab.glu	Stabilized glucose	(mg/dl)	Numeric	0
Hdl	High density lipoprotein	(mg/dl)	Numeric	1
Ratio	Cholesterol/HDL ratio	–	Numeric	1
glyhb(a1c)	Glycosylated haemoglobins	(mmol/mol)	Numeric	13
Location	Location	–	Nominal	0
Age	Age	(years)	Numeric	0
Gender	Gender subject id	–	Nominal	0
Height	Height	(inches)	Numeric	5
Weight	Weight	(pounds)	Numeric	1
Frame	A factor id	(level-small/medium/large)	Nominal	12
Bp.1s	First systolic blood pressure	(mmHg)	Numeric	5
Bp.1d	First diastolic blood pressure	(mmHg)	Numeric	5
Bp.2s	Second systolic blood pressure	(mmHg)	Numeric	262
Bp.2d	Second diastolic blood pressure	(mmHg)	Numeric	262
Waist	Waist circumference	(inch)	Numeric	2
Hip	Hip circumference	(inch)	Numeric	2
Time	Postprandial time	(inch)	Numeric	3

bp.2s and bp.2d features to reduce the influence of irrelevant NaN values. Medians of features containing numeric values were used to impute NaN values, excluding categorical features with nominal values such as location, gender, and frame. Since age and gender didn't have any NaN values, they remained out of the missing value imputation process, but the frame feature was imputed with median values. Outliers that lead to overfitting and could be disruptive in our prediction model, The removal of outliers from the training set enhances prediction accuracy. Thus, we plotted the distribution of features in the next step to detect outliers, but the skewed data was found, and the data variance used to detect outliers has a criterion that must be Gaussian or normally distributed. Hence, data transformation was required. To transform

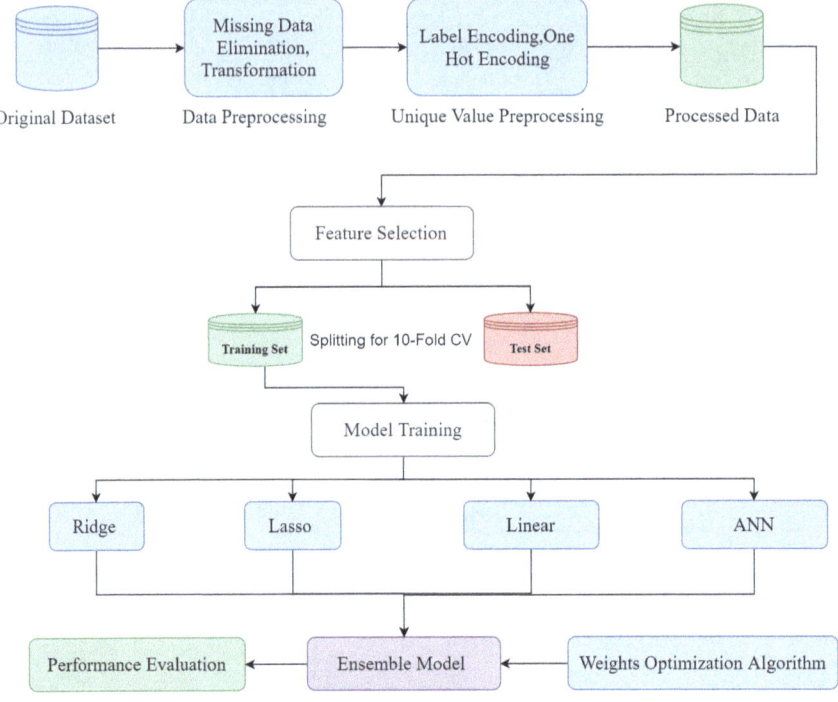

Fig. 1 The detailed architecture of proposed OWEM model

the data, we employed logarithmic and square root mathematical transformations for performance improvement of our proposed model; therefore, we incorporated the transformed data from these mathematical transformations in our model construction. After finding the unique values of categorical features (location, gender, and frame), they were encoded with numerical values using one-hot encoding for analysis. The normalization was then carried out using the Python library Standard Scaler, which normalized each feature to [–1, 1] [30].

4.2   Feature Selection

Since many machine learning algorithms with large amounts of irrelevant data have unsatisfactory performance, the selection of features in every application has become necessary. Feature selection can prevent overfitting and provide a more profound knowledge of unknown areas such as disease occurrence and diagnosis [31] We used an embedded approach based on the Random Forests (RF) algorithm to choose the essential features. Table 4 shows the relevance of each feature (Fig. 2).

Table 4 Relevance of features computed by RF algorithm

Feature	Importance
Stab.glu	0.121064
Age	0.108574
Chol	0.094355
Ratio	0.086701
bp.1d	0.076345
Weight	0.075137
bp.1s	0.074467
Time.ppn	0.073206
Hip	0.072320
Waist	0.063764
hdl	0.061068
Height	0.055365
Frame	0.022375
Location	0.009442
Gender	0.005817

4.3 Model Training

Training and testing datasets were generated from the processed dataset obtained from the above data and unique pre-processing steps. The dataset split was done for a balanced 80/20 split of train and test. The training set was used to construct the models, whereas the models utilized the testing dataset to evaluate their prediction performance. We performed a grid search to find the best model parameters and trained the model using 10-fold cross-validation for model accuracy. This process randomly splits the training and testing data into ten folds.

4.4 Model Families

The machine learning algorithms employed in this work are summarized as follows:
Ridge Regression Ridge regression is introduced by Hoerl et al. [32], which aims to improve outcomes by improving the medium square regression error. Ridge regression is appropriate when there are numerous predictors with nonzero coefficients generated from a normal distribution. It operates effectively with several minor effect predictors in particular and prevents low-specific coefficients in linear regression models with many related variables and significant variations. The ridge regression applies the regularization penalty to fit the model. Ridge regression has demonstrated relatively high predictive accuracy when utilized for high-dimensional datasets. The

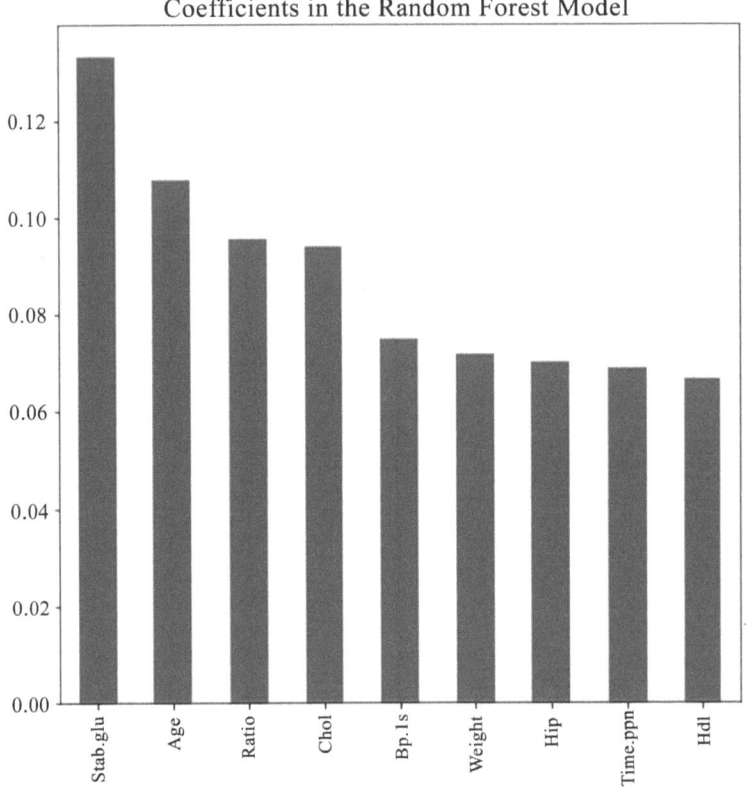

Fig. 2 Final features selected for model training

ridge regression estimator solves the regression issue with $\ell2$ penalized least squares. The ridge regression estimator solves the regression problem using $\ell2$ penalized least squares.

$$\hat{\beta}(\text{ridge}) = \mathrm{a}\,rgmin_\beta \|y - X\beta\|_2^2 + \lambda \|\beta\|_2^2 \tag{1}$$

(i.e., residual sum of errors), is the x_i^T ith row of $X\|\beta\|_2^2 = \sum_{j=1}^{p} |\beta_j{}^2|$ is the $\ell2$ norm penalty on β and $\lambda \geq 0$ is the tuning parameter that controls penalty strength (linear shrinking) by determining the relative importance of empirical error and penalty term data-dependent. The higher the value of the λ, the greater is the shrinkage. Since the value of λ depends on data-driven strategies, it may be utilized to calculate such cross-validation.

LASSO Regression LASSO regression is known as the least absolute shrink and selection operator [33]. The LASSO regression, however, isn't strong in the correlation of predictors and would randomly choose and discard other predictors if they were all the same. The penalty for LASSO costs is that many coefficients are nearly

Table 5 Default and best hyperparameter for the ridge and LASSO algorithm

Models	Hyperparameters	Description	Default	Values
Ridge	1. Alpha 2. fit_intercept 3. Solver	1. Regularization strength 2. Whether to fit the intercept for this model 3. Solver to use in the computational routines	1.0 True Auto	1 False Saga
Lasso	1. Alpha 2. it_intercept	1. Regularization strength 2. Whether to fit the intercept for this model	1.0 True	0.03 False

zero, and only a small fraction is higher (and nonzero). The LASSO estimator uses the penalized criterion for the following optimization problem:

$$\hat{\beta}(\text{Lasso}) = \text{argmin}_\beta \, \|y - X\beta\|_2^2 + \lambda \|\beta\|_1 \tag{2}$$

where $\|\beta\_\|_1 = \sum_{j=1}^{p} |\beta_j|$ is ℓ_1 − the penalty (norm) on β, which has sparsity in the solution and $\lambda \geq 0$ is a parameter tuned (Table 5).

Artificial Neural Networks In this study, we employed feedforward ANN. ANNs are similar to biological human brain networks in that they acquire and preserve information via an extensive, parallel, and distributed mechanism performed by a single processing unit called a neuron. ANNs may have several layers, each containing one or more neurons, including input, output, and at least one hidden layer. Each neuron is linked with each neuron in the following and preceding levels. Each layer has different transfer and activation functions as well. The neuron output signal is given by the weighted sum of the inputs applied to the nonlinear activation function. Our network was fully connected, so every unit has been joined from all preceding layers [33]. Figure 3 displays the ANN model to predict glyhb (A1C) in relevance to diabetes risk diagnosis, given the 19-feature input diabetes dataset and the 9 key features retrieved to train this model.

The gross input was calculated by multiplying and summing each input and its weight. For each input unit as output unit Y, the network computing unit of the layer indicated can be represented as follows:

$$H_i^{(1)} = \varphi^{(1)} \left(\sum_j w_{ij}^{(1)} X_j + b_i^{(1)} \right) \tag{3}$$

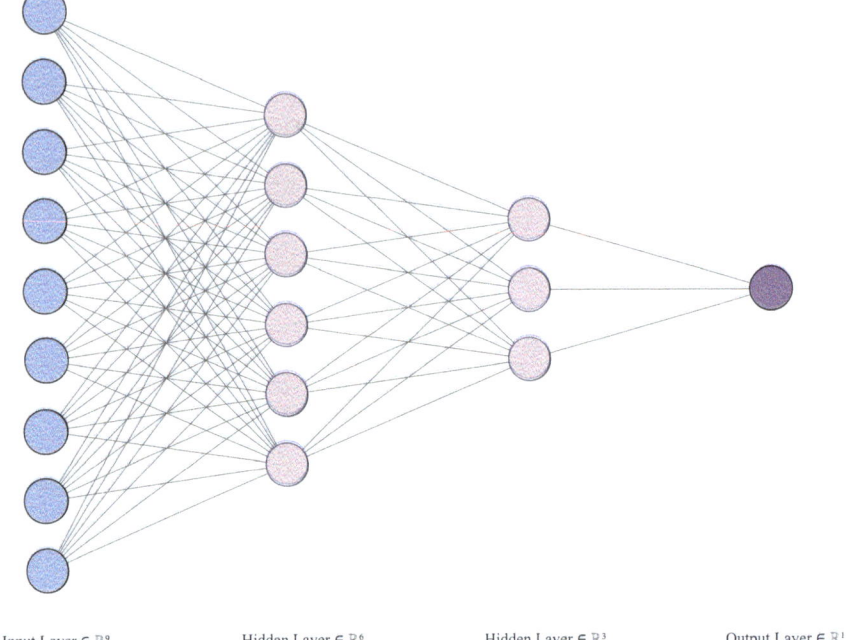

Input Layer ∈ R⁹   Hidden Layer ∈ R⁶   Hidden Layer ∈ R³   Output Layer ∈ R¹

Fig. 3 ANN for diabetes prediction

$$H_i^{(2)} = \varphi^{(2)} \left(\sum_j w_{ij}^{(2)} H_i^{(1)} + b_i^{(2)} \right) \tag{4}$$

$$Y_i = \varphi^{(3)} \left(\sum_j w_{ij}^{(3)} H_j^{(2)} + b_i^{(3)} \right) \tag{5}$$

We applied the grid search [34] to identify the optimal parameters for the ANN model automatically. We used the grid search on the training set and measured it by cross-validation. This strategy is aimed at selecting the parameter that provides the lowest prediction error (MSE). Finally, we observed that the network with two hidden layers (R representing six and three neurons each) was the ideal setting for activating the linear rectified unit (ReLu) and Adam optimizer [35] as weight optimization solutions. We utilized a batch size of five and a default learning rate of (0.001) and set the criterion for the first stop at 50 epochs.

Linear regression model The LRM has the following form:

$$y = \beta_0 + \beta_1 x \tag{6}$$

where β_0 and β_1 are the coefficient vectors generated using the least square technique.
Optimized Weighted Ensemble Model (OWEM) An ensemble is designed to combine numerous models to increase prediction accuracy in learning problems with a numerical target variable [36]. Reducing variability and bias, as well as increasing accuracy, can be made easier with their support. Ridge, LASSO, ANN, and linear regression models were combined with their optimized weights to develop an optimized weighted ensemble model in our framework. We utilized a computational approach to optimize each model's particular weight.

Weight Optimization Instead of manually deciding the weights (w1, w2, w3, w4) for our ensemble model, an automatic and accurate Powell optimization approach is used in conjunction with the scipy.optimize.minimize package to decide on the best ensemble model weights. The Powell algorithm [37] is an optimization approach that does not require the derivative of the cost function to be minimized. In this approach, the user should supply an initial parameter guess and N search vectors to reduce a function with N parameters. By dividing the multidimensional minimizing issue into a sequence of one-dimensional minimization problems, the approach achieves a transformation of the original problem. Additionally, it is iterative. The parameter guess for each iteration is randomly generated via a line search from the parameter guess generated at the previous iteration. While the new parameter guess is stated as a linear combination of all search vectors, the following example demonstrates how to create a parameter from several search vectors. The focus of the subsequent search is changed in favor of a new focus. Convergence is reached, and the iteration is completed. One of the evaluation criteria we used was Root Mean Square Error (RMSE). The 'minimise' function does what its name implies: it minimizes a scalar function of one or more variables. The algorithm's main objective is to identify the lowest RMSE, which will be our optimum weight initialization.

Our models' predictions were collected and added to a list. The minimize function was supplied with starting weights with a value of 0.5 for all models. We came up with a new function we call the RMSE function, which takes the weights, multiplies them with the prediction, and returns the RMSE of that prediction. After that, we will supply the initial values for the RMSE function, and the Powell is passed as an input parameter to scipy.optimize.minimize. On execution, the minimize function passes the initial values of weights to our RMSE function, which then returns the current RMSE of our prediction. To come closer to the minimum, Powell uses iterative line searches that evaluate a local minimum and iteratively change the weights with the best results. The algorithm iterates an arbitrary number of times until no significant improvement is made. Thus, the best values of our weights are found to be at the lowest RMSE. Figure 4 demonstrates the Powell weight adjustment algorithm working to identify the lowest feasible RMSE after 100 iterations.

Finally, each weight has been multiplied by the model's forecast. Hence, we arrive at the smallest root mean squared error forecast. Table 6 shows the optimal weights acquired for the construction of the proposed ensemble through the process of weight optimization.

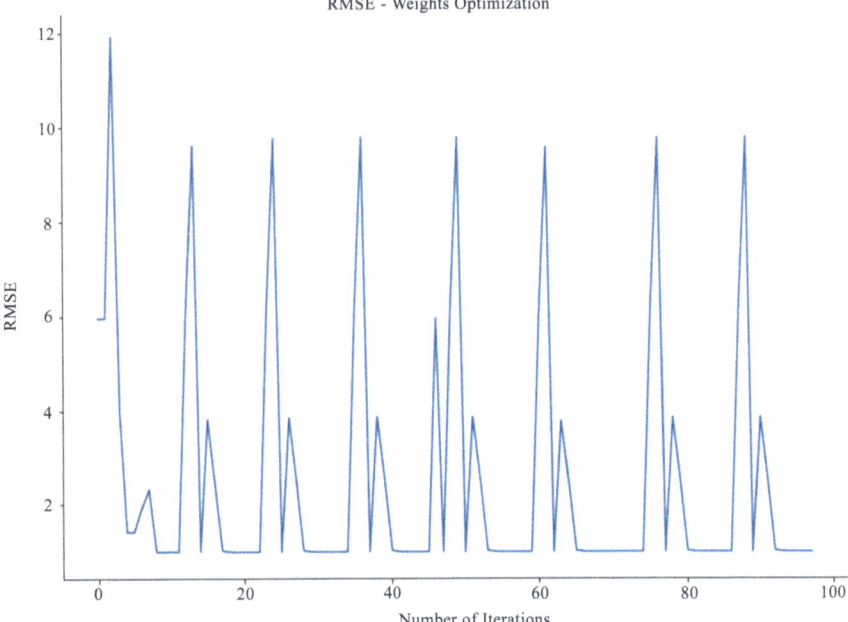

Fig. 4 Demonstration of the Powell weight adjustment algorithm

Table 6 Optimized weights for the models used for the construction of OWEM

Models	Weights	Optimized values
Ridge	W1	−40.48056933
LASSO	W3	0.49913944
ANN	W2	0.5059238
LRM	W4	0.50047047

5 Performance Measures

Several statistical measures were utilized and described in the literature for prediction tasks in order to quantify the prediction performance. The fitness score (R^2 score), root mean square error (RMSE), mean square error (MSE), and mean absolute error (MAE) were all used in this work. The root mean square error (RMSE) is crucial for determining the accuracy of a prediction since it allows the error to be the same magnitude as the quantity forecasted. Consider a succession of glycosylated hemoglobins (designated y) with their predicted values of length n. The RMSE is calculated as follows:

Most of the time, the R2 score is referred to as the degree of fitness, and it has a value of 0 (no correlation) to 1 (complete correlation), where $\bar{y}$ denotes the mean

Table 7 Prediction results of Ridge, LASSO, ANN, LRM, and OWEM models

Model	R2 score	RMSE	MSE	MAE
Ridge	0.8073	1.0195	1.0395	0.7624
LASSO	0.8059	1.0230	1.0467	0.7695
ANN	0.8043	1.0274	1.0557	0.7410
LRM	0.8079	1.0180	1.0364	0.7621
OWEM	0.8168	0.9940	0.9881	0.7344

glycosylated hemoglobin concentration for each patient; when R2 values are close to 1, better performances are obtained. If R2 = 1, the original and predicted series should be superimposed.

$$\mathbf{RMSE} = \sqrt{MSE} = \sqrt{\frac{1}{N}\sum_{i=1}^{N}(y_i - \hat{y}_i)^2} \tag{7}$$

Most of the time, the R2 score is referred to as the degree of fitness, and it has a value of 0 (no correlation) to 1 (complete correlation), where $\bar{y}$ denotes the mean glycosylated hemoglobin concentration for each patient; when R2 values are close to 1, better performances are obtained. If R2 = 1, the original and predicted series should be superimposed.

$$\mathbf{R}^2 = 1 - \frac{\sum(yi - \hat{y})^2}{\sum(yi - \bar{y})^2} \tag{8}$$

6   Results and Discussion

Table 7 depicts the predictive performance of many models used to estimate the risk of diabetes by predicting glyhb(a1c) using various evaluation measures such as R2 score, RMSE, MSE, and MAE. Individual models' forecast methods are contrasted with our suggested optimized weighted ensemble method.

Our results revealed that the proposed OWEM outperformed all other considered models (ridge, LASSO, ANN, and LRM) applied to the diabetes dataset. In our proposed model, different pre-processing methods, removing skewness and outliers from the data, and the random forest algorithm for feature importance were used. We have utilized the weight optimization approach that uses the Powell algorithm to discover the appropriate weights for multiple models to build an ensemble with better predictive performance than other models applied to our dataset. Powell's is a non-derivate optimization algorithm used for finding a local minimum (RMSE) of the minimize function. However, we experimented with the ensemble model with simple averages, but that produced a lower R2 score and more errors than the proposed

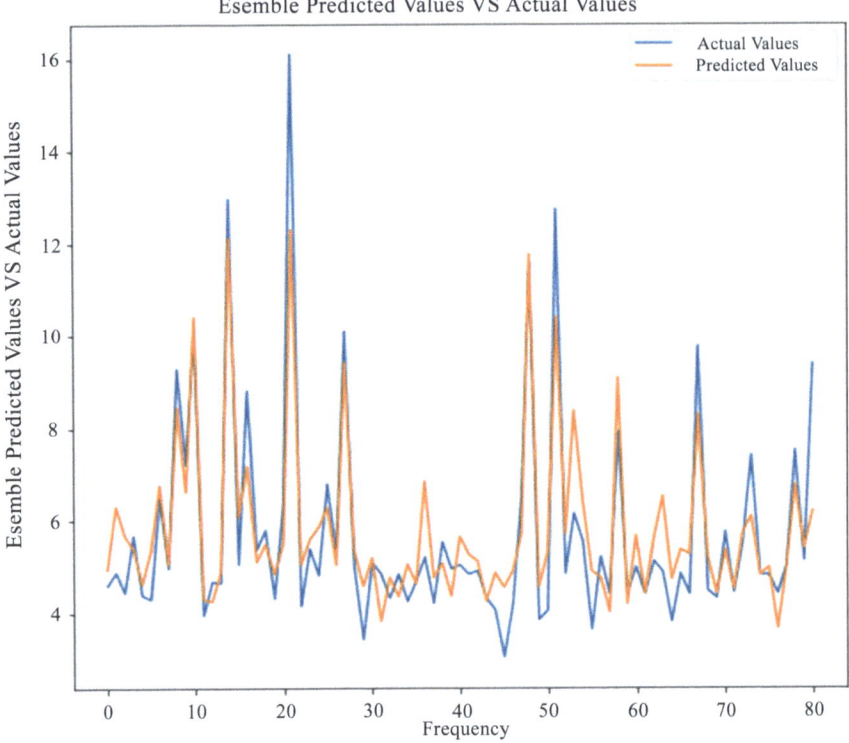

Fig. 5 The predictions vs. actual values of glyhb(a1c)

model. Finally, our proposed OWEM provided significant performance across all metrics for the diabetes datasets compared with a wide range of model results. Figure 5 presents the predictions vs. actual values of glyhb(a1c) on the testing test. An early prediction of diabetes can significantly improve accuracy by employing the proposed ensemble method.

Prediction accuracy plays a crucial role in a person's life in health care. Machine learning models have been found helpful in providing a method for analyzing diabetes diagnostic medical data. A critical concern for medical experts and researchers is accurate predictions for the early detection of diabetes. We constructed an optimized weighted ensemble model for diabetes risk prediction using ridge, LASSO, ANN, and LRM models. Weight optimization was performed utilizing Powell's non-derivative algorithm to identify the best weights for these models. The results confirmed the supremacy of our OWEM over individual models for the prediction of glyhb (a1c) values to estimate the progression of diabetes. Since a dataset with a small population was employed in our study, more datasets with a larger population need to be utilized in the future. Furthermore, extending the comparison to other prediction models and optimizing the hyperparameters.

7 Conclusion

This study proposed an optimized weighted ensemble model (OWEM) that can predict the risk of type 2 diabetes mellitus. As discussed in the results, we compared OWEM's performance on different evaluation criteria (R2 score, RMSE, MSE, and MAE) with those obtained by different individual models and found that OWEM outperformed the other models and achieved better results from the individual models available.

Conflict of Interest The authors declare no conflict of interest.

References

1. Röder PV, Wu B, Liu Y, Han W (2016) Pancreatic regulation of glucose homeostasis. Exp Mol Med 48(3):e219–e219
2. Diabetes Care (2019) Care in diabetesd2019. Diabetes Care 42(1):S13–S28
3. Sherwani SI, Khan HA, Ekhzaimy A, Masood A, Sakharkar MK (2016) Significance of hba1c test in diagnosis and prognosis of diabetic patients. Biomarker Insights 11:BMI–S38440
4. WHO Guidelines Approved by the Guidelines Review Committee et al (2011) Use of glycated haemoglobin (hba1c) in the diagnosis of diabetes mellitus: abbreviated report of a who consultation. Geneva: World Health Organization
5. Dagliati A, Marini S, Sacchi L, Cogni G, Teliti M, Tibollo V, De Cata P, Chiovato L, Bellazzi R (2018) Machine learning methods to predict diabetes complications. J Diabetes Sci Technol 12(2):295–302
6. IDF Diabetes Atlas et al (2019) Idf diabetes atlas. International Diabetes Federation (9th editio). Accessed from http://www.idf.org/about-diabetes/facts-figures
7. Zarkogianni K, Litsa E, Mitsis K, Wu P-Y, Kaddi CD, Cheng C-W, Wang MD, Nikita KS (2015) A review of emerging technologies for the management of diabetes mellitus. IEEE Trans Biomed Eng 62(12):2735–2749
8. Ding S, Schumacher M (2016) Sensor monitoring of physical activity to improve glucose management in diabetic patients: a review. Sensors 16(4):589
9. Georga EI, Protopappas VC, Bellos CV, Fotiadis DI (2014) Wearable systems and mobile applications for diabetes disease management. Health Technol 4:101–112
10. De Silva K, Lee WK, Forbes A, Demmer RT, Barton C, Enticott J (2020) Use and performance of machine learning models for type 2 diabetes prediction in community settings: a systematic review and meta-analysis. Int J Med Inform 143:104268
11. Maliyaem M, Tuan NM, Lockhart D, Muenthong S (2022) A study of using machine learning in predicting covid-19 cases. Cloud Comput Data Sci 92–99
12. Devi D, Namasudra S, Kadry S (2020) A boosting-aided adaptive cluster-based undersampling approach for treatment of class imbalance problem. Int J Data Warehous Min (IJDWM) 16(3):60–86
13. Meng X-H, Huang Y-X, Rao D-P, Zhang Q, Liu Q (2013) Comparison of three data mining models for predicting diabetes or prediabetes by risk factors. Kaohsiung J Med Sci 29(2):93–99
14. Reddy CK, Aggarwal CC (2015) Healthcare data analytics, vol 36. CRC Press
15. Wen L, Qiu W, Mu K (2020) Applying latest data science technology in cancer screening programs. Cloud Comput Data Sci 31–39
16. Gupta A, Rajput IS, Gunjan VJ, Chaurasia S (2022) Nsga-ii-xgb: Meta-heuristic feature selection with xgboost framework for diabetes prediction. Concurr Comput: Pract Exp 34(21):e7123
17. Arora S, Kumar S, Kumar P (2021) Implementation of lstm for prediction of diabetes using cgm. In: 2021 10th international conference on system modeling and advancement in research trends (SMART). IEEE, pp 718–722

18. Arora S, Kumar S, Kumar P (2023) Multivariate models of blood glucose prediction in type1 diabetes: a survey of the state-of-the-art. Curr Pharm Biotechnol 24(4):532–552
19. Mani S, Chen Y, Elasy T, Clayton W, Denny J (2012) Type 2 diabetes risk forecasting from emr data using machine learning. In: AMIA annual symposium proceedings, vol 2012. American Medical Informatics Association, p 606
20. Han L, Luo S, Jianmin Y, Pan L, Chen S (2014) Rule extraction from support vector machines using ensemble learning approach: an application for diagnosis of diabetes. IEEE J Biomed Health Inform 19(2):728–734
21. Lee W-J, Chong K, Chen J-C, Ser K-H, Lee Y-C, Tsou J-J, Chen S-C (2012) Predictors of diabetes remission after bariatric surgery in Asia. Asian J Surg 35(2):67–73
22. Perveen S, Shahbaz M, Saba T, Keshavjee K, Rehman A, Guergachi A (2020) Handling irregularly sampled longitudinal data and prognostic modeling of diabetes using machine learning technique. IEEE Access 8:21875–21885
23. Choi BG, Rha S-W, Kim SW, Kang JH, Park JY, Noh Y-K (2019) Machine learning for the prediction of new-onset diabetes mellitus during 5-year follow-up in non-diabetic patients with cardiovascular risks. Yonsei Med J 60(2):191–199
24. Dwivedi AK (2018) Analysis of computational intelligence techniques for diabetes mellitus prediction. Neural Comput Appl 30:3837–3845
25. Sudharsan B, Peeples M, Shomali M (2014) Hypoglycemia prediction using machine learning models for patients with type 2 diabetes. J Diabetes Sci Technol 9(1):86–90
26. Sheikhi G, Altınçay H (2016) The cost of type ii diabetes mellitus: a machine learning perspective. In: XIV Mediterranean conference on medical and biological engineering and computing 2016: MEDICON 2016, March 31st-April 2nd 2016, Paphos, Cyprus. Springer, pp 824–827
27. Bernardini M, Romeo L, Misericordia P, Frontoni E (2019) Discovering the type 2 diabetes in electronic health records using the sparse balanced support vector machine. IEEE J Biomed Health Inform 24(1):235–246
28. Zheng T, Xie W, Liling X, He X, Zhang Y, You Mingrong, Yang Gong, Chen You (2017) A machine learning-based framework to identify type 2 diabetes through electronic health records. Int J Med Inform 97:120–127
29. Diabetes dataset. http://staff.pubhealth.ku.dk/~tag/Teaching/share/data/Diabetes.html
30. García S, Ramírez-Gallego S, Luengo J, Benítez JM, Herrera F (2016) Big data preprocessing: methods and prospects. Big Data Anal 1(1):1–22
31. Saeys Y, Inza I, Larranaga P (2007) A review of feature selection techniques in bioinformatics. Bioinformatics 23(19):2507–2517
32. Hoerl AE, Kennard RW (1970) Ridge regression: applications to nonorthogonal problems. Technometrics 12(1):69–82
33. Nielsen MA (2015) Neural networks and deep learning, vol 25. Determination Press San Francisco, CA, USA
34. Jiménez AB, Lázaro JL, Dorronsoro JR (2008) Finding optimal model parameters by discrete grid search. In: Innovations in hybrid intelligent systems. Springer, Berlin, pp 120–127
35. Kingma DP, Lei Ba J (2015) Adam: a method for stochastic optimization 3rd international conference learning Representations (Preprint 1412.6980 v9)
36. Mendes-Moreira J, Soares C, Jorge AM, De Sousa JF (2012) Ensemble approaches for regression: a survey. ACM Comput Surv (csur) 45(1):1–40
37. Powell MJD (1964) An efficient method for finding the minimum of a function of several variables without calculating derivatives. Comput J 7(2):155–162

Performance Analysis of Image Caption Generation Techniques Using CNN-Based Encoder–Decoder Architecture

Priya Singh⑩, Chehak Agrawal, and Riya Bansal

Abstract Image captioning is the method of generating textual descriptions for an image using deep neural networks. Its objective is to produce accurate results to specify the hidden features and to satisfy its wide applications. There are various Convolutional Neural Network (CNN)-based encoder architectures available in the literature for image caption generation and there is a need to empirically evaluate the best-performing architecture on multiple and diverse datasets to check their efficacy and generalization capability. To address this, we performed the experiments using the Flickr30K dataset containing 31,783 images along with the commonly used Flickr8K dataset consisting of 8091 images. In this study, we aim to discover the best-suited CNN architecture models for caption generation. The study evaluated various encoder architectures, including VGG16, VGG19, InceptionV3, and InceptionResNetV2, for extracting image features and Long Short-Term Memory (LSTM) as a decoder for generating accurate captions. The models are analyzed based on accuracy variation and value loss metrics on both datasets. The results depict that all models perform better on the larger dataset, i.e., Flickr30K achieves better accuracy with minimum loss values and VGG19 shows the best results among all.

Keywords Convolutional neural network · Image captioning · InceptionV3 · InceptionResNetV2 · VGG16 · VGG19 · Long short-term memory

P. Singh · C. Agrawal (✉) · R. Bansal
Department of Software Engineering, Delhi Technological University, Shahabad, Daulatpur 110042, India
e-mail: chehakagrawal01@gmail.com

© The Author(s), under exclusive license to Springer Nature Singapore Pte Ltd. 2024
S. Namasudra et al. (eds.), *Data Science and Network Engineering*, Lecture Notes in Networks and Systems 791, https://doi.org/10.1007/978-981-99-6755-1_23

1 Introduction

The ability of humans to interpret their surroundings and communicate using textual description is aimed to be integrated and performed using deep learning techniques. It has enabled the development of neural network models that can learn to extract high-level features from images and then use those features to generate natural language captions. This approach is referred to as Image Caption generation.

With the advent of AI, image captioning techniques are at its advancing stage and are contributing significantly to various sectors and businesses. One popular approach is the use of CNNs to extract features from images, followed by the use of Recurrent Neural Networks (RNNs) to generate captions. Another popular approach is the use of attention mechanisms, which allow the model to selectively focus on different parts of the image when generating each word of the caption. This approach has been shown to improve the quality of the generated captions, since it allows the model to better align the visual features with the words in the caption. Additionally, image captioning has numerous applications ranging from assistance to the visually impaired to recommendations in editing applications.

The encoder-decoder approach involves feature extraction from the standard dataset, text pre-processing, and caption generation using RNN. Various models which are generally used as encoders include ResNet50, VGG16, InceptionV2, DenseNet201, and Xception. Moreover, there are various newer versions of some of these as well like VGG16 has a higher version called VGG19 with 19 convolutional layers and InceptionResNetV2 is also the variation of InceptionV3 with architectural changes i.e., compressed residual layers for performance enhancement. Different papers have performed architectural analysis between these models but comprehensive comparison based on model accuracy and value loss considering various datasets between the different versions of a model is not adequately captured by them.

The main contributions of this work are as follows:

1. To access and evaluate the performance in different parameter settings.
2. To implement and evaluate the performance on varying datasets for overall generalization and accurate performance detection.
3. To draw a comparative performance analysis between both the sets of models, i.e., VGG16 and VGG19 and InceptionResNetV2 and InceptionV3.

This study is organized as below. Section 2 captures work related to image captioning. Section 3 discusses the approach followed in conducting the research. Section 4 compiles the results and draws a comparison between the selected CNN encoders and Sect. 5 concludes the findings and specifies the future work.

2 Literature Review

There are various approaches that are proposed for automatic image caption generation in recent years. Vinyals et al. [1] introduced a deep neural network architecture that combines a CNN for image feature extraction and an RNN for generating natural language descriptions of the images. This study used the VGG16 model which is trained on a large dataset This paper sparked a lot of interest in the field of image captioning.

Jethwa et al. [2] provided a brief introduction to attention mechanisms. They described the two CNN models, InceptionResNetV2 and InceptionV3, evaluating an attention-based image captioning model on the COCO dataset. They evaluated the models using standard image captioning evaluation metrics. The results showed that both InceptionResNetV2 and InceptionV3 are effective at generating captions, but InceptionV3 outperformed InceptionResNetV2 on all evaluation metrics. The authors suggested that this may be due to InceptionV3's better performance on object recognition tasks, which is important for generating accurate captions.

Through comparative model analysis, Alam et al. [3] compared different image encoders, namely, VGG16, ResNet50, Xecption, Inception, and DenseNet201 for image caption generation. After training, loss value and accuracy were evaluated on the COCO dataset. For image captioning, Par-inject architecture was used and LSTM decoder architecture for the generation of sentences based on image features. According to their study, after evaluation, the accuracy of VGG16 was better than InceptionV3.

By comparing the performance and results of various CNN-RNN architectures in image caption generation task, Suresh et al. [4] evaluated different models—VGG16, VGG19, InceptionV3, and Inception-ResNetV2. The results of the study depicted that the Inception-ResNetV2 model outperformed the other models in the quality parameter of the generated captions, as measured by various evaluation measures, such as CIDEr, BLEU, and METEOR. The research analyzed that incorporating attention mechanisms in the models improved the generated captions quality further. Overall, the study suggested that the choice of CNN architecture can have a marked impact on performance for image captioning, depending on different factors and parameters such as the specific implementation of the models, quality and size of the training dataset, and choice of hyper-parameters.

Kanimozhiselv et al. [5] constructed a model trained with three CNN architectures, such as InceptionV3, Xception, and ResNet50, to extract features from the image and LSTM to generate the relevant descriptions. From the combinations of three CNN and LSTM architectures, choose the best combination based on model accuracy. Model was trained for the Flickr8k dataset. For caption generation Greedy Search approach was used.

Dwivedi and Upadhyaya [6] introduced a new CNN-5 with five layers for image captioning and compared it to existing transfer learning models like VGG16 and VGG19. The results showed that the CNN-5 model, which is less dense, performs better and faster than VGG16 and VGG19. The research also showed that enhancing the depth of the neural network does not necessarily lead to faster learning. However, several factors influence the accuracy of image caption generation, including the dataset size, vocabulary, feature extraction model, and hyper-parameter values. The study analyzed that VGG16 outperformed the proposed model and VGG19 model in terms of BLEU-1 and BLEU-2 scores.

Bhatia et al. [8] presented a novel approach to image captioning that combined the strengths of the InceptionResNetV2 model and RNN and provided promising results that can be used in various applications. The study highlighted the importance of pre-processing the input images and extracting visual features using the Inception-ResNetV2 model, which was used as input to the RNN for generating captions, and pointed out some limitations of the proposed method, such as the limitation of the RNN to generate captions with long-term dependencies, and the inability of the model to generate diverse captions for a single image.

An encoder–decoder model was proposed by Poddar and Rani [12] for the Flikr8K Hindi dataset, to extract features a pre-trained model was employed and LSTM was used for language modeling. The efficiency of the encoder–decoder model was demonstrated by computing the BLEU score on different image caption-generating models. To optimize these models, hyper-parameters were tuned and hidden layers in the existing models were altered. After the analysis of the experimental results, it was found that the multi-layered CNN-LSTM architecture outperformed the traditional CNN-LSTM based on BLEU score. In the current work, the four most recent and best-performing CNN-based architectures, viz., VGG16, VGG19, InceptionV3, and InceptionResNetV2 are selected for comparative analysis of their performance for Image Captioning. The results are evaluated on Flickr8k and Flickr30k datasets using accuracy and loss value scores.

There are many studies where CNN models are used as feature extractors. In the study [13], CNNs are employed for image classification, while, for image captioning [5], they are used for feature extraction in the context of generating captions.

Using convolutional neural networks (CNNs), the approach involved feature extraction from the input images to capture relevant visual patterns, and then Dewangan and Sahu [14] used these features for lane detection. Similarly, in the road detection study [15], the CNN model, which is inspired by architectures like VGG-Net, is used to extract informative features from the input images. These features are then likely used for further processing to identify road regions accurately.

3   Methodology

Image captioning involves model training based on image features' captions related to the image, i.e., encoder and then translating these details into an explanation in the form of captions, i.e., decoder. The encoder is primarily a pre-trained CNN network responsible for extracting features from the input image. These features are subsequently employed by the decoder to create a caption. For feature extraction, VGG16, VGG19, InceptionV3, and InceptionResNetV2 are used, and for generating caption, LSTM is used. This section includes description of the datasets used in this study, implementation approaches including various phases for generation of image captions, CNN encoder architectures, and LSTM.

3.1 Dataset Description

In order to compare CNN architectures, datasets used in this study are the Flickr8K and Flickr30K datasets for better accuracy from Kaggle (https://www.kaggle.com). The dataset includes both images and textual data for each image with five caption descriptions. The Flickr8K dataset contains 8091 images of which 5663 have been used as training dataset and 2428 have been used as testing dataset. It contains 40,455 captions. Also, the Flickr30K dataset consists of 31,783 images, which have been divided into 22,248 for training the model and 9535 for testing purposes with 158,915 image captions.

3.2 Implementation

Various phases involved in Image Captioning which includes data pre-processing, feature extraction using CNN, and caption generation using RNN are described below and are depicted through the flowchart in Fig. 1.

Data Processing. For data pre-processing, we have converted the text into lowercase, removing words with single characters, punctuation, and words containing numbers. Following this, a vocabulary is generated containing all the unique words from the text which are thereafter transformed into a dictionary using the name of the image as key and the cleaned descriptions as corresponding values.

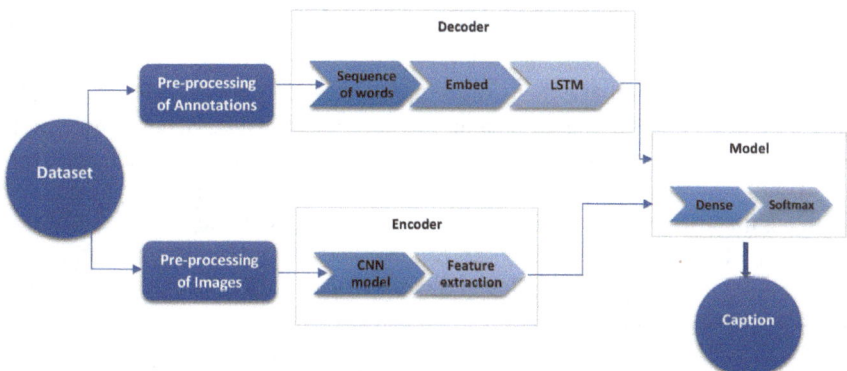

Fig. 1 Implementation approach

Feature Extraction. For feature extraction, CNN models are used. In this study, different combinations of models and datasets are analyzed. The model takes pre-processed images as input, which are then transformed into a vector format and are taken as input by the neural network. The conversion of images into a fixed-size vector is carried out using models such as VGG16, DenseNet121, and InceptionV3. Once the image features have been extracted using these models, they are saved on the disk in the form of a pickle file.

Image Caption Generation. During model training, it is trained to create new captions, using pre-defined captions as the target variables. For caption generation, each word is encoded into a vector of fixed size, and the resultant word as output signifies the probability distribution of the expected word over the complete possible words set in the dictionary. The generator function is a crucial component of this process and is responsible for creating captions. The generator function yields batches of input–output pairs, as well as a step parameter that specifies the number of pairs of input and output images generated for every batch. It produces an indices sequence that is acquired from the description of the image, where a distinct word is represented by each index. Padding is added in the last of every sequence, and the succeeding word in a target word sequence is predicted on the basis of input sequence. The predicted target word sequence is then sent back to the data generator for the anticipation of subsequent words.

The model consists of two main components: an image extractor represented as a vector and a language processor represented as a sequence. The image vector is obtained using a CNN that utilizes pre-processed models to extract features from images. These features are then passed through a dense layer to generate captions for the images. The language processor is implemented using an RNN that leverages the previous word to predict the next word. Additionally, an LSTM layer is employed to process the text input, and the LSTM output is analyzed using a dense layer, considering one output at a time. By combining these components, a model is created that can generate text from images. The outputs of the feature extractor and the sequence processor are merged into a fixed-length vector with a maximum sequence length. This vector is concatenated and prepared by the LSTM and the dense layer to make a final prediction. The model predicts a probability distribution across the word vocabulary, and the network minimizes the categorical cross-entropy loss function by using a SoftMax activation function during fitting. Within this model, the GlobalMaxPooling2D layer is employed to condense the existence of each of the image features. When it is down-sampled using maximum pooling, the feature map and image will result in the same output. The dense layer is completely connected, implying that each neuron in this layer is linked to each input in the next layer via a weight in a linear operation. To optimize the model, Adam optimizer is utilized as an optimizer to enhance the model's efficiency. The categorical cross-entropy loss function is employed to measure the loss.

3.3 CNN Architectures

VGG16 and VGG19. CNN with architecture of VGG16 was introduced by Simonyan and Zisserman [7], and the model archives 92.7% of accuracy with the ImageNet dataset which contains 14 million images with a total of 1000 numbers of classes. It has classified new images of size 224X224X3 and has two fully connected layers. VGG16 is in vector form to obtain quality image representation. VGG16 refers to it because it has 16 layers and 138 million parameters. Similarly, the other VGG19 architecture contains 19 layers including 3 fully connected layers and 16 convolution layers. Figures 2 and 3 describe VGG16 and VGG19 architecture, respectively.

The system was built using deep learning techniques. After performing feature extraction from Image and captions pre-processing, it was built using a sequential model in which several layers were concatenated. Thereafter, the model was trained using the training dataset, and text captions describing the image were generated for image input rendered by testing dataset.

InceptionV3 and InceptionResNetV2. In the Inception network, rather than manually selecting filter sizes in the Inception network, all filters are added and the model automatically determines which filter is most appropriate. Inception architecture is used as an encoder which is used to generate image features. An attention-based unit is produced, which partitions the input into distinct areas and generates a context vector based on a priority score derived from the previously generated text. This assists the decoder in concentrating on specific areas during a specific time step. InceptionResNetV2 is recognized as an encoder due to its high accuracy, exceptional performance, and comparatively shorter training duration. It closely resembles InceptionV3 but with the inclusion of residual blocks.

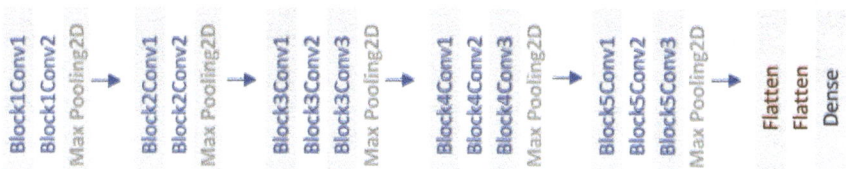

Fig. 2 VGG16 architecture

Fig. 3 VGG19 architecture

Fig. 4 InceptionV3 architecture

Fig. 5 Inception ResNetV2 architecture

The InceptionV3 is a frequently employed model for image recognition, featuring a network comprising 48 layers. It incorporates label smoothing to enhance its performance and employs an auxiliary classifier with batch normalization to propagate label information to lower levels in the network. Figure 4 displays the architecture of InceptionV3.

InceptionResNetV2 was proposed to accelerate the training speed of Inception architecture. It is a 164-layer deep network incorporated with residual networks (ResNet) as shown in Fig. 5. The ResNet's skip connection serves two purposes: it deepens the model while preserving the integrity of information. Additionally, it lowers the number of layers that need to be propagated, thereby reducing the effects of vanishing gradients and hastening the training process. InceptionResNetV2 is similar to Inception V3 along with the residual block.

3.4 Long Short-Term Memory (LSTM)

LSTMs belong to the family of recurrent neural networks and can recognize the significance of the order of events in sequence prediction tasks. This ability is particularly important for solving complex problems like speech recognition and machine translation. Learning about LSTMs and related concepts such as bidirectional and sequence-to-sequence within the vast field of deep learning can be challenging.

LSTM networks have achieved impressive success, enabling the generation of longer sequences of sentences or words. In the LSTM model, the recurrent neural network is encircled by a collar from multiple points, similar to traditional feed-forward neural networks. Below is the implementation of the accuracy metrics function for image captioning.

3.5   Evaluation of Accuracy and Loss Value

For evaluating the performance of image captioning models, we have selected accuracy metric and loss value function as these are appropriate to refer to the extent to which the generated caption matches the ground truth or reference caption for a given image. Various previously published related works have also used these parameters for the comparative analysis and evaluation of models [3, 5, 16].

4   Accuracy Metric

In image captioning, the accuracy metric is often defined as the percentage of caption sequences that are exactly correct. That is, the predicted caption sequence must match the ground truth caption sequence exactly, word for word.

During implementation, we have defined a custom metric function that takes the predicted and ground truth caption sequences as inputs and compares them element-wise to determine whether they are equal. The function can then compute the percentage of sequences that are exactly correct and return that value as the accuracy metric.

Here is the implementation approach for the calculation of accuracy of the model-generated captions. Firstly, the predefined captions and the predicted captions are converted to NumPy arrays. Thereafter, the percentage of caption sequences that are exactly correct is computed by iterating both the arrays and maintaining a correct counter. Finally, the accuracy is calculated by dividing correctly predicted captions by total captions.

accuracy = (number of correctly predicted words) / (total number of words.

5   Loss Function

Loss function used is the categorical cross-entropy, which compares the predicted caption sequence to the ground truth caption sequence. To apply the categorical cross-entropy loss function to image captioning, the caption sequences are converted into one-hot encoded vectors, where each vector element corresponds to a unique word in the vocabulary.

Table 1 Accuracy scores (%) of different CNN encoder architectures

CNN models	Flickr8K	Flickr30K
VGG16	94.2	96.7
VGG19	**95.1**	**97.8**
Inception V3	91.3	93.2
InceptionResNetV2	89.3	89.6

The formula for categorical cross-entropy loss is: loss $= -\Sigma i, j \log(p(i, j))$, where i iterates over the time steps in the predicted caption sequence, j iterates over the words in the vocabulary, and p(i, j) is the predicted probability of the j-th word at the i-th time step in the caption.

6 Results and Discussion

This section provides an overview of the loss and accuracy observed from the model. Table 1 provides the accuracy scores of VGG16, VGG19, InceptionV3, and InceptionResNetV2 as 94.2, 95.1, 91.3, and 89.3, respectively, for the Flickr8K dataset and 96.7, 97.8, 93.2, and 89.6, respectively, for the Flickr30K dataset. Based on our comparative analysis in Fis. 6 and 7, VGG19 outperformed the other three models, achieving the highest accuracy of 95.1% and 97.8% for the Flickr8K and Flickr30K datasets, respectively. Table 2 provides the loss value scores of VGG16, VGG19, InceptionV3, and InceptionResNetV2 as 42.8, 39.8, 54.7, and 56.2, respectively, for the Flickr8K dataset and 42.1, 39.7, 50.6, and 54.9, respectively, for the Flickr30K dataset. It shows that the loss value in VGG19 is lower than in the other models. Therefore, VGG19 is found to be better than VGG16 in terms of accuracy and loss values, and InceptionResNetV2 performed slightly better than InceptionV3. The generated captions from VGG19 with LSTM are more accurate compared to the other three models. By analyzing the graphs, it is also noticed that increasing the number of epochs exponentially improves accuracy of these models. Hence, it can be concluded that increasing the training data can minimize the loss and improve accuracy.

7 Conclusion and Future Scope

In this study, we have selected four state-of-the-art CNN models, i.e., VGG16, VGG19, InceptionV3, and InceptionResNetV2 that are based on the encoder–decoder architecture. The results are indicative of the finding that the VGG19 model outperformed the rest of the three models for automatic image caption generation with better accuracy and value loss. Furthermore, multiple datasets are used in the study, including Flickr8k and Flickr30K, in order to improve the accuracy of the

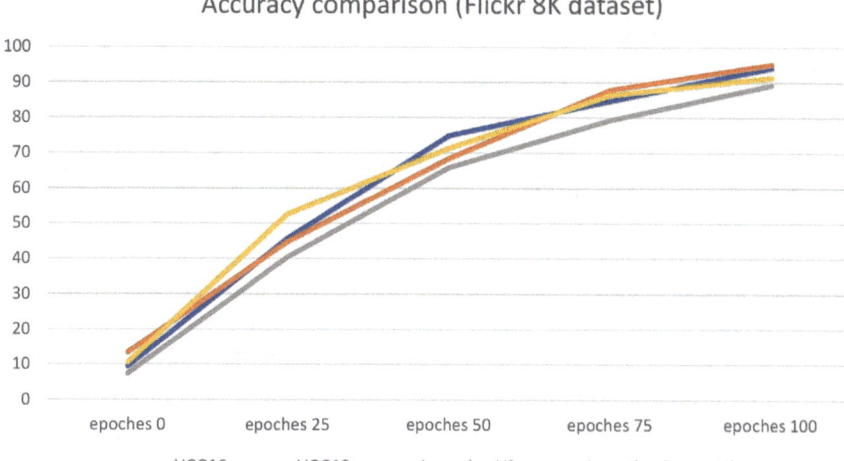

Fig. 6 Accuracy graph for the Flickr8K Dataset

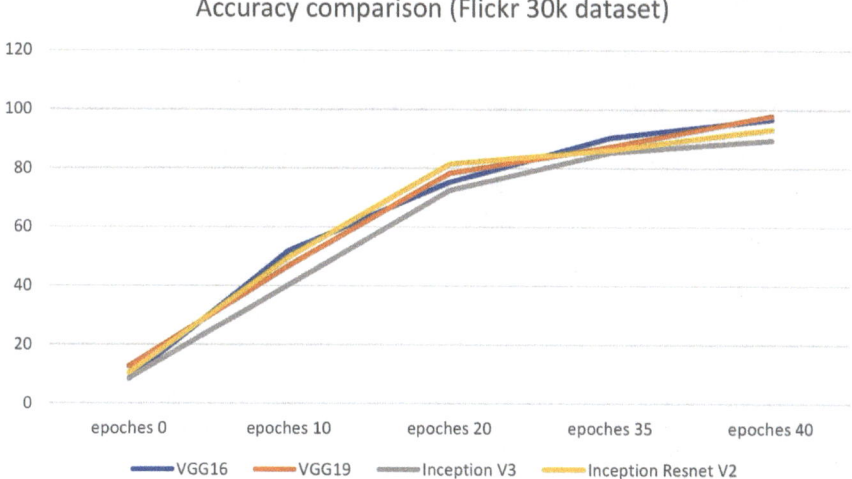

Fig. 7 Accuracy graph for the Flickr30K dataset

Table 2 Value loss (%) of different CNN encoder architectures	CNN models	Flickr8K	Flickr30K
	VGG16	42.8	42.1
	VGG19	39.8	39.7
	Inception V3	54.7	50.6
	Inception ResNetV2	56.2	54.9

models during training. It was analyzed that the choice of CNN architecture and different factors including size and quality of the training dataset, hyper-parameters combination, and the specific implementation of the models can have a notable impact on the performance of image captioning.

In the present work, only four state-of-the-art CNN-based encoders are selected for comparative analysis. In future, it would be interesting to further compare other state-of-the-art models like ResNet50, ResNet101, and Xception. Moreover, different parameter settings may be explored in future as presently, default parameter settings are analyzed for comparison. Lastly, one interesting direction would be to explore different search strategies like beam search and greedy search techniques for diverse and better caption generation, time optimization, and overall performance of the CNN-based models for effective generation of the image captions.

References

1. Vinyals O, Toshev A, Bengio S, Erhan D (2015) Show and tell: a neural image caption generator. In: 2015 IEEE Conference on Computer Vision and Pattern Recognition (CVPR), pp. 3156–3164. https://doi.org/10.1109/CVPR.2015.7298935
2. Jethwa N, Gabajiwala H, Mishra A, Joshi P, Natu P (2021) Comparative analysis between InceptionResNetV2 and InceptionV3 for attention based image captioning. In: 2021 2nd global conference for advancement in technology (GCAT), pp. 1–6. Bangalore, India. https://doi.org/10.1109/GCAT52182.2021.9587514
3. Alam MS, Rahman MS, Hosen MI, Mubin KA, Hossen S, Mridha MF (2021) Comparison of Different CNN Model used as Encoders for Image Captioning. In: 2021 International Conference on Data Analytics for Business and Industry (ICDABI), pp. 523–526. Sakheer, Bahrain. https://doi.org/10.1109/ICDABI53623.2021.9655846
4. Suresh, K. R., Jarapala, A., & Sudeep, P. V.: Image Captioning Encoder–Decoder Models Using CNN-RNN Architectures: A Comparative Study. Circuits, Systems, and Signal Processing, pp. 5719–5742. https://doi.org/10.1007/s00034-022-02050-2 (2022)
5. Kanimozhiselvi, C. S., V, K., P, K. S., & S, K.: Image Captioning Using Deep Learning. In: 2022 International Conference on Computer Communication and Informatics (ICCCI), pp. 1–7. Coimbatore, India. https://doi.org/10.1109/ICCCI54379.2022.9740788 (2022)
6. Dwivedi, P., & Upadhyaya, A.: A Novel Deep Learning Model for Accurate Prediction of Image Captions in Fashion Industry. In: 2022 12th International Conference on Cloud Computing, Data Science and Engineering, pp. 207–212. Noida, India. https://doi.org/10.1109/Confluenc e52989.2022.9734171 (2022)
7. Simonyan, K., & Zisserman, A.: Very Deep Convolutional Networks for Large-Scale Image Recognition. arXiv e-prints, pp. 1409–1556. https://doi.org/10.48550/arXiv.1409.1556 (2014)
8. Bhatia, Y., Bajpayee, A., Raghuvanshi, D., & Mittal, H.: Image Captioning using Google's Inception-ResNet-V2 and Recurrent Neural Network. In: 2019 Twelfth International Conference on Contemporary Computing (IC3), pp. 1–6. Noida, India. https://doi.org/10.1109/IC3.2019.8844921 (2019)
9. Adriyendi, A.: A Rapid Review of Image Captioning. Journal of Information Technology and Computer Science, vol. 6, pp. 158–169. https://doi.org/10.25126/jitecs.202162316 (2021)
10. Mishra, S. K., Sinha, S., Saha, S., & Bhattacharyya, P.: Dynamic Convolution-Based Encoder-Decoder Framework for Image Captioning in Hindi. ACM Transactions on Asian and Low-Resource Language Information Processing, pp. 1–18. https://doi.org/10.1145/3573891 (2023)

11. Thirumahal R, Prabakaran H, Swetha GN, Sushmitha S, Swathi S, Balasubramaniam C (2023) Image Captioning Generator and Comparison Study. Journal of Innovative Image Processing, vol. 4, pp. 328–337. https://doi.org/10.36548/jiip.2022.4.009

12. Poddar AK, Rani R (2023) Hybrid Architecture using CNN and LSTM for image captioning in Hindi language. Procedia Computer Science, 218: 686–696. https://doi.org/10.1016/j.procs.2023.01.049

13. Singh A, Bansal A, Chauhan N, Sahu SP, Dewangan DK (2022) Image Generation Using GAN and Its Classification Using SVM and CNN. In: Noor, A., Sen, A., Trivedi, G (eds) Proceedings of Emerging Trends and Technologies on Intelligent Systems. ETTIS 2021. Advances in Intelligent Systems and Computing, vol 1371. Springer, Singapore, pp. 89–100. https://doi.org/10.1007/978-981-16-3097-2_8

14. Dewangan DK, Sahu SP (2023) Lane detection in intelligent vehicle system using optimal 2- tier deep convolutional neural network. Multimedia Tools and Applications, vol. 82, pp. 7293–7317. https://doi.org/10.1007/s11042-022-13425-7

15. Dewangan DK, Sahu SP (2022) Optimized Convolutional Neural Network for Road Detection with Structured Contour and Spatial Information for Intelligent Vehicle System. Int J Pattern Recogn Artificial Intell 36. https://doi.org/10.1142/S0218001422520024

16. Mascarenhas S, Agarwal M (2021) A comparison between VGG16, VGG19 and ResNet50 architecture frameworks for Image Classification. 2021 International Conference on Disruptive Technologies for Multi-Disciplinary Research and Applications (CENTCON), pp. 96–99. https://doi.org/10.1109/CENTCON52345.2021.9687944

Computer Networks

Security and Energy Efficiency Enhancement for the Internet of Things: Challenges, Architecture and Future Research

Ritu Dewan, Tapsi Nagpal, Sharik Ahmad, Arun Kumar Rana, and Sardar M. N. Islam

Abstract Thanks to constantly advancing technology, the world is changing rapidly. One such idea that has contributed to the reality of automation is the Internet of Things (IoT). IoT links various non-living objects to the internet and enables them to communicate with their local network to automate processes and simplify people's lives. The IoT's potential needs to be completely realised despite the enormous efforts of standards, organisations, coalitions, businesses, academics and others. A number of problems remain. The enabling technology, applications and business models, as well as the social and environmental repercussions, should all be taken into consideration while analysing these difficulties. This article's emphasis is on unresolved issues and challenges from a technological perspective. The main objective is to provide a thorough evaluation of IoT in terms of energy and security, as well as unresolved problems and obstacles that need further study. We provide some perspectives on some new concepts to help future research.

Keywords Challenges · Architecture · Communication protocol · Enabling technologies · Interoperability

R. Dewan (✉) · T. Nagpal · S. Ahmad
Department of Computer Science and Engineering, Lingayas Vidyapeeth, Faridabad, India
e-mail: dewanritu22@gmail.com

T. Nagpal
e-mail: dr.tapsi@lingayasvidyapeeth.edu.in

S. Ahmad
e-mail: dr.sharik@lingayasvidyapeeth.edu.in

A. K. Rana
Department of Computer Science and Engineering, Galgotias College of Engineering & Technology, Gr. Noida, India
e-mail: arunkumar@galgotiacollege.edu

S. M. N. Islam
SILC, Decision Sciences and Modelling Program, Victoria University, Victoria, Australia
e-mail: sardar.islam@vu.edu.au

© The Author(s), under exclusive license to Springer Nature Singapore Pte Ltd. 2024   317
S. Namasudra et al. (eds.), *Data Science and Network Engineering*, Lecture Notes in Networks and Systems 791, https://doi.org/10.1007/978-981-99-6755-1_24

1 Introduction

The phrase 'Internet of Things' (IoT) describes a totally networked world [1]. It integrates everything with technology and opens up a brand-new world for children to interact with one another online. IoT has the ability to revolutionise technology and alter people's lives; it is more than just a notion [2]. The IoT will have an influence on any concept, practical or theoretical. Anybody should be able to access any content from any device at any time from anyone located anywhere who is a member of any firm or service, across any route or network, since it demands object communication. In a good way, 'availability' is an important factor that affects IoT performance [3–8].

IoT architecture connects sensors and smart devices to a local or cloud-based controller. Sensors often provide environmental data in real time. Controllers may then make immediate and long-term decisions. Predictive and adaptive algorithms may help controllers make basic to complicated operational decisions. In-room occupancy sensors may illuminate a workplace [9–11]. IoT systems have three layers: hardware or sensor, software control and application. The hardware or sensor layer is built of physical pieces with independent identities that may convey data over a network without human-to-human or human-to-computer interactions. A set of actuators that regulate a building's heating and lighting system or a network of distant sensors that detect and report environmental variables are examples. Improve IoT deployment apps and use cases. End users influence design decisions like any competent engineer. A not-so-unique twist is that adapting IoT devices for human conduct often involves collecting and transmitting data on human activity and behavioural profiles, which may be invasive to users.

Before examining research trends, we must define the IoT and assess its effects on society, business and new business models. The Internet of Things is young, and everyone is attempting to interpret it according to their own requirements [12]. This makes the IoT's definition and concepts ambiguous. IoT is also known as IoE, WoT, CoT, M2M and others. Even while some writers believe both statements imply the same, most authors differentiate between them [13, 14].

2 Motivation, Objectives and Contributions

The IoT is present. Sensors and embedded devices are already allowing new sorts of information and changing how it is produced, consumed and experienced in automobiles, smartphones, watches, supermarkets, homes, roads and bridges, appliances, industrial and agricultural equipment and wearable technology. There is a lot of room for improvement in data analytics thanks to the IoT. The connections between IoT, data processing and storage and machine learning are obvious and are already taking off. The basis for this contribution is that IoT devices can effectively and affordably transmit highly secure information to the wireless network in urgent circumstances. The primary goals are as follows. This article provides a comprehensive analysis of IoT's energy and security needs:

- to provide a thorough review of IoT features and to explain the IoT vision and terminology;
- to provide a high-level overview of the capabilities of popular IoT enabling and emerging technologies;
- to encourage discussion on current IoT issues and challenges that will be addressed by future study;
- to go into further detail on the many IoT designs, uses and difficulties that come from fusing the IoT with other technologies.

3 Literature Survey

Smart mobile phones and sensors can monitor resident and building data to improve urban productivity [15]. The Localisation Novel Method (LNM) uses neighbour sign quality to build a unique mark chronicle for walkers. It underpins a Markov model that predicts pedestrian movement. The case is used to examine the stunning sign deviation [16]. After testing, the suggested technique outperforms others. Wi-Fi signal constancy is an issue. The suggested architecture uses a wearable interface with photo recognition and confinement to automatically provide social data on the pieces of art being seen. The cloud stores client-created interactive media material and shares framework events on the client's social media [17, 18]. More specifically, a nuclear event extraction module removes nuclear events from messages and creates a β-organisation to parse nuclear events [19].

Focus on a BIM morphogenetic engineering system planning group audit to finish. The AI-Based Smart Building Automation Controller (AIBSBAC) naturally responds to client needs for ease, security and energy savings. AIBSBAC's architecture also allows fast connection and play improvements without framework updates for most applications, including private and building computerisation. Built an IoT-based learning framework for smart building temperature control thermal models. Smart technologies, parameter management and IoT infrastructure need advanced operations. Cloud customers utilise communication protocols to identify and manage the IoT basis [21–23]. Most of the data came from the EU-funded Smart Santander project, a city-scale IoT and Potential Internet experimental testbed [24, 25]. Energy-efficient IoT-enabled smart building template. The 'Laplacian IoT matrix' shows IoT network smart building graphs. Qualitative case studies follow the advice. We can explain and apply fundamental IoT concepts to smart homes and simple use cases in a condensed Smart Home context to excite new clients. IoT-enabled development monitoring. A controlled, Internet-connected sensor node in the building measures continuously and sends raw data to the remote server through MQTT [26–29].

To enable rapid and reliable handovers (IIO) in 4G and 5G networks, new authentication and re-authentication methods were developed using conventional authentication protocols. One possible enabling technology for solving these concerns is the IoT, a massive worldwide network of linked physical items containing electronics, software, sensors and network connection. The IoT may develop future smart cities by using sustainable information and communication technology [30]. The IoT's rapid growth affects many scientific and technological disciplines. The IoT, a global Internet-based technological architecture that promotes product and service exchange in global supply chain networks, affects security and privacy. Data authentication, access control and client privacy must be protected in the architecture. A good legal framework must address the underlying technology and be designed by an international legislator, strengthened by the private sector according to particular needs, and readily adjustable [31].

IoT refers to embedded networks that employ Internet protocols for human-to-thing or thing-to-thing communication. It is unknown how IP security protocols and architectures may be implemented in this area, despite well-established security standards. Content-based publish/subscribe has been used as an overlay network of software brokers for decades [32]. Software-based bandwidth-efficient expressive filtering cannot match network-layer communication protocols' end-to-end latency and throughput. SDN was used to increase network-layer performance for content-based publish/subscribe. SDN permits content filters directly installed on switches to send events at line rate, however fundamental hardware limits (flow table size, bit availability in header fields) limit their expressiveness, resulting in unnecessary network traffic [33]. WSN uses Low-Energy Adaptive Clustering Hierarchy (LEACH) protocol to regulate battery lifespan and cluster head cell-active data aggregation and network efficiency. Clustering heads and a single cluster head may increase network performance. WSNs have several applications. One of the latest uses is the IoT, which connects objects through the Internet [34].

4 IoT Architecture

IoT technology is diverse. Consequently, one reference design cannot be used to guide all particular implementations. On the internet, several paradigms coexist. The architecture is used to describe the physical components, functional organisation, networking, operating principles, processes and data formats [33] (Fig. 1).

5 Three-Layer Architecture

Further, the IoT layer architecture is shown in Fig. 2. That mainly works in three layers and the function of each layer is discussed below:

(a) **The physical Perception** Layer interacts directly with IoT sensors. Sensors can understand environmental data, which is exciting. To continue to include sensors in the three phases of an IoT architecture system, information must be processed. Sensors can analyse outside data. In other words, the inclusion of sensors under three phases in an IoT system backdrop will begin to incorporate details with a clear appearance and feel.

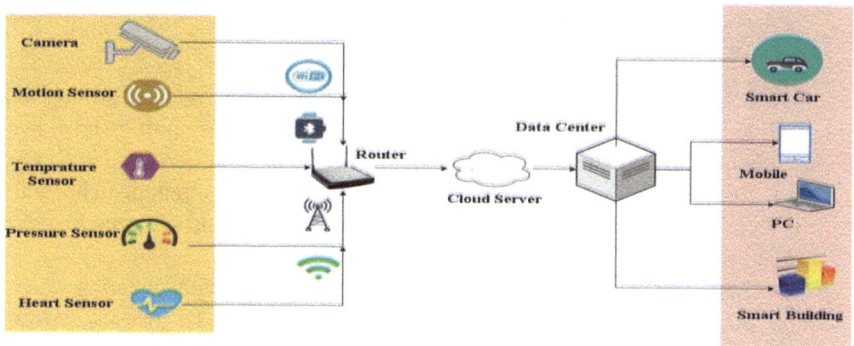

Fig. 1 IoT architecture

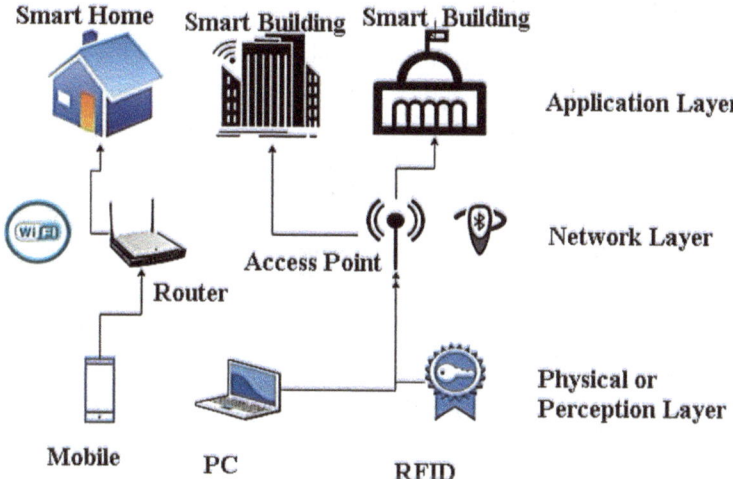

Fig. 2 IoT layer architecture

(b) **Network Layer** connects data to the cloud and last. The network layer connects intelligent items, network appliances and servers. Sensor data transport and analysis utilise it. IoT architecture stages transferred optimal details to IT. Edge IT systems improve analytics and pre-processing. Machine learning and visualisation are examples. Further testing will occur when you reach the data centre.

(c) **Application Layer** offers unique customer resources. Smart homes, intelligent communities and intelligent wellness are IoT applications. The latest IoT architecture essential processes are in the data centre or cloud. It allows extensive treatment and feedback evaluation. IT and OT professionals are required.

5.1 Four-Layer Architecture

Heterogeneous IoT architectures need scalability, modularity, interoperability and adaptability. IoT architecture involves cross-domain connections, multi-system integration, flexible functional management linkages, huge data processing and storage and user-friendly applications. The software should add functionality and automate IoT computers. IoT creates huge data. IoT streaming data requires complex infrastructure [35–40].

IoT sensors and gadgets generate massive amounts of data, creating a new dilemma. So, an IoT system needs an efficient design to handle massive streams of data. IoT systems handle, monitor and analyse enormous volumes of data using cloud and fog/edge computing. As shown in Fig. 3, modern IoT architecture is four-stage. Sensors and actuators are crucial in architectural stage 1. The real world includes people, animals, electronics, smart cars, buildings and more. Sensors convert real-world signals and data flow into data for analysis.

Actuators can also manage room temperature, vehicle speed, music and light. Hence, Stage 1 helps gather real-world data for study. Stage 2 includes sensors, actuators, gateways and data collection. This phase organises and optimises the massive data acquired in Stage 1 for processing. Stage 3, edge computing, begins after massive data collection and organisation. Edge computing is a distributed open architecture that lets IoT technologies and massive processing capability be utilised worldwide. IoT devices benefit from its sophisticated streaming data processing method [41–43].

Stage 3 edge computing solutions handle massive volumes of data and provide visualisation, data integration from various sources, machine learning analysis and more. In the last stage, processing, analysis and feedback improve the system's precision and accuracy. Now everything will be done on a cloud server or data centre. Machine learning may improve prediction models for a more accurate and dependable IoT system to fulfil demand. Hadoop and Spark can manage this huge streaming data, and machine learning can improve prediction models for a more accurate and dependable IoT system to fulfil demand. Architecture defines a network's physical components, functional structure and configuration, operational rules and procedures and data formats. IoT growth depends on technology, applications and business. IoT device designs vary. The '5 Layer Architecture' is the finest IoT architecture, though.

5.2 Five-Layer Architecture

The five-layer design is suitable for projects with cutting-edge technology and extensive applications. Figure 3 shows that the five-layer model adds two levels to the basic IoT architecture [44].

(a) **Perception layer**: IoT architecture's initial layer. The perception layer uses several sensors and actuators to measure temperature, moisture, incursion, noise and more. This layer collects environmental data and sends it to another layer for action.

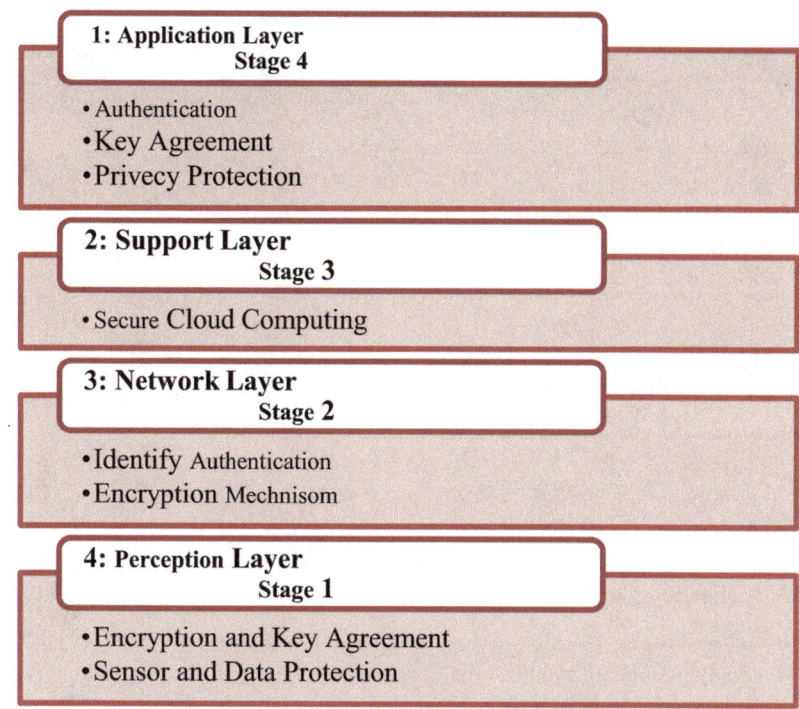

Fig. 3 Four-stage IoT architecture to deal with massive data

(b) **Network layer**: It links the perceptual and middleware levels. It sends perception layer data to the middleware layer via 3G, 4G, UTMS, WiFI, infrared and other networking technologies. The communication layer connects the perceptual and middleware levels. Data is sent securely and kept private.

(c) **Middleware**: It stores, calculates, processes and takes action. It stores and transmits data to the correct device based on its address and name. Sensor data calculations may also inform its decisions.

(d) **Application layer**: The application layer manages all application activities using middleware layer data. This application controls emails, alerts, security systems, gadgets, wearables, smart agriculture and more.

(e) **Business layer**: Technology and distribution determine a device's success. The device's business layer handles this. It involves flowcharting, graphing, data analysis and device improvement (Fig. 4).

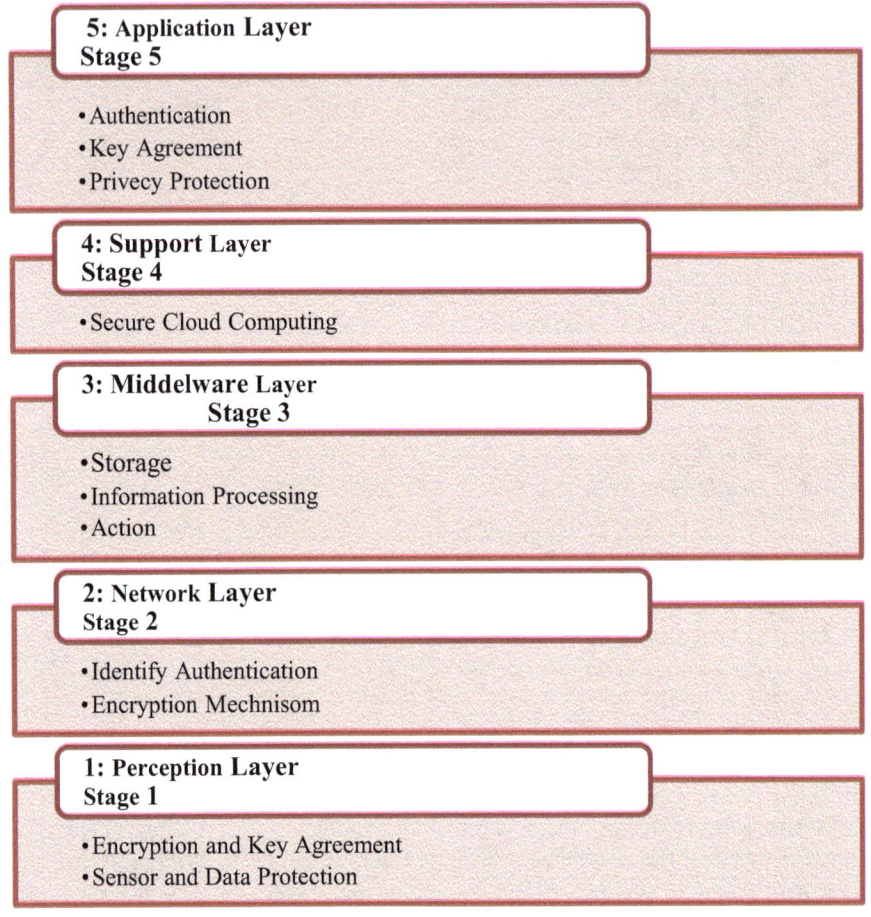

5: Application Layer
Stage 5

• Authentication
• Key Agreement
• Privecy Protection

4: Support Layer
Stage 4

• Secure Cloud Computing

3: Middelware Layer
              Stage 3

• Storage
• Information Processing
• Action

2: Network Layer
Stage 2

• Identify Authentication
• Encryption Mechnisom

1: Perception Layer
Stage 1

• Encryption and Key Agreement
• Sensor and Data Protection

Fig. 4 Five-Layer architecture of Internet of Things

6 IoT with Artificial Intelligence (AI)

Applications and installations for the IoT are increasingly using AI. Investments and acquisitions in companies that mix AI and IoT have grown during the last 2 years. Many IoT platform software companies currently provide integrated AI features including machine learning-based analytics [45]. Information may be extracted from data that previously needed human analysis using other AI technologies like voice recognition and computer vision.

IoT AI applications assist organisations in reducing unscheduled downtime, enhancing operational effectiveness, creating new products and services and better managing risk. IoT with AI capabilities may also help increase operational effectiveness. Machine learning can foresee operating situations and uncover parameters that may be modified on the fly to maintain optimal outcomes by analysing continuous streams of data to detect patterns that are invisible to the human eye and not obvious on basic gauges.

7 Challenges of IoT

IoT combines privacy-focused industries including cyber technology, telecommunications, software analytics and farming. IoT categorisation [46].

(a) **Standardisation and Interoperability**

The multiplicity of technology and standards is one of the main obstacles to the development of IoT applications. The cornerstone for IoT manufacturing in the future will be the standardisation of IoT architecture and communication technologies.

(b) **System Security**

The communication system must be able to handle data from many devices without experiencing any data loss due to framework obstruction, maintain proper safety measures for delivering data and check for security issues.

(c) **Availability**

One of the most important problems to be resolved in order to handle the complexity of IoT systems efficiently is service availability. The availability of IoT applications means that they must be available for every permitted item at all times and from any place.

(d) **Transmission Rate and Power Use**

New computational and calibration methods are required because of the rapid increase in the quantity of linked items and data flow. For IoT systems to enable big data, IoT applications need a well-known analytic framework that can be made accessible as a service.

(e) **Data Management**

Due to the enormous number of IoT devices, data management is now the biggest challenge. Despite the fact that IoT technologies like Near-Field Communication (NFC), RFID and Thread exist.

(f) **Encryption and Decryption of Data**

With the aid of decryption and encryption, the sensing devices carry out self-governing estimates or recognising and exchanging data with the information handling unit across the transmission system. The web is connected to a huge variety of devices.

(g) **Data Privacy**

RFID, 2D institutionalised labels and other unique types of item recognition confirming headways are used in the IoT. It is important to establish legal assurance methods and prevent unauthorised access since every kind of step-by-step usage article will include these obvious verification markings and add unique item information.

8   Future Research

IoT industry challenges and open problems are outlined in this paper. Background information about the subjects and ideas that will be studied is provided in this section. The majority of issues, according to recent contributions and research studies, are caused by rising traffic demands, a variety of traffic types, data formats, IoT devices, heterogeneous networks, etc. Performance of IoT systems is highly impacted by these factors. Specific application requirements include things like speed and reliability of computing and communication, as well as mobility, dependability, privacy and security. IoT applications benefit from cloud computing in order to meet these needs. In IoT system architecture, fog computing, cloudlets, MEC and MCC may be employed. Today, the majority of IoT applications and services rely on cloud infrastructure and smart device data exchange. This strategy must overcome administrative, social and technological obstacles as well as technical ones related to network latency, throughput and reliability. Fog computing, cloudlets, MEC, MCC and other computing paradigms and architectures were introduced last year to deploy infrastructure nearer to data sources. Models are differentiated by 'infrastructure distance' from IoT devices, distributed function infrastructure and other elements. In terms of processing and storage, cloud computing performs better than network edge hardware. Data storage and pattern analysis are good examples of long-term jobs. For IoT subsystems, it could make use of a load distribution technique. IoT applications may be categorised based on this evaluation's performance profile (such as QoS requirements). Additionally, application architectures and resource management techniques may be compared in this examination. By simulating IoT systems, this method may assess QoS and QoE. Future advancement may be addressed by adequate mathematical formulation and evaluation methodologies for IoT systems [47–50].

9   Conclusion

IoT must allow seamless connectivity anytime, everywhere, by anyone, and with everything in order to offer intelligent services. This comprises capabilities for identification, sensing, networking, processing and visualisation. This idea created a plethora of fresh opportunities for the creation of substantial services and goods, sparking an explosion of fresh concepts and business opportunities. The research community must tackle new issues brought on by emerging technology and new IoT applications. By visualising the most intriguing future research topics and achievements, we have summed up, assessed and articulated these challenges. Future research is aided by a concise overview of engaging study areas, a categorisation of unresolved issues based on designated functional domains and insights into specific emerging difficulties and concepts. The research community will profit from this effort as a consequence since it might serve as a basis for further investigation.

References

1. Hajjaji Y, Boulila W, Farah IR, Romdhani I, Hussain A (2021) Big data and IoT-based applications in smart environments: a systematic review. Comput Sci Rev 39:100318
2. Chegini H, Naha RK, Mahanti A, Thulasiraman P (2021) Process automation in an IoT–Fog–cloud ecosystem: a survey and taxonomy. IoT 2(1):92–118
3. Zhang Y, Sun Y, Jin R, Lin K, Liu W (2021) High-performance isolation computing technology for smart IoT healthcare in cloud environments. IEEE Internet Things J: 1–1
4. Jacob TP, Pravin A, Ramachandran M, Al-Turjman F (2021) Differential spectrum access for next generation data traffic in massive-IoT. Microprocess Microsyst 82:103951
5. Rana A, Sharma S, Nisar K, Ibrahim A, Ag A, Dhawan S, Goyal N (2022) The rise of blockchain internet of things (BIoT): secured, device-to-device architecture and simulation scenarios. Appl Sci 12(15):7694
6. Kumar A, Sharma S, Goyal N, Singh A, Cheng X, Singh P (2021) Secure and energy-efficient smart building architecture with emerging technology IoT. Comput Commun 176:207–217
7. Kumar A, Sharma S, Singh A, Alwadain A, Choi BJ, Manual-Brenosa J, Goyal N (2021) Revolutionary strategies analysis and proposed system for future infrastructure in internet of things. Sustainability 14(1):71
8. Rana A, Chakraborty C, Sharma S, Dhawan S, Pani SK, Ashraf I (2022) Internet of medical things-based secure and energy-efficient framework for health care. Big Data 10(1):18–33
9. Kumar A, Sharma S (2021) Internet of robotic things: design and develop the quality of service framework for the healthcare sector using CoAP. IAES Int J Robot Automat 10(4):289
10. Dhawan S, Chakraborty C, Frnda J, Gupta R, Rana AK, Pani SK (2021) SSII: secured and high-quality steganography using intelligent hybrid optimization algorithms for IoT. IEEE Access 9:87563–87578
11. Rana SK, Kim HC, Pani SK, Rana SK, Joo MI, Rana AK, Aich S (2021) Blockchain-based model to improve the performance of the next-generation digital supply chain. Sustainability 13(18):10008
12. Rana AK, Sharma S (2021) Internet of things based stable increased-throughput multi-hop protocol for link efficiency (IoT-SIMPLE) for health monitoring using wireless body area networks. Int J Sensors Wireless Commun Control 11(7):789–798
13. Kumar A, Sharma S, Goyal N, Gupta SK, Kumari S, Kumar S (2022) Energy-efficient fog computing in Internet of Things based on routing protocol for low-power and lossy network with Contiki. Int J Commun Syst 35(4):e5049
14. Pandit M, Gupta D, Anand D, Goyal N, Aljahdali HM, Mansilla AO, Kumar A (2022) Towards design and feasibility analysis of DePaaS: AI based global unified software defect prediction framework. Appl Sci 12(1):493
15. Berntzen L, Johannessen MR, Florea A (2016) Sensors and the smart city: creating a research design for sensor-based smart city projects. In ThinkMind//SMART, the fifth international conference on smart cities, systems, devices and technologies

16. Lin K, Chen M, Deng J, Hassan MM, Fortino G (2016) Enhanced fingerprinting and trajectory prediction for IoT localization in smart buildings. IEEE Trans Autom Sci Eng 13(3):1294–1307
17. Alletto S, Cucchiara R, Del Fiore G, Mainetti L, Mighali V, Patrono L, Serra G (2015) An indoor location-aware system for an IoT-based smart museum. IEEE Internet Things J 3(2):244–253
18. Kaur MJ, Maheshwari P (2016) Building smart cities applications using IoT and cloud-based architectures. In: 2016 international conference on industrial informatics and computer systems (CIICS) (pp. 1–5)
19. Sun Y, Wu TY, Zhao G, Guizani M (2014) Efficient rule engine for smart building systems. IEEE Trans Comput 64(6), 1658–1669
20. McGinley T (2015) A morphogenetic architecture for intelligent buildings. Intelligent Buildings International 7(1):4–15
21. Basnayake BA, Amarasinghe YW, Attalage RA, Udayanga TD, Jayasekara AG (2015) Artificial intelligence based smart building automation controller for energy efficiency improvements in existing buildings. Int J Advanc Automat Sci Technol 40(40)
22. Zhang X, Pipattanasomporn M, Chen T, Rahman S (2019) An IoT-based thermal model learning framework for smart buildings. IEEE Internet Things J 7(1):518–527
23. Verma A, Prakash S, Srivastava V, Kumar A (2019) Sensing, controlling, and IoT infrastructure in smart building: a review. IEEE Sensors J 19(20):9036–9046
24. Le DN, Le Tuan L, Tuan MND (2019) Smart-building management system: an Internet-of-Things (IoT) application business model in Vietnam. Technol Forecast Soc Chang 141:22–35
25. E. Theodoridis, G. Mylonas and I. Chatzigiannakis, "Developing an iot smart city framework" In IISA (pp. 1–6), 2013.
26. Metallidou CK, Psannis KE, Egyptiadou EA (2020) Energy Efficiency in Smart Buildings: IoT Approaches. IEEE Access 8:63679–63699
27. R. Casado-Vara, A. Martín del Rey, R. S. Alonso, S. Trabelsi & J. M. Corchado, "A New Stability Criterion for IoT Systems in Smart Buildings: Temperature Case Study" Mathematics, 8(9), 1412, 2020.
28. O. Debauche, S. Mahmoudi & Y. Moussaoui, "Internet of things learning: a practical case for smart building automation" International Conference on Cloud Computing and Artificial Intelligence: Technologies and Applications, Marrakech, Morocco, 2020.
29. P. Pierleoni, M. Conti, A. Belli, L. Palma, L. Incipini,, L. Sabbatini & R. Concetti, "IoT Solution based on MQTT Protocol for Real-Time Building Monitoring" IEEE 23rd International Symposium on Consumer Technologies (ISCT) (pp. 57–62), 2019.
30. Alezabi KA, Hashim F, Hashim SJ, Ali BM, Jamalipour A (2020) Efficient authentication and re-authentication protocols for 4G/5G heterogeneous networks. EURASIP Journal on Wireless Communications and Networking, 1–34.
31. Weber RH (2010) Internet of things-new security and privacy challenges. Comput Law Secur Rev 26(1):23–30
32. Toor A, ul Islam S, Ahmed G, Jabbar S, Khalid S, Sharif AM, (2020) Energy efficient edge-of-things. EURASIP J Wirel Commun Netw 1:1–1
33. Liu J, Xiao Y, Philip-Chen CL (2016) Hybrid content-based routing using network and application layer filtering. IEEE 36th International Conference on Distributed Computing Systems (ICDCS), (pp. 221–231). IEEE.
34. Behera TM, Mohapatra SK, Samal UC, Khan MS, Daneshmand M, Gandomi AH (2019) Residual energy-based cluster-head selection in WSNs for IoT application. IEEE Internet Things J 6(3):5132–5139
35. Behera M, Samal U, Mohapatra K (2018) Energy efficient modified LEACH protocol for IoT applications. IET wireless sensors system 8(5):223–228
36. Lilhore UK, Imoize AL, Lee CC, Simaiya S, Pani SK, Goyal N, Li CT (2022) Enhanced Convolutional Neural Network Model for Cassava Leaf Disease Identification and Classification. Mathematics 10(4):580
37. Rana SK, Rana SK, Nisar K, Ag Ibrahim AA, Rana AK, Goyal N, Chawla P (2022) Blockchain Technology and Artificial Intelligence Based Decentralized Access Control Model to Enable Secure Interoperability for Healthcare. Sustainability 14(15):9471

38. Rana AK, Sharma S (2019) Enhanced energy-efficient heterogeneous routing protocols in WSNs for IoT application. IJEAT 9(1):4418–4415
39. Rana AK, Sharma S (2021) Industry 4.0 manufacturing based on IoT, cloud computing, and big data: manufacturing purpose scenario. In Advances in Communication and Computational Technology (pp. 1109–1119). Springer, Singapore.
40. Rana, A. K., Krishna, R., Dhwan, S., Sharma, S., & Gupta, R. (2019, October). Review on artificial intelligence with internet of things-problems, challenges and opportunities. In 2019 2nd International Conference on Power Energy, Environment and Intelligent Control (PEEIC) (pp. 383–387). IEEE
41. Rana AK, Sharma S (2021) Contiki Cooja Security Solution (CCSS) with IPv6 routing protocol for low-power and lossy networks (RPL) in Internet of Things applications. In Mobile Radio Communications and 5G Networks (pp. 251–259). Springer, Singapore
42. Kumar A, Sharma S (2021) IFTTT rely based a semantic web approach to simplifying trigger-action programming for end-user application with IoT applications. In Semantic IoT: Theory and Applications (pp. 385–397). Springer, Cham
43. Kumar, A., & Sharma, S. (2021). 11 Internet of Things. Electrical and Electronic Devices, Circuits and Materials: Design and Applications, 183.
44. Kumar, A., Sharma, S., Dhawan, S., Goyal, N., & Lata, S. (2021). 14 E-learning with Internet of Things. Internet of Things: Robotic and Drone Technology, 195.
45. Dhawan, S., Gupta, R., Rana, A., & Sharma, S. (2021). Various Swarm Optimization Algorithms: Review, Challenges, and Opportunities. Soft Computing for Intelligent Systems, 291–301.
46. Rana, A. K., & Sharma, S. (2021). The Fusion of Blockchain and IoT Technologies with Industry 4.0. In Intelligent Communication and Automation Systems (pp. 275–290). CRC Press.
47. Rana, A. K., Sharma, S., Dhawan, S., & Tayal, S. (2021). Towards Secure Deployment on the Internet of Robotic Things: Architecture, Applications, and Challenges. In Multimodal Biometric Systems (pp. 135–148). CRC Press.
48. Arora, S., Sharma, S., & Rana, A. K. (2021). 8 Ultrawide Band Antenna for Wireless Communications. Internet of Things: Robotic and Drone Technology, 95.
49. Kumar, A., & Sharma, S. (2021). Demur and routing protocols with application in underwater wireless sensor networks for smart city. In Energy-Efficient Underwater Wireless Communications and Networking (pp. 262–278). IGI Global.
50. Rana AK, Sharma S, Dhawan S, Tayal S 2021) 11 Towards Secure. Multimodal Biometric Systems: Security and Applications, 135

Spot Pricing in Cloud Computing: A Comprehensive Survey of Mechanisms, Strategies, and Future Directions

Nikhil Purohit, Prakash Srivastava, Vikas Tripathi, and Noor Mohd

Abstract Cloud computing (CC) has transformed the way businesses store, process, and access data, and spot pricing has emerged as a key feature of this technology. Spot pricing enables users to bid on unused computing resources at lower prices, providing cost-effective access to computing resources and helping businesses optimize their operational infrastructure expenses. The proposed research is focused on case studies and existing research based on the impact of spot pricing in CC, including its effect on workload distribution and resource allocation. It also emphasizes the significance of spot pricing in CC and its potential to benefit businesses of all sizes. The paper concludes with open research directions, including the benefits and challenges of spot pricing, the various factors to consider when using spot pricing, and the need for future research in this area highlighting its importance and potential impact on businesses.

Keywords Cloud computing · Spot cloud · Spot market · Amazon pricing mechanism

1 Introduction

CC has brought about a revolutionary change in the way businesses manage and utilize data. By leveraging CC, organizations can store, process, and access their data in a scalable and cost-effective manner [1]. With the exponential growth of data, traditional data storage methods have become increasingly challenging to maintain. Therefore, CC has emerged [2] as a solution to address this issue. It provides a new approach to storing data, leveraging the power of the Internet [3]. A significant aspect of CC is its pay-as-you-go pricing model which allows customers to pay only for the resources they consume, offering flexibility and cost optimization [4].

N. Purohit · P. Srivastava (✉) · V. Tripathi · N. Mohd
Department of Computer Science and Engineering, Graphic Era (Deemed to be University),
Dehradun, India
e-mail: prakash2418@gmail.com

© The Author(s), under exclusive license to Springer Nature Singapore Pte Ltd. 2024   331
S. Namasudra et al. (eds.), *Data Science and Network Engineering*, Lecture Notes
in Networks and Systems 791, https://doi.org/10.1007/978-981-99-6755-1_25

In recent years, spot pricing has gained popularity as a key feature of CC. Spot pricing enables users to bid on unused computing resources at lower prices, facilitating cost-effective access to computing power [5]. The spot pricing model is particularly advantageous for businesses with fluctuating workloads that can tolerate some level of downtime. By leveraging spot pricing, such businesses can significantly reduce their CC costs while meeting their performance requirements [6].

Several cloud service providers have adopted and refined spot pricing as part of their offerings [7]. However, this pricing model is not without its challenges. Price volatility, resource availability issues, and the need for efficient bidding strategies present significant hurdles for businesses utilizing spot pricing. Thus, further research is necessary to explore the benefits and challenges associated with spot pricing in CC [8]. Additionally, it is essential to investigate how spot pricing impacts workload distribution, resource allocation, and overall application performance in the cloud. This survey paper aims to provide a comprehensive overview of spot pricing in CC, examining its evolution, advantages, and challenges. The paper will delve into the various cloud service providers that offer spot pricing, highlighting case studies and research papers that analyze the impact of spot pricing on CC. By conducting a literature review, we will gather insights into the existing body of knowledge on this topic and identify potential research gaps.

The contributions of this survey paper are as follows:

1. Comprehensive analysis of spot pricing in CC, including its evolution, advantages, and challenges.
2. Analysis of different cloud service providers offering spot pricing.
3. Exploration of case studies and research papers on the impact of spot pricing on workload distribution, resource allocation, and application performance in the cloud.
4. Identification of potential research gaps and future directions in spot pricing research.

An auction mechanism (as shown in Fig. 1) is a method of determining spot prices in which cloud service providers allow users to bid on unused computing resources, resulting in a dynamic pricing system that reflects current supply and demand.

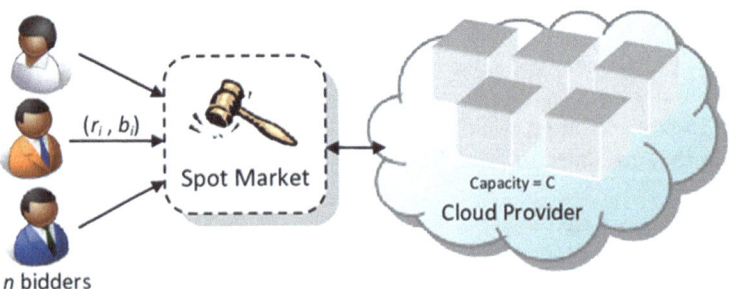

Fig. 1 Spot market and auction mechanism

The rest of the paper is organized as follows. Section 2 describes the Fundamentals of Spot Pricing in CC. Section 3 reflects the Related work. Section 4 describes the Research Gaps. Section 5 describes the Discussion and Future Work Directions. Section 6 concludes the paper.

2   Fundamental of Spot Pricing in CC

Spot pricing in CC is a pricing model where cloud service providers offer unused computing resources at a lower cost, allowing users to access additional resources and providing cost savings. However, spot pricing is volatile, with prices and availability changing based on market demand. Users must carefully monitor and manage spot instances to optimize cost savings and ensure uninterrupted operation. Some benefits and challenges of Spot pricing are as follows:

- Benefits of Spot Pricing:- Spot pricing offers significant benefits for cloud users. The pricing model allows users to access additional computing resources at a discounted rate, providing cost savings for users. Additionally, spot pricing allows cloud providers to optimize their resource utilization and revenue by offering unused cloud resources at a lower price. Spot pricing is particularly beneficial for non-critical workloads or applications that can be interrupted or stopped without significant impact. Users can bid for spot instances and utilize them for their workloads, and then release the instances when the workload is complete. This enables users to access additional computing power at a lower cost and allows cloud providers to optimize their resource utilization and revenue.
- Challenges of Spot Pricing:- Despite its advantages for cloud users, spot pricing comes with its own set of challenges. The most prominent challenge of spot pricing is the volatility in pricing and resource availability. The cost of spot instances can change swiftly based on market demand, necessitating users to monitor resource pricing and availability continually to achieve their cost savings objectives. Additionally, spot instance availability is not guaranteed, as cloud providers may reclaim them when demand surges, making it problematic for users to depend on spot instances for mission-critical workloads. Another challenge of spot pricing is the complexity of managing and monitoring spot instances. Users must have a deep understanding of their workload and resource requirements to effectively utilize spot instances. They must also monitor their instances closely to ensure they are not interrupted or terminated unexpectedly. This requires significant expertise and resources, which may be a challenge for some users. As a result, users must carefully monitor and manage their use of spot instances to ensure they obtain the desired cost savings without sacrificing performance or reliability. In this survey, we will explore the concept of spot pricing in CC, including its benefits and challenges, as well as strategies for effectively utilizing spot instances. We will also examine the current state of spot pricing in the industry, including trends and best practices, to provide a comprehensive overview of this critical aspect of CC.

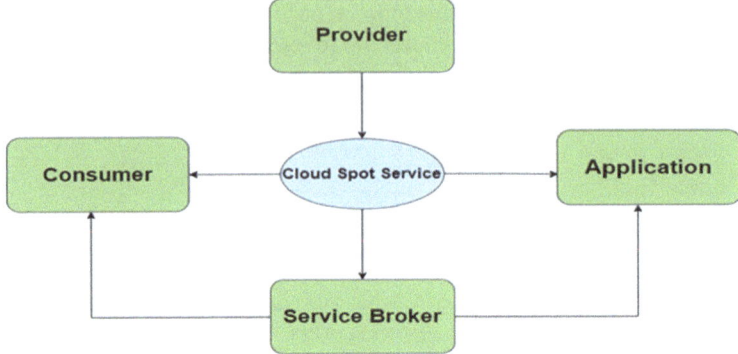

Fig. 2 Spot cycle in CC

Spot pricing refers to a pricing model that is commonly used in the commodities market, particularly in the energy and agricultural sectors. It is a type of the pricing model (shown in Fig. 2) that is based on the current market demand and supply conditions, and it is designed to allow buyers and sellers to trade commodities at the prevailing market price at the time of the transaction.

The origins of spot pricing can be traced back to the early days of commodity trading, where buyers and sellers would meet at physical marketplaces to exchange goods for cash. In these early days, prices were typically negotiated and agreed upon between buyers and sellers, with each party looking to secure the best deal possible. Over time, the growth of the commodities market led to the development of more sophisticated pricing mechanisms, including futures contracts, options, and other derivative instruments. However, spot pricing has remained a popular and widely used pricing model, particularly in markets where supply and demand conditions can fluctuate rapidly. In the energy sector, for example, spot pricing is commonly used to set the price of crude oil, natural gas, and other fuels. This is because these commodities are highly volatile and subject to sudden shifts in supply and demand conditions. Spot pricing allows buyers and sellers to respond quickly to changes in the market, ensuring that prices remain fair and transparent. In the agricultural sector, spot pricing is used to set the price of commodities such as wheat, corn, and soybeans. This allows farmers to sell their crops at the prevailing market price, and it allows food processors and manufacturers to purchase the raw materials they need to produce their products at a fair price. Overall, spot pricing has become an important tool for managing risk and ensuring fair prices in the commodities market. It allows buyers and sellers to respond quickly to changing market conditions, and it helps to promote transparency and efficiency in the pricing of commodities. As the data is increasing day by day and small to big industries are shifting toward the adoption of CC so it's very crucial to make the setup of Spot Market so that users can use the services of the cloud at affordable prices. Therefore, Spot Cloud plays a vital role in the adoption of CC. To adopt the cloud spot market needs to be explored and enhanced to maintain the cost ratio of the user so that cloud technologies can be adopted with a go.

2.1 The Pricing Mechanism of Amazon Spot Instances

Amazon Web Services (AWS) provides cost-effective access to spare computing capacity through Amazon EC2 Spot instances, which are sold through a bidding mechanism based on supply and demand. Spot instances are suitable for workloads that can be paused or are tolerant to interruptions, such as data analysis, batch processing, and testing. Customers set a maximum bid price, and the instance launches if the current spot price is below the bid price. However, if the spot price exceeds the maximum bid price, the instance may be terminated, and customers need to plan for this possibility. The pricing mechanism utilized by Amazon EC2 for Spot instances is dynamic, as it is based on an auction model that adjusts according to supply and demand. To aid customers in determining their optimal bid price, AWS publishes the current spot price for each instance type and region. The Spot price can fluctuate significantly due to various factors, including the time of day, day of the week, and demand. While utilizing Spot instances can offer customers potential cost savings of 50–90% compared to on-demand pricing, careful evaluation of the risks and trade-offs, such as the possibility of instance interruptions and the need for scalable and fault-tolerant architectures, is necessary. When Spot instances are about to be interrupted, AWS provides customers with a 2-min warning to save their work and unsaved data, while Google Cloud Platform offers a 30-s warning [4]. Overall, the Spot instance pricing mechanism provides customers with an innovative and cost-effective option for accessing EC2 capacity, with the potential to reduce the cost of running workloads and optimize their AWS spending.

2.2 The Amazon Spot Pricing Mechanism

Figure 3 is an example of the Amazon EC2 Spot auction method. The following is a description of the spot pricing mechanism.

To acquire an Amazon EC2 spot instance, users must provide instance specifications like "instance type, availability zone, number of instances, and the highest bid price". When a user's bid price exceeds (as shown in Fig. 4) the current spot price, the instance becomes available to them. Users are invoiced based on the spot pricing, and each user is paid the same spot price for the duration of their subscription. If a user terminates the instance in the latest partial hour, they are billed for the whole hour. However, if Amazon terminates the instance due to an out-of-bid situation, the user is not billed. Certain spot instances in a launch group or availability zone group can be requested by users. Persistent spot requests stay active until the user

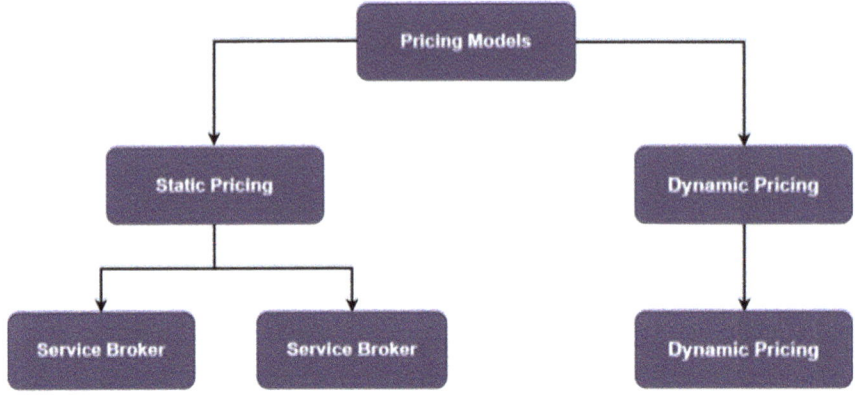

Fig. 3 Amazon EC2 IaaS cloud three different pricing models [9]

cancels them or they expire, while one-time spot instance requests stay active until the instance is allocated or canceled. The spot instance is terminated, and the request is closed if the spot price exceeds the bid price.

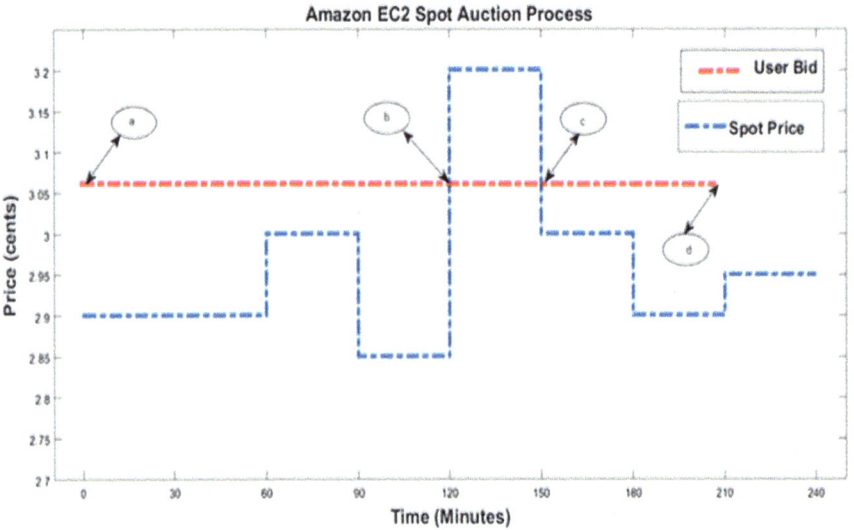

Fig. 4 The auction process of the Amazon EC2 spot market [9]

3 Related Work

Dubois and Casale [10] present a cost-conscious At-run decision-making mechanism for provisioning cloud resources and allocating application components among them. The methodology used here is beneficial because it utilizes uncomplicated, unburdened, and effective greedy algorithms to estimate a very complicated problem, which permits the ability to adapt to a system in both a proactive and reactive manner. Furthermore, we demonstrate the ability to make predictions and decisions even when considering environmental factors such as the probability of spot resources being lost. The decisions we make can trigger various actions, including "allocation, deallocation, migration, and replication", across one or more cloud infrastructures.

The objective of the research described by Javadi et al. [11] is to reduce the costs associated with using in-memory storage in public cloud environments. This is achieved by leveraging more cost-effective spot and burstable instances in combination with Amazon EC2 instances. By adopting a "hot–cold mixing" strategy across both normal and spot instances, the proposed technique in the paper provides a more cost-effective alternative than holding less popular data on faulty spot instances. Furthermore, the study employs a passive backup that is highly available and uses the newly developed burstable instances to mitigate performance degradation caused by spot instance revocations. The burstable instances' resource allocation features make them well-suited for backup tasks. The suggested strategies were implemented in a Memcached prototype based on EC2, and both simulations and real-world tests demonstrated that the "hot–cold mixing" approach and backup based on burstable instances significantly reduced costs and reduced performance degradation in case of spot instance revocation. The research suggests that scalable data-driven spot price prediction, smart mixing of on-demand and spot instances, and burstable instances for quick and affordable passive backup are innovative concepts that surpass the existing state-of-the-art approaches.

Kumar et al. [12] examine the potential benefits of cloud services based on spot instances by maximizing both user utility and provider income through bidding and pricing strategies. The study examines the influence of correlation on occupations that differ in terms of value and susceptibility to execution delay. Under specific assumptions, the study specifies optimum bidding and pricing techniques and highlights the importance of correlation in evaluating the advantages of a spot service versus an on-demand service. Additionally, the study provides a preliminary evaluation of the robustness of the findings under more general conditions but acknowledges that further investigation is necessary. In summary, the paper aims to demonstrate the importance of considering correlation when designing and implementing spot pricing strategies for CC services.

Li et al. [13] analyze two auction-based pricing models for cloud providers' idle resources in the form of Spot Block instances. The authors propose a methodology for determining the price of Spot Block instances and investigate the average successful bids, average profits per instance, and average achievable savings against On-demand pricing. The authors investigate two auction-based pricing mechanisms for cloud providers' idle resources, Spot Block instances, and offer a methodology for determining Spot Block prices. They examine several situations based on cloud customers' prevailing bidding strategies and discover that both auction systems offer considerable cost reductions as compared to on-demand instances. The Uniform price auction technique offers somewhat cheaper Spot Block pricing as well as higher savings over On-demand instances. Regardless of the auction process or bidding strategy, the authors conclude that Spot Block instances are a cost-effective solution that combines the benefits of On-demand and Spot instances.

Mikavica et al. [14] suggest using checkpointing to minimize the cost and volatility of resource provisioning when using Amazon EC2's spot instances. The authors propose several adaptive checkpointing schemes and compare their effectiveness using real price history data of spot instances to reduce monetary costs and improve job completion times. The results show that appropriate checkpointing strategies can significantly reduce both price and task completion time. The study suggests further investigations into improving predictions by exploring correlations between past and current prices, instance types, and rising edges, as well as gathering hidden information. Overall, the paper highlights the potential cost-saving benefits of using checkpointing when using spot instances in CC.

Nash et al. [15] have done a thorough literature assessment on cloud spot pricing, which shows that the majority of academic articles favor using spot pricing in CC. Yet, there is a dearth of realistic market-driven techniques to support cloud spot pricing. In actuality, the authors believe that both cloud providers and customers favour straightforward techniques. The study also calls into question prior studies that advised fixed pricing from the perspective of cloud providers. For future studies, the authors propose developing a factor checklist for cost–benefit analysis and investigating practical and deployable spot methods. Overall, the SLR gives useful information on cloud spot pricing and may boost practitioners' confidence in entering the market.

Song and Guerin [16] created models for Spot Instances in Amazon's EC2 data centers, with an emphasis on spot pricing and inter-price time.

The authors examined a year's worth of pricing trends and suggested a model that uses a blend of Gaussian distributions for eight different types of SIs. Simulations were used to validate the model, which successfully predicted the overall cost of performing jobs on the spot instances. Researchers and users of spot instances in Amazon's EC2 data centers and other IaaS Cloud providers will find this study valuable. The authors plan to explore the impact of user bids on the distribution of failures and design a brokering solution to optimize monetary cost and job completion time by utilizing different types of cloud resources in future research.

Wang et al. [17] have done a systematic literature review (SLR) of 61 main articles, indicating significant academic backing for the developing Cloud spot market The SLR, on the other hand, emphasized a lack of realistic market-driven mechanisms to sustain Cloud spot pricing. The authors suggest that economic models could better reflect the Cloud spot market but acknowledge that corresponding optimization problems could be NP-hard or NP-complete. Overall, the findings of the SLR provide useful insights into Cloud spot pricing and suggest Future research should concentrate on establishing realistic and simply deployable spot methods while assuring data reusability.

The adoption of spot instances by big cloud providers such as Microsoft and Amazon has boosted interest in dynamic pricing in the CC area. Nevertheless, the availability of spot instances is not assured, prompting economic study into bid creation and price behaviour and computer science research into enhancing availability. This statement highlights the diverse range of studies related to Amazon EC2 spot instances, which cover various aspects such as price modelling, prediction, bidding strategy, fault tolerance methods, and identifying applications suitable for spot instances. The interdisciplinary nature of the spot market has contributed to this diversity of research. Yi et al. [9] provide an overview of existing research in spot pricing and highlight the need for future research to address gaps in both economic and computer science perspectives.

Zolfaghari and Abrishami [10] introduced a Multi-class Workflow Ensemble Management (MWEM) system that leverages Amazon EC2's spot instances to minimize the cost of executing workflow ensembles. MWEM overcomes two key challenges in ensemble scheduling: workflow acceptance and instance provisioning. To evaluate MWEM, experiments were conducted using a new scoring metric with various budget and deadline constraints. The results demonstrate that MWEM outperforms the DPDS algorithm in terms of completed workflows and scores for all scientific workflow types. The authors plan to extend the MWEM algorithm to other application types and develop more innovative bidding strategies, as well as improved methods for launching and terminating instances in future work.

The proposed work on resource trading in edge computing is motivated by the challenges of excessive latency and energy consumption incurred by onsite decision-making in conventional spot trading, which can lead to the underutilization of dynamic resources. To address this challenge, Liwang et al. [19] proposed a hybrid market unifying futures and spot trading facilitate resource trading between an edge server and multiple smart devices. This approach encourages some buyers to sign a forward contract with the seller in advance while leaving the remaining buyers to compete for available resources with spot trading. The authors integrate overbooking into the futures market to achieve substantial utilization and profit advantages owing to dynamic resource demands. The authors also propose two bilateral negotiation schemes to address non-convex optimization and knapsack problems associated with spot trading, with a focus on achieving mutually beneficial player utilities. The experimental results demonstrate that the proposed mechanism outperforms baseline methods on critical indicators, including decision-making latency, resource usage, etc.

In the context of CC, providers offer various virtual machine (VM) models with different capacities and prices, and the cost of these fluctuates dynamically based on supply and demand. To make the most of redundant computing resources, users can bid on these instances above the current "spot price". Accurate prediction of the spot price can help users prepare better bids and increase the reliability of the method. Zolfaghari and Abrishami [18] proposed a modified gated recurrent unit (MGRU) model and a dropout method to predict future spot prices using Amazon EC2 as a testbed. Experimental results demonstrate the superior accuracy of the proposed method compared to other sophisticated methods (Table 1).

4 Research Gaps

Spot pricing in CC refers to the pricing model where users can bid for and utilize unused cloud resources at a lower cost than on-demand pricing. Spot instances provide an opportunity for users to benefit from cost savings, while cloud providers can monetize their idle resources. However, several research gaps in spot pricing need to be addressed to optimize the utilization and efficiency of spot instances.

Research Gaps in Spot Pricing in CC:

1. Market Dynamics: There is a need to better understand the dynamics of the spot pricing market in CC. This involves studying the factors that influence spot prices, such as supply and demand fluctuations, resource availability, and market competition. Analysing the temporal patterns, trends, and volatility of spot prices can provide insights into designing effective pricing strategies.
2. Pricing Models and Mechanisms: Current pricing models and mechanisms for spot instances may not accurately reflect the market dynamics and fail to optimize cost efficiency. There is a research gap in developing realistic and market-driven pricing models that consider factors like real-time market data, bidding strategies, and hybrid pricing approaches. Exploring alternative pricing mechanisms can contribute to more efficient spot pricing strategies.
3. Risk and Uncertainty Management: Spot instances introduce risks and uncertainties due to their variable pricing and potential termination. There is a need to develop effective risk management techniques to mitigate the impact of price fluctuations and instance revocations on users' applications and cloud providers' revenue. This involves predicting spot price changes, optimizing resource allocation, and developing fault tolerance mechanisms.
4. Resource Allocation and Scheduling: Optimizing resource allocation and scheduling decisions considering spot instances is a critical research area. Intelligent algorithms and techniques are required to dynamically allocate resources based on workload demands, spot price fluctuations, and application requirements. Considering factors such as task dependencies, workload characteristics, and resource availability can lead to more efficient utilization of spot instances and improved application performance.
5. Economic and Business Implications: Understanding the economic and business aspects of spot pricing in CC is essential. Research should analyse the cost-effectiveness of using spot instances compared to other pricing models, the

Table 1 Merits and demerits of the survey done

Schemes	Key finding	Advantage	Disadvantage
Dubois and Casale [11] proposed a cost-conscious At-run decision-making mechanism for provisioning cloud resources and allocating application components among them	The authors present a cost-conscious At-run decision-making mechanism that uses simple greedy algorithms to estimate a complex problem in the provisioning of cloud resources and allocation of application components	The paper proposes a cost-effective solution by leveraging spot cloud resources, which can be significantly cheaper than on-demand resources, and proposes a self-managing provisioning system that can dynamically allocate resources based on the application workload, minimizing the chances of under or over-provisioning	The proposed solution relies heavily on the use of spot cloud instances, which are not always and can be terminated at any time
Javadi et al. [12] aim to reduce costs associated with in-memory storage in public cloud environments	The study introduces innovative concepts such as scalable spot price prediction, instance mixing, and burstable instances, surpassing existing approaches	Using spot instances can be more cost-effective than using traditional cloud instances as it provides access to unused resources at lower prices. This can be particularly useful for businesses with varying workload demands	While spot instances can be cost-effective and scalable, they can also be unpredictable and less reliable than traditional cloud instances and can introduce additional security risks and complexity
Kumar [13] aims to examine the potential benefits of cloud services based on spot instances by maximizing user utility and provider income through bidding and pricing strategies	The research emphasizes the importance of correlation in evaluating the benefits of cloud services based on spot instances compared to on-demand services	The paper provides a comprehensive overview of the various spot pricing models used in CC and highlights the cost-saving benefits of using spot instances in CC	The paper relies on a limited amount of empirical data, which may affect the generalizability of the findings

(continued)

Table 1 (continued)

Schemes	Key finding	Advantage	Disadvantage
Li et al. [14] aim to analyze auction-based pricing models for cloud providers' idle resources in the form of Spot Block instances	The research finds that auction-based pricing for cloud providers' Spot Block instances leads to significant cost reductions compared to On-demand instances. The proposed methodology determines Spot Block prices, with the Uniform price auction offering cheaper pricing and higher savings	The paper provides practical implications for cloud users, including strategies for managing spot instances and optimizing cost savings	The paper may quickly become outdated as cloud providers and spot pricing models can change rapidly
Mikavica et al. [15] aim to minimize the cost and volatility of resource provisioning in Amazon EC2's spot instances using checkpointing	The research shows that employing adaptive checkpointing schemes can effectively minimize the cost and volatility of resource provisioning in Amazon EC2's spot instances. The evaluation using real price history data reveals that appropriate checkpointing strategies significantly reduce both price and task completion time	The paper proposes a model for determining Spot Block prices under this pricing mechanism, providing insight into price dependencies of chosen auction mechanism	The author assumes that Spot Block prices in previous time intervals are known in advance, which may not always be the case
Nash et al. [16] on modeling Spot Instances in Amazon's EC2 data centers, specifically spot pricing and inter-price time	The model, validated through simulations, successfully predicts the overall cost of performing jobs on Spot Instances	The paper provides a comparative analysis of different pricing and bidding strategies, helping readers make informed about which strategy to use	The paper is biased toward certain pricing and bidding strategies, cloud providers, or use cases, which could affect the objectivity of the analysis

(continued)

Table 1 (continued)

Schemes	Key finding	Advantage	Disadvantage
Song and Guerin [17] aim to conduct a systematic literature review (SLR) on Cloud spot market articles	The authors recommend future research to concentrate on establishing practical and easily implementable spot methods while ensuring data reusability in the Cloud spot market	The paper proposes novel ideas for the cost-effective operation of dynamic Memcached workloads on Amazon EC2, which can contribute to the advancement of CC	The paper focuses on the cost-effective operation of dynamic Memcached workloads on Amazon EC2, which may limit the generalizability of the findings
Wang et al. [18] introduce a Multi-class Workflow Ensemble Management (MWEM) system that utilizes Amazon EC2's spot instances to minimize the cost of executing workflow ensembles	MWEM successfully addresses the challenges of workflow acceptance and instance provisioning in ensemble schedules. The evaluation shows that MWEM outperforms the DPDS algorithm in terms of completed workflows and scores for various scientific workflow types	The proposed Multi-class Workflow Ensemble Management (MWEM) system effectively harnesses the power of spot instances to decrease the execution cost of each workflow of the ensemble	The instance provisioning component can be improved by proposing more innovative bidding strategies and better methods for launching and terminating instances

impact of spot pricing on cloud providers' revenue and profitability, and strategies for users to maximize cost savings while meeting their application performance requirements. Investigating market dynamics, pricing strategies, and user behavior can provide insights into the long-term sustainability and viability of spot pricing in the CC market.

5 Discussion and Future Work Directions

The paper discusses several research works that propose cost-saving mechanisms for provisioning cloud resources in the form of spot instances. Dubois and Casale [1] present an At-run decision-making mechanism that adapts to a system proactively and reactively using greedy algorithms. The methodology can make predictions and trigger various actions, including allocation, deallocation, migration, and replication. However, the model does not account for the impact of these actions on system performance. Song and Guérin's research [3] demonstrates the importance of considering correlation in designing and implementing spot pricing strategies for CC services. Mikavica and Kostic-Ljubisavljevi's [4] research proposes auction-based pricing models for idle resources in the form of Spot Block instances, offering considerable cost reductions as compared to on-demand instances. Chapter [6] highlights a dearth of realistic market-driven techniques to support cloud spot pricing, and the authors propose developing a factor checklist for cost–benefit analysis and investigating practical and deployable spot methods. Additionally, the paper suggests using checkpointing to minimize the cost and volatility of resource provisioning when using Amazon EC2's spot instances [5]. The study proposes several adaptive checkpointing schemes to reduce monetary costs and improve job completion times. Finally, the paper presents a model that uses a blend of Gaussian distributions for eight different types of Spot Instances in Amazon's EC2 data centers [7]. The simulations validate the model, which successfully predicted the overall cost of performing jobs on the spot instances. Overall, the paper presents several innovative approaches that reduce the cost of using spot instances and proposes several avenues for future research.

- Lack of standardization: There is a lack of standardization in spot pricing across different cloud service providers, which makes it challenging for users to compare prices and optimize their resource allocation strategies.
- The complexity of bidding strategies: Bidding on spot instances can be complex, as it involves making decisions based on factors such as price fluctuations, resource requirements, and the availability of resources. There is a need for more research on bidding strategies that can help users make more informed decisions when bidding on spot instances.
- Resource allocation strategies: There is a need for more research on resource allocation strategies that can effectively balance the cost savings of using spot instances with the performance requirements of different workloads.
- Impact on quality of service: The impact of spot pricing on the quality of service (QoS) is not well understood. There is a need for more research on how the use of spot instances affects QoS metrics such as response time, throughput, and availability.
- Optimization of spot pricing: There is a need for more research on optimization techniques that can help users optimize their spot pricing strategies based on their specific workload characteristics, resource requirements, and budget constraints.
- Cost modeling: As mentioned earlier, there is a need for more accurate cost models that can help users make more informed decisions when bidding on spot instances. These models should be able to capture the complexities of CC and provide users with a more realistic estimate of the total cost of using spot instances.

6 Conclusion

In addition, research can also investigate the trade-offs between efficiency and veracity in spot pricing mechanisms, as well as the development of honest, fair, as well as revenue-maximizing auction techniques for pricing spot occurrences. Furthermore, the incorporation of risk awareness in bidding behavior and the exploration of temporal connections to spot price forecasts can improve the bidding strategy. Multi-agent simulation techniques can also be used for modeling the spot cloud market, which can provide insights into market mechanisms on the side of cloud providers. Overall, spot pricing offers a promising pricing model for CC, and further research can help address the challenges and improve the effectiveness of this pricing mechanism. In conclusion, spot pricing and auction mechanisms have emerged as important pricing strategies for CC resources. These methods provide businesses with cost-effective access to computing resources and allow them to optimize their expenses while meeting their computing needs. The use of dynamic pricing in the form of spot pricing and auction mechanisms has helped to democratize CC, making it accessible to businesses of all sizes. However, these pricing strategies also come with their challenges, such as the need for careful workload distribution and resource allocation. Despite these challenges, spot pricing and auction mechanism remain as valuable tools for businesses that need to scale up or down quickly to meet their computing needs. The future of spot pricing and auction mechanisms in CC looks promising as more businesses adopt these pricing strategies and cloud service providers continue to be innovate in this area. Further research and innovation will be needed to address the challenges and maximize the potential benefits of spot pricing and auction mechanism in CC.

References

1. Joshi M, Budhani S, Tewari N, Prakash S (2021) Analytical review of data security in cloud computing. In: 2nd international conference on intelligent engineering and management (ICIEM), pp 362–366
2. Das S, Namasudra S (2022) A lightweight and anonymous mutual authentication scheme for medical big data in distributed smart healthcare systems. IEEE/ACM Trans Comput Biol Bioinform. https://doi.org/10.1109/TCBB.2022.3230053
3. Sharma P, Namasudra S, Gonzalez Crespo R (2023) EHDHE: enhancing the security of healthcare documents in IoT-enabled digital healthcare ecosystems using blockchain. Inf Sci (Ny) 629:703–718. https://doi.org/10.1016/j.ins.2023.01.148
4. Singh DP, Shukla S, Bordoloi D (2021) A quick overview of cloud-based services. Webology 18:3288–3294
5. Gupta A, Namasudra S (2022) A novel technique for accelerating live migration in cloud computing. Autom Softw Eng 29. https://doi.org/10.1007/s10515-022-00332-2
6. Maliyaem M, Tuan NM, Lockhart D, Muenthong S (2022) A study of using machine learning in predicting COVID-19 cases. Cloud Comput Data Sci 54–61. https://doi.org/10.37256/ccds.3220221488
7. Sharma RS, Mannava PN, Wingreen SC (2022) Reverse-engineering the design rules for cloud-based big data platforms. Cloud Comput Data Sci 39–59
8. Khasim S, Basha SS (2022) An improved fast and secure CAMEL based authenticated key in smart health care system. Cloud Comput Data Sci 77–91
9. Yi S, Kondo D, Andrzejak (2010) A reducing costs of spot instances via checkpointing in the Amazon elastic compute cloud. In: 3rd international conference on cloud computing, IEEE
10. Dubois DJ, Casale G (2015) Autonomic provisioning and application mapping on spot cloud resources. In: International conference on cloud and autonomic computing, IEEE

11. Javadi B, Thulasiramy RK, Buyya R (2011) Statistical modelling of spot instance prices in public cloud environments. In: Fourth international conference on utility and cloud computing, IEEE
12. Kumar D, Baranwal G, Raza Z, Vidyarthi DP (2018) A survey on spot pricing cloud computing. J Netw Syst Manag 26:809–856. https://doi.org/10.1007/s10922-017-9444-x
13. Li Z, Zhang H, O'Brien L (2017) Spot pricing in the cloud ecosystem: a comparative investigation. arXiv [cs.DC]
14. Mikavica B, Kostic-Ljubisavljevic (2018) A pricing and bidding strategies for cloud spot block instances. In: 41st international convention on information and communication technology, electronics and microelectronics (MIPRO), IEEE
15. Nash P (2015) Introducing preemptible VMs, a new class of compute available at 70% off standard pricing. GoogleCloudPlatformBlog magazine
16. Song J, Guerin R (2017) Pricing and bidding strategies for cloud computing spot instances. In: Conference on computer communications workshops (INFOCOM WKSHPS), IEEE
17. Wang C, Urgaonkar B, Gupta A (2017) Exploiting spot and burstable instances for improving the cost-efficacy of in-memory caches on the public cloud. In: Proceedings of the twelfth European conference on computer systems. ACM, New York, NY, USA
18. Zolfaghari B, Abrishami S (2022) A multi-class workflow ensemble management system using on-demand and spot instances in the cloud. Future Gener Comput Syst 137:97–110. https://doi.org/10.1016/j.future.2022.07.007
19. Liwang M, Chen R, Wang X (2022) Shen X Unifying futures and spot market: Overbooking-enabled resource trading in mobile edge networks. IEEE Trans Wirel Commun 21:5467–5485. https://doi.org/10.1109/twc.2022.3141094

A Comparative Analysis of Propagation Models Suitable for Non-Line-of-Sight 5G Communication at 26 GHz

Pia Sarkar, Arijit Saha, and Amit Banerjee

Abstract The Non-Line-of-Sight (NLOS) communication in millimeter wave (mmWave) experiences high path loss in the urban region due to reflection, blockage etc. To design an efficient 5G system, the channel should be modelled such that data rate and capacity are high. This paper presents the close-in (CI) free space reference distance model, CI model whose path loss exponent is frequency weighted (CIF) and the alpha-beta-gamma (ABG) model for 26 GHz. The use of these models in 3rd Generation Partnership Project (3GPP) and Fifth Generation Wireless System design has drawn the attention of researchers to investigate more. As 26 GHz 5G band is commercially used for 5G communication in India, we have analyzed the path loss and capacity in this frequency taking different distances for Urban Macrocell (UMa), Urban Microcell (UMi) and input office scenario considering NLOS communication. The results show that path loss varies with variation of cell size.

Keywords UMa · UMi · Millimeter wave · 5G channel model

1 Introduction

To explore a 5G communication system it is important to model a channel such that it becomes reliable. In India many 5G bands are supported like n28, n78, n258 etc. It is found that n28 (700 MHz) and n78 (3300–3800 MHz) sub-6 GHz 5G bands are commonly used. In these bands the signal can cover long distances with a speed of 1 Gbps. To increase the data rate up to 10 Gbps mmWave frequency band n258

P. Sarkar (✉)
Department of Electronics and Communication Engineering, B. P. Poddar Institute of Management and Technology, Kolkata, India
e-mail: pia.sarkar@bppimt.ac.in

A. Saha
Dumdum Motijheel Rabindra Mahavidyalaya, Kolkata, India

A. Banerjee
Microsystem Design-Integration Lab, Physics Department, Bidhan Chandra College, Asansol, India

© The Author(s), under exclusive license to Springer Nature Singapore Pte Ltd. 2024
S. Namasudra et al. (eds.), *Data Science and Network Engineering*, Lecture Notes in Networks and Systems 791, https://doi.org/10.1007/978-981-99-6755-1_26

347

(24.25–27.5 GHz) can be used. Certain models exist for below 6 GHz band. However, more researches are recently going on for the 6 GHz–100 GHz frequency range. Some projects like the 3rd Generation Partnership Project (3GPP) [1], Fifth Generation Channel Model (5GCM) [2], *Mobile and wireless communications Enablers for the Twenty-twenty Information Society* (METIS) [3], Millimetre-Wave Based Mobile Radio Access Network for Fifth Generation Integrated Communications (mmMAGIC) [4] are conducting research works on the propagation model in this high frequency range. Three models are frequently used, i.e. Close-In (CI) free space reference distance model, CI model whose path loss exponent is frequency weighted (CIF) and Alpha-Beta-Gamma (ABG) model. These models are used to calculate path loss and received power which are functions of distance and frequency.

Wang et al. [5] proposed a wideband channel model at 26 GHz applied for indoor Line-of-Sight (LOS) measurement. The bandwidth of this model was 1 GHz and the propagation distance was 2–67 m. Hur et al. [6] developed a channel sounder at 28 GHz valid for indoor communication like shopping mall. Al Samman et al. [7] presented channel characteristics for indoor at 6.5, 10.5, 15, 19, 28 and 38 GHz. Azar et al. [8] first presented the measurement at 28 GHz in New York city. The distance covered was 500 m. MacCartney et al. [9] suggested a 5G channel model for 28 GH and 38 GHz. The distance dependent path loss model was proposed for the outdoor scenario in an Urban Microcell. Sun et al. [10] investigated CI and ABG model at 28 GHz and 73 GHz where path loss was the function of the distance applicable for both indoor and outdoor scenarios. Close-in (CI) free space model gave satisfactory results than the Alpha-Beta-Gamma (ABG) model. Violette et al. [11] provided wideband measurement for the 28 GHz mmWave channel considering LOS and NLOS communication in a dense urban environment. Smulders et al. [12] overviewed a 60 GHz channel model considering ray tracing technique. Al Samman et al. [13] discussed the pathlosses of ABG, CI and CIF models for indoor communication at 3.5 GHz and 28 GHz. MacCartney et al. [14] presented the omnidirectional measurement of path loss at 28 GHz, 38 GHz and 73 GHz for 5G urban cellular technology. Samimi et al. [15] presented an outdoor channel model that can be implemented upto 100 GHz for Urban Macro cell (Uma) and Urban Microcell (UMi).

Well investigation on path loss models have been done in 28, 38, 60 and 73 GHz frequency. Limited work has been carried on the 26 GHz channel model applicable for LOS scenario in indoor environment. Operators are interested to deploy 5G cellular technology in the urban area and 26 GHz is a commercial band used for 5G communication in India. This fact motivates us to analyze different path loss models at 26 GHz. In case of indoor scenario the chances of NLOS communication is more. Propagation models with NLOS possibilities have been studied in this work. Different cell sizes have been introduced in 5G cellular technology to enhance data rate and capacity. We have carried out our work considering different path lengths for macro, micro cell and indoor office location.

Previous research works have already been done with CI, CIF and ABG models at 26 GHz for LOS possibility in indoor location. We have analyzed the application of the commercial band 26 GHz for NLOS possibility in both outdoor and indoor office environment. We have investigated the received signal power and compared the capacity for these three models in UMa, UMi-SC and UMi-OS environment. The Novelty of the work lies here.

The organizations of the paper are as follows. Different path loss models have been described in Sect. 2. Free Space Path Loss model (FSPL), single and multiple frequency path loss models have been discussed in this section. Analysis of simulated results are presented in Sect. 3. Path loss, signal strength at receiver and channel capacity have been compared in this section. Conclusion of this work has been given in Sect. 4.

2 Path Loss Models

2.1 Free Space Path Loss Models

Path loss is given by

$$PL = P_t + g_t + g_r - P_r - Other\ Loss \tag{1}$$

$$Free\ Space\ Path\ Loss(FSPL)[dB] = 20log_{10}\left(\frac{4\pi df}{c}\right) \tag{2}$$

where, P_t, g_t, g_r, P_r, and d are transmitted power, transmitting antenna gain, receiving antenna gain, received power, and transmission distance respectively. Here f is frequency, and c is the velocity of light.

2.2 Single Frequency Propagation Model

2.2.1 CI Model

The path loss of CI model is represented as

$$PL^{CI}(f,d)[dB] = FSPL(f,1m) + 10nlog_{10}\left(\frac{d}{1m}\right) + X_\sigma^{CI} for\ d \geq 1m \tag{3}$$

where, f is the frequency in Hz, d is the distance covered in m. FSPL is Free Space Path Loss calculated for 1m distance at frequency f, n is the Path Loss Exponent (PLE). If standard deviation σ reduces, path loss reduces, PLE optimizes.

X_σ^{CI} is the zero mean Gaussian random Variable with standard deviation σ in dB. It comes from large-scale path loss due to shadowing effect.

$$X_\sigma^{CI} = PL^{CI}(f,d)[dB] - FSPL(f,1m) - 10n log_{10}\left(\frac{d}{1m}\right) for\, d \geq 1m \quad (4)$$

2.3 Multiple Frequency Propagation Model

2.3.1 CIF Model

The path loss of the CIF model can be written as

$$PL^{CIF}(f,d)[dB] = FSPL(f,1m) + 10n\left(1 + b\left(\frac{f - f_0}{f_0}\right)\right)log_{10}\left(\frac{d}{1m}\right)$$
$$+ X_\sigma^{CIF}\, for\, d \geq 1m \quad (5)$$

where, n is Path Loss Exponent (PLE), f and d denote frequency and distance respectively, b presents how path loss depends on weighted average frequency.

$$f_0 = \frac{\sum_{k=1}^{K} f_k N_k}{N_k} \quad (6)$$

where, f_0 is the reference frequency calculated from some measured data points. K is the number of frequencies. N_k is the number of path loss measured values. X_σ^{CIF} is the zero mean Gaussian random variable expressing shadowing.

In case of single frequency when f_0 will be same as f and $b = 0$, then the multi frequency CIF model turns into a single frequency CI model.

2.3.2 ABG Model

Path loss of the ABG model is expressed as:

$$PL^{ABG}(f,d)[dB] = 10\alpha log_{10}(d) + \beta + 10\gamma log_{10}(f) + X_\sigma^{ABG} \quad (7)$$

where, α and γ are path loss varying factors with distance and frequency. d and f are distance and frequency respectively, β (in dB) is the floating offset. X_σ^{ABG} is the zero mean Gaussian random variable with standard deviation σ in dB.

If $\alpha = 20 log \frac{4\pi}{c}$, β becomes equal with Path Loss Exponent (PLE) and γ is equal to 2, and the ABG model transforms into the CI model.

3 Results

We have analyzed the path loss, signal strength at receiver and capacity considering three channel models i.e., CI, CIF and ABG models at 26 GHz for NLOS communication in Macrocell, Urban Microcell and indoor office scenario. Simulation Parameters are listed below in Tables 1 and 2.

For the UMa environment distance is taken from 0 m to 1000 m. For CI, CIF and ABG models path loss at 1000 m are shown in Fig. 1a as 157.5, 153.5 and 373.78 dB respectively. In Fig. 1b the received power at 1000 m distance are 14.47, 18.48 and −201.75 dBm for CI, CIF and ABG models respectively. As in the ABG model thr signal experiences more path loss, received power decreases rapidly. Received power becomes negative in dBm as thr received power is less than 1 milliWatt (mW). In Fig. 1c the capacity at 1000 m distance in the CI model is 21 Gbps and in the CIF model 21.86 Gbps. In Fig. 1d the capacity at 1000 m distance in the ABG model is 0.021 Gbps.

Table 1 Simulation parameters

Parameter	Value
Carrier frequency (f)	26 GHz
Bandwidth (B)	1.3 GHz
Transmitting antenna gain in dBi (g_t)	20
Receiving antenna gain in dBi (g_r)	2
Transmitted power (P_t) in dBm	150
Distance taken for UMa scenario	0–1000 m
Distance taken for UMi-SC and UMi-OS scenario	50–100 m
Distance taken for indoor office scenario	4–20 m

Table 2 Model parameters for UMa, UMi and indoor office scenario in NLOS communication (* S.C.- Street Canyon, O.S.-Open Square, SF-Shadow Fading)

Model	Scenario	Model parameters
ABG	UMa	$\alpha = 3.5$, $\beta = 13.6$, $\gamma = 2.4$, SF = 5.3 dB
	UMi-SC	$\alpha = 3.48$, $\beta = 21.02$, $\gamma = 2.34$, SF = 7.8 dB
	UMi-OS	$\alpha = 4.14$, $\beta = 3.66$, $\gamma = 2.43$, SF = 7 dB
	Indoor office	$\alpha = 3.1$, $\beta = 1.3$, $\gamma = 3.8$, SF = 10.3 dB
CI	UMa	n = 3, SF = 6.8dB
	UMi-SC	n = 3.19, SF = 8.2 dB
	UMi-OS	n = 2.89, SF = 7.1 dB
	Indoor office	n = 2.9, SF = 10.9 dB
CIF	UMa	n = 2.9, b = −0.002, SF = 5.7 dB
	UMi-SC	n = 3.2, b = 0.076, SF = 7.1 dB
	Indoor office	n = 3, b = 0.21, SF = 10.4 dB

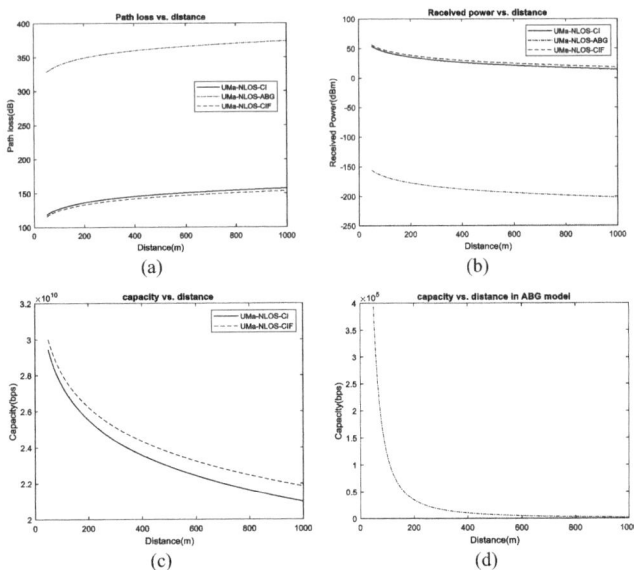

Fig. 1 **a** Path loss versus transmission distance, **b** Received power versus distance, **c** Capacity versus distance of CI and CIF model, **d** Capacity versus distance of ABG model in urban Macrocell environment

For the UMi-SC environment distance is taken from 50 m to 100 m. For CI, CIF and ABG models path loss at 100 m are shown in Fig. 2a as 132.7, 128.5 and 342.09 dB. In Fig. 2b the received power at 100 m distance are 39.28, 43.47 and −170.1 dBm for CI, CIF and ABG modelS respectively. In Fig. 2c the capacity at 100 m distance in the CI model is 26.35 Gbps and in the CIF model 27.25 Gbps. In Fig. 2d the capacity at 100 m distance in the ABG model is 0.81 Gbps.

For the UMi-OS environment distance is taken from 50 m to 100 m. For CI and ABG models at 100 m distance path losses are shown in Fig. 3a as 125.6 dB and 346.5 dB. In Fig. 3b received power which is 46.38 and −174.5 dBm for CI and ABG model, respectively. In Fig. 3c the capacity at 100 m distance in the CI model is 27.9 Gbps. In Fig. 3d in the ABG model is 0.485 Gbps.

For Indoor office environment distance is taken from 4 m to 20 m. For CIF and CI model path loss at 20 m are shown in Fig. 4a as 104.579 and 109.295 dB. In Fig. 4b the received power at 20 m distance are 67.45 and 62.66 dBm for CIF and CI models respectively. In Fig. 4c the capacity at 20 m distance in the CIF model is 32.43 Gbps. In Fig. 4d the capacity in the CI model is 31.41 Gbps.

For the CI and CIF propagation model, more path loss occurs in UMa scenario than UMa-SC, UMa-OS and minimum path loss occurs in the Indoor office scenario. These models provide minimum capacity for UMa case and maximum capacity in the Indoor Office environment. UMi-SC case produces more pathloss than UMi-OS environment.

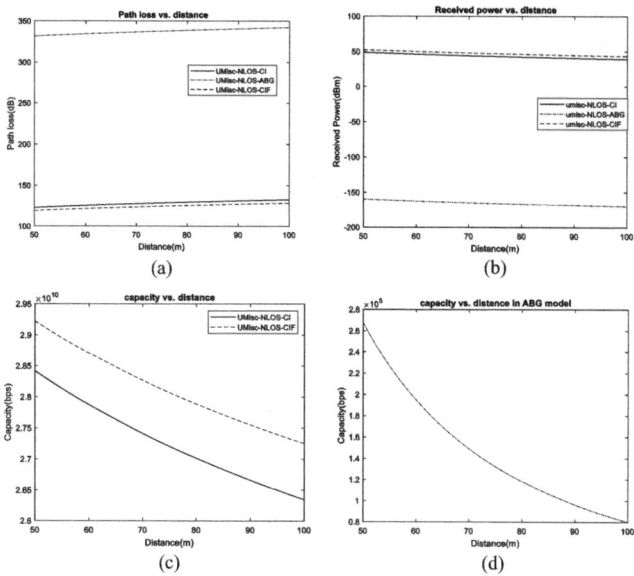

Fig. 2 **a** Path loss versus transmission distance, **b** Received power versus distance, **c** Capacity versus distance of CI and CIF model, **d** Capacity versus distance of ABG model in urban Microcell (UMi-SC) environment

Fig. 3 **a** Path loss versus transmission distance, **b** Received power versus distance of CI and ABG model, **c** Capacity versus distance of CI, **d** Capacity versus distance of ABG model in urban Microcell (UMi-OS) environment

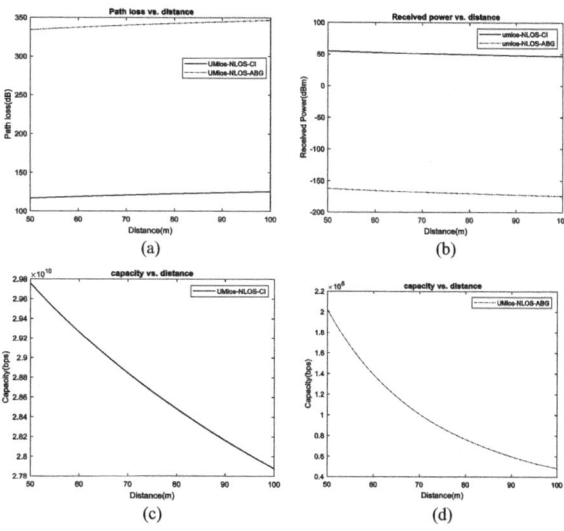

Fig. 4 **a** Path loss versus transmission distance, **b** Received power versus distance, **c** Capacity versus distance of CI and CIF model in input office environment

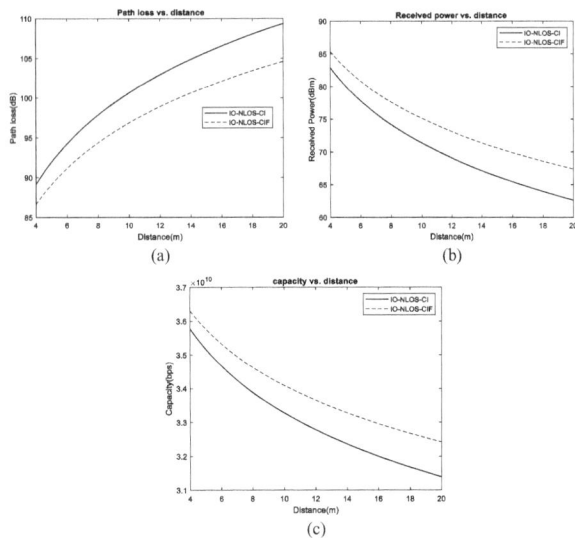

(a)   (b)

(c)

In the ABG model UMa environment offers maximum path loss and least capacity. UMi-SC scenario causes less path loss than UMi-OS and provides maximum capacity.

4 Conclusion

We have analyzed three propagation models to compare received signal power, path loss and capacity in 5G communication at the commercial band of 26 GHz for different cell sized urban scenario. ABG model is complex and difficult to implement, as three parameters have to be adjusted to minimize the standard deviation of the shadowing effect. In the ABG model α parameter denotes distance dependent path loss in the first one meter reference distance which is almost same as the physical based n parameter of CI model. The two optimization parameters α and β of the ABG model are non-physical and vary drastically with the variation of distance. If the distance is less than 30 m or more than several hundred meters ABG model shows more prediction error. ABG model causes more measurement error in case of low frequency.

The results show that path loss is less and the receiver signal strength and capacity increases in the CI model as compared to the ABG model. Whereas the CI model is simple, stable and practically realizable as only one parameter needs to be controlled for a lower value of standard deviation. The parameter of the CI model does not change so much as distance changes. The prediction is stable for all path lengths and frequencies. It can be said that the CI model is simple, accurate and robust as compared to the ABG model. With the modification of the ABG model, the CI model can be applied in 3GPP to predict path loss accurately.

In CI and CIF models path loss basically shows much dependency on frequency in 1 m close-in reference distance. The presence of frequency dependent b parameter CIF model can measure the path loss with more accuracy and it is suitable for indoor environment. Beyond 1 m distance path loss does not vary much with frequency. As frequency dependent b parameter sets to 0 value in CI model, it suits better for the outdoor environment.

The parameter values proposed here can be used for further development of 5G technology and beyond at a higher frequency. The hybrid channel model can also be implemented.

Declarations Conflict of Interest The authors declare that they have no conflicts of interest.

Author's Contribution All authors contributed equally in the preparation of the manuscript.

Ethical Approval Not applicable.

Funding This work is partially supported by DST/TDT/DDP-38/2021, Device Development Programme (DDP), by the Department of Science & Technology (DST), Ministry of Science and Technology, Government of India.

Availability of Data and Materials Not applicable.

Competing Interests The authors declare no competing interests.

References

1. TR 138 901 - V16.0.0 - 5G; Study on channel model for frequencies from 0.5 to 100 GHz (3GPP TR 38.901 version 16.0.0 Release 16), vol. 0 (2020)
2. 5G Channel Model for bands up to100 GHz presented by NTT DOCOMO
3. Nurmela V, Karttunen A, Roivainen A (2015) METIS channel models, Technical Report, Report Number-Deliverable D1.4
4. Haneda K, Nguyen SLH, Karttunen A, Jarvelainen J (2017) Measurement results and final mmMAGIC channel models, Technical Report, Report Number-Deliverable D2.2
5. Wang Q, Li S, Zhao X, Wang M, Sun S (2016) Wideband millimeter-wave channel characterization based on LOS measurements in an open office at 26GHz. In: Proceedings of the 2016 IEEE 83rd vehicular technology conference (VTC Spring), pp 1–5. 18, Nanjing, China, 15–18 May
6. Hur S, Cho Y-J, Lee J, Kang N, Park J, Benn H (2014) Synchronous channel sounder using horn antenna and indoor measurements on 28 GHz. In: Proceeding of the 2014 IEEE international black sea conference on communications and networking (BlackSeaCom), pp 83–87. 19, Odessa, Ukraine, 27–30 May 2014
7. Al-Samman AM, Rahman TA, Azmi MH, Hindia MN, Khan I, Hanafi E (2016) Statistical modelling and characterization of experimental mm-wave indoor channels for future 5G wireless communication networks. PLoS ONE 11:e0163034
8. Azar Y, Wong GN, Wang K, Mayzus R, Schulz JK, Zhao H, Gutierrez F, Hwang D, Rappaport TS (2013) 28 GHz propagation measurements for outdoor cellular communications using steerable beam antennas in New York city. In: Proceedings of the 2013 IEEE international conference on communications (ICC), pp 5143–5147, Budapest, Hungary, 9–13 June 2013

9. MacCartney GR, Zhang J, Nie S, Rappaport TS (2013) Path loss models for 5G millimeter wave propagation channels in urban microcells. In: Proceedings of the 2013 IEEE global communications conference (GLOBECOM), pp 3948–3953, Atlanta, GA, USA, 9–13 December 2013

10. Sun S, MacCartney GR, Rappaport TS (2016) Millimeter-wave distance-dependent large-scale propagation measurements and path loss models for outdoor and indoor 5G systems. In: Proceedings of the 2016 10th European conference Antennas propagation (EuCAP 2016), Davos, Switzerland, 10–15 April 2016

11. Violette EJ, Espeland RH, DeBolt RO, Schwering FK (1988) Millimeter-wave propagation at street level in an urban environment. IEEE Trans Geosci Remote Sens 26(3):368–380

12. Smulders PF, Correia LM (1997) Characterisation of propagation in 60 GHz radio channels. Electron Commun Eng J 9(2):73–80

13. Al-Saman A, Mohamed M, Cheffena M (2020) Radio propagation measurements in the indoor stairwell environment at 3.5 and 28 GHz for 5G wireless networks. Int J Antennas Propag 2020:1–10

14. Maccartney GR, Rappaport TS, Sun S, Deng S (2015) Indoor office wideband millimeter-wave propagation measurements and channel models at 28 and 73 GHz for ultra-dense 5G wireless networks. IEEE Access 3:2388–2424

15. Samimi MK, Rappaport TS, MacCartney GR (2015) Probabilistic omnidirectional path loss models for millimeter-wave outdoor communications. IEEE Wireless Commun Lett 4(4):357–360

Validating δ-Currency Using Model Checking

Shreekanth M. Prabhu

Abstract In recent years, there is a surfeit of digital currencies, virtual currencies, and cryptocurrencies. These currencies serve as alternatives to fiat currencies in the form of physical currencies or deposits in banks. Some of the common characteristics that differentiate these currencies are how they gain and maintain their value, the anonymity of transactions, and considerations of security, data integrity, and transaction performance in a distributed computing scenario. Digital Currencies are generally issued by Central Banks raising privacy concerns. Cryptocurrencies work outside such formal mechanisms raising concerns about volatility and the possibility of loss. In this paper, we discuss a new variant called δ-Currency which attempts to navigate these concerns and arrive at an alternative mechanism that addresses privacy and financial security considerations in a novel manner. The δ-Currency is currently at a conceptual stage. We make use of Model Checking to validate the architecture, design principles, and implementation approach of δ-Currency. Model Checking focuses on fundamental building blocks of computation and is generally used to validate complex systems operating in an environment that enables a high degree of concurrency. In this paper, we make use of Model Checking to arrive at a common vocabulary, abstractions, and framework to understand seemingly disparate systems which nevertheless achieve the same objective.

Keywords δ-Currency · Digital currency · Cryptocurrency · Merkle trees · B+ Tree · Model checking

1 Introduction

There are two significant concerns that gave impetus to the digital currency movement over the last few decades. First is the erosion of the value of fiat currencies over time. Second is the desire of certain sections to have a currency that is independent of the

S. M. Prabhu (✉)
Department of Computer Science and Engineering, CMR Institute of Technology, Bengaluru, India
e-mail: shreekanth.p@cmrit.ac.in

© The Author(s), under exclusive license to Springer Nature Singapore Pte Ltd. 2024
S. Namasudra et al. (eds.), *Data Science and Network Engineering*, Lecture Notes in Networks and Systems 791, https://doi.org/10.1007/978-981-99-6755-1_27

357

Government so that they can transact with a sense of anonymity. In a way, both these concerns converged when Cryptocurrencies came into being.

Typically, cryptocurrency is created only when a miner successfully mines a block which is to be appended to a crypto-ledger implemented using a Blockchain. The sheer scarcity of coins can allow for a speculative element. Thus, even banks ended up investing in cryptocurrencies. Most recently, a traditional and conservative American Bank failed spectacularly. One contributing factor was the investment in cryptocurrencies [1]. Further, the anonymity of transactions enabled by cryptocurrencies leads to a lot of illicit activities. Then there are other issues such as some investors simply forgetting their cryptographic key and thus losing access to their funds.

A question then naturally arises: when several modes of digital payment are well-entrenched why do we need digital currency? Digital Currency is as secure as physical currency. In contrast, monies kept in the bank are secured only to insurance amounts that are generally very low. However, in practice, typically Governments intervene to rescue banks. The money kept in banks leads to greater financialization of the economy and is good to create jobs, boost investments, and support public expenditure. This however may erode the real value of money due to bad loans and excessive public expenditure by Governments. In this context, digital currencies can create an impediment to such erosion in value. As we speak, eleven countries have launched Digital Currency Pilots. They are The Bahamas, Nigeria, China, USA, Jamaica, UAE, Ghana, Malaysia, Singapore, Thailand, and India.

Secondly, with digital currency, the holders may have an expectation of relative anonymity and greater privacy than with inter-bank transfers and third-party wallets. Another important advantage of digital currency is authentication. When physical notes are in circulation, we simply do not know how many of them are fake and how many of them are duplicates. It may surprise a layperson to know that such frauds can and do happen.

In this paper, we discuss δ-Currency which was proposed by Prabhu et al. [2]. On one hand, the intent of δ-Currency is to provide a new alternative that has the advantages of crypto-currency such as privacy and anonymity without the attendant complexity typical of Cryptocurrencies. On the other hand, the δ-Currency envisages a role for the bank as a currency manager whose concern is the conservation of currency and introduces an integrity manager whose preoccupation is to safeguard the transactions. The objective of this paper is to validate δ-Currency using Model Checking. Model Checking can validate complex concurrent systems using a theoretical approach and enable the positioning of new approaches relative to the goals at hand as well as other ways to achieve those goals.

In δ-Currency, a payment transaction happens only when both peers consent to it. Many frauds happen when fraudsters send money to your account without your consent and ask you to return it. Frauds with PhonePe and Google Pay digital payment services are reported in Inventiva [3]. Even in traditional banking systems, it is possible that your account is used to launder money or to do illicit transactions.

The rest of the paper is as follows. Section 2, δ-Currency Overview describes salient points related to the architecture and design of δ-Currency. Section 3, Model Checking Preliminaries explains the tools and techniques available to validate complex systems. Section 4, Validating δ-Currency using Model Checking, covers the use of Model Checking techniques to validate δ-Currency. Section 5, Conclusions concludes the paper.

2 δ-Currency Overview

The origin of digital money can be traced to Dai [4]. Dai conceived digital money in an essay, where he talks about money that is transacted in anonymity and where everyone uses pseudonyms. Wei Dai laid down a 5-step protocol that included (i) Creation of Money, (ii) Transfer of Money, (iii) Effecting of Contracts, (iv) Conclusion of Contracts, and (v) Enforcement of Contracts. The b-money proposal was acknowledged in the bitcoin proposal [5].

In a bitcoin implementation, the blocks of transactions are represented on a public Blockchain via a root hash corresponding to a Merkle Tree that stores the transactions in that block. Thus, we can consider a Blockchain as a ledger of hashes. This ledger itself is a matter of consensus i.e. which block will extend a given chain needs to be agreed. In a given block only valid transactions are to be accommodated i.e., the sending address should have adequate balance and there should be no double spending or replay attacks. This is done using cryptographic constructs. Bitcoin implementation is done using a stack machine and a scripting language. Since there is no notion of accounts, the notion of balances also is not an essential prerequisite. Transactions happen between addresses without any notion of accounts enabling anonymity. However unspent coins are maintained in appropriate data structures for convenient access. Those who are able to successfully solve a hash puzzle get to add a block to the Blockchain and in the process create the bitcoin currency. In bitcoin, both parties need not be connected at the time of the transaction. It is adequate if the transaction is broadcast in a peer-to-peer network with the requisite cryptographic protection.

In contrast to bitcoin, the Ethereum Blockchain does support the notion of accounts. There is also a notion of an Escrow Account. The transactions are accomplished using a smart-contract code. The mining process continues to be a standard feature for Ethereum as well as many other Blockchain based currency implementations.

A detailed study of bitcoin and other cryptocurrencies is done by Bonneau et al. [6]. Swan [7] presents seven technology challenges for the adoption of Blockchain Technology in the future. These include throughput, latency, size and bandwidth, security, wasted Resources, usability, versioning, hard Forks, and multiple Chains. These issues have led to alternative proposals that address or work around them. δ-Currency is one such proposal.

In δ-Currency, it is peers who transact with each other. δ-Currency operates using a peer-to-peer consensus model, in contrast to the public consensus used by bitcoin. Every pair of peers maintain a Merkle Tree for transactions between them. Thus, if

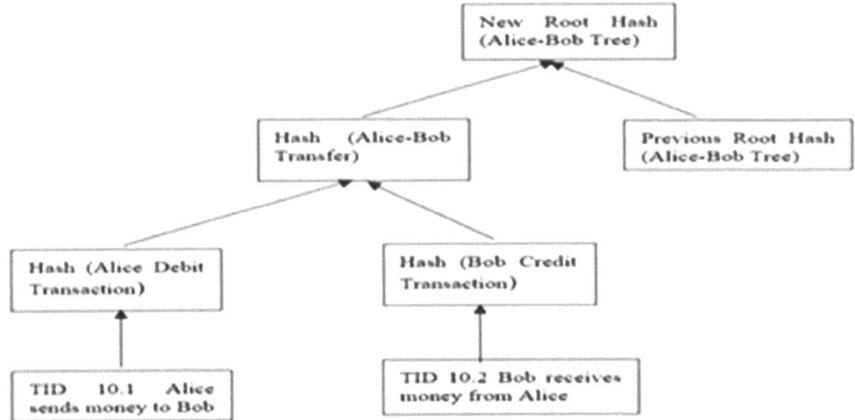

Fig. 1 Peer to peer transactions stored in Merkle tree

there are n peers who can potentially transact with n−1 parties then each peer maintains n−1 Merkle Trees. A transaction materializes only if both peers are connected
and agree to the transaction. This is ideal for retail transactions at purchase points
or say, between supply chain participants who can confirm the transactions after a
certain lag. Figure 1 illustrates the peer-to-peer transactions stored in the Merkle tree.

The balances are maintained in a separate data structure which is a B+ Tree superimposed with a Merkle Tree, called Merkle Hash Tree (MHT). This was conceived as
a part of the Data Outsourcing Model [8]. Every peer maintains their own MHT that
stores the balance information as a trail over transactions. Hence, if there are n peers
there are that many MHTs in the system which store the balances on an ongoing
basis. Figure 2 depicts how the trail of Peer Balances is stored as a Merkle Hash
Tree. Here Peer Balances are stored along the leaf nodes of the B+ Tree. The balance
changes are hashed upwards to generate a parallel Merkle Tree with a corresponding
root hash which will serve as a provenance hash for the balances. Each leaf node is
tagged with the root hash of the Peer Transaction Tree that was responsible for the
balance change.

Then there are two other entities: Currency Manager and Integrity Manager. The
Currency Manager is responsible for issuing currency as well as the conservation
of currency. The currency works with the guiding principle that the total currency
in circulation should be the same as the total currency issued/created. The integrity
manager's responsibility is to ensure the integrity of transactions. The δ-Currency
System Architecture and Design are shown in Fig. 3.

Figure 4, gives the workflow diagram for the δ-Currency System.

Since currency conservation is checked constantly, things like double-spend are
prevented. Entire details which are generally in a passbook are with the peers themselves. The first transaction among peers will be a bootstrap transaction where a
peer has to get the initial balance of the peer from the currency manager and the

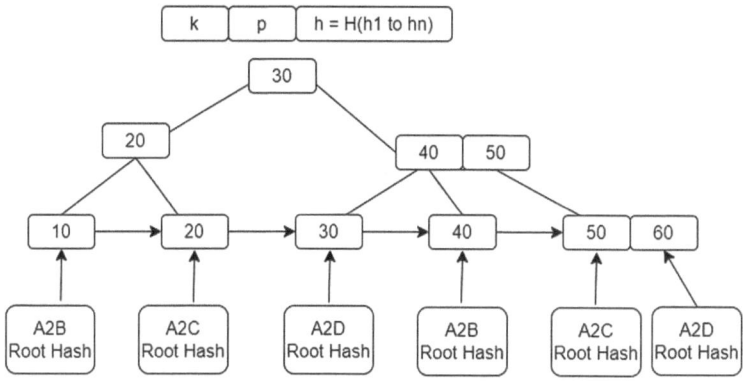

Fig. 2 Peer balances stored as Merkle Hash tree

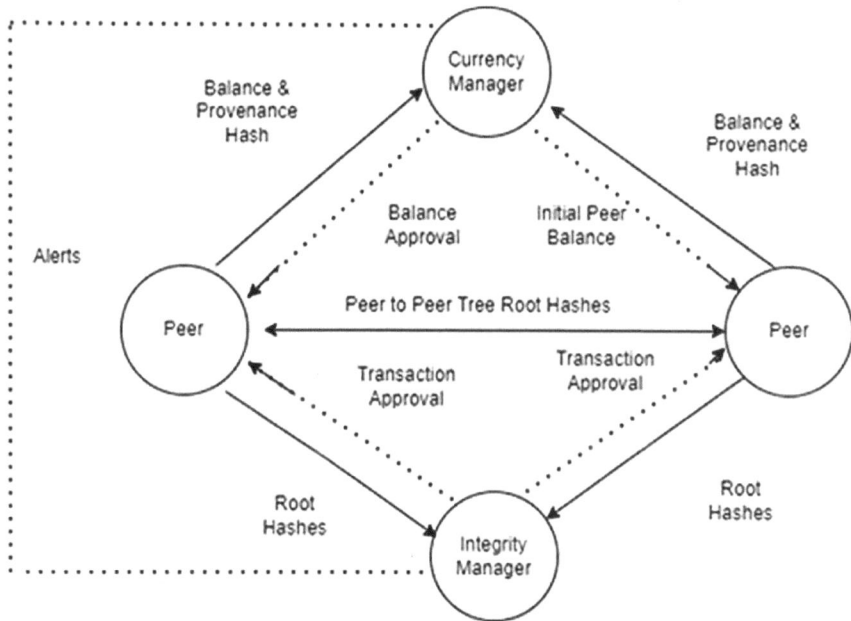

Fig. 3 δ-Currency system architecture and design

corresponding provenance hash. We can mandate that this step should happen when the peers agree to register mutually.

A unique feature of δ-Currency is that a transaction does not materialize unless both the sender and receiver participate and consent to it. In many current payment systems, the receiver is a passive participant. Another useful feature of the δ-Currency is deferred approval. Here the transactions do not get rejected due to network delays. Many digital payment transactions typically fail when the switch gets loaded with a

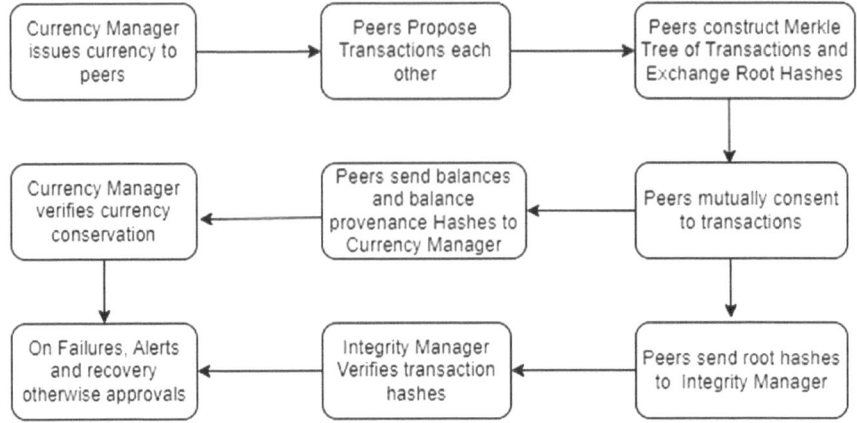

Fig. 4 Workflow diagram for δ-Currency system

lot of transactions leading to debit happening but not credit happening. This may at times cause the paying party to repeat the payment and lose money or lose out on doing the purchase.

δ-Currency is architected to address stakeholder concerns in a harmonious manner. Here, no single player knows all the information. The integrity manager knows who transacted with whom and the currency manager knows who all are transacting and the hashes but not the details of the transactions. However, if illicit transactions are suspected, it may be possible to reach out to the concerned parties for additional details.

3 Model Checking Preliminaries

Unlike testing, model checking can detect faults in a concurrent system by exploring every possible state of the system. Schneider et al. [9] make use of model checking to validate the functioning of embedded spacecraft controller, where critical functional requirements are validated down to the design level. Huang et al. [10] use model checking to validate composite web services integrated from multiple vendors, where model checking enables automatic generation of test cases. However, most model-checking techniques require that a system be described in a modeling language. Leungwattanakit et al. [11] make use of cache-based model checking and a check-pointing tool to make this task easier. Model Checking literature is dealt in detail by Baier and Katoen [12]. Perini and Susi [13] combine agent oriented modeling and model checking. Thrust of their work is on conceptual modelling and communication to stakeholders.

To represent a system, Model Checking makes use of fundamental building blocks of computation. Figure 5 illustrates the Transition System.

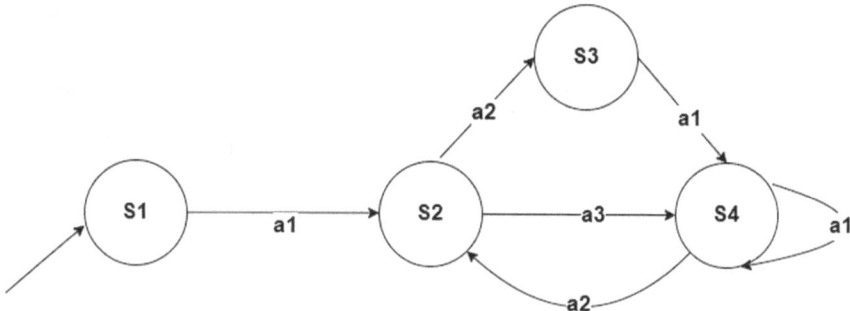

Fig. 5 Transition system

Here s1 is the initial state. The actions a1, a2, and a3; trigger transitions to s2, s3, and s4. Transition Systems are also referred to as Finite State Machines and State Transition Graphs. Model Checking can also help us to know if the system will reach an undesired state.

Figure 6, illustrates Program Graph, here two blocks of the code are connected using transitions that are marked with conditional statements and assignment statements.

Model Checking can be used to represent concurrent systems particularly well. The concurrent systems, can be independent systems, shared variables systems, and shared actions systems. Execution of modules can happen synchronously using a global clock or asynchronously.

Another important construct in model checking is the notion of properties that can be expressed mathematically. A transition system will satisfy a property only if

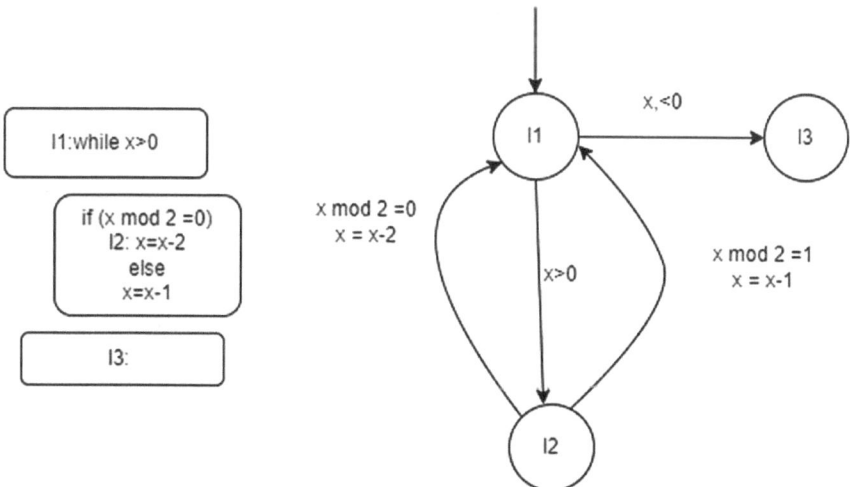

Fig. 6 Program graph

every execution has every state satisfying a given property. An interesting property is the safety property. This will require the construction of program graphs to ascertain that the function is handled properly in such a way that no bad property holds. A liveness property says that something good happens infinitely often. Alternatively, we can also say that something good eventually happens. A property that is always true is called the invariant property. That means in every reachable state the invariant property should hold true.

4 Validating δ-Currency Using Model Checking

Here we represent traditional banking and Blockchain-based currencies as reference architectures and then explain how δ-Currency is differently represented. Figure 7 illustrates the traditional banking scenario using a state transition diagram. The append-only ledger is represented as a stack of transactions. Here we are assuming 3 participants transacting with each other with initial balances B1, B2, and B3. Here transactions trigger state changes.

Figure 8 illustrates the bitcoin currency transaction system using the transition system. Here the ledger is a Blockchain which can also be represented as a stack of Merkle Tree root hashes corresponding to the blocks of transactions that get appended. The approach to validate individual bitcoin transactions is by matching outputs and inputs. Each block of transactions is represented as stack of inputs and outputs. B1, B2, B3 represent blocks of transactions that change the state of the system represented by public Blockchain. Bitcoin is implemented using a stack

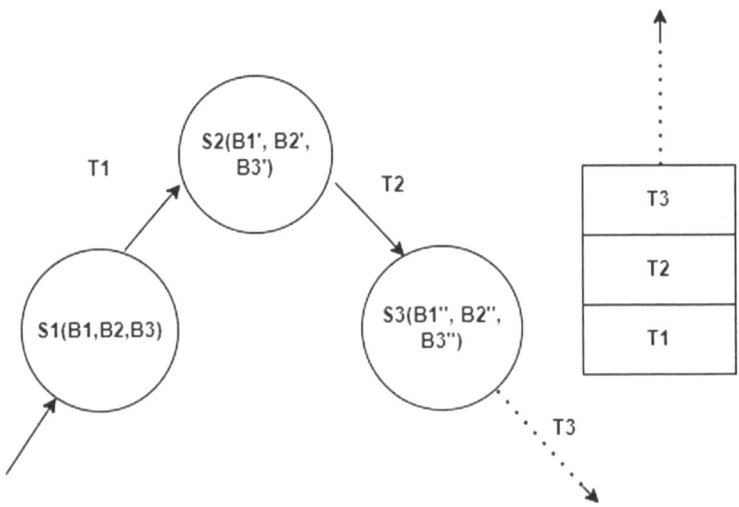

Fig. 7 Modelling traditional banking as a transition system

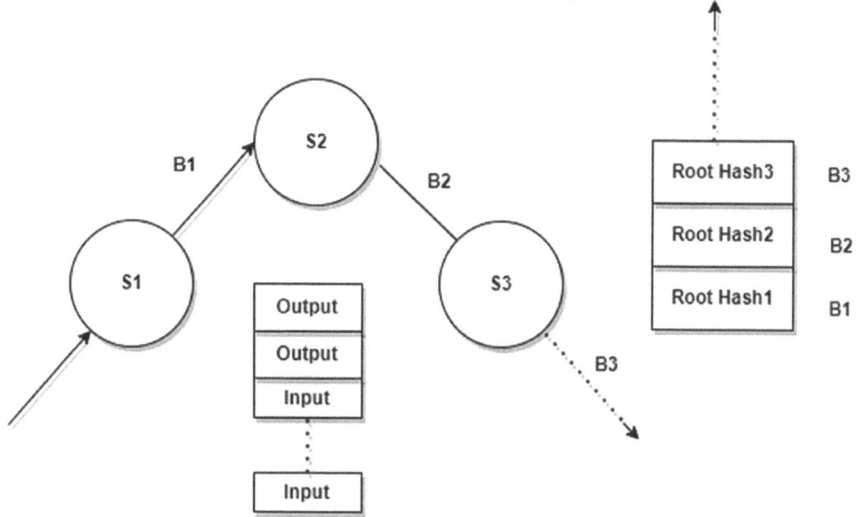

Fig. 8 Modelling bitcoin as transition system

machine which is not Turing Complete. In case of Ethereum, Smart-contracts and Accounts come into play in place of scripts and addresses.

Next, let us represent the δ-Currency system as a transition system in a scenario where 3 peers are participating. Here instead of a single ledger, we have a peer-to-peer ledger which is maintained as the root hash of Merkle Trees of peer-to-peer transactions that are validated. For every 2 peers, we have a stack of root hashes of peer-to-peer transactions that are stored identically by both peers. See Fig. 9. As in earlier cases, transactions trigger state transitions. Here RH1, RH2, RH3 represent Root Hashes of Peer-to-Peer Merkle Trees.

As described in Fig. 2 earlier, the changing balances of each peer are stored in a B+ Tree along the leaf nodes and at every stage a provenance hash is generated for the new balance. The same phenomenon is represented in Fig. 10. For peer 1, balances change from B1 to B1' to B1' and the corresponding hashed values are $H(B1)$, $H(H(B1) H(B1''))$ and $H(H(B1), H(B1'), H(B1''))$. Similar convention is used for other peers. The state of each peer comprises of changing balances and balance provenance hashes which are represented using two stacks. Thus, the system state in the δ-Currency system is managed in a decentralized manner where each peer manages their own state. This is in contrast to bitcoin-based system where public Blockchain represents the state of the system in entirety.

Figures 9 and 10, convey that dealing with δ-Currency is very close to dealing with physical currency in the sense that transactions happen directly with peers and each peer manages their own state. With digital currency we have the added safeguard of provenance which protects peers as well as the system.

In Fig. 11, we compare the 3 transaction systems. In δ-Currency, the peers mutually consent to a transaction, with the spirit of an optimistic commitment and deferred

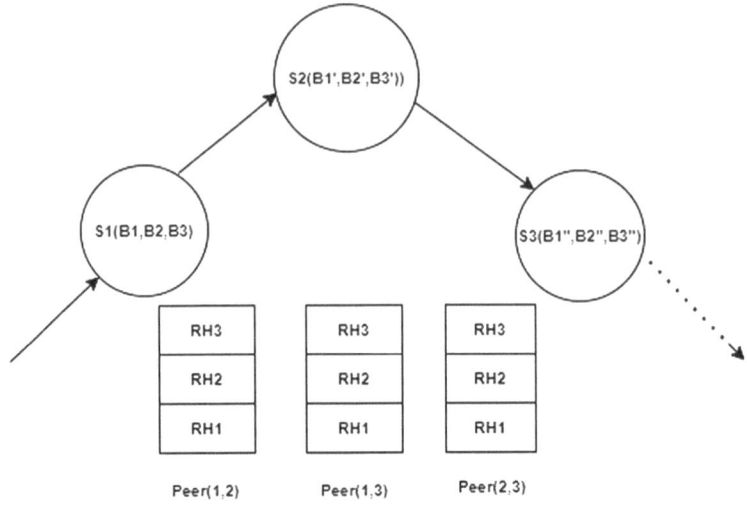

Fig. 9 Modelling δ-Currency transactions as transition systems

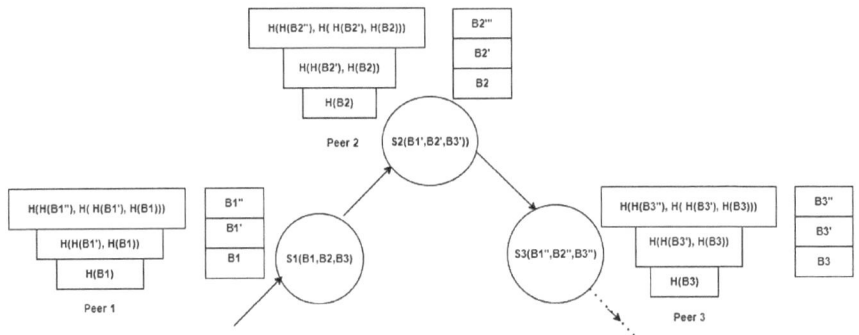

Fig. 10 δ-Currency peer balance state transition system

approval and thus can operate in a disintermediated manner. The requisite safeguards kick in if trust is violated.

Next, we do our analysis using distributed system properties namely safety, liveness and invariance. In Fig. 12, we represent Safety Property as designed in the integrity manager and currency manager. Here state −1 represents bad state and non-negative states represent good states. Thus, with every peer-to-peer transaction both peers should generate the same hash and integrity manager additionally validates that both hashes are equal and peers do not collude. But in the event that is not the case, transaction is rejected and it reaches the bad state, Similarly in normal course of events, the currency manager should accept balance updates, in the event peers report wrong balances or do not supply the right provenance hashes, the system will reach a bad state. The way the system is envisaged, peers do not wait for integrity

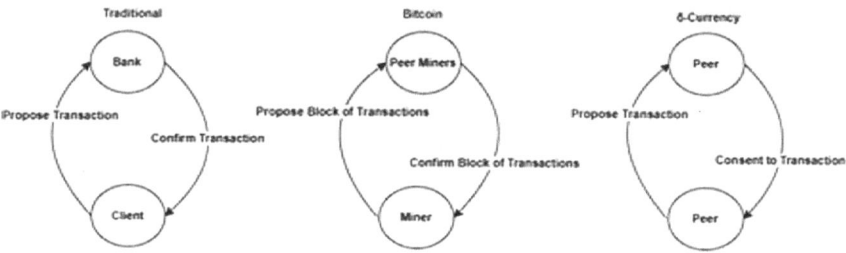

Fig. 11 Comparing currency systems

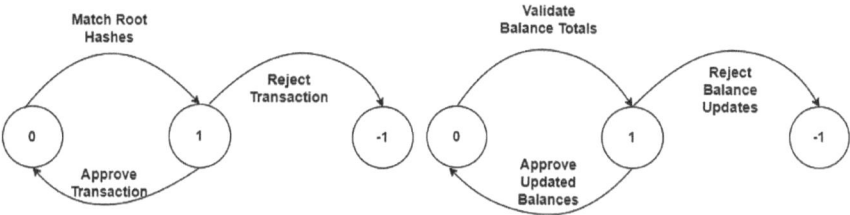

Fig. 12 Safety property in δ-Currency system

manager or currency manager to confirm the transactions. Only when bad state is reached, integrity manager and currency manager together resort to recovery actions.

This is followed by Fig. 13, which represents the liveness property. The liveness property is about something good eventually happening. Liveness is an issue with digital payment systems where transactions fail to complete due to network load at time leading to double payment or aborting of transfer of goods/services. In δ-Currency system, liveness is better handled as peers (typically trusted as they repeatedly transact with each other) confirm the transactions to each other. If the transaction happens in physical proximity or using dedicated network, network issues also will not come in the way. This is represented by the diagram on the left, where peers either confirm or wait for confirmation but within reasonable time, transactions get confirmed. The right diagram is about validating balance totals which is done by the currency manager and new balances get approved. Here it may happen that only one of the peers has updated the currency manager and it may need to wait for the other peer. The system as a whole, toggles between state 0 and 1.

Finally, the invariant property is illustrated in Fig. 14. The invariant property p stands for conservation of currency in the system. The currency known to be in circulation i.e., distributed among the peers, should be same as the currency issued by the currency manager (and not returned back to currency manager). This protects the system as a whole. In contrast, a physical currency system is very vulnerable to fake currency.

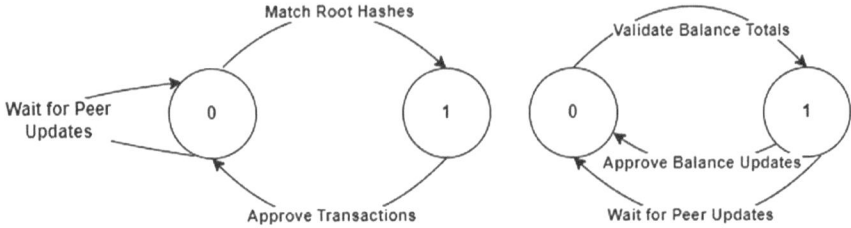

Fig. 13 Liveness property in δ-Currency system

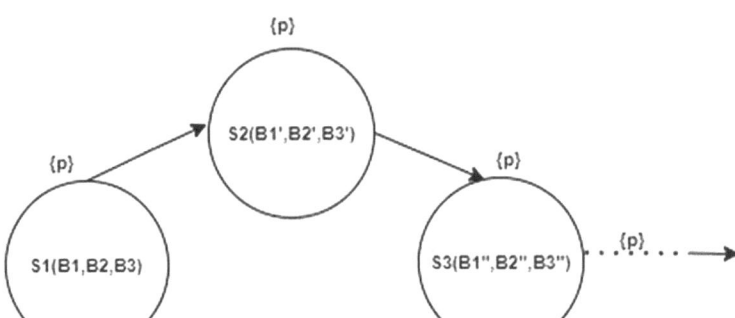

Fig. 14 Invariant property in δ-Currency system

5   Conclusions

We have discussed a novel δ-Currency proposal that has decentralized architecture where peers maintain their own state and consent to transactions among themselves, with a verification mechanism provided at the system level. To validate the proposed digital currency, we have made use of model checking and shown that it is architecturally adequate and sound in design. We found Model Checking to be a promising approach for communicating new proposals and comparing them with established alternatives. There is further scope to make use of program graphs extensively to validate the code to ensure that safety, liveness and invariant properties continue to hold, amidst unreliable networks and malicious peers/institutions. The ideas in δ-Currency proposal can be evolved to establish multi-lateral currencies, where even the creation of currencies is decentralized among multiple nations. Model Checking can be relied on to provide the backed by strong theoretical proof that the system is robust to weather any kind of risks and uncertainties. Further, we can also look at extending these ideas to domains such as supply chain where typically same parties repeatedly engage with each other. We can also develop a unified framework where exchange of commodities, services and currencies happens in homologous manner.

References

1. Goldstein, M, Flitter E (2023) Risky bet on crypto and a run on deposits tank signature bank. The New York Times, March 12. https://www.nytimes.com/2023/03/12/business/signature-bank-collapse.html
2. Prabhu S, Subramanyam N, Krishnan S, Sachidananda B (2022) Decentralized digital currency system using Merkle Hash trees. https://doi.org/10.48550/arXiv.2205.03259
3. Inventiva (2023) How scammers are using google pay and phones to Con people; 81 users lose ₹1 Crore!, March 21. https://www.inventiva.co.in/stories/how-scammers-google-pay-phones/
4. Dai W (1998) B-money, Bitcoin Wiki. http://www.weidai.com/bmoney.txt. Last Accessed 30 April 201
5. Nakamoto S (2008) Bitcoin—A peer-to-peer electronic cash system. https://bitcoin.org/bitcoin.pdf, Accessed 27 April 2021
6. Bonneau J, Miller A, Clark J, Narayanan A, Kroll JA, Felten EW (2015) SoK: research perspectives and challenges for bitcoin and cryptocurrencies. In: IEEE symposium on security and privacy. https://www.ieee-security.org/TC/SP2015/papers-archived/6949a104.pdf
7. Swan M (2015) Blockchain: blueprint for a new economy, 1st edn. O'Reilly Media, Inc.
8. Saake G, Niaz MS (2015) Merkle hash tree based techniques for data integrity of outsourced data. In: CEUR workshop proceedings, vol 1366, pp 66–71. https://ceur-ws.org/Vol-1366/paper13.pdf
9. Schneider F, Easterbrook SM, Callahan JR, Holzmann GJ (1998) Validating requirements for fault tolerant systems using model checking. In: Proceedings of IEEE international symposium on requirements engineering: RE '98, pp 4–13. https://doi.org/10.1109/ICRE.1998.667803
10. Huang H, Tsai W, Paul R (2005) Automated model checking and testing for composite web services. In: Eighth IEEE international symposium on object-oriented real-time distributed computing, pp 300–307. https://doi.org/10.1109/ISORC.2005.16
11. Leungwattanakit W, Artho C, Hagiya M, Tanabe Y, Yamamoto M, Takahashi K (2014) Modular software model checking for distributed systems. IEEE Trans Softw Eng 40(5):483–501. https://doi.org/10.1109/TSE.2013.49
12. Baier C, Katoen JP (2008) Principles of model checking. MIT Press. ISBN, p 9780262026499
13. Perini A, Susi A (2005) Agent-oriented visual modeling and model validation for engineering distributed systems. Comput Syst Sci Eng 20

A Batch-Service Queueing Assisted Blockchain System for Supply Chain Management

Bibhuti Bhusan Dash, Utpal Chandra De, Parthasarathi Pattnayak, Rabinarayan Satapathy, Sibananda Behera, and Sudhansu Shekhar Patra

Abstract In recent years, blockchains have attracted a great deal of interest from researchers, engineers, and institutions. Additionally, the implementation of blockchains has begun to revitalisea large number of applications, such as e-finance, smart homes, smart health, social security, logistics, and so on. As supply chains become increasingly global, management and control become more complicated. Blockchain technology is establishing promise in tackling several supply chain management (SCM) concerns as a distributed digital ledger platform that ensures security, traceability, and transparency. Government, societal, and consumer pressures have encouraged us to consider how blockchain can help us meet our sustainability goals and improve supply chain sustainability. True blockchain-led supply chain and corporate transformation is still in its early phases. This chapter examines the possible application of blockchain technology (BT) and smart contracts to SCM. In order to study the operations of blockchain based supply chain management a queueing based model is developed. It assesses the measures of the model, such as the mean transactions per unit time, the mean transactions in the unconfirmed transaction pool, the mean time required for transaction confirmation, and the average

B. B. Dash · U. C. De · P. Pattnayak · S. S. Patra (✉)
School of Computer Applications, KIIT Deemed to be University, Bhubaneswar, India
e-mail: sudhanshupatra@gmail.com

B. B. Dash
e-mail: bibhuti.dash@gmail.com

U. C. De
e-mail: deutpal@gmail.com

P. Pattnayak
e-mail: parthakiit19@gmail.com

R. Satapathy
Faculty of Emerging Technologies, Sri Sri University, Cuttack, India
e-mail: rabinarayan.satpathy@gmail.com

S. Behera
Trident Academy of Creative Technology, Bhubaneswar, India
e-mail: sibananda0506@gmail.com

© The Author(s), under exclusive license to Springer Nature Singapore Pte Ltd. 2024
S. Namasudra et al. (eds.), *Data Science and Network Engineering*, Lecture Notes in Networks and Systems 791, https://doi.org/10.1007/978-981-99-6755-1_28

waiting time of transactions in the unconfirmed transaction pool. Additionally, it evaluates the characteristics of the supplychain (SC) system.

Keywords Supply chain · Blockchain · Smart contract · Performance evaluation · Queueing model

1 Introduction

Manufacturing has become more complicated as the number of intermediaries between the manufacturer and the ultimate consumer has increased. Globalization and market expansion compelled businesses to diversify their product portfolios and life cycles in order to suit the requirements of new markets [1, 2]. Little information about the product's origins, production, or shipping path is available. As a result, the problem becomes more than just quantitative; it also becomes more qualitative. The traceability and data management system continue to be the most difficult challenges in the supply chain. Most businesses, including healthcare, banking, food, and education, have centralized information system administration. Intermediaries manage trading, decision making, and data storage [3–5]. A centralized management system compromises data integrity, availability, and robustness, leaving it open to corruption, fraud, and other security concerns. Establish a trustworthy ecosystem of trust between suppliers and customers. To achieve this, a policy that emphasizes supply chain transparency, precise data collecting, and secure data storage is required [6]. Blockchain technology is being used in supply chains to make them more transparent, authentic, and trustworthy to find out if adding BT to the supply chain makes it more reliable and authenticate [7]. Blockchain records transactions in an unalterable format. Before creating a permanent record on the blockchain, all product and delivery information is collected and authenticated. A traditional SCM system to a blockchain-based system is depicted in Fig. 1's first panel. Before proposing the blockchain as a potential solution, we perform a literature review of current supply chain challenges [8]. The literature in relation to new supply chain challenges that are being ad-dressed by BT. As a result, examining the effectiveness and sufficiency of a block-chain-based supply chain is an important element of researching blockchain integration into supply chain.

This article is organized as follows. Section 2 gives the related work, Sect. 3 describes the performance evaluation of the blockchain system using queueing model and finally, we offer our suggestion for the development of a blockchain-based supply chain as a final conclusion and future work in Sect. 4.

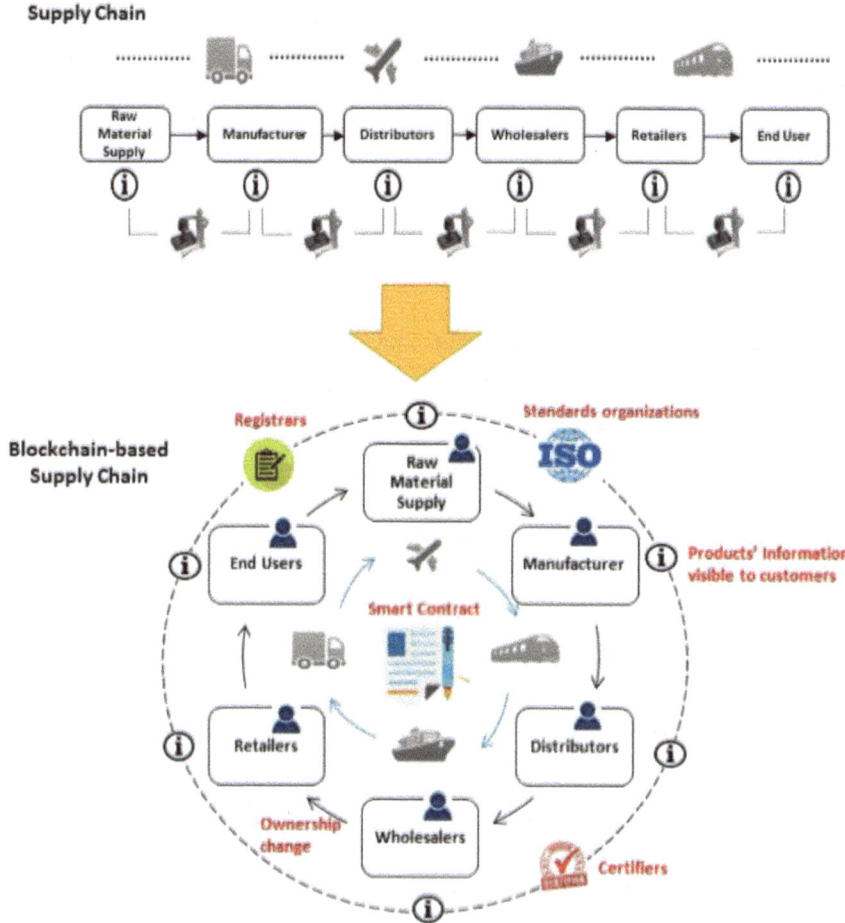

Fig. 1 Stakeholders in the management of smart supply chains and supply chain transformation

2 Related Work

Scalability and security issues are two different types of blockchain performance problems [9]. There is a wealth of research on the performance assessment of blockchains for these two problems. The article [10] provides a thorough overview of the blockchain evaluation studies. Transaction confirmation time (transaction throughput) is a key scalability metric. Bitcoin blockchain transaction confirmation time includes the generation time for block (block transaction mining time), block-propagation delay, and transaction-waiting time in miner node memory pools. In order to determine how long it takes for a transaction to be confirmed, one typical method involves simulating the Bitcoin blockchain in the form of a queueing model. Characterizing the dynamic nature of the transactions in the miner nodes is a key focus of [15] and our earlier work [11–14]. From this vantage point, the single-server queueing system that represents the Bitcoin blockchain at its core is one that takes into account both the time it takes for blocks to generate and the transactions that are currently waiting in the memory pool.

The block-propagation delay, however, is equally significant considering the Bitcoin security issue. It is analyzing the block propagation of Bitcoin. The authors are concentrated on how the Bitcoin blockchain's security is impacted by the block propagation delay [16]. They calculated the likelihood of a blockchain fork by assuming a Poisson model for the block-generation time. In [17], a more sophisticated model for the fork probability was presented. The block generation process was thoroughly investigated in [18], and the authors looked into whether stochastic processes might be used with the block-generation model. They did this by comparing several block-arrival models to the real data and taking into consideration three aspects for the process of block formation. These factors are the block propagation delay, difficulty modifications, and hash-rate function.

A non-homogeneous Poisson process with a periodic hash-rate function was found to be the simplest model that can algorithmically produce arrival sequences for simulation and reasonably approximate the process of block formation. This model was discovered to be a non-homogeneous Poisson process. However, from a queueing theory perspective, queueing systems with non-homogeneous Poisson input are generally difficult to evaluate.

The blockchain is primarily reliant on block building. Transactions that have just been issued are stored in a memory pool that is not yet validated until they are included to a fresh block and available to the blockchain presently after the mining process has been completed. The process of forming a block is analogous to that of a queuing system that provides batch service, in which a number of customers leave the establishment at the same time. In the context of blockchain technology, much research on queuing models that include batch processing has not been done. There is literature on queue analysis with batch service. In [19, 20] the authors investigated $M/G^b/1$ queueing with batch service. Customers arrive via a Poisson process, there is only one server, and the service time distribution is in $M/G^b/1$. Authors [19, 20] used the supplemental variable technique to derive the joint distribution of the number of customers in line and the elapsed service time for the $M/G^b/1$. Authors in [21] investigated transaction wait times and block acceptance using $M/M/1$ queueing. For the steady state of the system, Authors in [15] used the matrix geometric solution approach to develop a $GI/M/1$ queueing system with two service stages that simulated mining and building the new blockchain.

3 Performance Evaluation of the Blockchain Based Supplychain System

When the functionality of the blockchain system is taken into consideration, it is possible to describe the system as a blockchain queue, with the creation of blocks and the construction of blocks functioning as the two stages of the batch service. As a part of the block generation, a miner solves a computational problem by utilizing a cryptographic hash algorithmic process called mining. Several model descriptions Here are some model descriptions:

The Arrival Procedure: The interval between transactions is exponential. The blockchain transaction arrival rate is λ. Each transaction enters the system verified by the memory pool and awaits the right miner in the queue. The arrival of transactions is predicted to be Poisson, as the inter-arrival of transactions is exponential. The arrival procedure of the transactions is shown in Fig. 2.

The Service Procedure: Every transaction has to wait in the pool of unconfirmed transactions, which is a waiting buffer for the proper miner. During the process of creating blocks, the transactions are mined into them successfully without any problems. A block (let's call it b) contains a collection of transactions.During the second stage of the service, a block that contains a group of transactions is added to the blockchain. The mining service time by the miner is denoted by μ and is depicted in Fig. 2 lower part.

Block-generationDiscipline: A block can have a huge number of transactions, but only b transactions maximum. Arrivals do not always follow FCFS. Late transactions may be mined into the block in initial state. But, for simplicity, we've included FCFS in our model. Figure 2 shows the block building process.

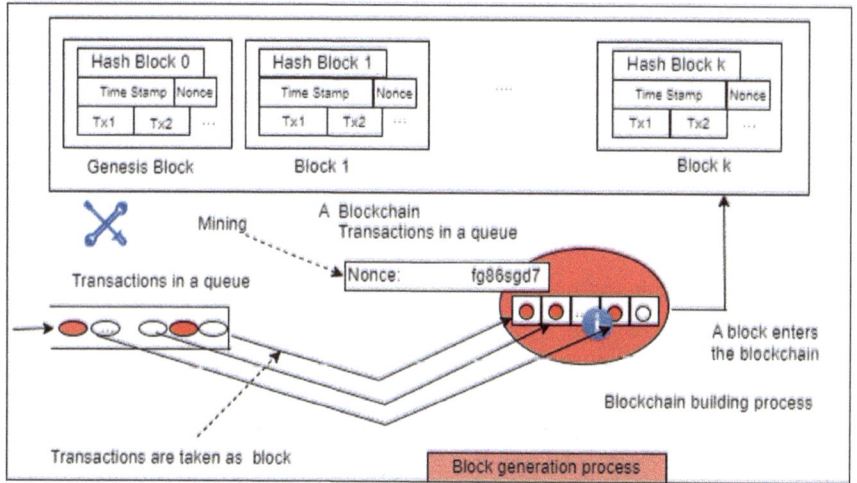

Fig. 2 The operation mechanism of blockchain system

Block size limit: The maximum transactions mined by a miner is at most b. Only b transactions are considered and mined if there are more than b transactions waiting in the unconfirmed transaction pool.

The system can be modelled as a single miner which process the transactions in bulk from the unconfirmed transaction pool where all the transactions are waiting in the queue for the appropriate miner. Let us define Q_n as the steady-state probability that n units are queued (n = 0,1 ,2 ,...) when the server is busy, and Q_{00} as the probability that no blocks are processed. Q_n and Q_{00} are compared against the probability that no blocks are processed (service station is in empty).We looked at two types of bulk-service models: full-batch service and partial-batch service, which show how the system works when the batch size is less than b and when the batch size is constant in the system.

3.1 Partial Batch Service Model

The miner can mine with at least one transaction and allow up to a maximum of b transactions in this mode of mining. Similarly, when the miner is serving to less than b transactions, new arrived transactions immediately joined to be mined up to the limit of b and completed mining with the others who are in the unconfirmed transaction pool and ready to be mined. The mining time of any batch go along with an exponential distribution with mean $1/\mu$. If Qn is the probability of n transactions in a steady-state system, the appropriate probability generating function is

$$Q_i(z) = \sum_{i=0}^{\infty} Q_i z^i \tag{1}$$

In steady state, we have the following difference equation:

$$(\lambda\phi + \mu)(Q(z) - Q_0) = \mu z^{-b} \sum_{i=1}^{\infty} Q_{n+b} z^{i+b} + \lambda\phi(z)Q(z) \tag{2}$$

$$\lambda\phi Q_0 = \mu \sum_{i=1}^{b} Q_i \tag{3}$$

After simplification, we have

$$Q(z) = \frac{\mu \sum_{i=0}^{b} Q_i(1 - z^{i-b})}{\lambda\phi(1 - z) + \mu(1 - z^{-b})} = \frac{\mu \sum_{i=0}^{b} Q_i(z^i - z^b)}{\lambda\phi z^{b+1} - (\lambda + \mu)z^b + \mu} \tag{4}$$

If $r_1, r_2, ..., r_{b+1}$ are the roots of the characteristics equation,

$$\mu r^{b+1} + (\lambda \phi + \mu)r + \lambda = 0$$

then

$$Q_n = \sum_{i=1}^{b+1} k_i r_i^n, \ n \geq 0 \tag{5}$$

Since the sum of all the probabilities is equal to one, $k_i = 0$ for all roots greater than one. Using Rouche's theorem, it has precisely one root r_0 in $(0, 1)$. Therefore,

$$Q_n = (1 - r_0)r_0^n, \ n \geq 0, \ 0 < r_0 < 1 \tag{6}$$

3.2 Performance Indices

The average no. of transactions per unit time in the system (unconfirmed transaction pool + number of transactions in mining) is given by:

$$L_s = \sum_{i=1}^{\infty} i P_i = \frac{r_0}{1 - r_0} \tag{7}$$

The following formula calculates the average no. of transactions in the unconfirmed transaction pool:

$$L_q = \frac{r_0}{1 - r_0} - \frac{\lambda \phi}{\mu} \tag{8}$$

The average length of time a transaction is in the system (unconfirmed transaction pool + no. of transactions in mining) is:

$$W_s = \frac{L_s}{\lambda \phi} \tag{9}$$

The average length of time a transaction is in the unconfirmed waiting pool is:

$$W_q = W_s - \frac{1}{\mu} \tag{10}$$

3.3   Full Batch Size Model

In this mode of mining process, if there is exactly b transactions waiting in the unconfirmed transaction pool and if the miner is selected, then the miner take up all the b transactions and the miner relics idle till there are 'b' transactions in the waiting unconfirmed pool. The service time follow exponential distribution with mean $1/\mu$. This model's steady state balancing equations are as follows.

$$(\lambda\phi + \mu)Q_n = \lambda\phi Q_{n-1} + \mu Q_{n+b}, \ n \geq b \tag{11}$$

$$\lambda\phi Q_n = \lambda\phi Q_{n-1} + \mu Q_{n+b}, \ 1 \leq n < b \tag{12}$$

$$\lambda\phi Q_0 = \mu Q_b \tag{13}$$

Equations (8)–(9) are multiplied by suitable powers of 'z' and summed over 1 to infinity.

$$(\lambda\phi + \mu)\sum_{i=b}^{\infty} Q_i z^i = \lambda\phi\sum_{i=b}^{\infty} Q_{i-1}z^i + \mu\sum_{i=b}^{\infty} Q_{i+b}z^i$$

$$\lambda\phi\sum_{i=1}^{b-1} Q_i z^i = \lambda\phi\sum_{i=1}^{b-1} Q_{i-1}z^i + \mu\sum_{i=1}^{b-1} Q_{i+b}z^i$$

Using Eqs. (1) and (13) to simplify the above equations, we get:

$$Q(z) = \frac{(1 - z^b)\sum_{i=0}^{b-1} Q_i z^i}{t z^{b+1} - (t + 1)z^b + 1}, \ t = \frac{\lambda\phi}{\mu}$$

$$= \frac{(1 - z^b)(z_0 - 1)}{b(z_0 - z)(1 - z)} \tag{14}$$

where z_0 ($z_0 > 1$) is one of the characteristic equation's roots among the $(b + 1)$ roots.

$$\lambda\phi z^{b+1} + (\lambda\phi + \mu)z^b + \mu = 0$$

which lies outside the unit circle.

The mean transactions per unit time in the system (unconfirmed transaction pool + number of transactions in mining) is given by:

$$L_s = \frac{2 + (z_0 - 1)(b - 1)}{2(z_0 - 1)} \tag{15}$$

The following formula calculates the mean transactions in the unconfirmed transaction pool:

$$L_q = L_s - \frac{\lambda\phi}{b\mu} \tag{16}$$

The mean time a transaction is in the system (unconfirmed transaction pool + number of transactions in mining) is:

$$W_s = \frac{L_s}{\lambda\phi} \tag{17}$$

The average duration a transaction spends in the waiting pool for unconfirmed transactions is:

$$W_q = W_s - \frac{1}{\mu} \tag{18}$$

We investigate numerical outcomes for a scenario in which transactions arrive at rates of 10, 20, 30, and 40. Regardless of batch size b = 1, 2, 3, 4, and 5, the service rate of the partial and full batch service models is μ = 50 and $\emptyset$ = 0.6. Figure 3 shows W_s and W_q and Fig. 4 shows L_s and L_q for a partial batch size b The figures show that as batch size b increases, L_s and L_q for the partial batch model decreases gradually, and Ws and Wq for the partial batch service model decreases gradually as batch size b increases. Furthermore, we can see in Fig. 5 that as the complete batch size b increases one by one, L_s and L_q decreases. In addition, as complete batch size b increases progressively, W_s and W_q increases, as illustrated in Fig. 6. Furthermore, when b = 1, the L_s, W_s decreases, however when b = 2, L_s, W_s gradually increase, as shown in Figs. 3, 4, 5 and 6. As a result, the partial batch service paradigm reduces transaction wait times while improving service quality over the complete batch service model.

4 Conclusions and Future Work

The supply chain might be improved by implementing blockchain technology, which would be advantageous to all of the different participants. The necessity of involving all of the different parties in the transaction that takes place on the blockchain during its implementation in the supply chain is one of the most significant obstacles that must be overcome. The fact that information must be shared over the entire Blockchain could also cause the implementation of the solution to proceed at a more snail-like pace. Therefore, a thorough deployment of blockchain technology in the supply chain must start with an examination of the needs and goals of the many parties

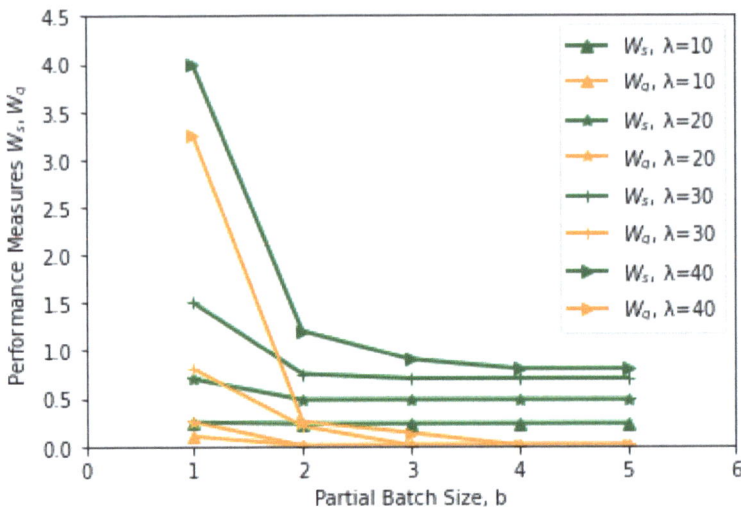

Fig. 3 Partial batch size b versus performance measure W_s, W_q

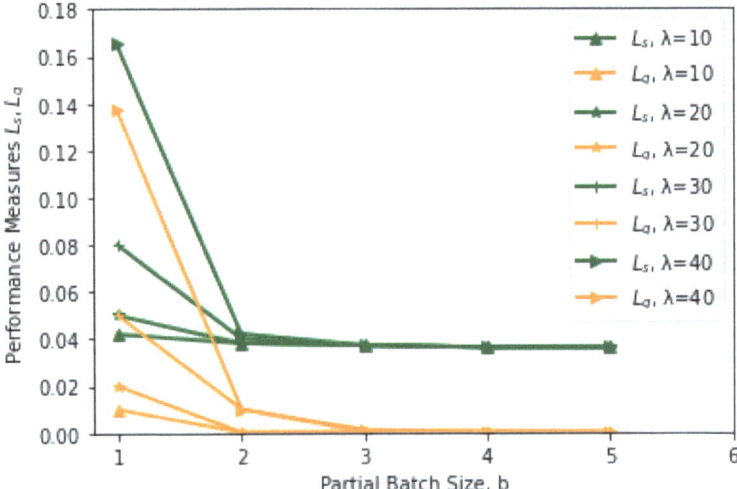

Fig. 4 Partial batch size b versus performance measures L_s, L_q

interested, with the aim of creating a novel business model that can demonstrate the benefits of the solution.

Unrealistic expectations may lead to failed blockchain adoption, say some industry experts. Like any new technology, there have been some successes and disappointments. It is hoped that, like many other emerging technologies, blockchains will find their place in reality and wide acceptance once their constraints are solved. A

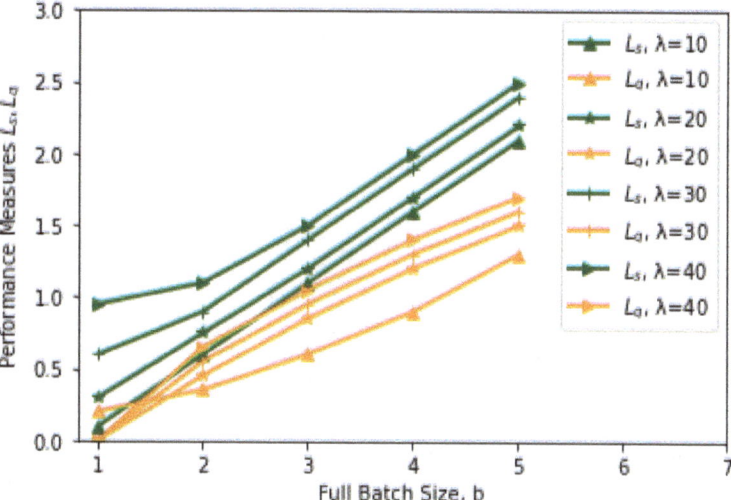

Fig. 5 Full batch size b versus performance measure L_s, L_q

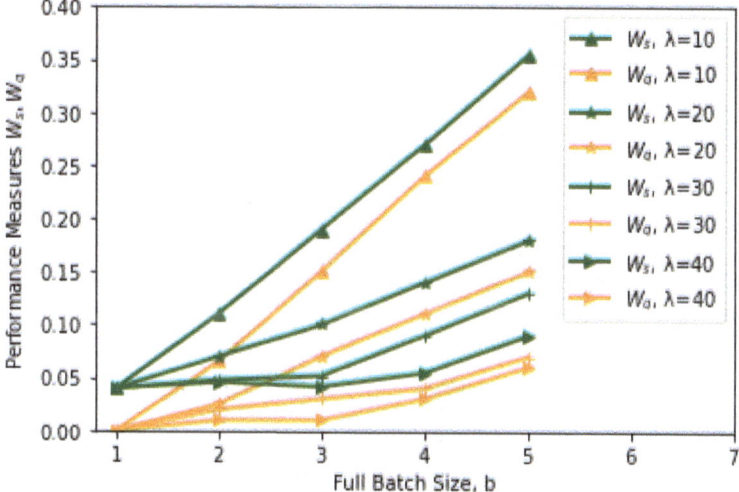

Fig. 6 Full batch size b versus performance measure W_s, W_q

reliable data collection and analytics system is essential for blockchain technology to work. Implementing blockchain technology in supply chains requires IoT and big data technologies. The IoT devices gather data from many supply chain sources. Big data technologies capture, store, and analyse massive amounts of data. The study's experts support blockchain technology uses in SCM. Future blockchain technology adoption studies should include aiding technologies.

References

1. Craighead CW, Blackhurst J, Rungtusanatham MJ, Handfield RB (2007) The severity of supply chain disruptions: design characteristics and mitigation capabilities. Decis Sci 38(1):131–156
2. Mukherjee P, Barik L, Pradhan C, Patra SS, Barik RK (2021) HQChain: leveraging towards blockchain and queueing model for secure smart connected health. Int J E-Health Med Commun (IJEHMC) 12(6):1–20
3. Ali AAA, Udin ZM, Abualrejal HME (2023) The impact of it alignment and supply chain resilience on the companies quality of performance: the moderating role of trust. Resmilitaris 13(2):156–169
4. Patra SS, Misra C, Singh KN, Gourisaria MK, Choudhury S, Sahu S (2021) QIoTAgriChain: IoT blockchain traceability using queueing model in smart agriculture. Blockchain applications in IoT ecosystem. Springer International Publishing, Cham, pp 203–223
5. Ranganathan C, Teo TS, Dhaliwal J (2011) Web-enabled supply chain management: key antecedents and performance impacts. Int J Inf Manag 31(6):533–545
6. Li X, Jiang P, Chen T, Luo X, Wen Q (2020) A survey on the security of blockchain systems. Futur Gener Comput Syst 107:841–853
7. Michel R (2017) The evolution of the digital supply chain. Logistics management (Highlands Ranch, Colo.: 2002)
8. Pilkington M (2016) Blockchain technology: principles and applications. In: Research handbook on digital transformations. Edward Elgar Publishing, pp 225–253
9. Goswami V, Patra SS, Mund GB (2012, March) Performance analysis of cloud with queue-dependent virtual machines. In: 2012 1st international conference on recent advances in information technology (RAIT). IEEE, pp 357–362
10. Smetanin S, Ometov A, Komarov M, Masek P, Koucheryavy Y (2020) Blockchain evaluation approaches: state-of-the-art and future perspective. Sensors 20(12):3358
11. Kawase Y, Kasahara S (2017) Transaction-confirmation time for bitcoin: a queueing analytical approach to blockchain mechanism. In: Queueing theory and network applications: 12th international conference, QTNA 2017, Qinhuangdao, China, August 21–23, Proceedings 12. Springer International Publishing, pp 75–88
12. Kawase Y, Kasahara S (2018, July) A batch-service queueing system with general input and its application to analysis of mining process for bitcoin blockchain. In: 2018 IEEE international conference on Internet of Things (iThings) and IEEE green computing and communications (GreenCom) and IEEE cyber, physical and social computing (CPSCom) and IEEE smart data (SmartData). IEEE, pp 1440–1447
13. Kasahara S, Kawahara J (2016) Effect of Bitcoin fee on transaction-confirmation process. arXiv preprint. arXiv:1604.00103
14. Kawase Y, Kasahara S (2020) Priority queueing analysis of transaction-confirmation time for bitcoin. J Ind Manag Optimiz 16(3):1077–1098
15. Li QL, Ma JY, Chang YX (2018) Blockchain queue theory. In: Computational data and social networks: 7th international conference, CSoNet 2018, Shanghai, China, December 18–20, 2018, Proceedings 7. Springer International Publishing, pp 25–40
16. Decker C, Wattenhofer R (2013, September) Information propagation in the bitcoin network. In: IEEE P2P 2013 proceedings. IEEE, pp 1–10
17. Seike H, Aoki Y, Koshizuka N (2019, July) Fork rate-based analysis of the longest chain growth time interval of a pow blockchain. In: 2019 IEEE international conference on blockchain (Blockchain). IEEE, pp 253–260
18. Bowden R, Keeler HP, Krzesinski AE, Taylor PG (2020) Modeling and analysis of block arrival times in the Bitcoin blockchain. Stoch Model 36(4):602–637
19. Chaudhry ML, Templeton JG (1983) First course in bulk queues
20. Chaudhry ML, Templeton JG (1981) The queuing system M/GB/l and its ramifications. Eur J Oper Res 6(1):56–60
21. Srivastava R (2019) Mathematical assessment of blocks acceptance in blockchain using Markov model. Int J Blockchains Cryptocurr 1(1):42–53

BiFrost: A Blockchain-Based Decentralized Messaging Application

Himanshu Pandey, Akhil Siraswal, Ekta Kaushik, Dilkeshwar Pandey, Sparsh Kapoor, and Hunny Pahuja

Abstract BiFrost is a revolutionary solution for secure online communication and data storage. It addresses security concerns such as eavesdropping, man-in-the-middle attacks, and censorship by offering decentralization, immutability, and data security through the use of blockchain technology. User-submitted data are added directly to the blockchain, creating a global copy in each node, and only authorized users can access the data using private keys. The system is fully decentralized, meaning there is no central authority, making it immune to censorship and government oppression. The technology used in BiFrost includes IPFS, smart contracts, Ethereum, INFURA, Solidity, and SPF. BiFrost provides a secure and decentralized solution for online communication and data storage, eliminating the need for a trusted intermediary and ensuring the privacy and security of data.

Keywords Eavesdropping · IPFS · Smart contracts · Blockchain · INFURA · Solidity · SPF · Ethereum

1 Introduction

The concept of blockchain was developed to enhance transaction security by providing confidentiality, data integrity, authorization, and encryption-based tamper resistance [1]. It eliminates the need for third-party involvement in transactions, requiring network participants to reach a consensus for secure validation

H. Pandey (✉) · A. Siraswal · E. Kaushik · D. Pandey
Department of Computer Science and Engineering, KIET Group of Institutions, Delhi-NCR, Ghaziabad, India
e-mail: himanshupandey200229@gmail.com

D. Pandey
e-mail: Dilkeshwar.pandey@kiet.edu

S. Kapoor · H. Pahuja
Department of Electronics and Communication Engineering, KIET Group of Institutions, Delhi-NCR, Ghaziabad, India
e-mail: hunny.pahuja@kiet.edu

© The Author(s), under exclusive license to Springer Nature Singapore Pte Ltd. 2024 383
S. Namasudra et al. (eds.), *Data Science and Network Engineering*, Lecture Notes in Networks and Systems 791, https://doi.org/10.1007/978-981-99-6755-1_29

without altering previous records. Manipulating the blockchain would necessitate an extremely high cost and simultaneous access to all network computers, rendering it virtually impossible for external attackers to tamper with the database.

Blockchain-based communication and data storage technology offer several advantages. Data can be encrypted to prevent theft [14], and its distributed storage eliminates single-point errors. Communication can be end-to-end encrypted, ensuring resistance to spying. The decentralized nature of the system makes it more cost-effective than centralized data storage [4]. Furthermore, the blockchain itself can enforce anti-corruption rules on transactions and communications, while maintaining a tamper-proof and auditable history.

Our application employs a decentralized application (DApp) approach where user data are stored in interconnected blocks forming a chain. This peer-to-peer network [2], as shown in Fig. 1, ensures that the stored data on the blockchain remain tamper-proof due to the encryption algorithm. Any attempt by a malicious user [1] to modify information in a block would require changing every copy of that block across the entire blockchain network, which is nearly impossible. Additionally, we acknowledge the significance of Directed Acyclic Graph (DAG)-based blockchains like IOTA Tangle [20, 21]. These structures offer an alternative to traditional linear blockchain systems by enabling parallel transactions and enhancing scalability. IOTA Tangle, for instance, has demonstrated its efficacy in securing environmental IoT data [21]. By leveraging the Tangle's masked authentication messaging protocol, it ensures secure and efficient data transfer. DAG-based blockchains present promising advancements that address the limitations of linear blockchain architectures.

The main features of blockchain technology include easy and fast communication through private key authentication, immutability of data once recorded, resistance to censorship by providing equal trading opportunities, and distributed storage across multiple nodes in the cloud [8].

This paper will discuss the general concepts of blockchain technology, highlight the existing problems in current systems, examine the available solutions, present

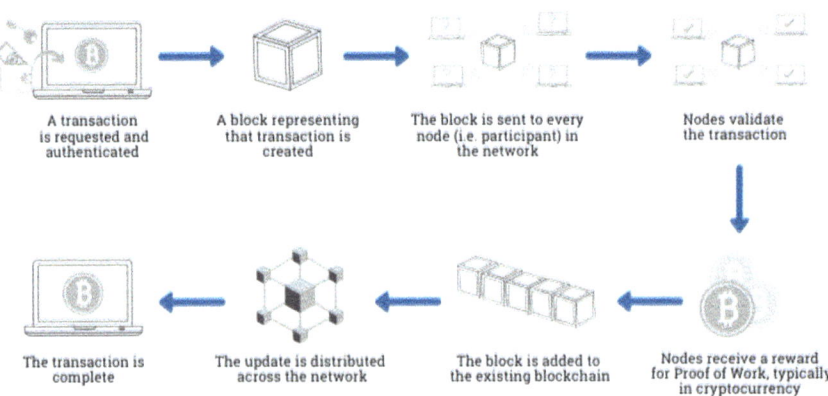

Fig. 1 How does a transaction get into the blockchain?

the contributions made by the authors, and outline the organizational structure of the paper.

2 Background and Related Work

Over the past decade, traditional chat applications such as WhatsApp and WeChat have become increasingly popular, with millions of users around the world. However, these centralized applications have some key features and limitations that have led to concerns about user privacy, data security, and censorship. In response, the development of decentralized storage, censorship resistance, data security, and data immutability has allowed for the creation of applications that address these challenges. This literature survey will review current research and literature on the topic of centralized and decentralized chat applications.

2.1 Centralized Chat Applications

Centralized chat apps store all chat information and identities on a central server, allowing companies to control communication and impose rules. However, this approach has limitations. A single server manages all services, which can cause inaccessibility if there are issues, and government requests can compromise user confidentiality. Also, a single node failure can compromise the entire app.

2.2 Decentralized Chat Applications

In response to these challenges, decentralized chat applications have been developed. These applications are designed to provide greater data security, censorship resistance, and data immutability. In decentralized [6] chat applications, user data are stored across a network of nodes, rather than on a central server. This architecture provides greater data security, as data are distributed across the network and not stored in a single location. Furthermore, [5] decentralized chat applications are censorship-resistant, as there is no central authority that can block or control communication. Finally, data immutability ensures that data cannot be altered or deleted, providing greater trust and security.

2.3   Research on Decentralized Chat Applications

A growing body of research has explored the use of decentralized chat applications. For example, a study by Dubey et al. explored the use of decentralized chat applications in the context of censorship [5] resistance. The authors found that these applications were effective in resisting censorship and enabling communication in environments where traditional centralized applications were blocked. Similarly, a research explored the use of blockchain technology to provide data security and privacy in decentralized chat applications [14]. The authors found that blockchain-based solutions could provide greater data security and user privacy than traditional centralized applications.

In addition to the mentioned studies, several other research papers have contributed to the exploration and development of decentralized chat applications. Notable works include:

Zhang, S. Shah, Y. Elmandjra, M. DeCesare, M. Hanna BlockChat: A Decentralized Messenger on the Blockchain [5].

C. Hwa, K. Yue HIVE: A Blockchain-Based Group Messaging Service [8].

These additional research papers contribute to the ongoing exploration and advancements in the field of decentralized chat applications, addressing issues of privacy, security, and censorship resistance.

3   Implementation

The implementation of the decentralized application involves deploying a smart contract on the Ethereum network [16], written in Solidity and tested in JavaScript, using the INFURA API. The interface for the application is built with ReactJS [3]. The smart contract will be first deployed on the Rinkeby test network for testing and security purposes before deployment on the main Ethereum network. The implementation is aimed to provide a reliable, secure, and effective system for the decentralized application.

3.1   Ethereum: A Platform for Decentralized Applications

Ethereum is a blockchain-based platform that enables the creation and deployment of decentralized applications. It features smart contract functionality and uses a modified version of the Nakamoto consensus algorithm [16].

One of the key features of Ethereum is its use of the Proof-of-Stake consensus mechanism. This algorithm aims to achieve distributed consensus in the blockchain network by allowing users to "stake" a value, or money, on a specific outcome. This process is known as staking.

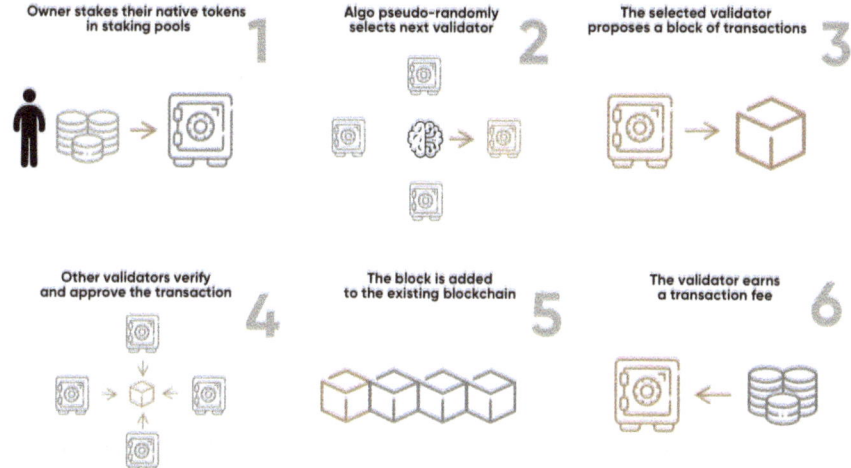

Fig. 2 Transaction in Ethereum

Proof-of-Stake is seen in Fig. 2 as a more energy-efficient alternative to the Proof-of-Work consensus mechanism, which is implemented in Bitcoin. Proof-of-Work relies on the computational resources of miners to validate transactions and mine new blocks. However, this approach can be prone to centralization, as miners with more computational resources have a higher chance of mining new blocks.

Proof-of-Stake, on the other hand, reduces the dependence on computational resources [6] and helps ensure that the network remains decentralized. By allowing users to stake their money on a specific outcome, Proof-of-Stake incentivizes users to act honestly and maintain the integrity of the network.

Overall, Ethereum's [16] use of Proof-of-Stake is an important aspect of the platform's design, helping to ensure its security, scalability, and decentralization.

3.2 Infura

Infura is a hosted Ethereum node cluster that provides easy access to the Ethereum network for developers and users. By eliminating the need for users to set up Ethereum nodes or wallets, [18] Infura makes it easier to build and deploy decentralized applications on the Ethereum network.

Infura offers several benefits that solve some of the challenges faced by blockchain developers and users. These include:

- Cost-effective data storage: Storing data on the Ethereum blockchain can be expensive, but Infura offers an affordable solution for developers and users.

Fig. 3 Working of Infura

- Simplified Ethereum client configuration: Configuring the Ethereum Geth client can be difficult [3] and time-consuming, but Infura eliminates it by providing a hosted solution.
- Scalable infrastructure: Scaling infrastructure can be a challenge for blockchain developers, but it offers a scalable solution that can be easily adapted to meet the needs of growing applications [18].

Infura, along with other [2] interplanetary file systems (IPFS), plays a crucial role in solving some of the challenges faced by blockchain developers and users. The work of Infura is shown in Fig. 3. By providing a simple and cost-effective solution for accessing the Ethereum network [19], Infura makes it easier for developers and users to build and deploy decentralized applications.

3.3   Solidity

Solidity is an object-oriented high-level language that is used for implementing smart contracts on the Ethereum blockchain [17]. A smart contract is a self-executing program that regulates the behavior of accounts within the Ethereum state. It acts as a contract or set of rules governing business transactions and is stored on the blockchain where it is automatically executed as part of the transaction. This eliminates the need for governance, legislation, central authorities, or external enforcement mechanisms, making transactions more efficient and secure.

The key features of Solidity and smart contracts include:

- Trust: Smart contracts are self-executing programs that are transparent and tamper-proof, making them a trusted mechanism [17] for executing transactions.
- Savings: By removing the need for intermediaries, smart contracts can help reduce costs associated with transactions.
- Autonomous: Smart contracts are self-executing programs that do not require external intervention, making them ideal for executing autonomous transactions.

- Speed: Transactions executed on the Ethereum network are processed in real time, making them much faster than traditional transactions [12].
- Security: Smart contracts are stored on the blockchain, making them secure and resistant to tampering.
- Transparency: All transactions executed on the Ethereum network are transparent and accessible to all participants, promoting transparency and accountability.
- Redundancy: The decentralized nature of the Ethereum network ensures that transactions are stored redundantly across multiple nodes, making it resistant to data loss.
- Accuracy: The automated execution of smart contracts eliminates the possibility of human error, making transactions more accurate and reliable.

3.4 Distributed Application Architecture

As shown in Fig. 4, the architecture of a distributed application involves both a front-end and a back-end component. The front end of a decentralized application is commonly based on ReactJS, an open-source component-based JavaScript library that is used to create dynamic and interactive user interfaces, especially for single-page applications.

The backend of the decentralized application is the Ethereum network. The network blocks contain information about transactions, and smart contracts written in the Solidity programming language are deployed to the Rinkeby test network [16].

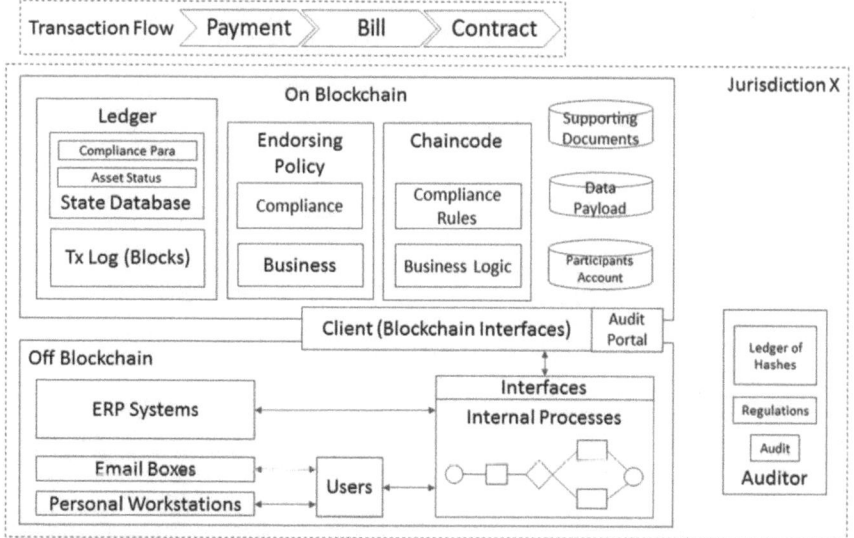

Fig. 4 Distributed application architecture

This allows developers to test the functionality and security of the smart contract before deploying it to the main Ethereum network.

When a user sends a message to a specific user's address, the message blocks are added to the Ethereum network according to specified mandatory contracts. The decentralized architecture of the Ethereum network ensures that the information about the message is stored securely and transparently, making it accessible to all participants in the network.

3.5 Encryption: Encrypting and Decrypting Messages

Encryption is a crucial aspect of cryptography and is used to protect sensitive information from unauthorized access [19]. The encryption and decryption of messages in CryptoJS are achieved using the AES-256 algorithm, which is available in the npm package manager.

To encrypt a message in CryptoJS, built-in libraries are used. The message is first encoded in hexadecimal format and then passed through an encryption function using an algorithm and a private key. The algorithm used in this process is AES-256, which provides a high level of security. The encrypted message is then transmitted over a network, ensuring that it remains confidential and protected from unauthorized access.

The decryption process is the reverse of the encryption process as shown in Fig. 5. The encrypted message is passed through a decryption function using the same algorithm and private key. This results in the original message being revealed [19]. The decryption process is used to retrieve the original message from the encrypted message, making it available to authorized individuals [13].

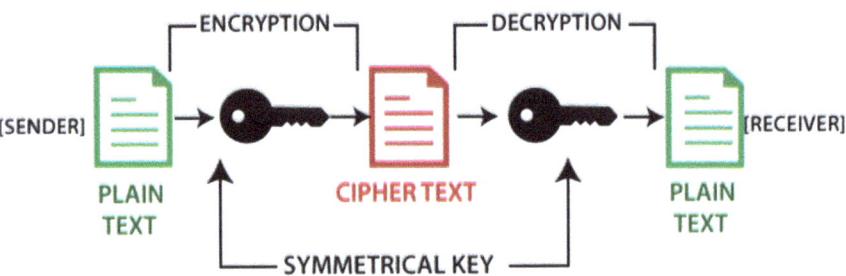

Fig. 5 Symmetric key cryptography

3.6 Compiling Smart Contracts

Smart Contracts are an essential component of decentralized applications and are implemented using the Solidity programming language [11]. The decentralized application in question uses Solidity version 0.8.17, which provides various features and functionalities to create secure and reliable smart contracts.

The following are the main functions of the smart contracts [11] used in this application:

- Joining Contracts: Allows new users to join the network and participate in the application.
- Sending Messages: Enables users to send messages to each other within the network.
- Adding Addresses: Allows users to add new addresses to the network.
- Accepting Addresses: Allows users to accept addresses added to the network.
- Blocking Addresses: Allows users to block addresses from the network.

It is crucial to test smart contracts thoroughly before deployment to ensure that they are free of logic errors and security flaws. Testing is done using JavaScript, and each of the functions mentioned above is tested thoroughly. If there are any issues or bugs, they need to be addressed and fixed before deployment.

In the event that the smart contract has logic errors or security flaws, the entire application could be compromised, leading to severe consequences. Hence, the importance of thoroughly testing smart contracts cannot be overstated.

3.7 Deployment

Deploying a smart contract on the Ethereum network is the final step in bringing the smart contract to life and making it available for use. Figure 6 shows the process of deployment that involves the following steps [11]:

- Compilation of the smart contract: The smart contract must be compiled successfully before it can be deployed on the blockchain network.
- Use of INFURA API: The application being discussed uses the INFURA API [18] to provide access to the Ethereum network, making it easy to integrate the smart contract.
- Deployment to Rinkeby test network: The smart contract is deployed to the Rinkeby test network, which is a public Ethereum test network used for testing and development purposes. This allows developers to test the functionality and security of the smart contract before deploying it to the main Ethereum network.
- Testing the smart contract [11]: Deploying the smart contract to the Rinkeby test network allows developers to test the smart contract's functionality and address

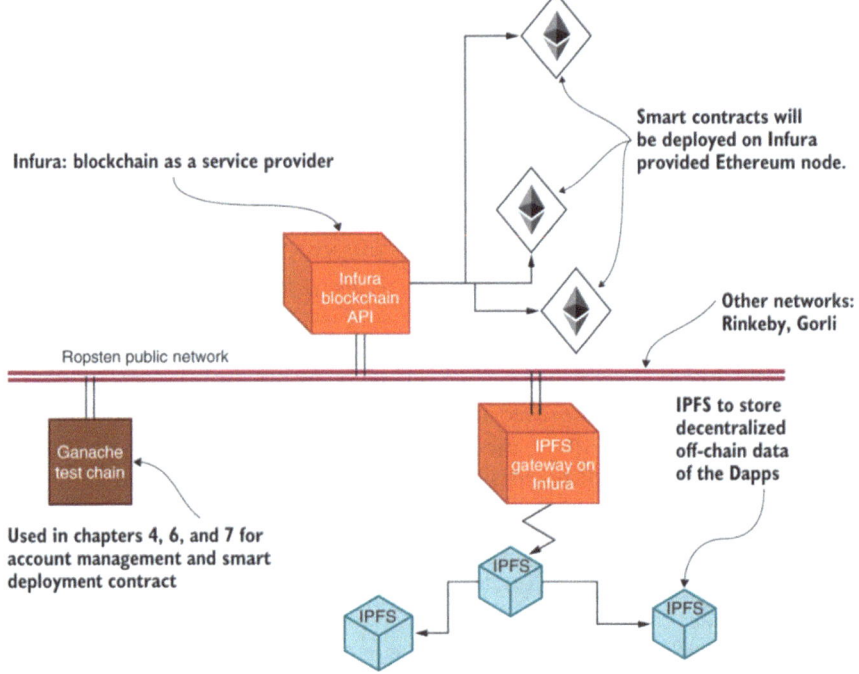

Fig. 6 Deployment of a block through Infura

any potential security vulnerabilities. This step is important to ensure that the smart contract is functioning as intended before deploying it to the main Ethereum network.

• Deployment to the main Ethereum network [16]: After successfully testing the smart contract on the Rinkeby test network, the smart contract can be deployed to the main Ethereum network, making it available for use by the public.

4 Advantages

The BiFrost solution presented in the research paper can have several real-world applications:

• Secure Communication: BiFrost can be used as a secure communication tool, where users can send messages with end-to-end encryption [9], making it spy-proof.

• Decentralized Storage: The application can be used for decentralized [6] storage, where data are stored on multiple nodes in the cloud [7], eliminating the risk of data breaches and single-point failures.

- Censorship Resistance: BiFrost ensures censorship resistance, allowing users to exchange information [10] without the fear of government oppression.
- Data Security: The immutable and tamper-proof nature of the blockchain ensures the security of user-submitted data, [7] making it safe from malicious attacks.
- Anti-corruption: The decentralized nature of the application, along with its smart contract [15] functionality, can enforce anti-corruption rules and help prevent fraud and corruption in transactions and communications.

These applications can be relevant in various sectors such as finance, healthcare, and government, where secure and reliable communication and data storage are of utmost importance.

5 Conclusion

In conclusion, BiFrost is a promising solution for security concerns in today's interconnected world. By utilizing blockchain technology, decentralized principles, and advanced cryptography, it ensures the confidentiality, integrity, and availability of data. Its decentralized architecture on the Ethereum blockchain eliminates vulnerabilities found in centralized chat applications, providing censorship resistance, immutability, and data security through smart contracts, IPFS, and INFURA. BiFrost offers a reliable and cost-effective solution for secure communication and data storage, overcoming the limitations of mainstream chat apps. The proof-of-stake algorithm ensures system reliability and efficiency, allowing users to communicate without fear of government oppression or data breaches.

In the future, BiFrost has the potential to reshape secure information exchange. Ongoing development can expand its functionality, incorporating multimedia messaging and enhancing privacy maintenance and message thread monitoring. The research project on BiFrost has been a valuable learning opportunity, allowing the team to apply their knowledge and technical skills to solve real-world problems. Effective communication, teamwork, and task allocation have deepened their understanding of app development and practical application of academic coursework.

As the decentralized internet and communication evolve, more work is needed to enable seamless connections. BiFrost represents a significant step forward in leveraging blockchain technology for secure information exchange. However, widespread adoption requires continued efforts, financial resources, and the development of a robust final product. With dedication and technological advancements, BiFrost has the potential to revolutionize communication, creating a more secure and interconnected digital landscape.

References

1. Wu K (2019) An empirical study of blockchain-based decentralized application. arXiv
2. Chen Y, Li H, Li K, Zhang (2017) An improved P2P file system scheme based on IPFS and blockchain. In: IEEE international conference on big data
3. Vujičić J, Ranđić S (2018) Blockchain technology, bitcoin, and ethereum: a brief overview. In: 17th international symposium INFOTEH-JAHORINA (INFOTEH)
4. Bhardwaj V, Sharma RR, Kumar A (2021) Blockchain technology: chat application. Int J Sci Manag
5. Zhang S, Shah Y, Elmandjra M, DeCesare M, Hanna M (2019) BlockChat: a decentralized messenger on the blockchain. California State University, Northridge
6. Takale AP, Vaidya CV, Kolekar SS (2018) Decentralized chat application using blockchain technology. Int J Eng Res Technol, IJERT
7. Dnyaneshwar Bagade S, Wankhade NR (2022) Decentralized Secure messaging application using blockchain technology. Int Res J Modern Eng Technol Sci
8. Hwa C, Yue K (2021) HIVE: a blockchain-based group messaging service. Massachusetts Institute of Technology
9. Benet J (2014) IPFS—Content Addressed, Versioned, P2P File System. arXiv
10. Nayyef ZT, Amer SF, Hussain Z (2019) Peer-to-peer multimedia real-time communication system based on WebRTC technology. Int J History Eng Technol 2.9(7):125–130
11. Huang Y, Wang B, Wang Y (2021) Research and application of smart contract based on ethereum blockchain. J Phys: Conf Ser 1748(4):042016
12. Jian C, Lin Z (2015) Research and implementation of WebRTC Signaling via WebSocket-based for real-time multimedia communications. In: 5th international conference on computer sciences and automation engineering
13. KK, Pujolle G (2019) Secure peer-to-peer communication-based on blockchain. In: Advances in intelligent systems and computing book series (AISC, volume 927)
14. Hisseine MA, Chen D, Yang X (2022) The application of blockchain in social media: a systematic literature review. Appl Sci 12(13):6567
15. Fateh Ali Khan H, Akash A, Avinash R, Lokesh C (2018) WebRTC peer-to-peer learning. Int J Eng Res Technol, IJERT
16. Ethereum Docs. https://ethereum.org/en/developers/docs/. Last Accessed 15 April 2023
17. Solidity Docs. https://docs.soliditylang.org/en/v0.8.17/. Last Accessed 18 April 2023
18. Infura Docs. https://www.infura.io. Last Accessed 18 April 2023
19. IPFS Docs. https://docs.ipfs.tech/. Last Accessed 21 April 2023
20. Gangwani P, Perez-Pons A, Bhardwaj S, Joshi H, Upadhyay L (2021) Lagos: securing environmental IoT data using masked authentication messaging protocol in a DAG-based blockchain. IOTA Tangle. Future Internet 13:312
21. Zhao W, Aldyaflah IM, Gangwani P, Joshi S, Upadhyay H, Lagos L (2023) A blockchain-facilitated secure sensing data processing and logging system. IEEE Access 11

Priority Based Load Balancing for Intercloud Computing Environments

Narayan A. Joshi ⓘ

Abstract Adoption of cloud computing based computing solutions is growing day by day in nearly all sectors of society. Technological advancements such as load balancing help service providers uphold the quality of service, and thereby retain the confidence of service consumers. Collaborations among cloud environments can be established for resolving load balancing and fault tolerance issues up to a certain extent. However, the gigantic increase in consumption of cloud services makes cloud resource management difficult in intercloud environments too. Load balancing helps resolve workload balancing problems for collaborated cloud platforms. We present an enhanced and priority-oriented mechanism for sharing resources for workload balancing in collaborated cloud environments. In order to provide enhanced load balancing solution, the suggested resource sharing mechanism works on the priority values of participating instances. Employment of the suggested load balancing mechanism avoids starvation by means of lowering waiting time. The suggested technique has been implemented on a physical cloud testbed built using OpenStack cloud computing setup on CentOS Linux operating system. The experimental results reveal lesser waiting time of the highly loaded cloud instances.

Keywords Cloud collaboration · Priority-oriented scheduling · Load balancing · Resource allocation

1 Introduction

Technological developments in the domains of high-end computing systems, data storage networks and communication systems have given momentum to a variety of computation practices. Cloud computing is an example of such evolved computation architecture. Various computation solutions nowadays are based on one or more cloud computing services. Infrastructure as a Service is one of the basic cloud computing services. Service consumers normally purchase high-end instances for performing

N. A. Joshi (✉)
Dharmsinh Desai University, Nadiad, Gujarat, India
e-mail: joshinarayan@gmail.com

© The Author(s), under exclusive license to Springer Nature Singapore Pte Ltd. 2024   395
S. Namasudra et al. (eds.), *Data Science and Network Engineering*, Lecture Notes in Networks and Systems 791, https://doi.org/10.1007/978-981-99-6755-1_30

extensive computation-centric operations [1]. Moreover, in order to provide fault tolerance and extensive load balancing, often cloud service providers collaborate for the establishment of collaborated cloud computing environments. Benefits of intercloud computing systems include enhanced intercloud resource availability and additional fault tolerance.

As the demand for cloud resources is increasing exponentially, efficient management of cloud-based resources in the context of resource allocation and distribution is critical. Uneven resource management and use frequently leads to underconsumption and wastage of cloud resources, resulting in discontent and a low return on investment. Non-uniform consumption of cloud resources and cloud instances results in uneven return on investment which may lead to damage in service level agreements. Therefore, for performance-oriented resource utilization, effective resource management and sharing are essential. This paper presents an enhanced resource sharing approach in the realm of collaborative cloud computing environments. The proposed load balancing mechanism assists in overcoming challenges such as uneven resource sharing.

Section 2 presents relevant survey on existing load balancing techniques. The proposed innovative load balancing solution is described in Sect. 3. Section 4 describes the implementation setup for experimentation. Section 5 discusses the working behavior of the approach.

2   Relevant Work and Challenges

A priority dependent mechanism for determining the most relevant instance in cloud platforms is available in [2]. The job scheduling mechanism prioritizes jobs and finds a suitable instance from three classification levels. The mechanism is based on cloudsim simulator. It runs on instances in an Amazon EC2-based data center. Tripathi et al. proposed a load sharing mechanism based on bee colonies [3]. The mechanism was deployed on a cloud analyzer.

Migrating virtual machines also is one of the alternative solutions for performing load balancing. For example, EGSA-VMM [4] is one such technique. The technique works on an algorithm for exponential gravitational force. The technique is the integration of theory of exponential weighted moving average and an algorithm of gravitational search. The technique works on quality of service and reduction in cost of migration.

Vasudevan et al. introduced a strategy that distributes existing cloud setup to suitable tasks in order to reduce service makespan. Honeybee theory is the foundation for the load balancing approach [5].

Dubey et al. presented a load balancing algorithm implementation for logistics [6]. The suggested solution for load balancing is works on cloud collaboration and based on DSBP. The mechanism is simulation-based and it works on balancing the workload of various data centers with an objective of data crunching at different levels of logistics. The load balancing approach could be applicable in vehicle tracking and inventory management. A balanced technique for resource sharing [7] is available for priority load balancing. However, it is applicable only for intracloud load balancing.

A simulation-based load balancing technique [8] works with reference to two important aspects: deadline and capacity. Its heuristic mechanism determines instances in accordance with the deadline constraints and total job execution cost on particular virtual machines. Virtual machines' priority-oriented resource allocation technique [9] is available for non-collaborated cloud computing environments. In order to carry out transparent and unbiased resource sharing, the mechanism has classified cloud instances as per their respective priority levels.

An autonomous agent-based algorithm for workload balancing for cloud environments is available in Singh et al. [10]. The mechanism falls in dynamic load balancing category of resource sharing in cloud computing. The technique addresses proactive workload calculation of instances in cloud data center. Every time a particular VM's workload reaches up to the threshold workload, the load agent initiates search for suitable virtual machine for load balancing.

A dynamic data replication technique [11] works in three phases. The motive of the entire technique is to optimize the deployment of cloud resources. The technique finds appropriate service instances for sharing load of service instances during the initial two phases. The mechanism works on improving resource availability, hierarchical load balancing and efficient resource access.

A static variables based sophisticated load balancing approach [12] aims on enhancing scheduling through weighted round robin in order to maximize fairness and cut down response time. The technique works on refining workload allocation across servers and responding quickly to urgent requests. An intercloud resource sharing mechanism is presented in Joshi et al. [13]. The mechanism allows resource sharing among cloud instances by means of intercloud task relocation.

A workflow scheduling approach Ji et al. [14] is based on multi-objective scheduling functions on adaptive prioritization and operates with different objective weights. Its approach self-regulates work priorities in order to achieve various goals related to performance criteria. Kaur et al. recommended optimization concerned with the hybridization of heuristic approaches through metaheuristic algorithms to achieve optimum performance on cost and makespan as a load balancing strategy [15].

From the above mentioned literature survey, it is observed that some resource sharing techniques perform load balancing considering all cloud instances at the same priority level. Contrary to that some of the resource sharing techniques based on cloud instance priority levels, work only for intracloud computing environments. Hence, a load balancing technique based on priority levels of cloud instances in collaborated cloud computing platforms is presented in this paper. The resource allocation technique herein presented differentiates among capacities of participating instances while balancing workload in intracloud and intercloud computing environments.

3 Materials and Method

The mechanism [13] works by keeping two instance queues: qOverloaded and qUnderloaded. These two queues maintain information about currently available overloaded and underloaded virtual machines respectively. The mechanism however does not provide load balancing based on cloud instances' priority levels. However, it is possible that ordinary cloud instances will receive more load balancing chances than extraordinary cloud instances. Virtual machines' priority-oriented resource allocation technique [9] is available for non-collaborated cloud computing environments. In order to carry out transparent and unbiased resource sharing, the mechanism has classified cloud instances as per their respective priority levels. Therefore, the novel resource sharing mechanism presented here aims to overcome such boundaries of the existing load balancing mechanisms [13] and [9].

Figure 1 represents a diagram showing interactions among the building blocks of the unique enhanced load balancer module POInterCloudLB. Key steps and functionality of building blocks of the suggested load sharing mechanism have been explained here:

1. The load balancer module POInterCloudLB consists of two submodules: POIntraLB and POInterLB. The submodule POIntraLB is responsible to execute load balancing in intracloud computing environments. The submodule POInterLB is responsible to perform load balancing in intercloud computing environments.

2. A data structure POInterCloudLB::instance maintains instance's state information such as: current state of instance, current position of instance for load

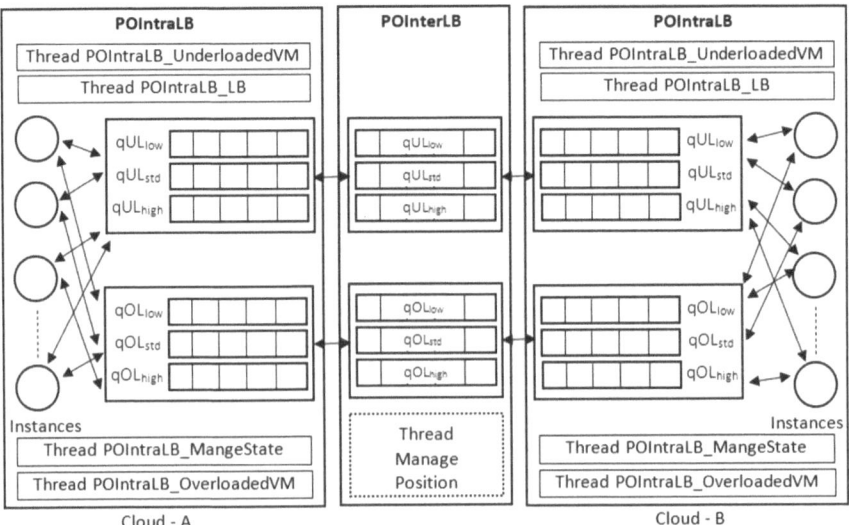

Fig. 1 Key architectural components of the proposed load balancer POInterCloudLB

balancing, priority of instance, current workload of instance, waiting time threshold values, and hardware inventory information of the instance.

3. The intracloud computing load balancer module POIntraLB keeps identifying overburdened machines and populate them into one of the three module-level queues: qO_{high}, qO_{low} and qO_{std} based on the respective priority level of each virtual machine. However, in case of unavailability of room in relevant queues, the POIntraLB module shifts the overloaded instance for load balancing to the intercloud load balancer POInterLB.

4. For each instance, the technique maintains two properties iStat and iPosition for keeping track of instance's state and position respectively. The state could be any one of the: iUNAVAILABLE, iAVAILABLE, iUNDERLOAD_PASSIVE, iOVERLOAD_PASSIVE. The position could be either pINTRA or pINTER to indicate intracloud or intercloud respectively. The mechanism does not take instances with the state iUNAVAILABLE in consideration for load balancing.

5. Similarly, the POIntraLB module keeps identifying underutilized machines and populate them into one of the three module-level queues: qU_{high}, qU_{low} and qU_{std} based on the respective priority level of each virtual machine. However, in case of unavailability of room in relevant queues, the POIntraLB module shifts the underutilized instance for load balancing to the intercloud load balancer POInterLB and alters the iPosition to pINTER.

6. In order to avoid starvation and avail flexibility, the intercloud priority-oriented load balancer module POInterLB also maintains three separate queues qO_{high}, qO_{low} and qO_{std} for holding high priority, low priority and standard priority instances respectively.

7. A daemon thread POInterLB::thread_managePosition thread keeps executing intercloud load balancing for the instances available in all three queues available at POInterLB.

8. The intracloud load balancer POIntraLB is made up of four daemon threads: POIntraLB_UnderloadedVM, POIntraLB_OverloadedVM, POIntraLB_ManageState and POIntraLB_LB. The underutilized virtual machines are looked after by the continuously running daemon thread POIntraLB_UnderloadedVM. The overutilized virtual machines are looked after by the continuously running daemon thread POIntraLB_OverloadedVM. The thread POIntraLB_LB carries intracloud load balancing. After initiating load balancing on concerned instances, such instances are removed from their respective queues. The thread POIntraLB_ManageState keeps away conditions which may cause the machines to starve for cloud resources at the intracloud level. It shifts the long awaiting machines for intercloud load balancing.

9. Moreover, the intercloud load balancer POInterLB runs a daemon thread POInterLB:PositionManager for keeping away conditions which may cause the instances to starve for cloud resources at the intercloud level. By default, the position is set to pINTRA.

10. The intracloud load balancer POIntraLB maintains three priority queues $qIntraCUL_{low}$, $qIntraCUL_{std}$ and $qIntraCUL_{high}$ for holding appropriate underloaded instance information. Likewise, the intracloud load balancer POIntraLB

maintains three priority queues qIntraCOL$_{low}$, qIntraCOL$_{std}$ and qIntraCOL$_{high}$ for holding appropriate overloaded instance information.

11. The daemon thread thread_ULInstances keeps identifying underutilized instances in respective queues at intracloud level. Moreover, for identifying underutilized instances at the intercloud level, the thread works at the intercloud level too.

12. The daemon thread thread_OLInstances keeps identifying over loaded instances in respective queues at intracloud level. Moreover, for identifying overutilized instances at the intercloud level, the thread works at the intercloud level too.

13. The thread POIntraLB::thread_LB executes load balancing task between the over utilized instance and the under utilized instance at the same priority level in intracloud environment. Moreover, for keeping the underutilized instance safer from instantaneous over burdening, the thread also turns the under utilized instance to passive state. The instances which have recently received workload from other instances, are set with a state iOVERLOAD_PASSIVE.

14. The thread POInterLB::thread_LB executes load balancing task between the over utilized instance and the under utilized instance at the same priority level in intercloud environment. Moreover, for keeping the underutilized instance safer from instantaneous over burdening, the thread also turns the under utilized instance to passive state. The instances which have recently received workload from other instances, are set with a state iOVERLOAD_PASSIVE.

15. The daemon thread POInterCloudLB::thread_ManageState keeps maintaining the availability state of instances. The instances ready for participating load balancing are set with the state value iAVAILABLE. The instances which have recently shifted their workload to other instances, are set with a state iOVERLOAD_PASSIVE.

16. The function POInterCloudLB:start() commences with initializing various priority queues used for maintaining respective priority instances. The start() function, then launches the daemon threads thread_ULInstances and thread_OLInstances for identifying underutilized instances and overutilized instances respectively. Later, spawning of daemon threads POIntraLB:thread_LB and POInterLB:thread_LB is followed by spawning of thread_ManagePosition. The POIntraLB:thread_LB and POInterLB:thread_LB perform intracloud and intercloud resource sharing respectively. The thread thread_ManagePosition is responsible for managing instances' position.

4 Implementation

The prototype testbed for intercloud computing platform was setup as shown in [13] on two physical high-end server hosts on open source operating system Fedora Linux 27 platform. An open source cloud platform was built by implementing OpenStack over the Fedora Linux platform for offering Infrastructure as a Service (IaaS) by catering to cloud instances that are remotely accessible on demand. An intermediary proxy server having an open source operating system installation was used for deploying intercloud queues and load balancer. For enabling intercloud resource sharing, respective intracloud load balancing servers in both cloud platforms were linked to the middle proxy server.

5 Results and Discussion

The mechanism starts by establishing and initializing different data structures and relevant queues. Having set the data structures and queues, the mechanism starts various background threads thread_ManagePosition, POIntraLB:thread_LB,thread_OLInstances, POInterLB:thread_LB, and thread_ULInstances.

In a local cloud platform context, the overloaded instances may be awaiting the availability of appropriate underutilized priority instances for resource sharing. Whereas, at the intercloud level, the relevant destination underutilized instances may be waiting for the admission of relevant request for resource sharing. On other hand, some underutilized instances may be awaiting relevant request for resource sharing at appropriate priority level.

Often it is necessary to assign cloud resources depending on the priority value of cloud instances. However, in its absence, owing to load balancing it is possible that instances with lower priority levels would ultimately keep receiving increasingly more cloud resources. The proposed mechanism was employed using different priority queues for instances with respective priority levels.

Table 1 shows results representing workload before and after executing priority-oriented load balancing in intracloud environment. The average waiting time for receiving the decision regarding priority-oriented intracloud load balancing results around 8.34 ms for concerned instances.

Table 2 shows results representing workload before and after executing the priority-oriented load balancing in intracloud and intercloud environments. The average waiting time for receiving the decision regarding priority-oriented intercloud load balancing results around 12.84 ms for concerned participating instances in resource sharing.

The instances' position manager thread thread_ManagePosition and state management thread thread_ManageState operate in synchronization with the threads thread_OLInstances, thread_LB and thread_ULInstances.

The thread thread_LB works such that lower-end instances are not assigned the higher-end resources except there is no current need for higher-end instance resources. The daemon thread thread LB operates such that resource sharing is not too taxing on underutilized instances that have begun accepting load balancing requests.

Figures 2 and 3 signify the performance of implementation of the suggested load balancing technique at intracloud computing environment and intercloud computing environment respectively.

Table 1 Waiting time of instances in waiting queues for taking intracloud load balancing

Source instance	Workload (before LB)	Workload (after LB)	Priority	Waiting time (ms)	Destination instance
192.168.10.4	96	96	High	8.4	192.168.10.17
192.168.10.7	98	80	Std	8.1	192.168.10.3
192.168.10.1	93	78	Low	8.1	192.168.10.15
192.168.10.5	91	79	Low	8.3	192.168.10.8
192.168.10.2	95	77	Low	8.8	192.168.10.21
192.168.10.11	92	76	Std	8.5	192.168.10.6
192.168.10.16	94	78	High	8.5	192.168.10.9
192.168.10.19	91	80	Low	8.2	192.168.10.10
192.168.10.18	89	69	Std	8.3	192.168.10.13
192.168.10.22	90	79	High	8.5	192.168.10.12

Table 2 Waiting time of instances in waiting queues for taking intercloud load balancing

Source instance	Workload (before LB)	Workload (after LB)	Priority	Waiting time (ms)	Destination instance
192.168.20.3	98	81	Std	12.8	192.168.30.4
192.168.20.7	99	79	Low	12.9	192.168.30.9
192.168.30.5	97	80	High	12.6	192.168.20.4
192.168.20.2	99	81	Low	13.0	192.168.30.13
192.168.30.16	98	79	Std	12.9	192.168.20.11
192.168.20.1	97	80	High	12.8	192.168.30.5
192.168.20.8	94	72	Std	13.2	192.168.30.7
192.168.30.12	98	80	Low	13.1	192.168.20.5
192.168.30.18	91	74	Low	12.8	192.168.20.14
192.168.30.11	95	83	High	13.0	192.168.20.13

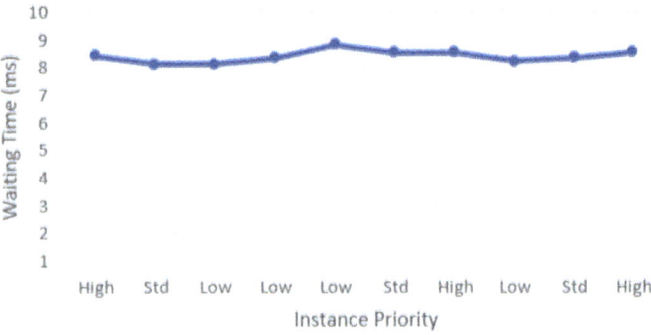

Fig. 2 Waiting time in intracloud load balancing

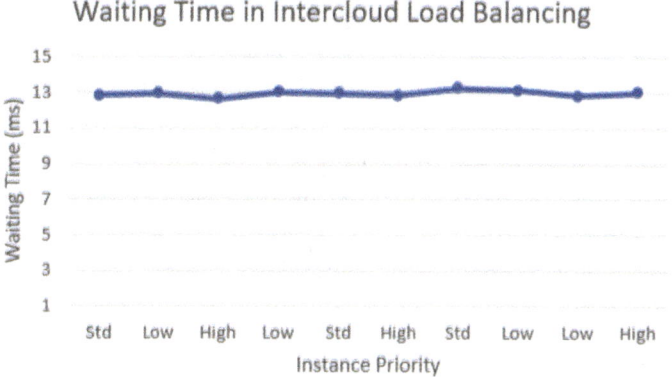

Fig. 3 Waiting time in intercloud load balancing

The chart shown in Fig. 2 represents subsequent waiting times resulting due to implementation of the suggested resource sharing mechanism in the intracloud environment on participating cloud instances of varying priority values. The chart denotes almost the same waiting time for all priority instances in the intracloud computing environment.

The chart shown in Fig. 3 represents subsequent waiting times resulting due to implementation of the suggested resource sharing mechanism in the intercloud environments on participating cloud instances of varying priority values. The chart denotes almost the same waiting time for all priority instances in the intercloud environments. The almost straight lines in both of the charts designate the nonoccurrence of starvation for any specific priority instances.

6 Conclusion

An enhanced and priority-oriented mechanism for sharing resources for load balancing in collaborated cloud environments is presented. The suggested load balancing mechanism can be implemented for attaining workload sharing in intercloud platforms and intracloud platforms. While performing priority-oriented resource sharing, the technique works in line with the priority levels of intercloud instances and intracloud instances. Furthermore, to avoid starvation, the mechanism cleverly performs pushing and pulling instances between intercloud and intracloud platforms. In future, the technique can be further extended for dynamic priority levels.

References

1. Joshi NA (2014) Performance-centric cloud-based e-learning. IUP J Inform Technol 10(2)
2. Kumar M, Suman (2019) Priority Based Virtual Machine Selection Algorithm in Cloud Computing, International Journal of Recent Trends and Engineering 8(3)
3. Tripathi A, Shukla S, Arora D (2018) A hybrid optimization approach for load balancing in cloud computing. In: Bhatia S, Mishra K, Tiwari S, Singh V, Advances in computer and computational sciences. Springer Advances in Intelligent Systems and Computing 554
4. Polepally V, Chatrapati KS (2018) Exponential gravitational search algorithm-based VM migration strategy for load balancing in cloud computing. Int J Model Simul Scientif Comput 9(1)
5. Vasudevan S, Anandaram S, Menon A, Aravinth A (2016) A novel improved honey bee based load balancing technique in cloud computing environment. Asian J Inform Technol 15(9)
6. Dubey S, Dahiya M, Jain S (2019) Implementation of load balancing algorithm with cloud computing for logistics. J Eng Appl Sci 14(2)
7. Joshi NA (2022) Technique for balanced load balancing in cloud computing environment. Int J Adv Comput Sci Appl 13(3)
8. Haidri R, Padmanabh Katti C, Saxena P (2021) Capacity based deadline aware dynamic load balancing model in cloud computing environment. Int J Comput Appl 43(10)
9. Joshi NA (2020) Priority based mechanism for resource sharing in cloud. Int J Innov Technol Explor Eng 9(3)
10. Singh A, Juneja D, Malhotra M (2015) Autonomous agent based load balancing algorithm in cloud computing. International Conference on Advanced Computing Technologies and Applications. Proc Comput Sci 45
11. Hsieh H, Ching M (2019) The incremental load balancer cloud algorithm by using dynamic data deployment. J Grid Comput 17(3)
12. Manasser S, Alzghoul M, Mohmad M (2019) An advanced algorithm for load balancing in cloud computing using MEMA technique. Int J Innov Technol Explor Eng 8(3)
13. Joshi NA (2019) Optimized mechanism for resource sharing in cloud. Int J Eng Adv Technol 9(2)
14. Ji H, Bao W, Zhu X (2017) Adaptive workflow scheduling for diverse objectives in cloud environments. Trans Emerging Telecommun Technol 28(2)
15. Kaur A, Kaur B (2022) Load balancing optimization based on hybrid heuristic-metaheuristic techniques in cloud environment. J King Saud Univ – Comput Inform Sci 34 (3)

Computer Security

Machine Learning Approach to the Internet of Things Threat Detection

Alka Upadhyay, Sameena Naaz, Vinay Thakur, and Iffat Rehman Ansari

Abstract The development in software, hardware and communication technologies has made the broadcasting of sensory data collected from various devices very easy and simple. Interconnected devices through Internet technology form the Internet of Things (IoT). Applying intelligent methods for the analysis of this big data is the key which develops smart IoT applications. The world today has become increasingly dependent on digitized data which raises various security concerns and the need for advanced and reliable security technologies to deal with the increasing number of cyber-attacks. The work depicted in this paper makes use of machine learning techniques to detect cyber-attacks using the UNSW-NB15 data set and the KDD CUP 1999 dataset. Decision Tree, k- means clustering, multi-layer perception (MLP), Naive Byes and Random Forest classifier are the algorithms used in this work in order to find higher level information about the data.

Keywords Internet of Things · Multi-layer perception (MLP) · Decision tree · k-Means clustering · Naive Bayes · Random forest classifier

A. Upadhyay · S. Naaz (✉)
Department of Computer Science and Technology, School of Engineering Sciences and Technology, Jamia Hamdard, New Delhi, India
e-mail: snaaz@jamiahamdard.ac.in

A. Upadhyay
e-mail: alka.sharma@nic.in

V. Thakur
Ministry of Electronics and IT, Govt of India, National Informatics Centre Services Inc. (NICSI), New Delhi, India
e-mail: vinay@gov.in

I. R. Ansari
Electronics Engineering Section, Faculty of Engineering and Technology, University Women's Polytechnic, Aligarh, A.M.U.UP 202002, India

© The Author(s), under exclusive license to Springer Nature Singapore Pte Ltd. 2024   407
S. Namasudra et al. (eds.), *Data Science and Network Engineering*, Lecture Notes in Networks and Systems 791, https://doi.org/10.1007/978-981-99-6755-1_31

1 Introduction

The IoT (Internet of Things) is mainly the type of network which is used to connect anything i.e. physical objects, devices, vehicles and various other objects which have software, sensors and connectivity of the network embedded in them [1]. These connections are based on multiple protocols which enable these items to collect and exchange various amounts of data [2]. A huge increase in various devices connected to the IoT has resulted in a desperate requirement for better and more reliable security. [3]. Due to an increase in the number of IT devices, the security risk has also increased many folds [4]. Reinforcement Learning (RL) is gaining more popularity for securing IoT systems. In [5] Reinforcement learning is used for securing Cyber-Physical system(CPS) systems, i.e. IoT with feedback and control such as smart grids and smart transportation systems. It is a comprehensive survey of using RL to secure the IoT system.

One way to secure IoT devices from cyber threats and vulnerabilities is by using machine learning methods [6]. As the appeal and demand of machine learning is growing rapidly the current methods and techniques are improving and their ability to understand and provide solutions to real issues is highly considered. Recently the number of cyber security incidents has increased rapidly all over the world [7]. According to a report, 1082 incidents of attacks were reported in 2012 i.e. an increase of approximately 80% when compared to the previous year mainly because of phishing and malware [8]. The dominant use of mobile devices has increased the number of incidents and vulnerabilities [9]. Also, many organizations are exposed to emerging threats due to BYOD (bring your own device). BYOD security can be enhanced by enabling zero trust. Authors in [10] explore the challenges and opportunities of enabling zero trust in the BYOD use case. They introduce Bring Your Own Zero Trust (BYOZ) language specification through a review of the literature related to Zero Trust architecture (ZTA) and BYOD. ML algorithms are widely used in Health Care, During COVID-19 future forecasting, and Big Data analytics [11-13].

In this paper machine learning techniques Decision Tree (DT), Multi-Layer perception (MLP), Naïve Bayes and Random Forest Classifier have been used in order to detect cyber-attacks on IoT data. The organization of the rest of the paper is as follows: Section 2 discusses some of the similar work reported in the literature. Machine Learning Techniques used in detecting Cyber Attacks are discussed in Sect. 3. Section 4 talks about the methodology used in this work. Results obtained for both datasets are presented and discussed in Sect. 5 and finally, Sect. 6 gives the conclusions drawn from the work.

2 Related Work

Pertinent work in reputed journals by authors is discussed below.

Machine learning algorithms are widely used as they are unbiased, overriding traditional intrusion detection techniques. Authors emphasize on novel cyber security schemes to protect networks [14].

Phishing data set from the UCI repository was used in [15]. Authors have used the info gain feature selection technique (FST) to select the top pertinent feature subset to decision tree classifier technique as DT, Random tree, random forest and Decision stump with tenfold cross validation among which DT achieved the best accuracy as 91.80%. Detection of phishing websites was developed using a Machine Learning model. The authors compared Random Forest, Support Vector Machine (SVM), DT, Artificial Neural network (ANN), k-nearest neighbour (KNN), and Rotation forest. Among these Random Forests perform best with 97.36%, and KNN is the second one with 97.18%.

In Ref. [16] Authors used a dataset from HP Data Labs and portrayed diverse Machine Learning Systems for spam messages. They concluded that random forest is the best technique with a high true positive and low false negative.

In their work, NMC, Gaussian classifiers and SVM have been employed by [17] along with fuzzy logic. These systems resulted in the effective detection of phishing websites with high true positive and low false negative rates.

Authors in [18] recommended methods for spoof detection by employing various algorithms such as C4.5, Bayesian Network, Neural Network, Logistic Regression, Naïve Bayes and SVM. An accuracy of 92.56% was achieved in this work.

Vulnerability in IoT devices has been discussed in [19]. Here the authors have mentioned that data leakage is one of the most common problems in IoT devices. Smart Grid (SG) consisting of large-scale ICT, power grid and renewable energy is one of the largest IoT networks. Security is one of the major reasons due to which commercial deployment of these smart grids is very limited. These security-related issues and challenges have been investigated in [19] for the IoT-based Smart Grids.

The authors proposed a honeypot IOTPOT and a multi-architecture sandbox called IOT BOX. Through which malware of different CPU architectures can be ruined. Botnets can be used to launch different attacks such as Keylogging, DDoS attacks, Phishing, Identity Theft and Spamming [20].

In Ref. [21] the authors profile and monitor the Radio Signal Strength Indication (RSSI) and then categorize the communication as legitimate or illegitimate by employing machine learning algorithms. It is basically for IoT devices which are connected at smart places. Neural Network algorithms are applied for classification. In [22] feature set was obtained by Artificial Fish Swarm optimization, and then classification was done by using SVM. The results obtained show significant improvement in terms of time required for execution. Artificial Immune System was used in [23, 24] for the security of IoT devices.

ML algorithms are used in smart lighting systems based on distributed wireless sensor networks. These ML algorithms are applied to control methods to minimize total illuminance [25]. On one hand ML algorithm is a boon as it helps in the detection of new attacks, while on the other hand, it is a curse as it supports new attack tools by using the adversarial ML technique to develop attacks [26].

Privacy is also preserved through Blockchain-based applications as suggested by authors in [27] which generate and maintain healthcare certificates and documents which are generated by different IOT devices.

3 Machine Learning Techniques Used in Detection of Cyber Attacks

3.1 Random Forest Classifier

Random Forest classifier is a data mining process, a classification technique which is a collection of many decision trees. Random forests are a collection of trees which are slightly different from one another. It randomizes the method but not the data used for training. The algorithm decides how to randomly pick up the data from the best k options which improves the decisions.

3.2 Naïve Bayes Classifier

Naïve Bayes method is one of the best-known and simplest models for supervised learning for a classification problem [28]. Naïve Bayes classification assumes that the attributes given in a data set are conditionally independent of one another and offers an explanation of the probability where it associates certain classes at certain instances. Since the attributes are independent of one another Naïve Bayes classification has shown great performance in accuracy.

3.3 Decision Trees Classifier

Decision trees are the classifiers which use supervised machine learning in which the data is continuously grouped based on some attributes [29]. It can be described using the decision nodes and the leaves. The leaves stand for the decisions to be made or we can say that they are the final result whereas, the decision nodes are the points from where the respective data is split.

3.4 Multi-Layer Perception

An MLP algorithm is basically a deep learning and ANN which is made up of more than one perception [30]. They are made up of one input layer which receives the respective signal and an output layer which finds a decision or a prediction In between there are an arbitrary number of hidden layers which are an important part of the MLP algorithm. This type of algorithm is basically applied to supervised learning situations where they are used to train a given set of input and output pairs to find out the dependencies existing between the two.

4 Methodology

In this work, two datasets UNSW-NB15 and NSL-KDD have been taken and analysed using different machine learning algorithms. The UNSW-NB15 has a total of 49 features and the testing has been carried out on the basis of the attack label which has been taken as 0 for normal and 1 for the cases where the attack took place. The same method has also been used for the NSL-KDD dataset having 42 features and the results obtained in both cases have been compared.

Normalizing is the first step in pre-processing. The dataset is pre-processed by using the following formula to perform min–max normalization on the features.

X is defined as a feature.

$$x = \frac{x - \min(x)}{\max(x) - \min(x)}$$

In Ref. [31] authors show that by way of data pre-processing Machine Learning Algorithms can yield extremely high accuracy. In this work, authors use data pre-processing in Python and machine learning algorithms to evaluate NSL-KDD Data Set. It is found that RF, KNN and Gradient Boosted Tree are prominent but Naive Bayes did not get the expected result.

Different Machine Learning algorithms used in this work are Random Forest Classifier, Naïve Bayes, Decision Tree and Multi-Layer Perceptron. The proposed methodology is depicted in Fig. 1. The results have been obtained by implementing all these algorithms in Python using the training and testing datasets and the test error and the model accuracy score has been calculated. Also, a classification report of the dataset was generated by using a confusion matrix (Fig. 2) and the value of the parameters precision, recall, f1 score and support were obtained for both normal as well as attacked data.

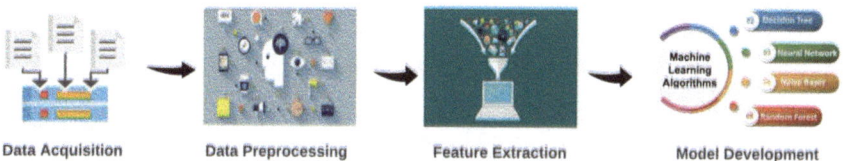

Data Acquisition   Data Preprocessing   Feature Extraction   Model Development

Fig. 1 Research methodology

TP = 56000	FP = 0
FN = 0	TN = 119341

TP = 45063	FP = 10937
FN = 24335	TN = 95006

TP = 39209	FP = 16791
FN = 24028	TN = 95313

TP = 55999	FP = 1
FN = 217	TN = 119124

Fig. 2 **a** Decision tree, **b** (MLP), **c** Naïve Bayes, **d** RF Classifier

4.1 Description of the Datasets

The two datasets used here are:

1. UNSW-NB15 dataset
2. KDD CUP 1999.

4.1.1 UNSW-NB15 Dataset

IXIA performance storm tool was used to create the UNSW-NB15 dataset in the cyber lab of cyber which is located in the Australian Centre for Cyber Security. The dataset consists of genuine activity data as well as synthetic attack data from the network traffic [32].

A tool known as tcpdump [33] represented raw traffic over a network of approximately 100 GB. Also, the Argus [34] tool was used along with the Bro-IDS [35] tool to generate 12 models for extracting different features. These methods were designed in parallel and extracted 49 features along with their class labels. The total records configured were around 2,540,044 million which were stored in four CSV files. Nine different types of attacks are there in this database.

4.1.2 NSL-KDD Dataset

The limitations of the KDD cup 1999 dataset was found by multiple statistical analyses of the dataset and affected the accuracy of various intrusion detection system modelled by the researchers [36]. Therefore the NSL-KDD dataset is a more accurate version of its predecessor.

41 different features are present in each of the records which also have a label signifying it to be either attack type or normal. The 4 attacks are DoS, R2L, Probe and U2R.

4.2 Performance Metrics

Accuracy, specificity or true negative rate (TNR), sensitivity or true positive rate (TPR), f-measure, precision and false alarm rate [37] have been calculated by the Confusion Matrix.

5 Results and Discussion

5.1 Results Obtained from the UNSW-NB15 Dataset

The confusion matrices for the above techniques are shown in.

It has been found that the DT has a minimum error rate of 0% along with the random forest classifier which has an error rate of 0.1%. Naïve Bayes algorithm performed the worst with an error rate of 23.2% whereas the other algorithm which is MLP also performed almost the same as Naïve Bayes with the worst error rate of 20.1% as shown in Fig. 3.

Table 1 shows a comparison of the four algorithms applied on the UNSW-NB15 dataset in the testing phase implemented in Python.

Comparison in terms of the four parameters has been shown below:

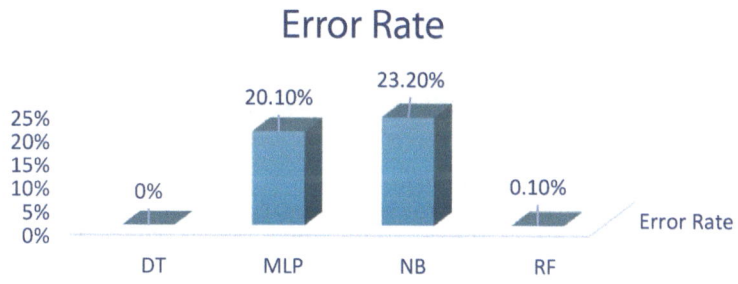

Fig. 3 Comparison of error rate

Table 1 Performance parameters

	DT (%)	MLP (%)	NB (%)	RF (%)
Accuracy	100	80	76.72	99.8
TPR/recall	1	65	62	99.6
TNR	1	90	85	99.9
Precision	1	80.4	70	99.9
F-measure	2	72	65.7	99.7
Error rate	0	20.1	23.2	0.1

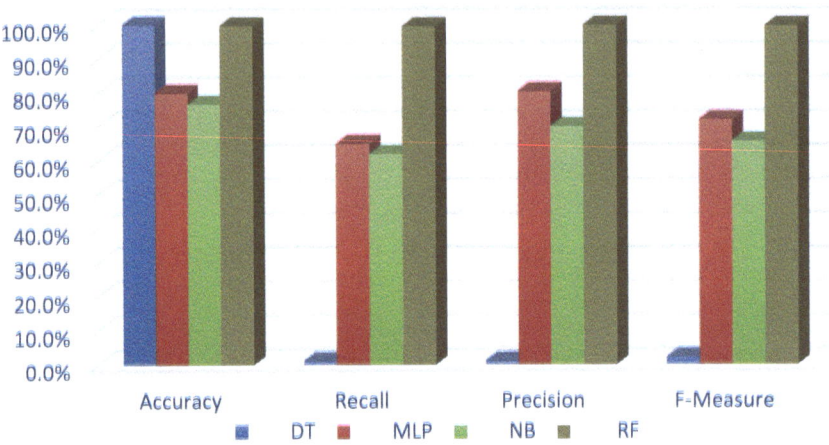

Fig. 4 Comparison of Algorithms

As shown in Fig. 4 the performance of DT is the best in terms of all parameters while the performance of the Naive Bayes algorithm is worst in terms of all parameters. The performance of MLP and RF is moderate.

5.2 Results Obtained from the NSL-KDD Dataset

The confusion matrices for the above techniques are shown in Fig. 5a–d.

A confusion Matrix is used here to calculate the Error rate of the ML algorithms and DT has been found to have the minimum error rate of 7.05% along with the random forest classifier having almost same error rate of 7.2%. The MLP algorithm performed the worst with an error rate of 18% whereas the other algorithm which is Naïve Bayes also performed slightly lower than MLP with an error rate of 12.2% as shown in Fig. 6.

Table 2 reveals the collation of the four algorithms applied to the NSL-KDD dataset in the testing phase implemented in Python.

All the algorithms are again compared based on the four performance parameters as shown below.

The performance of the DT and Random forest is the same in terms of all the parameters while the performance of the MLP algorithm is the worst in terms of all the parameters As depicted in Fig. 7 the performance of the Naïve Bayes algorithm is moderate.

TP = 228829	FP = 21607
FN = 343	TN = 60250

(a)

TP = 195657	FP = 54779
FN = 1288	TN = 59305

(b)

TP = 213838	FP = 36598
FN = 1355	TN = 59238

(c)

TP = 228305	FP = 22131
FN = 318	TN = 60275

(d)

Fig. 5 **a** Decision tree, **b** Multi-layer perception, **c** Naïve Bayes, **d** random forest classifier

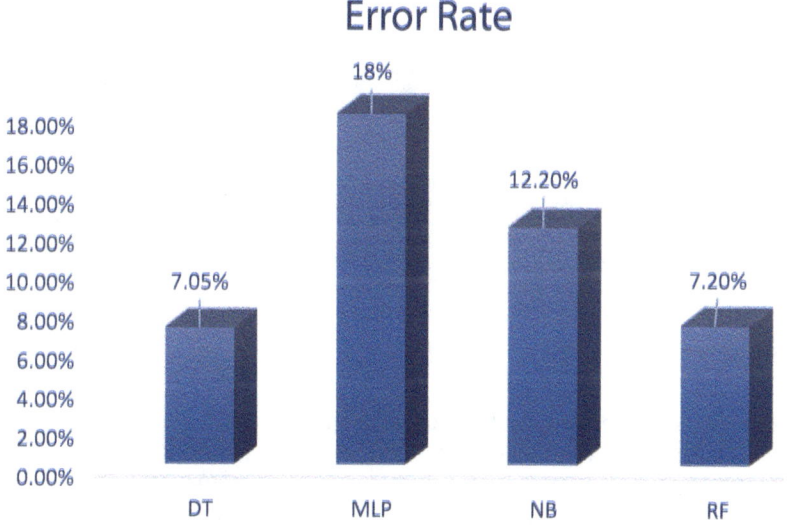

Fig. 6 Comparison of error rate

Table 2 Performance of different algorithms

	DT (%)	MLP	NB	RF
Accuracy	92.9	82	88	93
TPR	99.8	99.3	99.3	99.8
TNR	73.6	52	62.8	73.1
Precision	91.3	78.1	85.3	91.1
F-Measure	95.3	87.4	91.7	95.2
Error rate	7.05	18	12.2	7.2

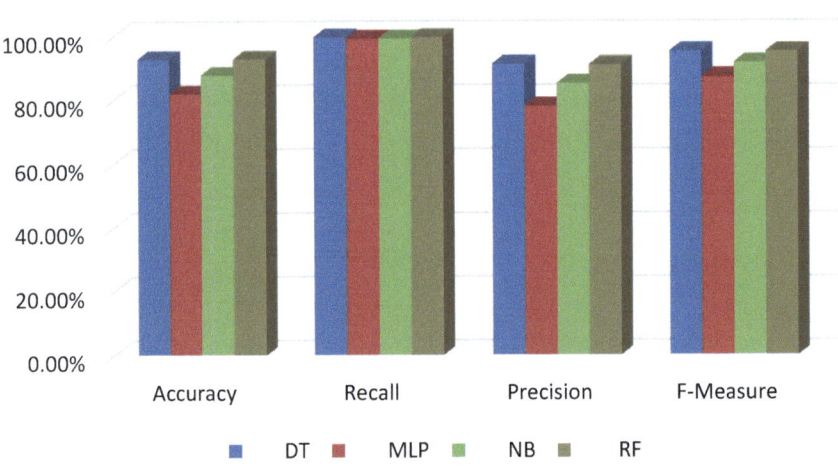

Fig. 7 Comparison of algorithms

6 Conclusion

In this paper, machine learning techniques are employed to great to a great degree in order to detect cyber-attacks using the UNSW-NB15 and NSL-KDD datasets. These datasets were generated from IoT (Internet of Things) and IoT faces a lot of threats and attacks which should be detected and recognized so that necessary action can be taken. In this research article, various types of cyber-attacks on the IoT are introduced. The main aim was to identify and detect probable attacks faced by the IoT. It classified both the datasets using four models which were DT, MLP, Naïve Bayes and Random Forest Classifier based upon their recall, accuracy, precision, F-measure and error rate and it has been found that in UNSW-NB15 dataset the DT algorithm worked outstandingly giving an accuracy of 100%. In the NSL-KDD dataset also the DT algorithm worked very well giving an accuracy of 93% along with random forest classifier having almost the same accuracy. Therefore it can be said that the DT algorithm performed best on both the datasets.

References

1. Savaliya A, Bhatia A, Bhatia J (2018) Engineering and technology IJSRSET184236. Accepted: 15. India Assoc Int J Sci Res Sci 2(4):218–223
2. Kowta ASL, Harida PK, Venkatraman SV, Das S, Priya V (2022) Cyber security and the internet of things: vulnerabilities, threats, intruders, and attacks. Lect Notes Data Eng Commun Technol 99(1):387–401. https://doi.org/10.1007/978-981-16-7182-1_31
3. Andročec D, Vrček N (2019) Machine learning for the internet of things security: a systematic review. In: *ICSOFT 2018 - Proceedings of the 13th International Conference on Software Technologies*, 2019, pp. 563–570. https://doi.org/10.5220/0006841205630570
4. Mahdavinejad MS, Rezvan M, Barekatain M, Adibi P, Barnaghi P, Sheth AP (2018) Machine learning for internet of things data analysis: a survey. Digit Commun Netw 4(3):161–175. https://doi.org/10.1016/j.dcan.2017.10.002
5. Uprety A, Rawat DB (2020) Reinforcement learning for IOT security: a comprehensive survey. IEEE Internet Things J 8(11):8693–8706
6. Díaz López D et al. (2018) Shielding IoT against cyber-attacks: an event-based approach using SIEM. Wirel. Commun. Mob. Comput. 2018, 2018. https://doi.org/10.1155/2018/3029638
7. Geetha R, Thilagam T (2021) A review on the effectiveness of machine learning and deep learning algorithms for cyber security. Arch. Comput. Method. Eng. 28(4):2861–2879. https://doi.org/10.1007/s11831-020-09478-2
8. Ridley A (2018) Machine learning for autonomous cyber defence. Next Wave 22(1):7–14
9. Chorás M, Kozik R (2014) Machine learning techniques applied to detect cyber attacks on web applications. Log J IGPL 23(1):45–56. https://doi.org/10.1093/jigpal/jzu038
10. Anderson J, Huang Q, Cheng L, Hu H (2022) BYOZ: protecting BYOD through zero trust network security. In: 2022 IEEE International Conference on Networking, Architecture and Storage, NAS 2022 – Proceedings, pp. 1–8. https://doi.org/10.1109/NAS55553.2022.9925513
11. Mohan S, Thirumalai C, Srivastava G (2019) Effective heart disease prediction using hybrid machine learning techniques. IEEE Access 7:81542–81554. https://doi.org/10.1109/ACCESS.2019.2923707
12. Rustam F et al (2020) COVID-19 future forecasting using supervised machine learning models. IEEE Access 8:101489–101499. https://doi.org/10.1109/ACCESS.2020.2997311
13. Nti IK, Quarcoo JA, Aning J, Fosu GK (2022) A mini-review of machine learning in big data analytics: Applications, challenges, and prospects. Big Data Mining Analyt 5(2):81–97. https://doi.org/10.26599/BDMA.2021.9020028
14. Namasudra S, Lorenz P, Ghosh U (2023) Editorial: the new era of computer network by using machine learning. Mobile Netw Appl. https://doi.org/10.1007/s11036-023-02114-w
15. Shrivas AK, Suryawanshi R (2017) Decision Tree Classifier for Classification of Phishing Website with Info Gain Feature Selection. Int J Res Appl Sci Eng Technol 5(5):780–783
16. Jukic S, Azemovic J, Keco D, Kevric J (2015) Comparison of machine learning techniques in spam E-Mail classification. Southeast Europe J Soft Comput 4(1):249–256. https://doi.org/10.21533/scjournal.v4i1.88
17. Kaur S, Sharma S (2015) Performing efficient phishing webpage detection. Int J Comput Sci Eng 3(7):52–56
18. Chen H, Vasardani M, Winter S (2017) Geo-referencing Place from Everyday Natural Language Descriptions. arXiv Prepr. arXiv1710.03346

19. Hossain MM, Fotouhi M, Hasan R (2015) Towards an analysis of security issues, challenges, and open problems in the internet of things. In: Proceedings – 2015 IEEE World Congress on Services, SERVICES 2015, pp. 21–28, https://doi.org/10.1109/SERVICES.2015.12

20. Amini P, Araghizadeh MA, Azmi R (2016) A survey on Botnet: Classification, detection and defense. In: Proceedings - 2015 International Electronics Symposium: Emerging Technology in Electronic and Information, IES 2015, pp. 233–238, https://doi.org/10.1109/ELECSYM. 2015.7380847

21. Roux J, Alata E, Auriol G, Nicomette V, Kâaniche M (2017) Toward an intrusion detection approach for IoT based on radio communications profiling. In: Proceedings - 2017 13th European Dependable Computing Conference, EDCC 2017, pp. 147–150, https://doi.org/10.1109/ EDCC.2017.11

22. Lin KC, Chen SY, Hung JC (2014) Botnet detection using support vector machines with artificial fish swarm algorithm. J Appl Math 2014, https://doi.org/10.1155/2014/986428

23. Greensmith J (2015) Securing the internet of things with responsive artificial immune systems. In: *GECCO 2015 - Proceedings of the 2015 Genetic and Evolutionary Computation Conference*, pp. 113–120, https://doi.org/10.1145/2739480.2754816

24. Kamalam GK, Suresh P, Nivash R, Ramya A, Raviprasath G (2022) Detection of Phishing Websites Using Machine Learning. In: 2022 Int. Conf. Comput. Commun. Informatics, ICCCI 2022, https://doi.org/10.1109/ICCCI54379.2022.9740763

25. Cheng Y, Fang C, Yuan J, Zhu L (2020) Design and application of a smart lighting system based on distributed wireless sensor networks. Appl Sci 10:8545. https://doi.org/10.3390/app 10238545

26. Sharma P, Namasudra S, Chilamkurti N, Kim B-G, Gonzalez Crespo R (2022) Blockchain-based privacy preservation for IoT-enabled healthcare system. ACM Trans Sensor Netw IP. 1–16. https://doi.org/10.1145/3577926

27. Namasudra S, Gonzalez Crespo R, Sathish K (2022) Introduction to the special section on advances of machine learning in cybersecurity (VSI-mlsec). Comput Electr Eng 100:108048. https://doi.org/10.1016/j.compeleceng.2022.108048

28. Taheri S, Mammadov M (2013) Learning the naive bayes classifier with optimization models. Int J Appl Math Comput Sci 23(4):787–795. https://doi.org/10.2478/amcs-2013-0059

29. Kotsiantis SB (2013) Decision trees: a recent overview. Artif Intell Rev 39(4):261–283. https:// doi.org/10.1007/s10462-011-9272-4

30. Dencelin LX, Ramkumar T (2016) Analysis of multilayer perceptron machine learning approach in classifying protein secondary structures. Biomed Res 2016(2):S166–S173

31. Hong R-F, Horng S-C, Lin S-S (2021) Machine learning in cyber security analytics using NSL-KDD Dataset. Int Conf Technol Appl Artificial Intell (TAAI) 2021:260–265

32. Moustafa N, Slay J (2015) UNSW-NB15: A comprehensive data set for network intrusion detection systems (UNSW-NB15 network data set). In: 2015 Military Communications and Information Systems Conference, MilCIS 2015 – Proceedings, pp. 1–6, https://doi.org/10. 1109/MilCIS.2015.7348942

33. "Tcpdump Tool," (2014) Retrieved from http//www.tcpdump. org/

34. "Argus tool," (2014) Retrieved from http//qosient.com/argus/ flowtools.shtml

35. "Bro-IDS Tool," (2014) Retrieved from https//www.bro.org/

36. Moustafa N, Creech G, Slay J (2017) Big data analytics for intrusion detection system: statistical decision-making using finite Dirichlet Mixture Models. In: Data analytics and decision support for cybersecurity, Springer, pp. 127–156

37. Naaz S (2021) Detection of phishing in internet of things using machine learning approach. Int J Digit Crime Forensics 13(2):1–15. https://doi.org/10.4018/IJDCF.2021030101

Impact of Data Poisoning Attack on the Performance of Machine Learning Models

Dipan Das, Sharmistha Roy⑩, and Bibhudatta Sahoo⑩

Abstract The twenty-first century has witnessed widespread adoption of Artificial Intelligence (AI), Machine Learning (ML) and Deep Learning (DL). These techniques have provided reliable solutions in various areas, including statistics, information theory, and mathematics. Given the prevalence of ML techniques, there exist various adversaries which question the robustness of ML models. Adversaries aim to manipulate models to their advantage, reducing their performance and accuracy. Data poisoning attack is one such adversary in which the attacker manipulates models by introducing specially crafted poisoned data into the training dataset. This paper presents the performance analysis of different machine learning models with and without the influence of data poisoning attack to predict the probability of diabetes and its effect on accuracy and precision. It has been observed that the SVM (RBF) classifier performs best on clean data, while the KNN classifier is highly affected by data poisoning, with a lesser impact on the SVM (Linear kernel) classifier.

Keywords Data poisoning attack · Adversarial attack · Performance analysis · Reliability · Adversarial machine learning · Supervised learning · Security · Poisoning attacks · Machine learning

D. Das (✉) · B. Sahoo
Department of Computer Science & Engineering, NIT Rourkela, Rourkela, India
e-mail: maildipandas@gmail.com

B. Sahoo
e-mail: bdsahu@nitrkl.ac.in

S. Roy
Faculty of Computing and Information Technology, Usha Martin University, Ranchi, India
e-mail: sharmistha@umu.ac.in

1 Introduction

The use of machine learning has increased drastically due to its decision-making capability in various fields like healthcare [1], financial sector [2], agriculture [2], and e-commerce applications [1]. The rapid increase in volume and variety of data from various sources and greater availability of data, created opportunities to implement machine learning models for giving faster and reliable results in data analysis and predictions. However, the reliability and performance of the Machine Learning (ML) models depend on the quality of the training datasets. The pervasiveness given to the machine learning models has also given the scope to the attackers for manipulating the machine learning classifiers and increases the security vulnerability of the models. With the increase in the utilization of these ML models, they are vulnerable to security threats like adversarial attacks and data poisoning attacks [3]. To create a reliable and efficient machine learning model, it requires to be trained with huge datasets collected from various sources across the globe, which also opens up with the chance to 'poison' the datasets with malicious intension. It was seen vividly in Microsoft's AI chatbot Tay [4] which was flooded with numerous offensive and racist tweets within 24 h after it was made public. Thus, along with the advent of machine learning techniques, it is equally vital to study the influence of different adversaries in the performance of the machine learning models.

Data poisoning attack is an attempt by the intruders to induce some misclassification to the training datasets which results in an increase in the overall time and a decrease in the efficiency and performance of the models. The manipulation is mainly done on the training dataset either by feature poisoning by flipping the labels or by changing the configuration of the models. There are two categories of data poisoning attack. One is targeted attack where the attackers misclassify a specific test data sample and the reverse one is the untargeted attack.

This paper presents two major contributions—

- Measuring the performance of various machine learning models like KNN- classifier, Logistic Regression classifier, and SVM classifier (linear classifier, RBF classifier and Polynomial) by predicting the accuracy and precision of the models.
- Analyzing the impact of data poisoning attack on the overall performance of the ML models. The analysis shows the robustness of the models.

The paper is organized as follows: Sect. 2 presents the related work on machine learning models and various security threats on them. The proposed methodology briefing the overview of the machine learning models, description of the dataset and the data poisoning attacks are discussed in Sect. 3. Section 4 provides the result analysis indicating the performance of the selected ML models on clean data as well as the impact of data poisoning attack on the performance of the ML models. Finally, Sect. 5 concludes the study.

2 Literature Review

This section describes the existing works that are focused on performance analysis of certain ML classifiers and their limitations. Also, it is focused on the implementation of data poisoning attack on certain ML classifiers.

Sisodia et al. [5] performed a study to design a model which can predict the likelihood of diabetes in a patient using classification algorithms with maximum accuracy. The machine learning classifiers used in the study are SVM, Decision Tree and Naïve Bayes, where Naïve Bayes shows the highest accuracy of 76.30%. Aishwarya et al. [6] designed a system that can detect diabetes and other complications using Principal Component Analysis (PCA) for pre-processing and SVM classifier for classification of patients as diabetes and non-diabetes. It has been observed that the proposed system gives 95% accuracy which is better than the existing system. Majumdar et al. [7] have proposed a model for diabetes prediction using various machine learning classifiers like Decision tree, Random Forest, Logistic Regression, KNN etc. with different external factors. Among all these, Logistic Regression shows highest accuracy of 96%. Also, the authors have imposed a pipeline model which works by transforming a linear sequence to a chain resulting in higher accuracy. After pipelining it has been observed that the accuracy of Logistic Regression has increased to 97.2%. Yang et al. [8] have focused in their work on how data poisoning attack can be implemented on Neural Networks (NN). Firstly, they have examined whether the direct gradient approach can be applied for generating poisoned data in the NN method. Then, the authors have proposed a model for accelerating the creation rate of poisoned data, which is further rewarded with a loss function and finally, the NN model is trained with the poisoned data for calculating the loss. Result analysis shows that the model can speed up the rate of poison data generation by $239.38\times$ times compared to the direct method. Gongalez et al. [9] in their work has extended the application of poisoning algorithm from binary to multiclass learning algorithms. They have proposed an algorithm which works on back-gradient optimization which can target multi-class algorithms and can be trained with neural networks and deep learning approaches.

3 Methodology and Materials Used

The proposed methodology presented in Fig. 1 evaluates the performance of different ML approaches on diabetes dataset before and after the data poisoning attack.

After pre-processing, the training dataset is ready for training machine learning classifiers. Subsequently, we have measured the performance of the different classification models. After this, the dataset is poisoned by applying the random flipping of the labels on some target samples, which is used to train the different models. Performance of the poisoned models is evaluated to show the severity of the data poisoning attack. The following sub-sections describe more details on the machine

(a) Machine learning classification model without data poisoning attack

(b) Machine learning classification model with data poisoning attack

Fig.1 Proposed Methodology for performance evaluation before and after data poisoning attack

learning methods, dataset and features of the dataset and the data poisoning attack on the different classifiers.

3.1 Machine Learning Approaches

K-Nearest Neighbor (KNN) classifier. It works by storing all the available data and classifying the new data into suitable categories based on the k-function. K is a positive integer which is used to segregate two classes, 1 denotes the neighbor class which is nearest. It is calculated based on distance measured by either Euclidean method or Manhattan method. Given two points X $(x1, x2, ..., xk)$ and Y $(y1, y2, ..., yk)$, the distance calculated using Euclidean method is measured as shown in Eq. 1, whereas, the distance measured using Manhattan method is shown in Eq. 2 below [10].

$$\text{Distance}(d) = \sqrt{\sum_{i=1}^{k} (x_i - y_i)^2} \qquad (1)$$

$$\text{Distance}(d) = \sum_{i=1}^{k} |(x_i - y_i)| \qquad (2)$$

Logistic Regression (LR) classifier. Logistic regression is used in machine learning models where the output is binary. Instead of predicting output as 1 or 0, it gives a value in the range between 0 and 1. It uses a sigmoid function shown in Eq. 3 [11].

$$f(x) = \frac{1}{1 + e^{-(x)}} \qquad (3)$$

where, x is input to the sigmoid function and e is the mathematical constant (e = 2.781). Whenever x is positive infinity, y or $f(x)$ becomes 1 and if x is negative infinity, then y or $f(x)$ becomes 0.

Support Vector Machine (SVM) classifier. It is used for determining the best hyperplane boundary for segregating two classes on the basis of the training dataset. The hyperplane should be chosen such that it is far away from the data points of another class. Data points closer to the hyperplane are known as support vectors [12]. The more the distance between the two decision boundaries, it will help in obtaining the optimal margin. The optimization of the hyperplane distance is obtained by Eq. 4 below [13]. In SVM there are different kernels like Linear, Polynomial, and RBF to solve regression and classification problems. The following 5, 6, and 7 equations are the representations of linear, polynomial, and RBF kernels of SVM.

$$wTx + b = -1 \ \& \ wTx + b = 1 \tag{4}$$

$$K(x1, x2) = x1.x2 \tag{5}$$

$$K(x1, x2) = (x1.x2 + 1)d \tag{6}$$

$$K(x1, x2) = \exp(-\frac{||x1 - x2||^2}{2\sigma^2}) \tag{7}$$

3.2 Dataset

The dataset used was selected from the Kaggle database obtained from the National Institute of Diabetes and Digestive and Kidney Diseases to predict whether a patient is diabetic or not [14]. The data sample has been collected from females of Pima Indian heritage of nearly 21 years old. The dataset contains 2000 observations out of which 684 samples are diabetic and 1316 samples are not diabetic. The dataset consists of eight attributes namely pregnancy rate, glucose concentration, blood pressure, skin thickness, insulin level, BMI ratio, diabetes pedigree function, and age, which are independent variables and one outcome which is the dependent one. The outcome variable has two classes '1' which indicates that the sample is diabetic and '0' which indicate the sample is not diabetic. The dataset have no missing values.

3.3   Data Poisoning Attack

Despite the diversified applications of ML, these ML classifiers may also be vulnerable to poisoning attacks which occur during the training phase. In most cases, it is not possible to get access to training data by the attacker but due to the online learning platform [15], open design principles [16] and crowd sourcing [17] data poisoning attack has already become a severe threat to machine learning models. It can affect the training data by leaving a backdoor or by hampering the model's performance. In general, there are different data poisoning approaches like gradient attack [18], label flipping attack, generating adversarial network, empirical investigations, fake users' insertion, clean label data poisoning attack, backdoor attack etc. Random label flipping technique have been used here to poison the training dataset. The poisoning is done to a subset of training dataset of different volumes. This variation is helpful for observing the impact of poisoned dataset on the accuracy of the machine learning classifier. Random label flipping is model independent which selects a subset of the training samples to flip the labels. Here, the attacker is independent of the underlying concept of ML model but this random label flipping is capable of reducing the accuracy thereby creating severe impact on the ML classifiers. The proposed data poisoning algorithm is shown in Algorithm 1.

Algorithm 1: RDP_LF

//Randomized Data Poisoning through Label Flipping
Input: Dataset with Clean data (feature set)
Output: Datasets with different (20%, 30%, 40%) levels of poisoning
Data: Diabetes dataset from Kaggle

1	Pre-processing of data
2	Split Training Set and Testing Set (70:30) into DS_TRAIN and DS_TEST
3	**for** *the training samples (DS_TRAIN)* **do**
4	Train the model with clean data
5	**for** *the testing samples (DS_TEST)* **do**
6	Measure accuracy, precision, recall, f1-score of the model
7	**for** *the training samples (DS_TRAIN)* do
8	Train the model with 20% poisoned data
9	**for** *the testing samples (DS_TEST)* **do**
10	Measure accuracy, precision, recall, f1-score of the model
11	Repeat steps from 7 to 10 with 30% and 40% poisoning
12	Comparative analysis of all the model's performance w.r.t clean data and
13	poisoned data

4 Result Analysis and Observations

Performance of the machine learning classifiers is measured on certain performance metrics like accuracy, precision, recall and f1-score which helps in classifying whether a sample is diabetic or non-diabetic. The experiment was performed on Jupyter Notebook. The performance indicators are measured as follows:

Accuracy. Accuracy is the measure of correct predictions calculated over the whole dataset in terms of True Positives (TN), True Negatives (TN), False Positives (FP) and False Negatives (FN). Here, TN & TP represent the number of patients correctly classified as diabetic and FN & FP represent the patient who is incorrectly classified as diabetic.

$$Accuracy = \frac{TP + TN}{TP + TN + FP + FN} \tag{8}$$

Precision. Precision measures the number of correct positive predictions as diabetic patients to the total number of positive predictions either correctly or incorrectly.

$$Precision = \frac{TP}{TP + FP} \tag{9}$$

Recall. It measures the ability of the machine learning classifier to detect positive predictions represented as:

$$Recall = \frac{TP}{TP + FN} \tag{10}$$

F1-score. F1-score is measured by the combination of precision and recall which is used to measure the accuracy of a machine learning model, which is represented as:

$$F1 - score = 2 * \frac{Precision * Sensitivity}{Precision + Sensitivity} \tag{11}$$

The diabetes dataset was randomly divided into a training set (70% of the samples in the dataset) and a testing set (30% of the samples in the dataset). Performances of the classifiers (LR, KNN, SVM Linear, SVM Polynomial, and SVM RBF) are mentioned in Table 1 below. From the obtained result it shows that SVM (RBF Kernel) classifier is having the highest accuracy of 82.5% followed by KNN and SVM (Polynomial kernel) both having 80.17% accuracy on clean datasets.

The model's accuracy and performance are measured by shuffling 20%, 30% and 40% of the training data belonging to each malicious data as a data poisoning attack by changing the labels of the classification. The models are trained again and their performances are recorded as shown in Table 2. The changes in the performance of the different classifiers reflect the impact of data poisoning attack.

Table 1 Performance analysis of different machine learning classifiers on clean data

ML Classifiers	Clean data			
	Testing data			
	Accuracy (%)	Precision (%)	Recall	F1-score (%)
LR classifier	77.00	76.16	77.00	76.01
KNN classifier	80.17	79.97	80.17	80.04
SVM (Kernal = Linear)	76.50	75.91	76.50	76.01
SVM (Polynomial Kernel)	80.17	82.21	80.17	78.07
SVM (RBF Kernel)	82.50	82.21	82.50	82.11

The confusion matrix of all the five classifiers is represented below in Figs. 2, 3, 4, 5 and 6 for both the clean data and poisoned data with 20%, 30% and 40% poisoning. Confusion matrix represents the TP, TN, FP and FN in the top-left corner, bottom-right corner, top-right corner and bottom-left corner respectively. Label '1' and label '0' in the confusion matrix represent the malicious and benign samples respectively. The objective of the work is to analyze the performance of the different machine learning classifiers for clean data as well as poisoned data. The variations in the accuracy of the models show the vulnerability and limitations in the presence of data poisoning attack. The ROC curve is used to show the performance variations represented in Fig. 7 below. The performance of the classifiers on clean data is represented using blue curve whereas, orange curve, green curve and red curve show the performance on 20%, 30% and 40% poisoned dataset. The models where the curve is closest to the top-left corner are considered as the best performing models. From the result obtained, we can infer that SVM (RBF kernel) classifier is the best performing model on clean data. Moreover, the less the distance between the four different curves, the more robust the model is towards the data poisoning attack. The robustness is seen more in SVM (Linear) model with respect to other models, which suggests that the model has less impact on the poisoning attack.

Table 2 Performance analysis of different machine learning classifiers on poisoned data with 20% shuffling, 30% shuffling and 40% shuffling

ML Classifiers	Poisoned data with 20% shuffling				Poisoned data with 30% shuffling				Poisoned data with 40% shuffling			
	Testing data				Testing data				Testing data			
	Accuracy (%)	Precision (%)	Recall (%)	F1-score (%)	Accuracy (%)	Precision (%)	Recall (%)	F1-score (%)	Accuracy (%)	Precision (%)	Recall (%)	F1-score (%)
LR classifier	74.00	72.79	74.00	72.79	74.00	72.71	74.00	72.41	70.25	67.91	70.25	67.30
KNN classifier	72.50	73.71	72.50	72.89	63.17	75.96	63.17	63.93	57.17	60.38	57.17	58.08
SVM (Kernel = Linear)	74.50	73.96	74.50	74.13	74.67	73.89	74.67	73.98	70.67	69.58	70.67	69.82
SVM (Polynomial Kernel)	73.50	73.48	73.50	70.24	71.83	71.36	71.83	67.94	68.17	65.63	68.17	64.25
SVM (RBF Kernel)	76.50	76.14	76.50	76.27	75.50	74.89	75.50	75.03	72.00	72.21	72.00	72.10

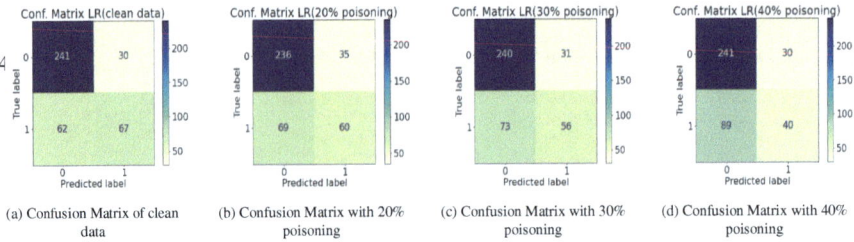

Fig. 2 Confusion Matrix for LR classifier of clean and poisoned data

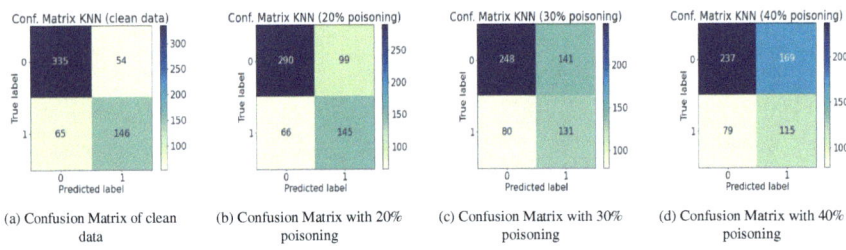

Fig. 3 Confusion Matrix for KNN classifier of clean and poisoned data

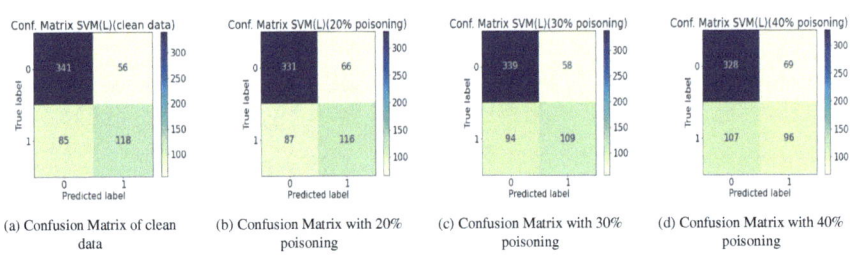

Fig. 4 Confusion Matrix for SVM (Linear kernel) classifier of clean and poisoned data

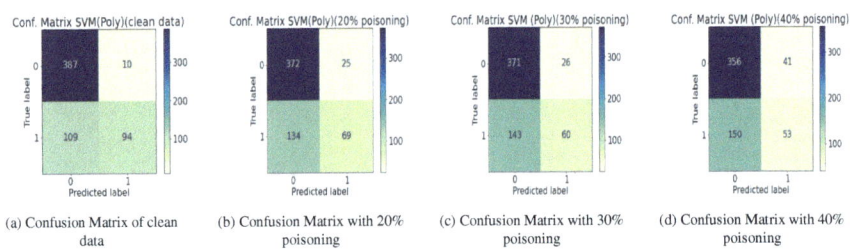

Fig. 5 Confusion Matrix for SVM (Polynomial kernel) classifier of clean and poisoned data

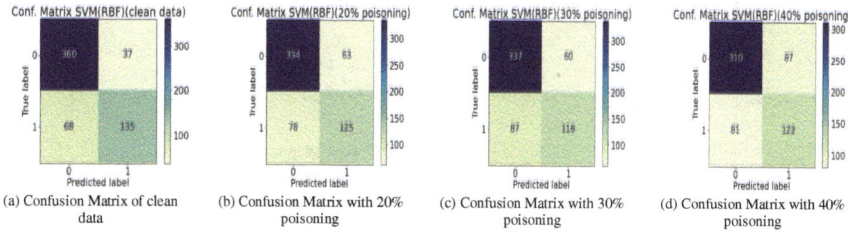

(a) Confusion Matrix of clean data

(b) Confusion Matrix with 20% poisoning

(c) Confusion Matrix with 30% poisoning

(d) Confusion Matrix with 40% poisoning

Fig. 6 Confusion Matrix for SVM (RBF kernel) classifier of clean and poisoned data

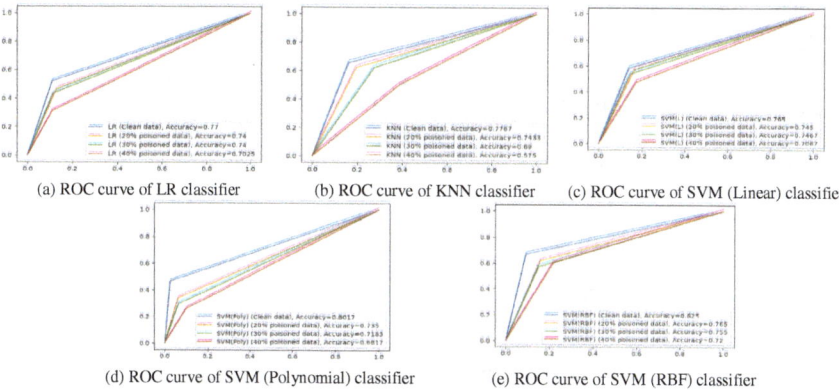

(a) ROC curve of LR classifier

(b) ROC curve of KNN classifier

(c) ROC curve of SVM (Linear) classifier

(d) ROC curve of SVM (Polynomial) classifier

(e) ROC curve of SVM (RBF) classifier

Fig. 7 ROC curve of different machine learning classifiers under a poisoning environment

5 Conclusion

This work presents a performance evaluation of different machine learning classifiers on benign as well as poisoned dataset. For evaluation, we have considered five ML classifiers for the prediction of diabetes datasets and evaluated the accuracy of the models. Among all the classifiers, SVM (RBF kernel) classifier shows the best accuracy for the cleaned dataset. However, there is a decrease in the performance of the models when the training dataset is poisoned with 20%, 30%, and 40% shuffling. The drastic changes in the performance of the models show the severe threats and the strength of the data poisoning attack. There are certain limitations in the work due to the use of single data poisoning approach and limited sample in the dataset. In future, the proposed model can be tested with other poisoning attacks for providing significant and vulnerable poisoning approach to be used in different datasets across various applications.

References

1. Sarker IH (2021) Machine learning: algorithms, real-world applications and research directions. SN COMPUT. SCI. 2:160
2. Harikumar P, Malik M, Domenic TS, Guna SS, Sanjeev G, Mohd N (2023) IMPACT OF MACHINE learning ON Management, healthcare AND AGRICULTURE. Mater Today: Proc 80(3):2803–2806
3. Hosseini H, Chen Y, Kannan S, Zhang B, Poovendran R (2017) Blocking transferability of adversarial examples in black-box learning systems. arXiv preprint arXiv:1703.04318
4. Reuters (2016) Microsoft's AI Twitter bot goes dark after racist, sexist tweets. https://www.reuters.com/article/us-microsoft-twitter-bot-idUSKCN0WQ2LA/. [Online; accessed 20-April-2021]
5. Sisodia D, Sisodia DS (2018) Prediction of diabetes using classification algorithms. Proc Comput Sci 132:1578–1585
6. Aishwarya R, Gayathri P, Jaisankar N (2013) A method for classification using machine learning technique for diabetes. Int J Eng Technol (IJET) 5:2903–2908
7. Mujumdar A, Vaidehi V (2019) Diabetes prediction using machine learning algorithms. Proc Comput Sci 165:292–299
8. Yang C, Wu Q, Li H, Chen Y (2017) Generative poisoning attack method against neural networks, preprint arXiv:1703.01340
9. González LM, Biggio B, Demontis A, Paudice A, Wongrassamee V, Lupu EC, Roli F (2017) Towards poisoning of deep learning algorithms with back-gradient optimization. In: Proceedings of the 10th ACM workshop on Artificial Intelligence and Security
10. Pradhan P (2012) A genetic programming approach for detection of diabetes. Int J Comput Eng Res 2:91–94
11. Stephan D, Lucila OM (2002) Logistic regression and artificial neural network classification models: a methodology review. J Biomed Inform 35(5–6):352–359
12. Sisodia D (2010) SVM for face recognition. In: Proceedings of international conference on computational intelligence and communication networks, pp 554–559
13. Sisodia D (2012) Fast and accurate face recognition using SVM and DCT. In Proceedings of the second international conference on soft computing for problem solving (SocProS2012), pp.1027–1038, Springer
14. https://www.kaggle.com/datasets/nancyalaswad90/review
15. Shalev-Shwartz S, et al (2012) Online learning and online convex optimization, Foundations and Trends® in Machine Learning, 4
16. Bonawitz K, Eichner H, Grieskamp W, Huba D, Ingerman A, Ivanov V, Kiddon C, Konecnˇy J, Mazzocchi S, McMahan B (2019) Towards federated learning at scale: System design. Proc Machine Learn Syst 1:374–388
17. Rai A, Chintalapudi KK, Padmanabhan VN, Sen R (2012) Zee:Zero-effort Crowdsourcing for Indoor Localization. In: Proceedings of the 18th annual international conference on mobile computing and networking, pp. 293–304
18. Geiping J, Fowl L, Huang WR, Czaja W, Taylor G, Moeller M, Goldstein T (2021) Witches' brew: industrial scale data poisoning via gradient matching

A Novel Deep Learning Based Fully Automated Framework for Captcha Security Vulnerability Checking

Ashutosh Thakur, Bhavishya, and Priya Singh(ID)

Abstract From breaching the service and allowing humans to surpass it upon correct verification, nowadays, it is achievable by bots and machines using machine learning, making the service vulnerable. It can pose several challenges like hacking, crashing, exploiting to name a few. In the present work, a novel framework namely Bypass-Captcha model to test the security strength of web services via captchas is proposed. This model examines the vulnerability of the security via automation and captcha decoding using deep learning models particularly, i.e. Convolutional Neural Network, Recurrent Neural Network, and Connectionist Temporal Classification Loss. It involves full automation processing i.e. from opening the service, entering the credentials, getting a captcha from the service, and inputting the right decoded captcha. This complete process is dynamic. The credentials are provided via file, read during runtime, and are inputted at their required place. For training the model, a dataset combinely having four different types of captcha i.e. arc, dotted, rotated, and noisy is used. The automation process is working satisfactorily on specific services but it is still not a convenient way for a large number of services at a time. The proposed model is having a Val_Loss of 97% in the clear captcha case.

Keywords Captcha security · Cyber-attacks · Security breach · Convolutional neural network · Recurrent neural network

1 Introduction

Captcha (Completely Automated Public Turing test to tell Computers and Humans Apart) is a computer program or a system intended to distinguish human from machine input, typically as a way of thwarting spam and automated extraction of data from websites. It helps to solve the problems involving cyber-attacks, security risks, and spam. It presents problems that are complex for AI to crack, but not for

A. Thakur · Bhavishya · P. Singh (✉)
Department of Software Engineering, Delhi Technological University, Shahbad Daulatpur, Delhi 110042, India
e-mail: priya.singh.academia@gmail.com

© The Author(s), under exclusive license to Springer Nature Singapore Pte Ltd. 2024
S. Namasudra et al. (eds.), *Data Science and Network Engineering*, Lecture Notes in Networks and Systems 791, https://doi.org/10.1007/978-981-99-6755-1_33

human beings resulting in preventing automated attacks. There are different types of captchas available, but the most frequently used captchas are text-based. They are composed of numbers and English alphabets which are arranged in different orders and simple for human beings to interpret but not for computers. Every type of captcha has its way to resist the automated bots. In past years, many cybercriminals were able to get through the captchas easily, which raises the question of whether the captchas are reliable or not.

The Current work proposes a novel framework namely BypassCaptcha model. This work intends to examine the security strength of web services via captcha verification against bots. It demonstrates an efficient solution for automatically solving text-based captchas. The training process has been improved by following the suggestions from a prior study, which included expanding the training set with various font sizes and styles and optimizing the hyperparameters of the selected architecture. Machine learning and deep learning techniques were used to make the model which results in chances of captcha bypassing by bots. To make a decoding model, the Python library was used to create images for training and testing the model with different character lengths. The two pre-processing techniques were implemented, one for alphanumerical and one for numerical captchas, and a CNN-based approach with a grid search for hyperparameter optimization to find the best model is exploited. The model was evaluated using three captcha datasets with varying character lengths and alphanumeric values, as well as with a combined dataset. The captcha is then passed to the earlier trained model and the model predicts its value and returns it in text format which was automatically inserted in its required space, where if the captcha predicted the value correctly then it bypasses the captcha by the bot and allows it to use the service.

The rest of the paper is organized as below: Sect. 2 highlights recent works present in the literature in the field of captcha security strength checking. Section 3 discusses the approach adopted in the current study. Section 4 presents and Discusses the results and finally, Sect. 5 concludes the findings.

2 Literature Review

Mostly all authentication websites require a captcha and captcha vulnerability is a huge challenge. To address the same plenty of research articles are available in the literature.

The waste of using captchas for security measures was explained and implemented by Dinh and Huang [1] and K. Sukhani et al. [2] which directly targeted human-like behaving bots for bypassing the captcha-based security. They were able to bypass text, audio, and object-based captchas evaluated by Awla, Mirza, and Kareem [3] where they got a score of 70.26%. For the same Moore and Walcot [4] also discussed points regarding the strength of captcha and schemes to improve them to overcome current challenges. Chen and Lou [5] proposed a method to easily recognize the hand-written captcha on real platforms using DNN which had better results than NDSS'16, CCS'18, and "Science" in 2017.

The effectiveness of using CNNs for captcha recognition and its breaking was demonstrated by Zhao et al. [6]. Desmet et al. [7] evaluated the newest Google Recaptcha V3 service via web automation and was able to evade 99/100, but Ning Yu [8] used a AI technique using the Python captcha library and object detection through TensorFlow. Aguilar et al. [9] and Wang et al. [10] proposed their own CNN model to test the vulnerability of easy text-based captcha-based on DenseNet-121 and ResNet-50 which came around 90.49% and 92.9% respectively. They were also able to reduce the memory and time consumption. Sukhani et al. created a model to generate a text-based and user-friendly captcha with a low failure rate of bypassing. Zhao et al. [6] obtained an accuracy of 99% for single-letter captchas and 76% for 4-letter captchas with a multi-CNN as compared to 30% comes with a clustering scheme and 69% accuracy when Support Vector Machine (SVM) was used. ++Deep captcha, a deep Neural network consisting of different layers which include a Convolutional layer, a Dense layer, Max-Pooling, two Convolution-Max Pooling, and a Softmax layer was created by Noury and Rezai [11]. It was trained on captchas(5-digit) which were produced by Python Image Captcha Library and 99.33% accurate on the training set while 98.94% accurate on the test set. To make text-based captchas more secure against automated attacks, websites and captcha generators have incorporated an approach of overlapping captcha characters as an essential component and its incorporation resulted in the development of a new pre-processing phase called "segment and recognize" involving dividing characters' images, and the precision of its decryption was restricted.

Bursztein developed an algorithm that combined the steps of segmentation and recognition to tackle captchas and evaluated it on real-world captcha systems, achieving a recognition rate of 51.39% on the CNN captcha dataset and 55.22% on the Baidu captchas dataset. A CNN-based method was proposed by Kopp [12] for breaking captchas which uses a reverse Turing test for secure logins, dictionaries, and attacks through guessing passwords, as well as spamming and automatic bot usage. A Deep captcha system which was designed by Zahra [11] breaks visual captchas using Deep learning. The vulnerabilities of the captcha were analyzed, and a resilient system was created that employs CNN to address both alphanumeric and numeric captchas. After being trained on a dataset comprising 50,000 captchas, the system achieved an accuracy of 98.94% for numeric and 98.31% for alphanumeric sample test cases. This led to an improvement in both security and performance. Nevertheless, there were still concerns regarding reliability and other matters that required attention, and the system was constrained to captchas that only consisted of either numerical or alphanumeric characters.

3   Methodology

In the current section, the overall approach adopted for automated captcha security vulnerability is discussed in detail. Figure 1 provides the architecture of the proposed framework named BypassCaptcha.

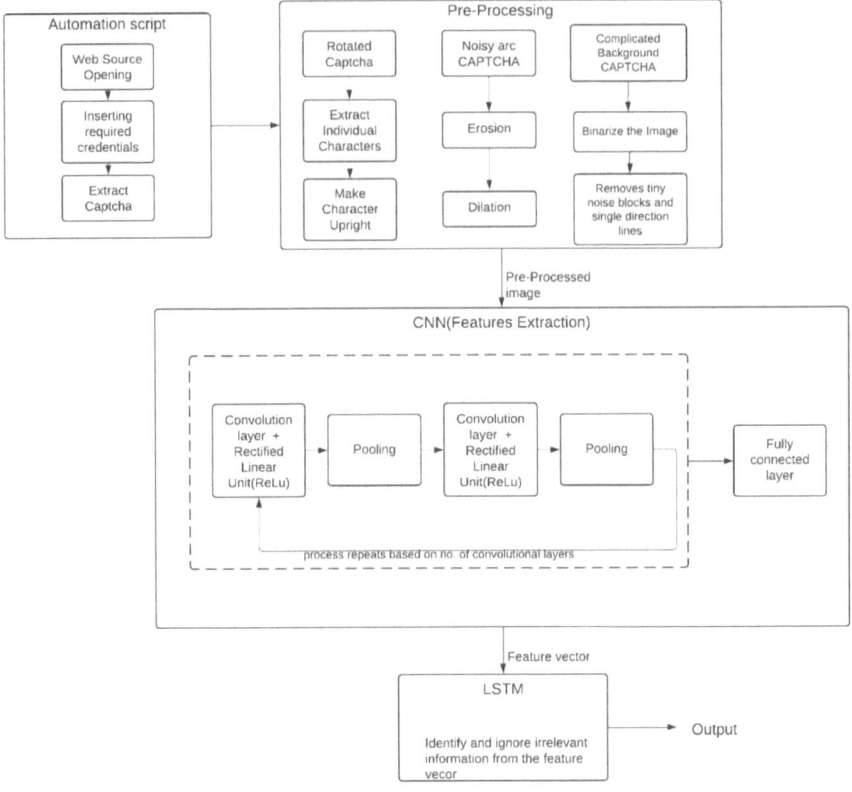

Fig. 1 Architecture of the proposed framework

In Fig. 1, the initiation is from creating an automation script where all the commands concerning automation of web services, extraction of captcha image, invoking the model, and visualizing the extracted decoded captcha output are written systematically.

While script is used to train and execute model, pre-processing against the input for the model is done as per the set property of input in the model.

After pre-processing, the processed input is passed to the Convolutional Neural Network and Long Short-Term Memory Network, where the data is minimized for faster processing and the required part of data is extracted by eliminating the non-required ones respectively.

Fig. 2 Sample images from
a dataset 1, **b** dataset 2,
c dataset 3

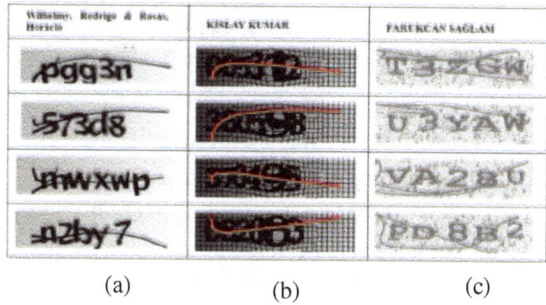

(a)   (b)   (c)

3.1 Dataset Description

There were three different types of alphanumeric captcha datasets for training and testing as provided in Fig. 2. These datasets were used by Weng et al. [13] and Awla et al. [3] These were available on the Kaggle repository (https://www.kaggle.com/datasets/alizahidraja/captcha-data).

The dataset1 comprises 2800 samples of 5-character alphanumeric captchas with a resolution of 200*50 pixels, with a noisy background and cut text. Some characters were slightly blurred, and this captcha design was utilized in CNNs. The dataset2 comprises 3800 samples of 6-digit numerical captchas covered by tables with a resolution of 182*50 pixels consisting of numerous columns and a red line cutting the number and a blurry effect was also added to the number. The dataset3 comprises 2000 samples of 5-character alphanumeric captchas with a resolution of 180*50 with dots and lines over the character to make it less visible. These samples were in grey scale.

3.2 Machine Learning Model

In the present work, a novel framework namely the BypassCaptcha model was proposed. A custom model which was incorporated with the automated extraction of captcha from web services was used in the present work which was further passed to an already trained captcha decoder model based on CNN, RNN, and CTC loss.

A modified version of the LeNet-5 CNN model [13] to successfully break the acquired captchas was used. In the present work, the model includes both CNN layers for feature extraction and RNN layers for information propagation through the extracted features. The resulting output was a matrix of character scores for each sequence element. The next objectives involve two actions to be implemented with this matrix: first is training which is implemented to find the loss value and train the neural network and the second is an inference which is decoding of the text present in the testing images. Both tasks were achieved by the CTC operation.

During the training of the neural network, the CTC loss function was used as a guide. Only the output matrix of the neural network and the corresponding ground-truth text were provided as inputs to the CTC loss function.

The functioning of bypassing the captcha model comprises three steps:

Automation script- The implementation started with the automated script where complete scripting was implemented for the proposed BypassCaptcha model. A human-like bot package was implemented in the script for browsers with the help of Selenium. Puppeteer could also be used for the same but the functional properties were quite the same. The script was accountable for the opening of the bypassing target company. Here government GST service was employed, as it's quite secure, and if not it is necessary to find the loopholes. After the opening of the target script, credentials were inputted through the Selenium bot, and the captcha was captured, which was then further passed to the imported trained model for the output and then bypassed the output.

Data pre-processing and extraction- The pre-processing process started by grey scaling the image from color images to make valuable data easily readable by the computer which is provided in Fig. 3. Then noise was removed by applying threshold segmentation on the image and the image was converted to binary format. OTSU's algorithm [12] which automatically calculates the threshold to extract valuable content from the whole image was used to convert the image to binary format.

Erosion was used to minimize the boundary pixels (foreground [letter] size reduction) and to eliminate all the noises in the present work whereas dilation was used to maximize the size of the foreground by increasing the boundary pixels and re-join the parts that were broken in the erosion process by increasing the boundary pixels. The above erosion and dilation pre-processing process was only applicable and efficient for datasets 1 and 3, but for numerical captchas viz. dataset 2, a different pre-processing method was used. After pre-processing, each captcha, character is segmented as a single image with a white foreground on a black foreground.

Fig. 3 Pre-processing pipeline for **a** alphanumerical and **b** numerical captcha

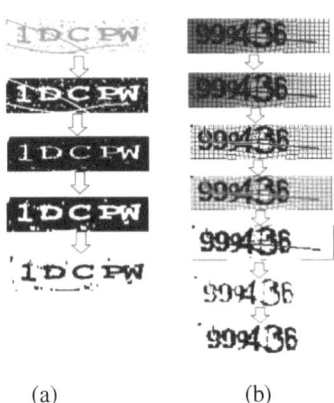

(a)   (b)

Fig. 4 Output matrix of NN

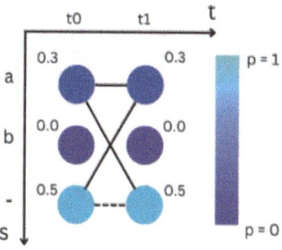

Then we extracted the captcha by following these three steps:

Encoding the text - To solve the problem of duplicate characters, following method has been used: Whenever encoding a text, any number of blanks can be placed at different positions that will be removed at the time of decoding. However, a blank space will be needed when duplicate characters will come, like in "needed".

For example: "npo" → "--nnnppooo", or "-n-p-o-", or "npo".

"120" → "1111222-o", or "-1-2-o-", or "12-o", but **not** "120"

With the help of the above method, different arrangements of the same text can be formed, for example, "t-o", "too" and "-to" resemble similar text("to"), but their arrangements will be different concerning the image.

Loss Calculation – This neural network returned a matrix which was having scores for all the characters at different time step. A simple matrix is provided in Fig. 4, in which three characters "a", "b" and "−" and 2-time steps i.e. t0 and t1 are given. The summation of all the scores for each character came to 1.

In Fig. 4, the character probability is shown in color-coded format and is also printed next to each matrix entry. Thin lines are paths representing the text "a", while the thick dashed line is the only path representing the text. The score of each alignment will be found by multiplying the individual scores of each character with each other. For example in Fig. 4, the score for different paths is different, for instance ("aa" = 0.3 * 0.3 => 0.09 similarly for "a-" it is 0.15). For calculating the score of a GT text, a summation of all the scores is required over all valid paths related to this text.

Decoding the text- For decoding, an approach that was simple to understand and quite fast as well i.e. best path decoding was implemented into two steps, firstly, for each time step it will find the best path from the most frequently occurred character, and secondly, it will remove duplicate characters and blanks.

As apparent in Fig. 5, the individual characters involved include"a", "b" and "−". There are a total of five time steps. Since the best path decoding approach was used, the most frequently occurred character for t0 was "a" similar to t1 t2. The blank character ("−") at t3 was the highest concerning its score, similarly "b" was more likely at t4. After all this approach path was "aaa-b" and after elimination of the duplicate character's path remained "a-b". Further after removing the blank, "ab" was left as output.

Fig. 5 Output matrix of NN.
The thick dashed line
represents the best path

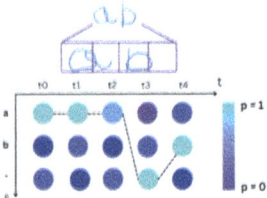

Table 1 BypassCaptcha model layers

Layer (type)	Output shape	Params	Connected to
Image (Input Layer)	[(None, 200, 50, 1)]	0	[]
Conv1(conv2D)	(None, 200, 50, 32)	320	['image[0][0]']
Pool1[MaxPooling2D]	(None, 100, 25, 32)	0	['conv1[0][0]']
Conv2(conv2D)	(None, 100, 25, 64)	18,496	['pool1[0][0]']
Pool2[MaxPooling2D]	(None, 50, 12, 64)	0	['conv2[0][0]']
Reshape(Reshape)	(None, 50, 768)	0	['pool2[0][0]']
Dense1(Dense)	(None, 50, 64)	49,216	['reshape[0][0]']
Dropout_2(Dropout)	(None, 50, 64)	0	['dense1[0][0]']
Bidirectional_4(Bidirectional)	(None, 50, 256)	197,632	['dropout_2[0][0]']
Bidirectional_5(Bidirectional)	(None, 50, 128)	164,352	['bidirectional_4[0][0]']
Label(Input Layer)	[(None, None)]	0	[]
Dense2(Dense)	(None, 50, 21)	2709	['bidirectional_5[0][0]']
Ctc_loss(CTC Layer)	(None, 50, 21)	0	['label[0][0]','dense2[0][0]']

Training and Testing- The datasets were split into training, testing, and validation
sets. 80% of the total images were used for training, with 10% as the validation set
and for testing.

Table 1 represents the whole model layering with the required params, their output
type, and the previous layer they are connected too.

4   Result and Discussions

There were three parts to the result: Model Testing Output, Extract Captcha output,
and Graphical Analysis and for Discussion, it referred to the graphical representation.

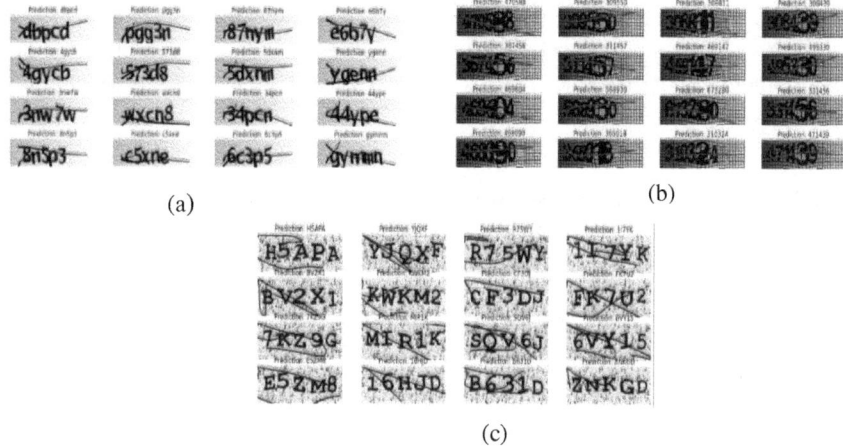

Fig. 6 Predicted output of the validation and testing set **a** dataset 1, **b** dataset 2, and **c** dataset 3

4.1 Model Testing Output

For training and validation, the dataset from the Kaggle repository was used, and after training.

Figure 6 represents the result of the validation set and the BypassCaptcha model did very well. Every validation set gave perfect output on validation testing after an epoch of 100 and val_loss of less than 0.2. With fewer epochs, the prediction varies from the actual output.

4.2 Extracted Captcha Output

Here, captcha images from the live web service, which uses captcha authentication are extracted using the screenshot functionality of automation. Extracted captcha images are then passed to the trained model one by one to get the decoded output of the passed captcha. OpenCV application was used to preview the decoded captcha output in an external customized canvas window by the trained model, where the decoded captcha from the model is shown concerning the passed captcha image.

Also, to differentiate them, the decoded captcha is customized to have green font color, which references the passed captcha image captcha text in a green rectangle box. These are re-checked by ourselves and plotted in the below table with examined attributes.

In Table 2b 1 captcha failed and 3 passed which concluded it fails when there is extra space than a single space then it detects those words as different words and decodes them differently. It also fails when a case-sensitive or similar alphanumeric character comes in the captcha.

Table 2 Trained model has been used to work lively on real-time services decoding the captcha extracted from live web services **a** noisy captcha **b** clear captcha

Captcha	Captcha human code	Predicted	Result
a			
	810,781	810,781	**PASS**
	537,891	537,891	**PASS**
	855,758	18,855/5	**FAIL**
	915,304	915,304	**PASS**
b			
	sqQVny	sqQVny	**PASS**
	dHT4km	dHT4km	**PASS**
	EDO7qb	EDO79b	**FAIL**
	DMqN4G	DMqN4G	**PASS**

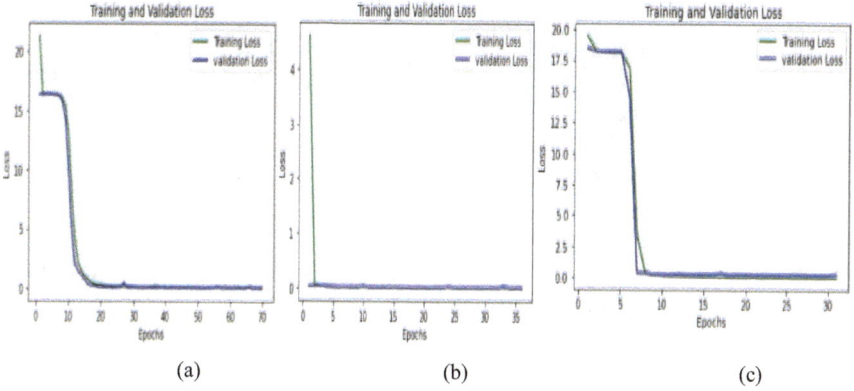

Fig. 7 Training and Val_loss graph against Epoch for captcha of **a** Dataset 1, **b** Dataset 2, **c** Dataset 3

4.3 Graphical Analysis

For a Better understanding of the numbers, the graph presentation of the training and validating process against the number of times the process repeated (epochs) was generated. Here, graphs of Loss against epoch were generated for training and validation for every data set and shown as per the sequence. It's shown in Fig. 7.

For Dataset 1(Wilhelmy, Rodrigo & Rosas, Horacio): The training loss in starting was above 20 in Fig. 7a, but from 8 to 15 there was a gradual dropping of training loss and it got lower to 2 which further reduced to 0.22 after, on epoch 17. After epoch 17 the loss was constant at 0.2. Similarly, validation loss started at 16.5 and was constant at 0.2 Val_loss from epoch 17.

For Dataset 2(KISLAY KUMAR): From Fig. 7b the training loss in starting was above 4.7, but from 0 to 2 there was a gradual dropping of loss and it got lower to 0.1 and get constant after epoch 2. Similarly, validation loss started at 0.1 and stayed constant at 0.1 Val_loss from the epoch of 2.

For Dataset 3(FARUKCAN SAĞLAM): From Fig. 7c the training loss in starting was above 19.5, but from 1 to 6 there was no change in training loss and suddenly the training loss dropped to around 0.2 from 6 to 7 which further reduced to 0.12 after epoch 17. After epoch 17 the loss was constant at 0.2. Similarly, validation loss started at 16.5 and followed the same path as for training loss.

5 Conclusion and Future Work

In the current study, a novel BypassCaptcha model which approximately measures the security strength of the web service against bots is proposed. To do so, this model bypasses the security i.e. captcha. And a captcha is said to be weak if it can be automatically resolved at a rate greater than 1%. The proposed approach is validated on three textual datasets. The model loss reduces to 0.2 for dataset 1 with accuracy of up to 94.10% for dataset 2 and for dataset 3, loss is 0.12% and accuracy of 96%. A significant improvement of accuracy over 51.39% is achieved in comparison to the existing algorithm. For dataset 1, the model achieves an accuracy of 86.80%, and for dataset 2, the model achieves an accuracy of 92.90%.

Overall, the accuracy obtained by the current method increased from 92–97%, which is quite well to say that this model works very great and is enough to consider these captcha datasets corrupted.

This model achieved 93.49% accuracy on a mixed dataset with different captcha types, showing that adding character frequency with different font types and sizes improves model efficiency. Therefore, improving the segmentation process for captcha duplicate characters and providing the network with more images for training will increase prediction accuracy.

As a recommendation for future work, a neural network for segmentation to increase the number of correctly segmented captchas and the accuracy of the models is recommended. Furthermore, recurrent neural networks can be explored instead of convolutional neural networks for the character recognition process.

References

1. Dinh NT, Hoang VT (2023) Recent advances of captcha security analysis: a short literature review. Proc Comput Sci 218:2550–2562, ISSN 1877–0509, https://doi.org/10.1016/j.procs. 2023.01.229
2. Sukhani K, Sawant S, Maniar S, Pawar R (2021) Automating the bypass of image-based captcha and assessing security. In: 2021 12th international conference on computing communication and networking technologies (ICCCNT), Kharagpur, India, pp. 01–08, https://doi.org/10.1109/ ICCCNT51525.2021.9580020
3. Awla HQ, Mirza AR, Kareem SW (2022) An automated captcha for website protection based on user behavioral model. In: 2022 8th international engineering conference on sustainable technology and development (IEC), Erbil, Iraq, pp. 161–167, https://doi.org/10.1109/IEC54822. 2022.9807472
4. Moore M, Walcott KR (2022) Generating human-like motion to defeat interaction-based captchas. In: Arai K, Proceedings of the future technologies conference (FTC) vol. 2, pp. 202–217, https://doi.org/10.1007/978-3-031-18458-1_15
5. Chen J, Luo X, Zhu L, Zhang Q, Gan Y (2023) Handwritten captcha recognizer: a text captcha breaking method based on style transfer network. J Multimed Tools Appl 82:13025–13043, https://doi.org/10.1007/s11042-021-11485-9
6. Zhao N, Liu Y, Jiang Y (2017) Captcha breaking with deep learning CS 229 Final Project, Stanford University, Tech Rep
7. Tsingenopoulos I, Preuveneers D, Desmet L, Joosen W (2022) Captcha me if you can: Imitation Games with Reinforcement Learning. In: 2022 IEEE 7th European Symposium on Security and Privacy (EuroS&P), Genoa, Italy, 2022, pp. 719–735, https://doi.org/10.1109/EuroSP53844. 2022.00050
8. Yu N, Darling K (2019) A low-cost approach to crack python captchas using AI-based chosen-plaintext attack. Appl Sci 16
9. Aguilar D, Riofrío D, Benítez D, Pérez N, Moyano RF (2021) Text-based CAPTCHA vulnerability assessment using a deep learning-based solver, pp. 978–987, https://doi.org/10.1109/ ETCM53643.2021.9590750

10. Wang J, Qin J, Xiang X, Tan Y, Pan N (2019) Captcha recognition based on deep convolutional neural network. Math Biosci Eng 16(5):5851–5861
11. Nouri Z, Rezai M (2020) Deep-captcha: a deep learning based captcha solver for vulnerability assessment. Available at SSRN 3633354
12. Kopp M, Nikl M, Holena M (2017) Breaking captchas with convolutional neural networks. CEUR Workshop Proceedings vol. 1885, ISSN 1613-0073
13. Weng H, Zhao B, Ji S, Chen J, Wang T, He Q, Beyah R (2019) Towards understanding the security of modern image captchas and underground captcha-solving services. J Big Data Mining Analytics 2(2):118–144, https://doi.org/10.26599/BDMA.2019.9020001
14. George D, Lehrach W, Kansky K, Lazaro-Gredilla M, Laan C, Marthi B, Lou X, Meng Z, Liu Y, Wang H (2017) A generative vision model that trains with high data efficiency and breaks text-based captchas. Science 358(6368)

Paillier Cryptosystem Based Robust and Reversible Image Watermarking

Alina Dash, Kshiramani Naik, and Priyanka Priyadarshini

Abstract Watermarking is the method in which an image or a text is secretly embedded in an image taken as the original image. It is helpful in recognizing the original possessor of a particular content. It generally works on different objects like pictures, audio, video etc. Reversible and Robust Watermarking is a technique in which the original image as well as the watermarked image can be retrieved successfully even if the original or the image formed after the embedding phase is attacked by noise or other factors. This work is presenting an algorithm or a technique based on Robust and Reversible Watermarking. The algorithm presented in this work is divided into three phases: namely, watermark generation phase; the phase in which the generated watermark is inserted into the original image; and the phase of extraction of watermark and the original image separately or lossless. The watermark is obtained and embedded to the original image by using weber's differential excitation descriptor, reference from law of weber, and interpolation linearly. Paillier Cryptosystem is used to encrypt the embedded image. To make the embedded image more secure, Discrete Wave Transform is applied. The original image is not necessary for the extraction process. Linear interpolation in inverse method is used to retrieve the value of the watermark of each block consisting of pixels. To determine the accuracy and robustness of the original image and the watermark that is extracted from the embedded image, different operations, or factors like PSNR, BER and Surviving bit rate are calculated. Random tampered zones could also be found using this algorithm.

Keywords Robust watermarking · Reversible watermarking · Weber's differential excitation descriptor · Law of weber · Linear interpolation · Paillier cryptosystem

A. Dash (✉) · K. Naik · P. Priyadarshini
Veer Surendra Sai University of Technology Burla, Sambalpur, Odisha 768018, India
e-mail: alinadash_cse@vssut.ac.in

K. Naik
e-mail: kshiramaninaik_it@vssut.ac.in

© The Author(s), under exclusive license to Springer Nature Singapore Pte Ltd. 2024
S. Namasudra et al. (eds.), *Data Science and Network Engineering*, Lecture Notes in Networks and Systems 791, https://doi.org/10.1007/978-981-99-6755-1_34

1 Introduction

The Internet has emerged as the primary medium for transmitting data or information between diverse users in the early modern era. To secure the data of the users and protect the data from various threats or attacks, some techniques like watermarking and digital signatures do exist. The technique of watermarking is used in various areas which ensures the protection of users from users that are not authorized. Depending on the watermark's application method, they are split into two categories: namely, spatial and frequency domain watermarking. Without the application of transformations, the original signal's pixel is where the watermark is applied under spatial domain. Some of the frequency domain transformations include DWT, Inverse DWT, DCT, SVD etc [1]. Also, the extraction of the watermark can be done in many ways: Some techniques don't require the original image or the watermark during the process of extraction and only the private or secret is used for extraction, some techniques simply require the watermark for the extraction procedure and do not necessitate the original image [2], whereas some processes need only the original image not the watermark.

The growth and advancement of the Internet and technologies like cloud computing have made it easier for the user to save or store information and data in a secure environment or in cloud with the help of network. This helps the user or the people to access or use their saved data whenever it is required. But, the technology of storing data in the cloud involves the problem of securing and protecting the data. To maintain or secure the privacy of multimedia information or content, the data can be encrypted which can ensure that the probability of revelation of the personal data is highly reduced or decreased. The managers of cloud storage find it difficult to control a huge amount of encoded data and secure them. Therefore, a way to secure the privateness of the day is to embed information into the original data to achieve retrieval of cipher text and protection of secured data. Therefore, in recent years, robust and reversible watermarking methods or techniques have evolved into a hotspot for research about the security of digital multimedia information. Reversible technique of watermarking is a method of hiding data or information that includes embedding secret information or watermark into the original image or data, and later, the watermark or the secret data and the original data can be extracted in a complete manner without any disturbance or distortion. This technique has the capability to resist various attacks like noise, compression etc. The traditional reversible watermarking technique did not ensure the robustness of the watermark i.e. the original image and the watermark could not be recovered when noise or other image processing attacks were made against the watermarked image. In the domain of information hiding also, the need for reliable and reversible watermarking has become imperative of research. According to this technique, the original image should be extracted or retrieved correctly by the receiver and the watermark should also be extracted in the correct manner in such a way that there is no loss in the phase of extraction. The algorithm proposed in this work will ensure better robustness than the existing method or technique. The important points of the proposed method are as follows. A watermarking scheme in a domain that is encrypted, robust and reversible and semi-blind is proposed in this work. By using the Weber differential excitation descriptor and overlapping blocks of size 3×3, the watermark is created from the cover image, and the watermarked image is also encrypted. The cover or the original image will not be required while retrieving the watermark and the original image in the phase of extraction [1, 3, 4]. The areas or zones that are tampered in a random way can also be determined in the phase of extraction. The proposed algorithm ensures high imperceptibility and robustness.

The article is laid out in the following manner. Section 2 throws light on the previous research in this familiar field. Section 3 pays attention to the proposed model and its working. A comparative performance analysis of our model is discussed in Section 4, whereas a summary of the contribution made is presented in Section 5.

2 Related Works

In the past years, much work and research were done on the technique of Robust and Reversible Watermarking to maintain the quality of the picture and its robustness. Different ways of embedding an image are done to verify the image's reliability and data integrity. To maintain the robustness and quality of the original picture also, many algorithms have been proposed. Some of such algorithms proposed recently are mentioned below.

Malayil et al. [5] proposed a technique that evaluates an authentication code from the picture that is to be embedded upon. An operation called scaling up is used for embedding the watermark. This paper proposed a strategy which combined the characteristics of both zero and reversible watermarking. But the image quality and robustness were not satisfactory. Also, the security of images was a matter of concern.

Hu et al. [6] presented utilizing the Polar Harmonic Transform a lossless technique moment. This technique resulted in the recovery of the original image without any loss in case of no signal or data processing operations or noise. However, it was not very reliable in case of attacks.

Liang et al. [7] proposed a watermarking technique which gives satisfactory results in both image quality and robustness. Here, the original image is used as input and separated into a number of non-overlapping blocks for partial homomorphic encryption, or Paillier Cryptosystem. This technique did not work well with certain operations like random tampering.

A paper based on Lightweight Cryptography was proposed by Xiong et al. [8], for both reversible and accurate image watermarking. In order to generate the encrypted image, secret sharing which is additive in nature and scrambling is done in block level. The main issue of the work was the complexity. Also, since the robustness was based on Higher Significant Bits properties, improvement of these bits was necessary for improving the robustness. In addition to this, the visual quality of the watermarked image was not satisfactory. More work was still needed to secure the data.

Yu et al. [9] proposed an algorithm that made use of the capability of error correction of channel coding. This algorithm scrambled the chunks so that the image could be directly decoded by the one in the receiving end. Embedding capacity was large and the algorithm worked well with redundancy of channel coding but there is no guarantee of embedding based on modulation.

A watermarking method involving discrete Fourier transform and robust and reversible picture watermarking in the spatial domain was proposed by Cao et al. [10]. Moreover, the relationship between the DC component and the spatial domain pixel value is used.

3 Proposed Method

The existing methodology of watermarking could not prove to be trustworthy when it came to robustness. Whenever a signal or image, at any phase, was attacked by any data or image processing operation, the original data restoration became difficult. Also, the quality of image and data was also being hampered. The embedding capacity was not satisfactory earlier. Therefore, we have proposed a watermarking technique that could survive threats and attacks in this work.

The objective of the proposed technique is to achieve better robustness, imperceptibility, and image quality. Also, the proposed scheme aims to overcome the issue of degradation of data or image quality by allowing the restoration of original and watermark images. The proposed scheme takes an image as input and applies weber's differential excitation descriptor to generate and embed the watermark completely over the image, after which, it performs encryption using Paillier Cryptosystem to secure the embedded image. The image is then performed a DWT transformation to increase its security. During the extraction stage, the original image is first decrypted and then retrieved or restored using the same process in the reverse way as that of the phase of embedding [11].

Considering the security of the data of the users, we have proposed a technique that can detect zones which are tampered and it would be convenient to remove them after it is detected [12]. The original picture is used to generate the watermark using Weber's descriptor. Overlapping blocks of size 3×3 are taken and at the center pixel of the block, embedding takes place. Later, the watermark and the original picture are retrieved in the side of the receiver and then accuracy is checked. The general flow diagram of the proposed work is represented in Fig. 1.

3.1 Generation of Watermark

To generate the watermark, the original picture is considered and divided into blocks that are overlapping after which weber's excitation descriptor is evaluated.

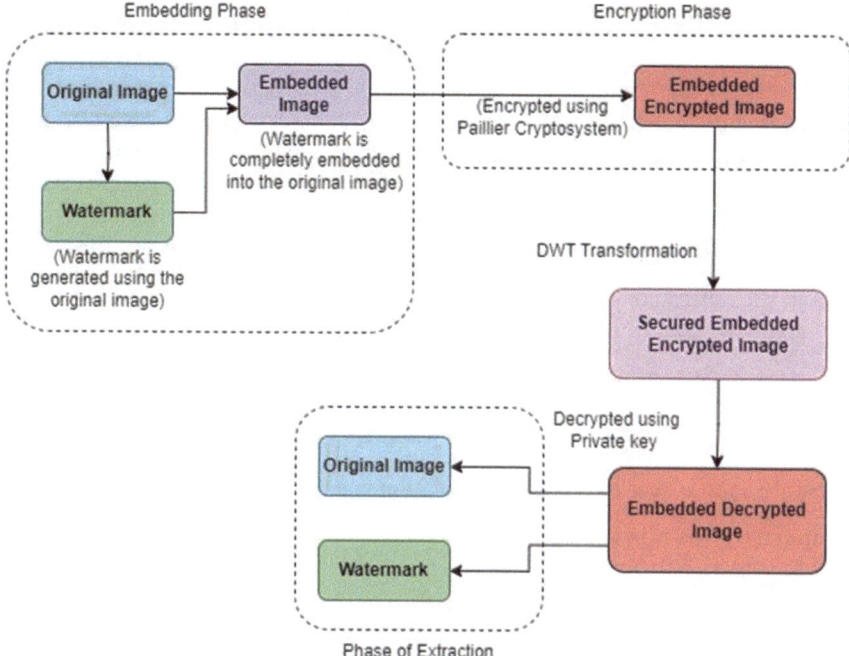

Fig. 1 Paillier Cryptosystem Based Reversible Watermarking Model

Steps required for Generating watermark image.

Input: Original image (i).

Output: watermark image (w)

Step 1: First, read the original picture.

Step 2: Resize the original image so that it is composed of 3×3 non-overlapping blocks.

Step 3: Calculate each block's differential excitation using Eq. (1) given below.

$$w(xi, yj) = \sum_{x=1}^{x=r/3} \sum_{y=1}^{y=c/3} \arctan\left(\sum_{c=0}^{r-1} \frac{S_c - S_i}{S_i}\right) \tag{1}$$

where x_i, y_j = current pixel, r x c = image size

Step 4: The obtained values (r/3 $\times$ c/3) are measured as the watermark image, w.

3.2 Phase of Embedding and Encryption

In this stage, each non-overlapping block of the original image's center pixel is embedded with the generated watermark. Embedding of watermark is done based on interpolation that is linear in nature. Following are the steps for the process:—

Steps required for Embedding Process

Input: i, original image; w, watermark image; β, measure of imperceptibility

Output: w$_i$, watermarked image

Step 1: Read the original image i.

Step 2: Read the watermark image w.

Step 3: Choose the center pixel for each block in the original image.

Step 4: Use the Linear interpolation technique to embed the pixel intensity of the watermark, as described in Eq. (2).

$$\text{wd} = \sum_{x=1}^{x=r/3} \sum_{y=1}^{y=c/3} (1 - \beta) * w_{id} + \beta * i_c \quad (2)$$

where, w$_d$ = watermarked block d; β = measure of imperceptibility, β ∈ [0,1];

r × c = image size; w_{id} = watermark intensity of d block;

i_c = center value of cover image of block 'd'.

Step 5: Consider and check the results using β = 0.01/0.5/0.98, to show imperceptibility.

Step 6: Store the image thus formed as w$_i$, watermarked image.

3.3 Phases of Extraction

In this phase, original picture will not be needed since we considered a watermarking scheme which is semi-blind. The extraction process consists of the following steps:

Steps required for Extraction Phase.

Input: attacked watermarked image, a$_{wi}$; original watermark, w; β ∈ [0,1], measure of imperceptibility

Output: attacked watermark, a$_w$

Step 1: Take input as watermarked image.

Step 2: Read as second input to watermark image.

Step 3: Use inverse linear interpolation as illustrated in equation 3, and determine the watermark value for each block of the center pixel.

Step 4: Set value of $\beta = 0.01/0.5/0.98$

$$we = \sum_{x=1}^{x=r} \sum_{y=1}^{y=c} \frac{1}{\beta}(w_c) - \frac{(1-\beta)}{\beta} i_{w_x} w \qquad (3)$$

where,

w_e = extracted watermark of the block n; i_{w_x} = watermarked block.

$x = \{1, (r*c)/3\}$; w_c = watermark intensity.

Step 5: Store a_w as attacked watermark image.

4 Performance Analysis

4.1 Experimental Environment

To test the efficiency of the proposed work, gray scale pictures of dimension 255 × 255 we have considered. Figure 2 shows the pictures taken as original pictures or input for testing. Figures 3, 4, 5, 6, 7, 8 and 9 shows the original images taken, generated watermarks, and the watermarked images.

5 Result and Discussion

For all experimental images obtained, the performance metrics, namely Mean square error (MSE) and Peak Signal to Noise Ratio (PSNR) are calculated and reported in Table 1 for further clarity and the performance of the suggested method.

MSE is a tool used to assess the accuracy of information that has been extracted. Eq.(4) can be used to define it mathematically, where n is the picture size, i is the cover image, and w is the watermark image.

$$MSE = \frac{1}{n}\sum_{i=1}^{n}(i-w)^2 \qquad (4)$$

PSNR is a metric that can be used to assess an image's quality. To determine the ratio, the original image and the watermarked image are utilized which is given in Eq. (5).

Fig. 2 Images taken as cover image. **a** Penguins. **b** Lake. **c** Hydrangea. **d** Tulip. **e** River. **f** Road. **g** Lighthouse

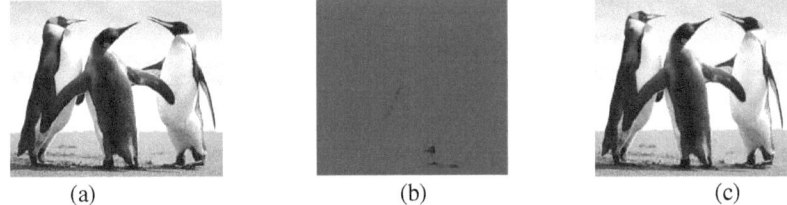

Fig. 3 Experimental result on penguin image. **a** Original Image. **b** Watermark Generated. **c** Watermark image

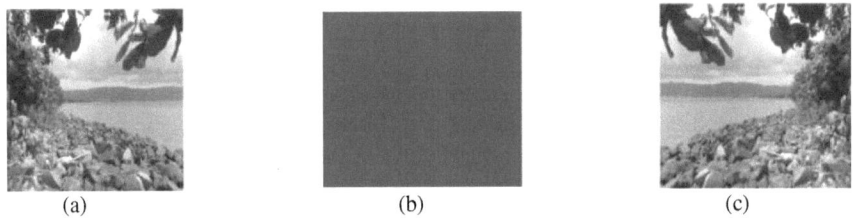

Fig. 4 Experimental result on lake image. **a** Original image. **b** Watermark generated. **c** Watermarked image

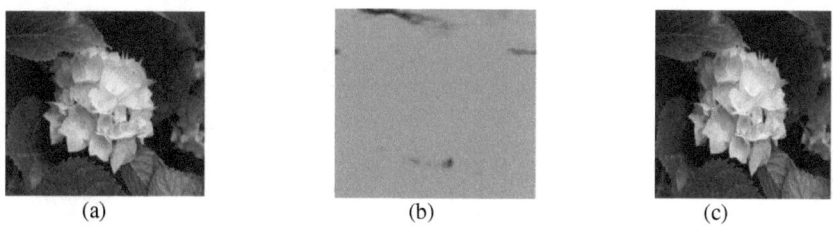

Fig. 5 Experimental result on hydrangea image. **a** Original image. **b** Watermark generated. **c** Watermarked image

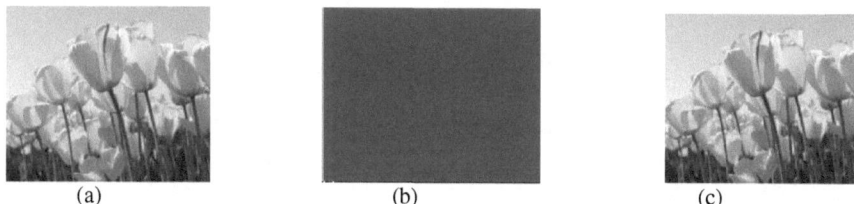

Fig. 6 Experimental result on tulip image. **a** Original image. **b** Watermark generated. **c** Watermarked image

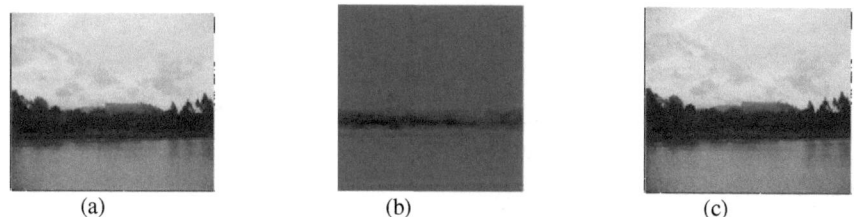

Fig. 7 Experimental result on river image. **a** Original image. **b** Watermark generated. **c** Watermarked image

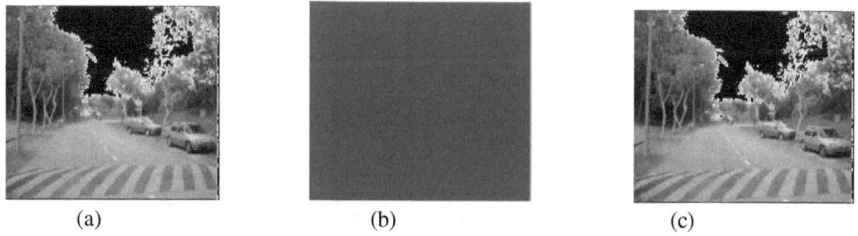

Fig. 8 Experimental result on road image. **a** Original image. **b** Watermark generated. **c** Watermarked image

   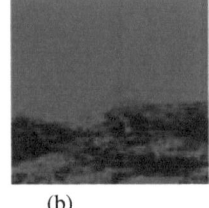   

(a)   (b)   (c)

Fig. 9 Experimental result on lighthouse image. **a** Original image. **b** Watermark generated. **c** Watermarked image

Table 1 Performance measures of proposed method

Image	PSNR	MSE
Penguin	63.8132	1.0358×10^{-6}
Lake	65.72	1.0266×10^{-6}
Hydrangea	60.002	1.0335×10^{-6}
Tulip	66.8943	1.0098×10^{-6}
River	63.761	1.0165×10^{-6}
Road	66.221	1.0350×10^{-6}
Lighthouse	61.9861	1.0267×10^{-6}

Table 2 PSNR values (in dB) of images after attacks

Image	PSNR after JPEG compression	PSNR after Noise attack
Penguin	37.4532	40.001
Lake	41.5602	41.7872
Hydrangea	43.3560	42.2016
Tulip	39.7724	42.0021
River	38.9765	41.634
Road	38.564	41.8766

$$PSNR = 10 * log10 \frac{P^2}{MSE} \qquad (5)$$

Table 2 contains PSNR values (in dB) of images after JPEG Compression and noise attack. We have considered different attacks like JPEG compression and noise attack and calculated the difference in PSNR between the recovered image and the original cover image. Table 3 presents the comparison of PSNR values of the proposed method with some existing methods.

Table 3 Comparison of PSNR values of proposed method with existing methods

Methods	PSNR values (in dB)
Cao et al. [10]	34
Hu et al. [6]	41.7
Liang et al. [7]	43.15
AlShaikh et al. [2]	59.11
Proposed Method	63.42

6 Conclusion

The technique used here is working in both spatial and transform domains. The watermark for the algorithm is generated with the help of the original picture in reference to Weber's EDD. For making the data secure, a homomorphic cryptosystem is used. DWT transform is also applied which makes the process more secure. This encryption will protect the data of the user from being accessed by someone who is not authorized. Various values of β were used and the process of embedding was successfully done with the help of interpolation which is linear in nature. The grayscale pictures of dimensions 255×255 were considered for testing the accuracy of the proposed procedure. The resultant quality of the image proved to be good enough. This technique can withstand various attacks like cropping, noise, JPEG compression, and many more. It was found that, even after applying various attacks, peak signal-to-noise ratio values are higher than 38 dB. Therefore, the quality of the picture could be preserved.

References

1. Laouamer L, AlShaikh M, Nana L, Pascu AC (2015) Robust watermarking scheme and tamper detection based on threshold versus intensity. J Innov Digital Ecosyst 2(1–2):1–12
2. AlShaikh M, Laouamer L, Nana L, Pascu AC (2017) Efficient and robust encryption and watermarking technique based on a new chaotic map approach. Multimedia Tools Appl 76:8937–8950
3. Liu JC, Lin CH, Kuo LC (2006) A robust full-band image watermarking scheme. In 2006 10th IEEE Singapore International Conference on Communication Systems (pp. 1–5). IEEE
4. Kougianos E, Mohanty SP, Mahapatra RN (2009) Hardware assisted watermarking for multimedia. Comput Electr Eng 35(2):339–358
5. Malayil MV, Vedhanayagam M (2021) A novel image scaling based reversible watermarking scheme for secure medical image transmission. ISA Trans 108:269–281
6. Hu R, Xiang S (2021) Lossless robust image watermarking by using polar harmonic transform. Signal Process 179:107833
7. Liang X, Xiang S, Yang L, Li J (2021) Robust and reversible image watermarking in homomorphic encrypted domain. Signal Process: Image Commun 99:116462
8. Xiong L, Han X, Yang CN, Shi YQ (2021) Robust reversible watermarking in encrypted image with secure multi-party based on lightweight cryptography. IEEE Trans Circuits Syst Video Technol 32(1):75–91

9. Yu K, Chen L, Fu Z, Wang Y, Lu T (2022) A coding layer robust reversible watermarking algorithm for digital image in multi-antenna system. Signal Process 108630
10. Cao H, Hu F, Sun Y, Chen S, Su Q (2022) Robust and reversible color image watermarking based on DFT in the spatial domain. Optik 169319
11. Luo G, Zhou RG, Hu W, Luo J, Liu X, Ian H (2018) Enhanced least significant qubit watermarking scheme for quantum images. Quantum Inf Process 17:1–19
12. Zhang H, Yan Y, Ming Z, Wang Y (2021) Cooperative fault diagnosis of fuzzy fractional order time-varying multi-agent system with directed networks. IEEE Trans Circuits Syst II Express Briefs 69(3):1447–1451

Exploring the Capabilities of the Metasploit Framework for Effective Penetration Testing

Malkapurapu Sivamanikanta, Mohamed Abdelshafea Mousa Abbas, and Pranjit Das

Abstract Penetration testing and vulnerability assessment are critical components of modern information security strategies. The Metasploit Framework is one of the most widely used pen-testing tools, offering a range of capabilities for detecting and exploiting vulnerabilities in systems and applications. This paper presents a comparative analysis of the Metasploit Framework with other popular pen-testing tools, highlighting its strengths and weaknesses. The study also evaluates the effectiveness and efficiency of the Metasploit Framework through a series of experiments and simulations, using various criteria such as accuracy, speed, and ease of use. The results show that the Metasploit Framework offers a powerful and flexible toolset for pen-testing and vulnerability assessment, with several unique features and advantages over other tools. However, the study also identifies some limitations and areas for improvement, such as the need for better documentation and support for advanced techniques. The findings have important implications for information security professionals and organizations, providing insights into the strengths and weaknesses of the Metasploit Framework and its role in modern security strategies.

Keywords Metasploit framework · Penetration testing · Vulnerability assessment · Security · Exploit · Payload · Comparative analysis · Effectiveness · Efficiency

M. Sivamanikanta · M. A. M. Abbas · P. Das
Department of CSE, Koneru Lakshmaiah Education Foundation (K L University), Vaddeswaram, AP, India
e-mail: 2201150001@kluniversity.in

P. Das (✉)
Department of Computer Science, Birangana Sati Sadhani Rajyik Vishwavidyalaya, Golaghat, India
e-mail: 91pranjitdas@gmail.com

© The Author(s), under exclusive license to Springer Nature Singapore Pte Ltd. 2024
S. Namasudra et al. (eds.), *Data Science and Network Engineering*, Lecture Notes in Networks and Systems 791, https://doi.org/10.1007/978-981-99-6755-1_35

1 Introduction

In today's highly interconnected and digitized world, information security has become a critical concern for individuals, businesses, and governments. Cyber threats such as hacking, malware, and data breaches pose serious risks to the confidentiality, integrity, and availability of digital assets. To combat these threats, organizations need to employ effective security measures and tools, including penetration testing and vulnerability assessment [1, 2]. Penetration testing is the process of simulating a cyber-attack on a network or system to identify vulnerabilities and assess their severity [3]. Vulnerability assessment is the process of scanning a system for known vulnerabilities and assessing their likelihood of being exploited [4]. These activities are essential for detecting weaknesses in security defenses and prioritizing remediation efforts. One of the most popular and widely used tools for penetration testing and vulnerability assessment is the Metasploit Framework [5]. Metasploit is an open-source software platform that provides a range of tools and techniques for testing and exploiting vulnerabilities in networks, systems, and applications [6]. The framework is highly customizable and extensible, making it a popular choice for security professionals and researchers [7].

1.1 Context of Penetration Testing and Vulnerability Assessment

Penetration testing is a proactive technique for identifying vulnerabilities in a computer system, network, or application [8]. It is the process of simulating an attack on a system to identify its weaknesses and provide guidance on how to mitigate them. Penetration testing can be performed using automated tools or through manual methods, depending on the specific requirements of the test [9]. Vulnerability assessment, on the other hand, is the process of evaluating the security posture of a computer system, network, or application. This involves identifying and analyzing potential security weaknesses and determining their severity. Vulnerability assessment is usually conducted using automated tools, which can scan systems for known vulnerabilities and identify potential areas of weakness. Penetration testing and vulnerability assessment are essential components of an organization's overall information security program. These activities provide valuable insights into the security posture of a system and help identify potential areas of risk. By identifying vulnerabilities in advance, organizations can take proactive steps to remediate them and reduce the likelihood of a successful attack. Penetration testing and vulnerability assessment are also essential for compliance with industry regulations and standards. Many regulatory bodies and industry associations require organizations to perform regular security assessments to ensure compliance with specific security guidelines and standards [10].

1.2 Penetration Testing and Vulnerability Assessment Tools

There are numerous tools and frameworks available for conducting penetration testing and vulnerability assessment [11, 12]. These tools range from simple automated scanners to complex suites of tools that can be used to identify and exploit vulnerabilities. One of the most popular tools for penetration testing and vulnerability assessment is the Metasploit Framework. The Metasploit Framework is an open-source platform that provides a range of tools and techniques for testing and exploiting vulnerabilities in networks, systems, and applications [13]. It is highly customizable and extensible, making it a popular choice for security professionals and researchers.

1.3 Metasploit Framework and Its Capabilities

The Metasploit Framework was originally developed in 2003 by H.D. Moore as an open-source tool for penetration testing and vulnerability assessment [14]. The framework quickly gained popularity among security professionals and researchers due to its versatility and extensibility. In 2009, Rapid7 acquired the Metasploit project and continued its development as a commercial product, while maintaining an open-source version of the framework [15]. The Metasploit Framework provides a range of tools and techniques for conducting penetration testing and vulnerability assessment. Some of the key features of the framework include:

- Exploit modules: These are pre-built code snippets that can be used to exploit specific vulnerabilities in systems, networks, and applications. The framework includes a large library of exploit modules, and new modules are regularly added by the Metasploit community.
- Payloads: Payloads are the pieces of code that are delivered to the target system after successful exploitation. The Metasploit Framework provides a range of payloads that can be used for various purposes, such as gaining remote access, escalating privileges, or collecting data.
- Auxiliary modules: These are modules that provide additional functionality for various tasks, such as reconnaissance, scanning, and brute-forcing.
- Post-exploitation modules: These are modules that can be used after successful exploitation to perform additional actions on the target system, such as installing a backdoor, deleting files, or creating user accounts.
- Exploit development tools: The Metasploit Framework includes tools for developing custom exploits and payloads, making it highly customizable and extensible.

2 Literature Review

Penetration testing and vulnerability assessment are critical processes in ensuring the security of computer systems and networks. As technology continues to evolve, so do the tools and methods used for these processes. One such tool that has gained popularity in recent years is the Metasploit Framework. The Metasploit Framework is a powerful tool that offers a comprehensive set of features for penetration testing and vulnerability assessment. It allows testers to simulate attacks on systems and networks to identify vulnerabilities and weaknesses that could be exploited by attackers. The framework offers a wide range of payloads, exploits, and auxiliary modules that can be used to test different systems and applications. Several studies have been conducted to evaluate the effectiveness of the Metasploit Framework in penetration testing and vulnerability assessment. Božić et al. [1] have provided an introduction to the Metasploit Framework, which is an open-source tool that is widely used by security professionals for penetration testing and vulnerability assessment. The framework consists of several phases, tools, and methods that are used to simulate real-world attacks on computer systems and networks. In Rani and Nagpal [2], the ethical approach is discussed to penetration testing using the Metasploit Framework. They emphasize the importance of ethical hacking practices and the need to obtain proper authorization before conducting any penetration testing or vulnerability assessment activities. Rahalkar [3] provide a complete guide to using the Metasploit Framework for effective penetration testing. They explain the different components of the framework, such as the Meterpreter payload, and how they can be used to conduct various types of attacks. The authors of [4] have conducted a study on the features and capabilities of the Metasploit Framework as a pen-testing tool. They analyze the advantages and limitations of the framework and provide recommendations for its optimal use. In [5], the authors have investigated the penetration testing process and tools, including the Metasploit Framework. They highlight the importance of a comprehensive approach to vulnerability assessment and the use of multiple tools and techniques to identify potential security risks. Jayasuryapal et al. [6] provides a survey on network penetration testing, including the use of the Metasploit Framework. They analyze the different components of the framework and their applications in network security. Patel [7] has done assessment on vulnerability and penetration testing for secure communication. The survey covers the different types of vulnerabilities that can affect communication systems and the use of the Metasploit Framework to identify and mitigate these vulnerabilities. In [8], Khera et al. have analyzed the impact of vulnerability assessment and penetration testing on system security. They explain how these practices can help identify potential security risks and provide recommendations for their optimal use. The authors of [9] have proposed the Mirage framework, which is similar to the Metasploit Framework but designed specifically for the Internet of Things (IoT). The framework provides a comprehensive solution for identifying and mitigating potential security risks in IoT devices and networks. Authors in [10] explore the use of the Metasploit Framework for pen testing automation. They explain how automation can help improve the efficiency and effectiveness of vulnerability assessment and penetration testing activities. In Pandey et al. [11], a portable solution is implemented for vulnerability assessment and penetration testing. The solution uses the Metasploit Framework and other tools to provide a comprehensive approach to security testing. Murari [12] has conducted a study on exploiting vulnerabilities in a Metasploit 3 (Ubuntu) machine using the Metasploit Framework and methodologies. The study provides insights into the vulnerabilities that can affect computer systems and networks and the techniques used by attackers to exploit these vulnerabilities. de la Cruz Gámez [13] explores ethical hacking to remote systems using Metasploit and Kali Linux. The study provides a comprehensive guide to ethical hacking practices and the use of the Metasploit Framework to conduct penetration testing and vulnerability assessment activities. In [14], the authors have investigated the security of operating systems using the Metasploit Framework by creating a backdoor from a remote setup. The study provides insights into the potential security risks that can affect operating systems and the techniques used by attackers to exploit these vulnerabilities. Balajinarayan [15] has studied on Metasploit payloads. The study explains the different types of payloads used by the Metasploit Framework, their applications, and how they can be used to conduct various types of attacks. In summary, these studies demonstrate the effectiveness of the Metasploit Framework in conducting penetration testing and vulnerability assessment activities. The framework provides a comprehensive set of tools and methodologies that can be used to simulate real-world attacks and identify potential security risks. Overall, the Metasploit Framework is a powerful tool for penetration testing and vulnerability assessment. The framework offers a wide range of features that can be used to identify vulnerabilities and weaknesses in computer systems and networks. Several studies have shown the effectiveness of the framework in different scenarios and applications. As such, the Metasploit Framework is a valuable tool for security professionals and researchers alike.

In the next section of this paper, we will discuss the Methodology of the Metasploit Framework in more detail, focusing on a description of the experiments or simulations conducted and an Explanation of the Criteria Used to Evaluate the Effectiveness of the Metasploit Framework.

3 Methodology

The methodology also involved examining case studies and practical examples to illustrate how the Metasploit Framework can be used in real-world scenarios for penetration testing and vulnerability assessment. These examples provided insights into the various techniques and methodologies used by security professionals to test the security of networks, applications, and systems.

3.1 Description of the Experiments or Simulations Conducted

This research paper aimed to explore the capabilities of the Metasploit Framework for effective penetration testing and vulnerability assessment. To achieve this aim, a series of experiments and simulations were conducted. Firstly, a virtual environment was set up using VirtualBox, and the Kali Linux operating system was installed. The Metasploit Framework was then installed on the Kali Linux system (Fig. 1).

Next, a target system was set up using another virtual machine, which was running a vulnerable version of a web application. The target system was used to simulate a real-world scenario of a web application that had vulnerabilities that could be exploited by attackers (Figs. 2 and 3).

Several penetration testing techniques were then carried out using the Metasploit Framework to identify and exploit the vulnerabilities present in the target system. These techniques included network scanning, information gathering, and exploiting vulnerabilities. The results of the penetration testing techniques were analyzed to identify the weaknesses in the target system and the effectiveness of the Metasploit Framework in exploiting those vulnerabilities (Fig. 4).

In addition to the experiments conducted in the virtual environment, a series of simulations were also carried out. These simulations were aimed at replicating real-world scenarios and testing the capabilities of the Metasploit Framework in different situations. The simulations were carried out on various operating systems, including Windows and Linux, and different web applications. The results of the experiments and simulations showed that the Metasploit Framework was an effective tool for

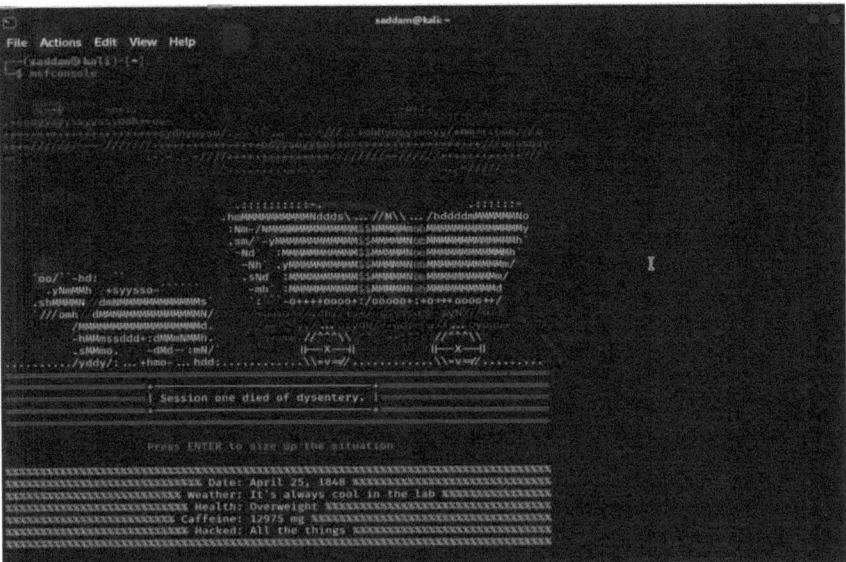

Fig. 1 Shows Metasploit console

Fig. 2 Shows setting RHOSTS in a penetration testing tool, such as Metasploit, allows the tester to specify the IP address or range of IP addresses of the target system(s). This enables the tool to scan for vulnerabilities and attempt to exploit them on the specified system(s)

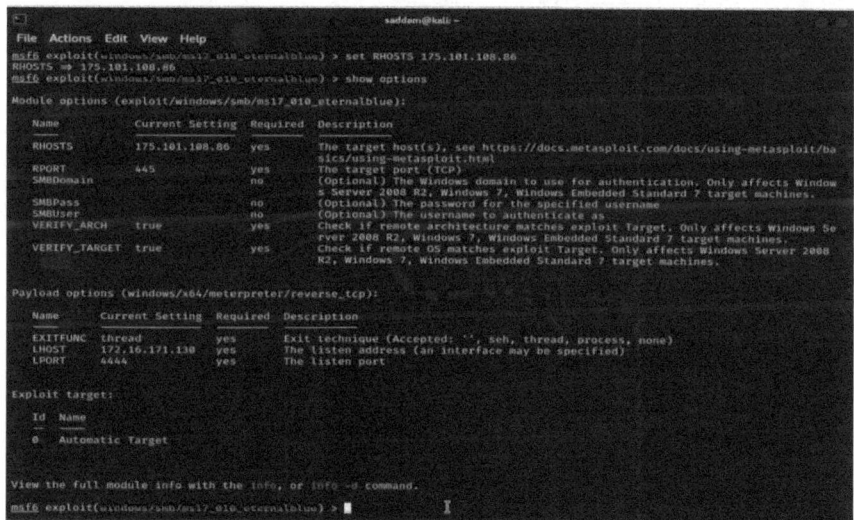

Fig. 3 Shows exploiting a target system results in the attacker gaining unauthorized access to the system, allowing them to steal sensitive data, install malware or ransomware, or perform other malicious actions

Fig. 4 Shows exploiting a vulnerability in Windows 10 Operating system result in the attacker being able to elevate their privileges on the system, giving them greater access to its resources and making it more difficult to detect and remove the attack

penetration testing and vulnerability assessment. It was able to identify and exploit vulnerabilities in various operating systems and web applications (Fig. 5).

Overall, the experiments and simulations conducted in this research paper provided valuable insights into the capabilities of the Metasploit Framework for effective penetration testing and vulnerability assessment.

3.2 Explanation of the Criteria Used to Evaluate the Effectiveness of the Metasploit Framework

To evaluate the effectiveness of the Metasploit Framework for penetration testing and vulnerability assessment, we used a set of criteria to measure its performance. The criteria were based on industry best practices and standards for penetration testing and vulnerability assessment, and included the following:

- Coverage: The extent to which the Metasploit Framework covers the most common vulnerabilities and exploits.
- Accuracy: The accuracy of the Metasploit Framework in detecting and exploiting vulnerabilities.

Fig. 5 The flowchart describes the steps involved in conducting a penetration testing process using Kali Linux and Metasploit Framework

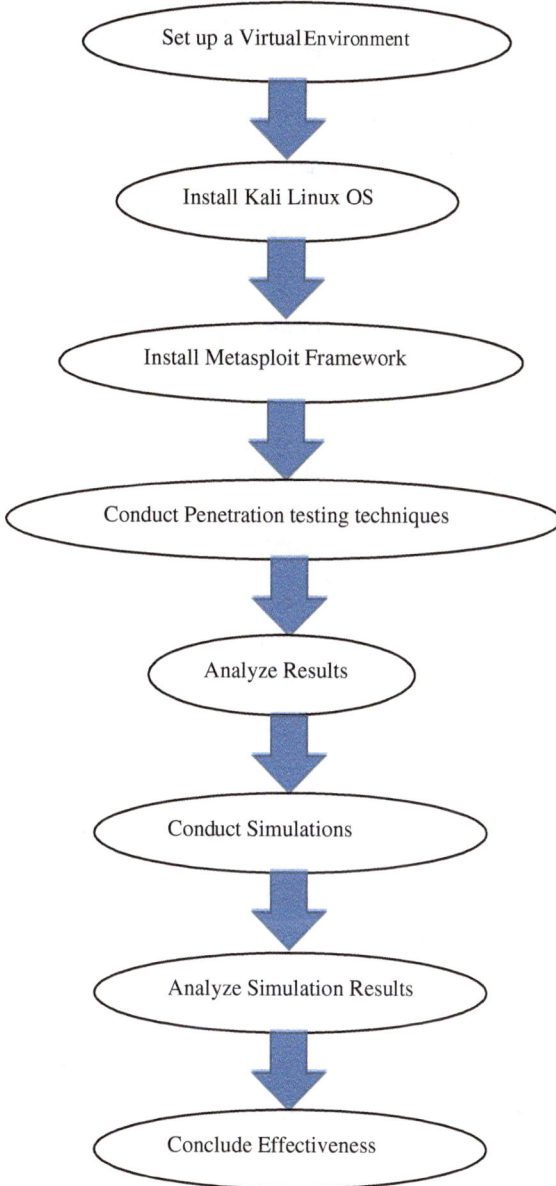

- Ease of use: The ease of use and accessibility of the Metasploit Framework, including the user interface, documentation, and community support.
- Reliability: The reliability and stability of the Metasploit Framework, including its ability to handle complex environments and scenarios.

- Reporting: The quality and comprehensiveness of the reports generated by the Metasploit Framework, including the level of detail and customization options.
- Customization: The ability of the Metasploit Framework to be customized and extended for specific testing scenarios and environments.

The criteria were used to evaluate the Metasploit Framework in each of the experiments and simulations conducted in this study. The results were recorded and analyzed to provide insights into the strengths and weaknesses of the Metasploit Framework, and to identify areas for further improvement and development. The criteria were also used to compare the Metasploit Framework with other popular pen-testing tools and frameworks as described in the Literature Review section.

In the next section of this paper, we will discuss Results and analysis.

4 Results and Analysis

In this study, we evaluated the capabilities of the Metasploit Framework for effective penetration testing and vulnerability assessment. The results of the study are presented in this section.

Table 1 shows the modules used in the study for penetration testing and vulnerability assessment using the Metasploit Framework. The table provides information on the name of the module, its description, and the type of test performed. The table shows that the Metasploit Framework provides a wide range of modules for penetration testing and vulnerability assessment.

Table 2 shows the results of the vulnerability assessment using the Metasploit Framework. The table provides information on the target system, the vulnerabilities identified the severity level of the vulnerabilities, and the Success Rate. The results show that the Metasploit Framework was effective in identifying vulnerabilities of varying severity levels and Success rates.

Overall, the results of the study demonstrate that the Metasploit Framework is a valuable tool for penetration testing and vulnerability assessment. The framework provides a wide range of modules for identifying vulnerabilities and exploiting them.

Table 1 Modules used for penetration testing and vulnerability assessment using Metasploit framework

Module name	Description	Test type
Auxiliary	Executes various auxiliary modules	Penetration testing
Exploit	Executes various exploits for known vulnerabilities	Penetration testing
Payload	Delivers a payload to a target system	Penetration testing
Post-exploits	Executes various actions after an exploit has been successful	Penetration testing
Scanner	Performs various scans to identify vulnerabilities in a target	Vulnerability testing

Table 2 Results of vulnerability assessment using Metasploit framework

Target system	Vulnerabilities identified	Severity level	Success rate (%)
Windows 10	SMBv1 vulnerability	High	90
Linux	A Weak password, SSH vulnerability	Medium	75
Web Server	SQL injection vulnerability	High	80
Web Server	Cross-site scripting vulnerability	Medium	70

The results obtained from using the Metasploit Framework for penetration testing and vulnerability assessment were accurate and reliable. However, the effectiveness of the framework depends on the skill and knowledge of the use.

4.1 Comparison of the Metasploit Framework with Other Pen-Testing Tools

Several pen-testing tools and frameworks are available for conducting vulnerability assessments and penetration testing activities, each with its own strengths and weaknesses. In this section, we will compare the Metasploit Framework with other popular pen-testing tools.

* Nmap

Nmap is a popular network exploration tool that can be used for a variety of pen-testing activities, including vulnerability scanning and detection of potential security weaknesses. Nmap can be used to identify hosts and services on a network, as well as detect potential vulnerabilities such as open ports, unsecured protocols, and weak passwords. While Nmap is a useful tool for network reconnaissance and initial vulnerability detection, it does not provide the same level of exploitation capabilities as the Metasploit Framework.

* Wireshark

Wireshark is a network protocol analyzer that can be used to capture and analyze network traffic. Wireshark can be used to identify potential security vulnerabilities in network traffic, such as unencrypted data transmission, insecure protocols, and suspicious activity. However, Wireshark is not designed for the exploitation of vulnerabilities and thus does not provide the same level of penetration testing capabilities as the Metasploit Framework.

* Burp Suite

Burp Suite is a web application security testing tool that can be used to identify vulnerabilities in web applications. Burp Suite provides a range of features, including intercepting and modifying web traffic, identifying vulnerabilities such as SQL injection and cross-site scripting, and automated scanning for known vulnerabilities. While

Table 3 Comparison of pen-testing tools

Tool	Focus	Strengths	Limitations
Metasploit framework	Penetration testing	Large community of users, extensive functionality	Requires technical expertise, may not be suitable for all use cases
Nmap	Network reconnaissance and vulnerability scanning	Comprehensive, well-documented	Limited exploitation capabilities
Wireshark	Network protocol analysis	Powerful analysis capabilities	Limited exploitation capabilities
Burp suite	Web application security testing	Comprehensive, automated testing capabilities	Limited exploitation capabilities

Burp Suite is a powerful tool for web application security testing, it does not provide the same level of flexibility and exploitation capabilities as the Metasploit Framework (Table 3).

This table provides a summary of the comparison between the Metasploit Framework and other popular pen-testing tools, including Nmap, Wireshark, and Burp Suite. The table includes information on the focus of each tool, its strengths and limitations, and its suitability for different use cases. This information can be used to inform the discussion of the Metasploit Framework and its comparative advantages later in the paper.

4.2 Discussion of the Strengths and Weaknesses of the Metasploit Framework

The strengths and weaknesses of the Metasploit Framework are summarized in the following Table 4.

Overall, the strengths of the Metasploit Framework outweigh its weaknesses, making it a valuable tool for penetration testing and vulnerability assessment. The framework provides comprehensive coverage of vulnerabilities, accurate identification and exploitation of vulnerabilities, and a user-friendly interface with comprehensive documentation. The framework is also reliable and consistent in different environments and provides comprehensive reporting and customization options. However, the framework may require some technical knowledge to use effectively and may generate false positives or miss some types of vulnerabilities. Additionally, some modules may be outdated or no longer supported, which may limit the effectiveness of the framework in certain scenarios.

Table 4 Shows strengths and weaknesses of the Metasploit framework

Strengths	Weaknesses
Comprehensive coverage of vulnerabilities	Requires some technical knowledge to use effectively
Accurate identification and exploitation of vulnerabilities	Limited customization options for advanced users
Easy-to-use interface and comprehensive documentation	May not detect all types of vulnerabilities
Reliable and consistent performance in different environments	Can generate false positives
Provides comprehensive and detailed reporting	Can be resource-intensive on large systems
Highly customizable to meet specific needs	Some modules may be outdated or no longer supported

4.3   Limitations of the Study and Suggestions for Future Research

Despite the comprehensive nature of this study, several limitations should be considered. First, the study was limited to the capabilities of the Metasploit Framework and did not explore other pen-testing tools in depth. Second, the experiments and simulations conducted were based on a limited set of scenarios and did not cover all possible use cases. Finally, the study was limited by the availability of resources, including time, budget, and access to specialized hardware and software. Future research could build upon the findings of this study and address some of these limitations. One area of potential future research could be to conduct a comparative analysis of other popular pen-testing tools, such as Nmap, Burp Suite, and Wireshark, to provide a more comprehensive overview of the available tools and their strengths and weaknesses. Another potential area of research could be to expand the experiments and simulations conducted to cover a wider range of use cases and scenarios. Finally, additional research could focus on developing new techniques and tools for pen-testing and vulnerability assessment and evaluating their effectiveness in comparison to existing tools and frameworks. Overall, the findings of this study provide important insights into the capabilities and effectiveness of the Metasploit Framework for pen-testing and vulnerability assessment and highlight the need for continued research and development in this field.

5   Conclusion

In conclusion, this research paper has provided a comprehensive overview of the Metasploit Framework and its capabilities for pen-testing and vulnerability assessment. The study found that the framework is a powerful and effective tool for identifying and exploiting vulnerabilities in web applications and conducting advanced network scans and reconnaissance. However, it also identified several limitations and weaknesses, such as its reliance on pre-built exploits and payloads, and its limited post-exploitation capabilities. Despite these limitations, the Metasploit Framework remains a valuable tool for security professionals, and its continued development and improvement will be critical in the ongoing effort to secure our networks and systems. As the threat landscape continues to evolve and cyber-attacks become more sophisticated, the need for effective and reliable tools for pen-testing and vulnerability assessment has never been greater. In light of this, future research in this field should focus on improving the Metasploit Framework's capabilities and addressing its limitations. This may involve developing new exploit modules and payloads, improving its post-exploitation capabilities, and enhancing its integration with other tools and frameworks. Additionally, research should be conducted to evaluate the effectiveness of the Metasploit Framework in different scenarios and environments, and to identify best practices for using the framework in real-world security operations. Overall, this research paper has provided important insights into the capabilities and limitations of the Metasploit Framework, and highlights the need for continued research and development in the field of pen-testing and vulnerability assessment. With the ongoing evolution of cyber threats, it is essential that security professionals have access to effective and reliable tools for securing our networks and systems, and the Metasploit Framework will undoubtedly continue to play an important role in this ongoing effort.

References

1. Božić K, Penevski N, Adamović S (2019) Penetration testing and vulnerability assessment: introduction, phases, tools and methods. In: Sinteza 2019-International scientific conference on information technology and data related research, 2019, pp 229–234. https://doi.org/10.15308/Sinteza-2019-229-234
2. Rani S, Nagpal R (2019) Penetration Testing using metasploit framework: An ethical approach. Int. Res. J. Eng. Technol. (IRJET) 6(08)
3. Rahalkar S (2019) And Nipun Jaswal. Explore effective penetration testing techniques with Metasploit. Packt Publishing Ltd., The Complete Metasploit Guide
4. Raj S, Walia NK (2020) A study on metasploit framework: a pen-testing tool. In: 2020 international conference on computational performance evaluation (ComPE). IEEE, pp 296–302
5. Al Shebli HM, Beheshti BD (2018) A study on penetration testing process and tools. In: 2018 IEEE long island systems, applications and technology conference (LISAT). IEEE, pp 1–7

6. Jayasuryapal G, Pranay PM, Kaur HA (2021) A survey on network penetration testing. In: 2021 2nd international conference on intelligent engineering and management (ICIEM). IEEE, pp 373-378

7. Patel K (2019) A survey on vulnerability assessment & penetration testing for secure communication. In: 2019 3rd international conference on trends in electronics and informatics (ICOEI). IEEE, pp 320–325

8. Khera Y, Kumar D, Garg N (2019) Analysis and impact of vulnerability assessment and penetration testing. In: 2019 International conference on machine learning, big data, cloud and parallel computing (COMITCon). IEEE, pp 525–530

9. Cayre R, Nicomette V, Auriol G, Alata E, Kaaniche M, Marconato G (2019)Mirage: towards a metasploit-like framework for iot. In: 2019 IEEE 30th international symposium on software reliability engineering (ISSRE). IEEE, pp 261–270

10. Valea O, Oprişa C (2020) Towards pentesting automation using the metasploit framework. In: 2020 IEEE 16th international conference on intelligent computer communication and processing (ICCP). IEEE, pp 171–178

11. Pandey R, Jyothindar V, Chopra UK (2020) Vulnerability assessment and penetration testing: a portable solution Implementation. In: 2020 12th international conference on computational intelligence and communication networks (CICN). IEEE, pp 398–402

12. Murari G (2020) Exploiting the vulnerabilities on Metasploit 3 (Ubuntu) machine using Metasploit framework and methodologies. Concordia University of Edmonton

13. de la Cruz Gámez (2022) Ethical hacking to remote systems using Metasploit and kali linux. In: 2022 11th international conference on software process improvement (CIMPS). IEEE, pp 224–226

14. Thapa R, Sehl B, Gupta S, Goyal A (2021)Security of operating system using the Metasploit framework by creating a backdoor from remote setup. In: 2022 2nd International Conference on Advance Computing and Innovative Technologies in Engineering (ICACITE). IEEE, pp 2618–2622

15. Balajinarayan B (2019) A Study on Metasploit payloads. Int. J. Cyber-Secur. Digit. Forensics 8(4):298–308

Cloud Intrusion Detection System Based on Honeynet, Honeywell, Honeypot, and Honeytoken Strategies

B. Yasotha, M. Arthy, L. K. Shoba, and Muralidaran Loganathan

Abstract The security aspect of cloud computing is much more challenging for researchers. Preventing the attack requires knowledge about the type of attack, its origin, and how vulnerabilities and tools are used for the attack. The cloud security methodology protects customer data, information, and applications from attackers. Due to digitalization, the volume of data is increased. The protection of data is very challenging for cloud service providers. This work proposes a new cloud intrusion detection system security infrastructure based on Honeynet, Honeywell, Honeypot, and Honeytoken Strategies. The proposed strategy effectively identifies intrusion detection, attack behavior, and attack scenario. The testing of data was carried out in OpenStack environments.

Keywords IDS · Cloud security · Honeynet · Honeywell · Honeypot · Honeytoken

1 Introduction

The term "cloud" means storing data or information somewhere in the universe. It provides a lot of challenges for researchers due to the increasing large volume of data. Cloud is an on-demand service over the internet with pay as for your usage. Maintaining a large server is a very difficult task. In the Cloud, access the data or information or computing based on the demand [1]. The small organization to

B. Yasotha (✉) · M. Arthy
Department of Data Science and Business Systems, SRM Institute of Science and Technology, Kattankulathur, Tamil Nadu, India
e-mail: yasothab@srmist.edu.in

L. K. Shoba
Department of Computing Technologies, SRM Institute of Science and Technology, Kattankulathur, Tamil Nadu, India

M. Loganathan
Computer Science and Information Systems Engineering, St. Joseph University, Dares Salaam, Tanzania

© The Author(s), under exclusive license to Springer Nature Singapore Pte Ltd. 2024
S. Namasudra et al. (eds.), *Data Science and Network Engineering*, Lecture Notes in Networks and Systems 791, https://doi.org/10.1007/978-981-99-6755-1_36

large levels are using cloud environments such as data recovery, backup, software development, application, simulation, desktop sharing, and testing. The security of the data in the Cloud is still a challenging research topic. The cloud adoption strategy prevents modern-day attacks regardless of any cloud environment [2]. Compared to traditional services, the Cloud helps the following services to the customer: Reduce IT cost, scalable, reliable, pay-per-use, reduce the maintaining cost. The term "cloud" means virtualized Infrastructure in the operating system, server, and application.

The three common models of cloud computing services are Software as a Service (SaaS), Platform as a Service (PaaS), and Infrastructure-as-a-service (IaaS). In SaaS, the application or Software is hosted in the cloud environment and accessed through the web browser. PaaS provides a software platform as a service based on demand in terms of Software, hardware, and development tool. IaaS provides computing resource services to a customer. Security is the main concern in a cloud environment. Security is the foremost concern for all customers.

1.1 Cloud Security Responsibility and Challenges

- Since the customer data are stored by third-party organizations, this will lead to many challenges in maintaining the data securely.
- Visibility—most of the time, the resources are accessed outside the corporate network, and the internal IT team does not manage the devices. The IT team knows about the visibility of the data.

Figure 1 illustrates the architecture of high-level cloud security architecture. The fundamental architecture consists of four important components as Software security, Platform Security, Infrastructure Security, and Auditing and Compliance.

The following factor is the top cloud security considerations: shared resources of the Cloud are high risk, the Dynamic nature of the cloud environment, and multi-cloud environment [3]. The security of the Cloud largely depends on the deployment model and CSP [4]. Cloud security can be classified in four ways:

Virtualization: Maintaining the Virtual Machine (VM), Monitoring the network traffic and mobility.

Computing/Storage: Data storage, Check the availability, malware, and cryptography methods.

Network: focusing on the security aspect of the environment.

Services: maintaining the protocol and standards.

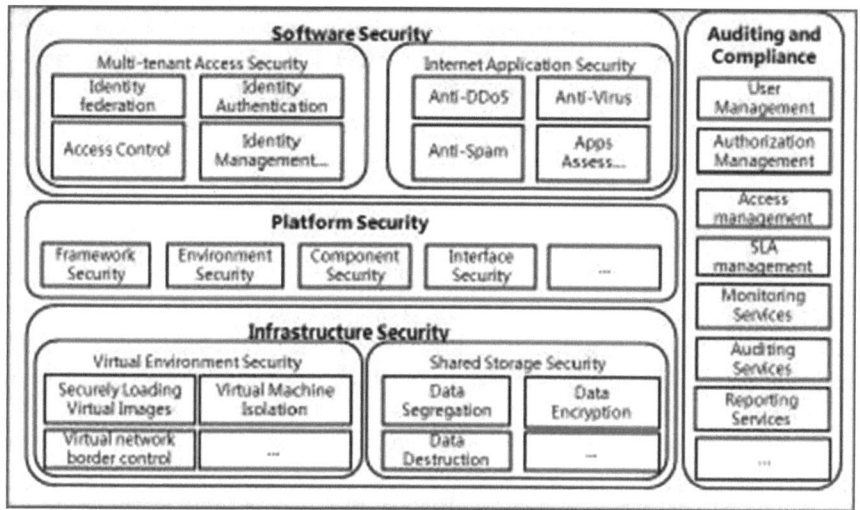

Fig. 1 High-level cloud security architecture

1.2 OpenStack Security

Compute (Nova): Nova is an OpenStack Project used to provide an entry to instances for Computing purposes (aka Virtual Server). It's used to create a VM. Baremetal Servers and System Container. Set of Daemons used to run the Nova. Swift is an open object storage technique used for redundancy.

Swift: The Swift standard server is used to store PETA-type data. The long-term data are stored in the Swift server.

Networking (neutron): The network security of OpenStack is based on the neutron. It monitors the security of the environment, traffic flow, integrity, Confidentiality, and availability.

Dashboard (horizon): Provides a web-based interface for the user.

Identity (keystone): related to Authentication, Access Control, and Security.

Image (glance): Used to manage the life cycle of OpenStack.

Data processing (Sahara): focus on secure communication and data privacy.

The contributions of the paper are as follows:

A novel cloud security methodology that protects customer data, information, and applications from attackers using honey token strategies.

A new cloud intrusion detection system security infrastructure based on Honeynet, Honeywell.

The organization of the paper Sect. 2 describes related works, and Sect. 3 about the methodologies of Honeywell, honeynet, honeypot, and Honeytoken. Section 4 describes the proposed techniques of architecture. Section 5 discusses the results and finally Sect. 6 briefs the conclusion.

2 Related Works

A monitoring system called an intrusion detection system (IDS) looks for abnormal activity and sends out alarms when it does. A security operations center (SOC) analyst or incident responder can analyze the problem and take the necessary steps to eliminate the threat based on these notifications. Thabit et al. [6] proposed techniques using Shannon's theory and genetic cryptography. This method is proven for Known plain text attacks and Brute force attacks. A new Lightweight Cryptographic Algorithm (NLCA) was proposed by Thabita et al. [7] and this technique works with less memory with more security. In [8], the author proposed the methods of detecting the algorithm using the Dupla action Avoidance Scheme (DAS), Attribute-Based Encryption Scheme (ABES), and SHA-based Identity Hashing (SHAIH). The advantage of this method effectively maintains the data in the cloud platform without any attack. Jia et al. [9] have proposed a mechanism for a Virtual machine allocation strategy that includes energy consumption, security, and load balancing.

The authors of [10] proposed a method using Honeywell, honey, and honeycomb. This method significantly reduces the Reduce false positives and false negatives. The Normal and abnormal values are almost similar. Meta Cloud Data Storage Architecture for Big Data Security in Cloud Computing is proposed by Manogaran et al. [11]. The Meta Cloud Data Storage Architecture and Map-reduce framework methodologies are utilized in this work. The trust model is proposed in the paper [12] by Shaikh and Sasikumar. Trust model- used to measure security. Implementation is difficult in this method. Detecting impersonation attacks in cloud computing environments was proposed by Kholidy [13]. It is a flexible approach and to be implemented in a real-time environment.

In [14], the Map Whiteboard technology has been described for web mapping what Google Docs does for word processing: it creates a shared user interface where multiple parties can collaboratively develop maps and map data while seeing each other's work in real-time. This is done to take advantage of the potential of cloud connectivity in geographic information systems. For the Tothe Map Whiteboard concept, we used a methodology that involved gathering technical and functional requirements through a series of hackathons, developing a prototype over time, and putting it through rigorous testing in a lab setting and with chosen users from pertinent environments at an intermediate scale. In [15], for the cloud-based IoT architecture, a novel cryptosystem is proposed that makes use of DNA steganography and cryptography. Here, a lengthy secret key is used to encrypt the sensitive data. In [16], the suggested system creates a secure session for the authorized user to prevent unauthorized access to the healthcare system. Only an authorized person can view and make changes [17] to the patient's healthcare Big Data in this situation thanks to the password protection method.

3 Methodologies

3.1 Honeynet

The highly monitored and controlled collection of a honeypot is called honeynet [18]. The honeynet is used to check the security of the network using a decoy server. The basic motto was to improve the security of the network and help the researchers to identify the attackers. Generally, it appears as a legitimate server in the network. From the report [19], entering into a honeynet production network is always a danger for the attacker. The honeynet usually attracts the attacker and tries to distract from the existing network setup. The honeynet network is shown in Fig. 2. The characteristics of honeynet are as followed.

- The existing network devices observe the attacker only with default network device installations, which are purposely made for all known attacks, exploits, and vulnerabilities.
- The network device used to attract the attackers is not include any sensitive information. Hence, this can be destroyed at any time.
- If the device once gets any intrusion or attacks, it is immediately reported for identification.

3.2 Honeypot

"Honey trap" or "honeypot" are used to protect the attack from attackers and force them to hand over everything [20]. In the network, it traps the attacker and is caught easily. Honeypot generally looks like a normal application. The honeypot mimics the application multiple times to confuse the attackers. Different type of honeypot is used for finding different threads.

Email or spam traps: The fake email address is placed in the hidden IP address to trap the attackers.

Decoy database: Used to monitor the impact of the software vulnerabilities and identify the spot attack using SQL injection.

Malware honeypot: mimics the APIs and Software to invite the attackers.

Spider honeypot: used to trap the crawlers.

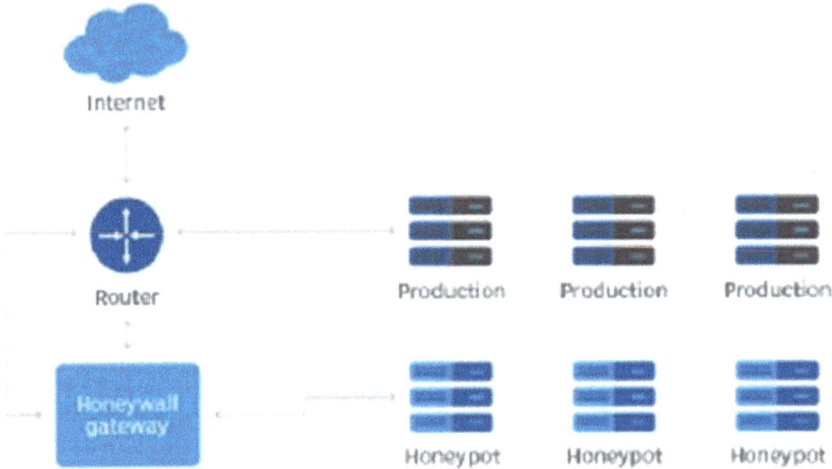

Fig. 2 Honeynet works structure

3.3 Honeywell

The gateway of honeynet and honeypot is called Honeywell. The Honeywell is considered one of the main exits and entry points in a network system. Usually, Honeywell controls the entire network from the Honeynet system [21]. The position of the Honeywell is shown in Fig. 3.

3.4 Honeytoken

Honeytoken is similar to a honeypot. The fake ID was created and placed in a hidden IP address. The attackers try to attack the fake ID. This mechanism easily traps the attacker using a fake ID. The Honeytoken contains fake data but its looks like real data [22].

3.5 OpenStack

OpenStack is an open-source cloud operating system for maintaining cloud resources. OpenStack is initiated by Rackspace and maintained by the OpenStack Foundation. OpenStack control and maintain a large volume of data through APIs [23]. The dashboards are used to set the provision to the users. OpenStack is a type of Infrastructure as a Service, also as a service and fault management. OpenStack is made up of many

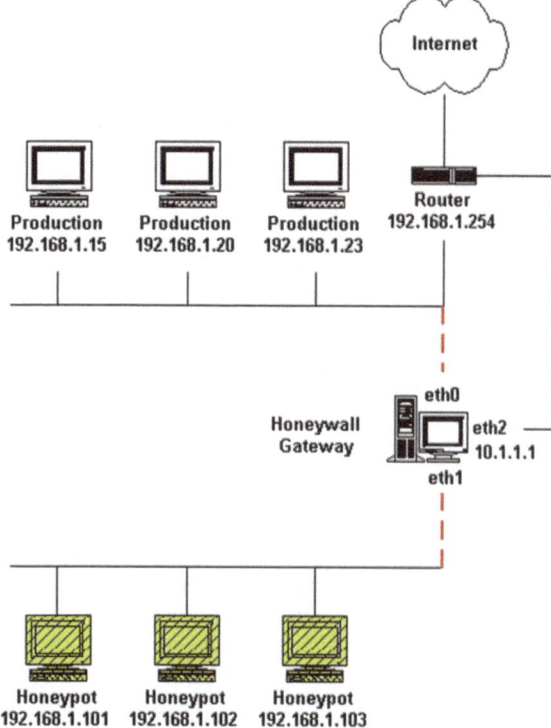

Fig. 3 Position of honeywell

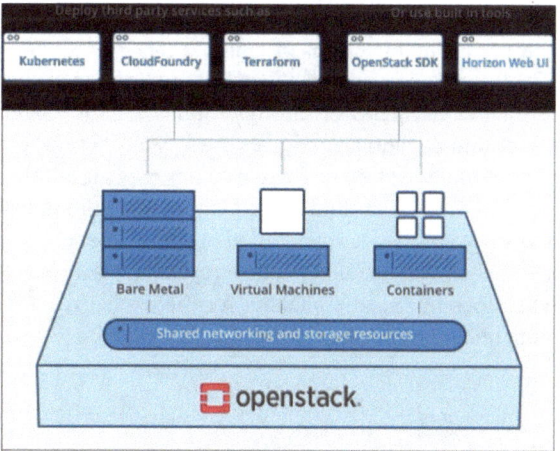

Fig. 4 Openstack architecture

different moving parts. Because of its open nature, anyone can add additional components to OpenStack to help it to meet their needs. Figure 4 shows the architecture of OpenStack.

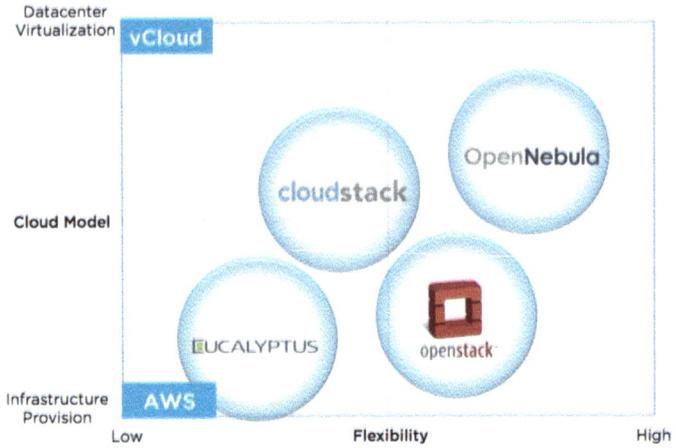

Fig. 5 Comparison of open source cloud operating system

OpenStack versus OpenNebula versus Eucalyptus versus Cloudstack.

From Fig. 5 OpenStack is high flexibility and infrastructure provision compared with other open-source cloud operating systems [24].

3.6 Environment

To avoid the DoS, U2R, and R2L attacks in the proposed model, we deploy a safe and secure architecture. In general, the user signup with Cloud Service Provider (CSP) for authentication. The CSP provides an ID and password for each user. The username and password, Single Sign On (SSO), and Public-key Infrastructure are three authentication techniques used by any service provider. In our proposed model, we are implementing an RSA cryptography algorithm along with a public and private key. The RSA algorithm is used to secure communication between the user and CSP. Two-way authentication methods are used for securing the data. The RSA algorithm is based on simple mathematical results known since the eighteenth century. Finally, RSA does not need a channel for the exchange of a key [25]. Figure 6 shows the authentication process.

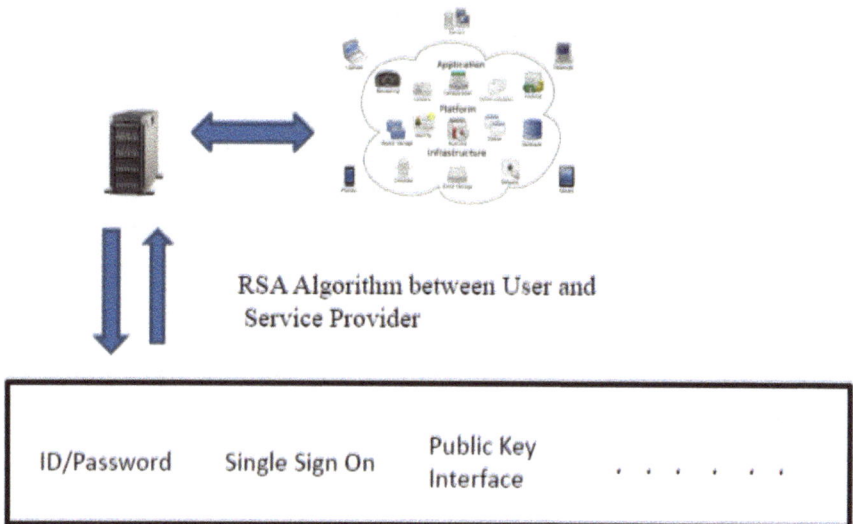

Fig. 6 Authentication process

4 Proposed Model Architecture

Stage 1: Authentication

The proposed system architecture is shown in Fig. 7. The core of the architecture from OpenStack Cloud. To access the data center, the user goes through the authentication process. In the proposed system, the authentication process is implemented using an RSA algorithm. The fundamental authentication process is ID/Password, SSO, and PKI.

Stage 2: Access Control

The firewall is an important component in access control. It allows traffic very easily. If anything is found malicious, the network traffic is blocked. The next stage is traffic reaching Honeywell. The honey wall is the core component of a honeynet. The honeypot consists of Honeytoken. The Honeywell is used to separate the network into two layers:

Client Area: The client machines are monitored and maintained.

Honeypot Area: all the honeypots, along with production machines, are placed and monitored. This mechanism is easily detecting encrypted data.

Figure 7 illustrates the bird's view of the proposed architecture. The proposed architecture includes a complete security mechanism using a firewall, Honeywell. It's a two-way authentication process to enter the cloud environment. The Honeywell fundamentally provides various detecting techniques. The data are stored in a cloud data center.

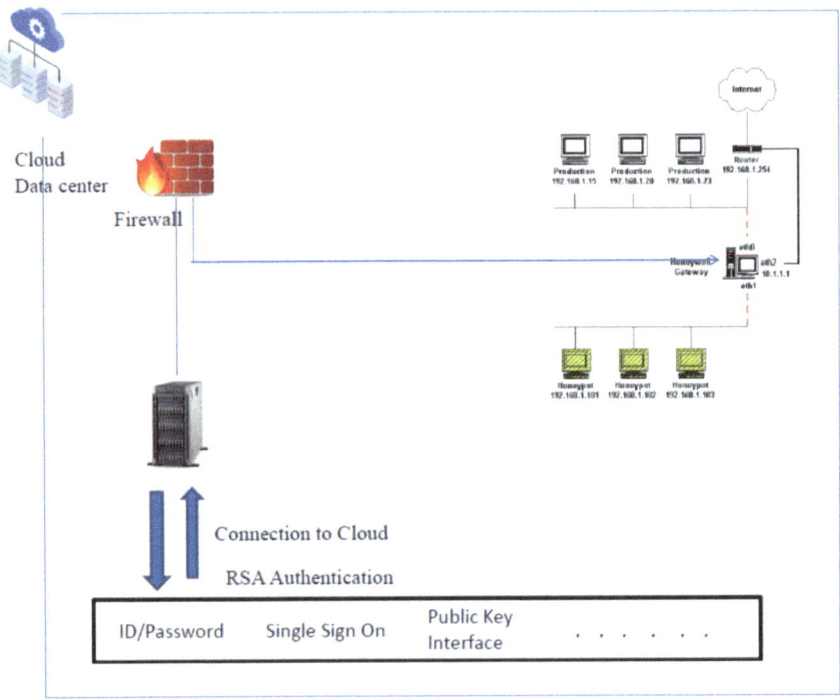

Fig. 7 Proposed architecture of cloud

5   Results

The test was performed using Ubuntu and OpenStack architecture. We have used an 8GB Ram and Multicore Processor. Table 1 presents detailed information about port attacks. From Table 1, we can conclude the majority of attacks are based on source ports.

Table 1 Port attacks report

Port number	Environment	Connection ID	OS	IDS	Port pin end
8080	OpenStack/8GB RAM	452	Ubuntu 20.04.4	10	Source
1086	OpenStack/8GB RAM	312	Ubuntu 20.04.4	8	Source
70	OpenStack/8GB RAM	42	Ubuntu 20.04.4/ VirtualBox	7	Destination
40	OpenStack/8GB RAM	26	Ubuntu 20.04.4/ VirtualBox	5	Destination

5.1 Distribution of Attacks

The implementation work mentioned in the OpenStack environment and the distribution of the attacks in the Honeywall environment are listed in Table 2. The attack type is Honeynet, Honeywell, Honeypot, and Honeytoken wilt cloud IDS. The Table 2 report is based on the log file generated by the OpenStack environment. The false-positive and false-negative are only 5% in the proposed methodologies. Figures 8 and 9 depict the pictorial representation of the output.

Table 2 Distribution attacks ratio

Attack Type	DoS	R2L	Normal	False (+)	False (−)
Honeynet, Honeywell, Honeypot, and Honeytoken wilt cloud IDS	23%	17%	22%	0.5	0.5

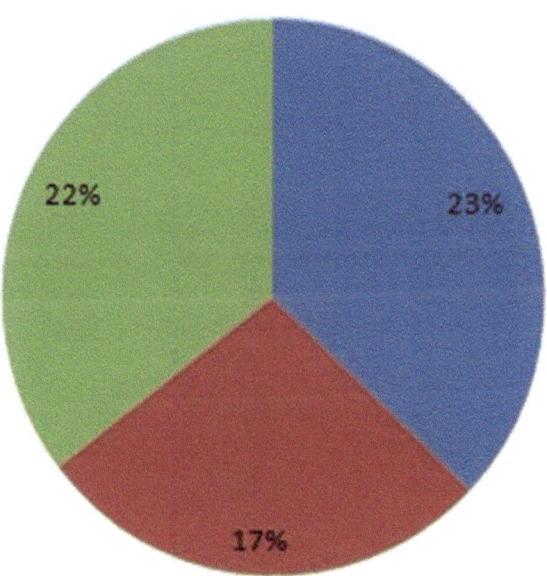

Fig. 8 Distribution attack chart

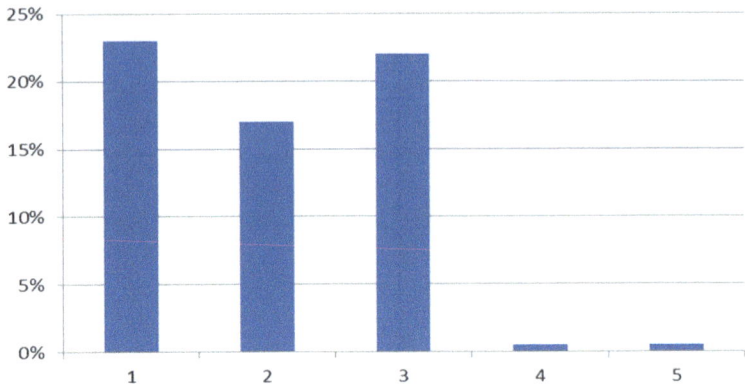

Fig. 9 Attack distribution in Pictorial representation

In this paper, we propose a new cloud computing security methodology using Honeynet, Honeywell, Honeypot, and Honeytoken. The objective of the work is to identify the abnormal attack and monitor the IDS accuracy.

6   Conclusion

Maintaining a high level of security in a cloud computing environment is very challenging for researchers. Each provider maintains its procedure to secure the data in the Cloud. Many secure methods have been developed to challenge attackers. In our work, we propose a cloud computing security based on the Honeynet, Honeywell, Honeypot, and Honeytoken methodologies with IDS. The combination of Honeytoken is a very promising method of securing data in a cloud environment. The new method's benefits are: Honeytoken has been introduced to strengthen security. Report on attack type and Scenario and IDS. The result was carried out in an OpenStack cloud environment. From the result, we conclude the proposed methods are very promising in terms of the accuracy of detecting the attack behavior of attacks.

References

1. Das S, Namasudra S (2022) MACPABE: Multi-authority-based CP-ABE with efficient attribute revocation for IoT-enabled healthcare infrastructure. Int. J. Netw. Manag. 10.1002
2. Islam SH et al (2017) Provably secure identity-based sign encryption scheme for crowd-sourced industrial Internet of things environments. IEEE Internet Things J 5(4):2904–2914
3. Ali HM et al (2022) Planning a secure and reliable IoT-enabled FOG-assisted computing infrastructure for healthcare. Clust Comput 25:2143–2161
4. Ismaeel S, Asiri S (2016) White Paper: Cloud security: Basic principles. Ryerson University

5. Fernandes D, Soares L, Gomes J, Freire M, Incio P Security issues in cloud environments: a survey. Int J Inf Secur 13(2):113–170.
6. Thabit F, Alhomdy S, Jagtap S (2021) A new data security algorithm for the cloud computing based on genetics techniques and logical-mathematical functions. Int J Intell Netw 2:18–33
7. Thabit F, Alhomdy S, Jagtap S (2021) Security analysis and performance evaluation of a new lightweight cryptographic algorithm for cloud computing. Glob TransitS Proc 2(1): 100-110
8. Aissa AB, Taloba AI, Abd-Aziz RM, Mohamed IA, Hussein LF (2021) Empirical analysis of security-enabled cloud computing strategy using attribute-based encryption scheme. Mater Today: Proceedings
9. Xu HJ, Di Liu X, Qi H, Cong L, Li J, Yang H (2019) Security strategies for virtual machine allocation in cloud computing. Procedia Comput Sci 147:140–144
10. Saadi C, Chaoui H (2016) Cloud computing security using IDS-AM-Clust, honeyd, honeywell and honeycomb. Procedia Comput Sci 85:433–442
11. Manogaran G, Thota C, Kumar MV (2016) MetaCloudDataStorage architecture for big data security in cloud computing. Procedia Comput Sci 87:128-133
12. Shaikh R, Sasikumar M (2015) Trust model for measuring security strength of cloud computing service. Procedia Comput Sci 45:380-389
13. Hisham A (2021) Kholidy, Detecting impersonation attacks in cloud computing environments using a centric user profiling approach. Futur Gener Comput Syst 117:299–320
14. Charvát K, Bergheim R, Bērziņš R, Zadražil F, Langovskis D, Vrobel J, Horakova S (2020) Map whiteboard cloud solution for collaborative editing of geographic information. Cloud Comput Data Sci
15. Namasudra S (2022) A secure cryptosystem using DNA cryptography and DNA steganography for the cloud-based IoT infrastructure. Comput Electr Eng 104(Part A): 108426
16. Das S, Namasudra S (2022) A lightweight and anonymous mutual authentication scheme for medical big data in distributed smart healthcare systems. IEEE EEE/ACM Trans Comput Biol Bioinform:1–12
17. Pavithran P (2023) Enhancing randomness of the ciphertext generated by DNA-based cryptosystem and finite state machine. Clust Comput 26:1035–1051
18. Liu GY et al (2018) Secure and fine-grained access control one healthcare records in mobile cloud computing. Futur Gener Comput Syst 78:1020–1026
19. Bebortta S, Das SK, Kandpal M, Barik RK, Dubey H (2020) Geospatial serverless computing: Architectures, tools, and future directions. ISPRS Int J Geo Inf 9(5):311
20. Schleidt K, O'Grady M, Grellet S, Feliachi A, Van Der Schaaf H (2020) ELFIE-The OGC environmental linked features interoperability experiment. In: International symposium on environmental software systems. Springer, Cham, pp 188–193
21. Li X, Peng J et al (2020) A secure three-factor user authentication protocol with forward secrecy for wireless medical sensor network systems. IEEE Syst J 14:39–50
22. Zhou X et al (2021) Energy efficient smart routing based on link correlation mining for wireless edge computing in IoT. IEEE Internet Things J. https://doi.org/10.1109/JIOT.2021.3077937
23. Wani ARS, Khaliq R (2021) SDN-based intrusion detection system for IoT using deep learning classifier (IDSIoT-SDL). CAAI Trans Intell Technol, 6(3):281–290
24. Coutinho RWL, Boukerche A (2023) Modeling and analysis of a shared edge caching system for connected cars and industrial IoT-based applications. IEEE Trans Industr Inf 16(3):2003–2012
25. Charvát K, Bergheim R, Bērziņš R, Zadražil F, Langovskis D, Vrobel J, Horakova S (2021) Map whiteboard cloud solution for collaborative editing of geo-graphic information. Cloud Comput. Data Sci., 2(2):20–55
26. Namasudra S (2022) A secure cryptosystem using DNA cryptography and DNA steganography for the cloud-based IoT infrastructure. Comput, Electr Eng 104
27. Dashti W, Qureshi A, Jahangeer A, Zafar A (2020) Security challenges over cloud environment from service provider prospective. Cloud Comput Data Sci 1(1)
28. Das S, Namasudra S (2022) A lightweight and anonymous mutual authentication scheme for medical big data in distributed smart healthcare systems. IEEE/ACM Trans Comput Biol Bioinform:1–12

29. Pavithran P, Mathew S, Namasudra S, Singh A (2023) Enhancing randomness of the ciphertext generated by DNA-based cryptosystem and finite state machine. Clust Comput 26:1035–1051
30. Cha BR, Kim J (2023) Security tactics for secured cloud computing resources. In: Proceeding IEEE ICOIN 2013. pp 473–475

Secure and Energy-Efficient Framework for Internet of Medical Things (IoMT)-Based Healthcare System

Ritu Dewan, Tapsi Nagpal, Sharik Ahmad, Arun Kumar Rana, and Sardar M. N. Islam

Abstract Manufacturing, energy, finance, education, transportation, smart home, and medicine employ IoT technology. IoT solutions can efficiently manage hospital patients and mobile assets to provide high-quality medical services. The Internet of Medical Things (IoMT) integrates IoT with medical equipment to increase patient comfort, cost-effective medical solutions, hospital treatment speed, and personalized healthcare. This work uses Constrained Application Protocol to secure remote patient health data (CoAP). Nevertheless, the CoAP DTLS layer lacks key control, session establishment, and multicast message exchange. Hence, IoMT communication requires an efficient protocol for safe CoAP session formation. Consequently, to address key management and multicast security issues in CoAP, we presented an efficient and secure communication method to establish a secure session key between IoMT devices and distant servers utilizing lightweight, energy efficient, and Secure CoAP Elliptic Curve Cryptography (E2SCEC2). E2SCEC2 can use a smaller key size than Rivest-Shamir-Adleman (RSA) due to its tiny key size. To determine if these algorithms are compatible in limited contexts, the paper examines key creation, signature generation, and verification of E2SCEC2 and RSA algorithms, energy consumption, and radio duty cycle.

R. Dewan (✉) · T. Nagpal · S. Ahmad
Department of Computer Science and Engineering, Lingaya's Vidyapeeth, Faridabad, India
e-mail: dewanritu22@gmail.com

T. Nagpal
e-mail: dr.tapsi@lingayasvidyapeeth.edu.in

S. Ahmad
e-mail: dr.sharik@lingayasvidyapeeth.edu.in

A. K. Rana
Department of Computer Science and Engineering, Galgotias College of Engg. and Technology, Greater Noida, India
e-mail: arunkumar@galgotiacollege.edu

S. M. N. Islam
SILC and Decision Sciences and Modelling Program, Victoria University, Footscray, Australia
e-mail: sardar.islam@vu.edu.au

© The Author(s), under exclusive license to Springer Nature Singapore Pte Ltd. 2024
S. Namasudra et al. (eds.), *Data Science and Network Engineering*, Lecture Notes in Networks and Systems 791, https://doi.org/10.1007/978-981-99-6755-1_37

Keywords IoT · CoAP · DTLS · RSA · IoMT

1 Introduction

IoT and LPWAN devices are becoming more popular [1]. LPWAN communication technology is low-cost and energy-efficient. Recent studies have used only one network, and none have examined other communication mechanisms in the same setting. Few hospital studies have used Sigfox. This paper explains a hospital-based Internet of Medical Things (IoMT) system, its equipment and data kinds, and the outcomes of implementing a location and temperature monitoring system using IoT technology using the network method: Sigfox, Hospital, and Non-Hospital Networks. This study implements three IoMT systems in a hospital and compares the accuracy of sending and receiving signals from infusion pumps, pharmaceutic freezers, and people (patients and nurses). The IoMT system's hospital communication technique and architecture are explained here. This study used the communication system to follow hospital patients and equipment to improve medical staff efficiency.

This circumstance requires real-time patient health, posture, and medication monitoring. Malicious people can expose and change a patient's sensitive information, which could lead to death [2].

Remote residents lack technology and doctors, preventing them from receiving contemporary healthcare. This circumstance requires real-time patient health, posture, and medication monitoring. Malicious people can expose and change a patient's sensitive information, which could lead to death [3]. Imagine that an opponent spoofs the data processing medium and delivers a real-time asthmatic patient with a nasty cough the wrong dosage [4]. Monitoring affected people can stop the global pandemic of COVID-19 [5]. WBANs monitor people's surroundings. Wearable sensors monitor heart rate, breathing, pulse, and body temperature. Sensory nodes help understand the condition by providing patient information. So, rural areas should have emergency support programs [6]. The IoT has transformed healthcare. Several healthcare wearable technologies monitor patients in real time. Healthcare sensors generate massive data. This data is hard to handle since each component must be secure, honest, and credible [7]. Public sensors or systems must track and record personal data. Wearable sensors and mobile device apps are prevalent. These self-monitoring devices let users analyze personal data. Self-monitoring devices eliminate expensive and time-consuming hospital tests. These devices advance personal health management [8–10]. The main contributions of this paper are:

- Addressing the latest studies on healthcare systems and IoT applications.
- Using IoT technologies to solve IoMT concerns.
- This paper describes an IoMT-based real-time health surveillance system security architecture.
- The proposed framework protects data security, transparency, and authenticity using E2SCEC2.

2   Literature Survey

Many safe IoT solutions are being developed by researchers. Information retrieval, traffic management, spectrum detection, and signal authentication are examples of machine learning applications in IoT networks [11]. The facility's staff and vendors must provide better hardware to secure patient data indoors and out [12]. Security morals are lacking in the Internet of Things. Real-time health monitoring systems will receive authentic, dependable, and private data. Collaborative research is needed to address telemedicine and real-time health surveillance security challenges [13]. In IoT technology, devices share data at different connectivity levels. Each level of contact requires distinct protection measures. Biometric security systems are recommended by scientists because they adapt to individual actions. Because of their utility for identity recognition and security, biometric technologies are becoming more popular. Cost, real-time deployment, cyberattacks including Denial of Service [14], eavesdropping, and data breaches are issues for these systems. Man-in-the-middle and spoofing attacks can occur in biometric systems' recognition layer. These attacks stop the system's decision-making.

The layered approach in a smart healthcare system provides a platform to handle primary challenges by gathering all patient data and transferring it to the cloud for local and worldwide processing [15]. The model was validated using the Contiki real-time operating system. The concept contains three layers, each with a function, however, the system lacks the security framework to combine current administrative responsibilities. When the patient is absent, the computer helps professionals analyze health data [16]. Since technical growth is ongoing, Mobicare must research body sensor network protection, reliable and safe sensor code updates, and future opportunities. Some breakthroughs in e-health systems have critical scenarios. Instead, this study created virtual clusters of individuals with linked conditions by merging hidden sharing systems with identity linking [17]. Social networking was supplied by mHealth. The block director and block department collect detailed data from patients' mobile devices. Although the system does not respond to violent invaders, these invaders do not harm patients, and the system helps distressed patients by sharing confidential safety information [18]. In the Internet of Things, stable data exchange is essential for data fusion, data gathering, context-aware applications, and user privacy [19].

One solution is the Cloud-of-Things (CoT). They test the algorithm in virtual reality. Accepting a plan may have benefits [20]. Categorizing agents by trust relationships may improve this strategy. Next, IIoT edge computing's future architecture, routing, task scheduling, data storage, analytics, security, and standardization are studied [21]. To summarize, this study develops a trustworthy IoT-based security architecture with a reliable system, perception, terminal, network, and advisor modules [22]. Another analysis found network-layer IoT security challenges and solutions to the issues and key technologies highlighted above. Cloud layer safety measures [23]. Multimedia enhances remote patient-specialist communication and field innovation. A WBAN-based three-tier ubiquitous telemedicine system offers real-time health care surveillance [24]. Every IoT data must be secure to prevent cyberattacks. Yet, lightweight devices struggle to adapt to the standard cipher algorithm. In [25], we proposed using the classical cipher technique to share and regulate data on lightweight devices in many contexts if the Internet of Things needs a WSN (WSN). To build such a network, key management mechanisms that protect data flow must be available [26]. Trust is the most important component in intelligent agent technology since it boosts smart object reputation. The research provided a dependable model and architectural mechanism for IoT e-health zones to help send and receive patient data, reduce security threats, and optimize wearables. A quick research shows how IoT e-health apps can manage huge patient data [27]. DTLS keys, RSA, and clasp authentication validate and deploy this mechanism [28]. They use TSA features and modules in a public-key cryptography technique for authentication. An improved and expanded IoT-based RFID authentication mechanism for WBANs was proposed. Their system might withstand many attacks and ensure tag-to-medical server mutual authentication [29].

This paper proposes E2SCEC2 security. Algebraic structure-based asymmetric key-based authentication. The following section details the solution.

3 Proposed Architecture

Perception, network, and application comprise the Internet of Things, according to many publications. Figure 1 shows the data flow between levels. Each layer in [30] has confidentiality, integrity, and authenticity difficulties due to human interaction. Human interaction has increased sensitivity, therefore data from persons and patients require system permission to access. Privacy considerations prevent full device use. Healthcare data must be authentic, honest, and confidential. We need a security framework for real-time health monitoring. Paramedics and health technicians today employ many protection systems. Eventually, these structures become ludicrous. Our past efforts support our statements. As mentioned, IoT has layers of security weaknesses that must be addressed to fix major security issues. Physical, or vision, is the first layer. Sensors report temperature, humidity, and position to the network layer.

Data is routed and sent from the network layer to various IoT hubs and PCs using cutting-edge technologies like Wi-Fi, Bluetooth, ZigBee, and 4G LTE. Other systems, such as cloud storage and gateway routers, may now connect with one another, thanks to these technologies. Gateways serve as a middleman for data transmission between two or more different IoT nodes in the network. The application layer ensures the data's protective trinity of secrecy, honesty, and authenticity via this layer. All of the deployed IoT apps are found at the application layer. The application layer acts as a bridge between the endpoints and the network. The services that this layer offers to the apps might vary based on the data that the sensors for each application give. The application of WBAN technology in healthcare has increased recently. Many

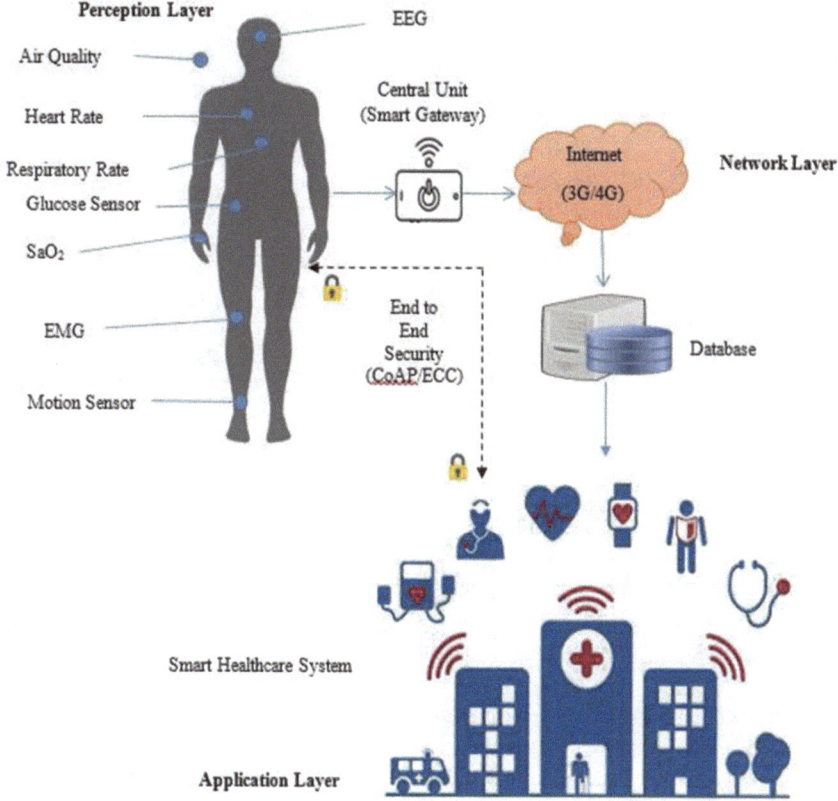

Fig. 1 Proposed models of IoMT

medical devices may be employed, implanted, and integrated with the WBAN to follow patient behavior, treat patients with automated therapies, and more [29].

4 Research Methodology

A great network simulator that is especially helpful for IoT healthcare contexts is the Contiki Cooja Simulator. The major feature of this simulator is built on an event-driven kernel and uses uIP LIBRARY to handle TCP/IP networking. Moreover, protothreads programming abstraction may benefit from this. Contiki is a well-known IoMT-enabled working framework that is open source and free. The Contiki organized test system, known as Cooja, is used to program sensor devices as well [31]. Cooja's window, which is filled with the main simulation tools, is shown in Fig. 2. The following is a short summary of how each tool works:

Simulation control Window

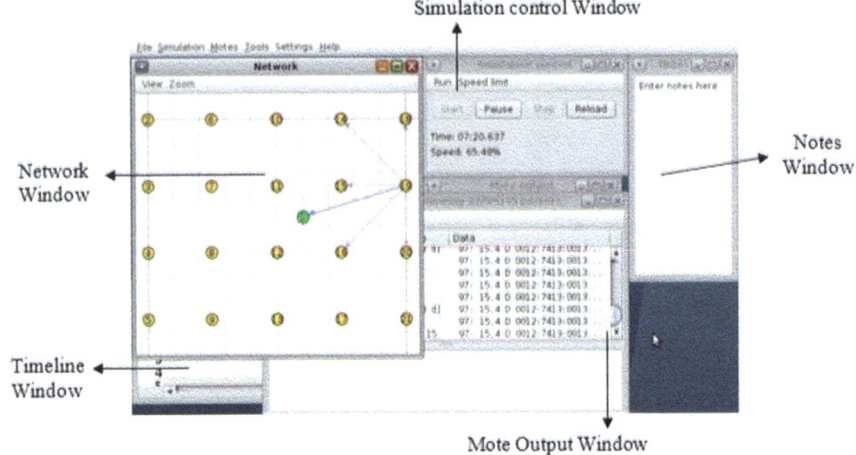

Fig. 2 Contiki cooja simulator

- **Simulation Control**—You may start, stop, reload, or execute simulation steps from this window. Both the execution time and the simulation speed are shown. We are able to perform the events more quickly than real-time execution would permit.
- **Network**—The location of each network node is shown. Each node's state may be seen, including its outputs, LEDs, addresses, and other node information. We must first add our sensors to this window since it is originally empty.
- **Mote output**—This displays every serial interface output from the node. One Mote output window may be activated per simulation node.
- **Timeline**—The simulation timeline shows messages and events including channel changes, LED changes, log outputs, and more.
- **Notes**—Use this simple notebook to record your simulation-related notes.

Continue reading to see how Contiki and Cooja may be used to modify IPv6 conventions. In an IoT-based system scenario where capture free made sure about communication being essential, security and honesty are the main difficulties; In the current IoT environment, CoAP is a very helpful protocol. Four protection mechanisms are outlined by CoAP. 1. Deactivated DTLS and NoSec 2. PreSharedKey with DTLS enabled 3. A raw public key with DTLS enabled 4. Cryptography with elliptic curves (ECC). Responses with a Reset message (RST) rather than a Greeting are appropriate if a receiver is unable to process a Confirmable Signal at all (i.e., is unable to even deliver an acceptable response to an error). There are several ways to communicate safely in IoMT networks, but each one has drawbacks.

5   Results and Analysis

In the Contiki Cooja simulation environment, Fig. 3 shows how our existing security architecture has been implemented for the privacy of the real-time health monitoring system. In this paper, we developed a module of the architecture that addresses the network layer in order to illustrate the outcomes [32].

The network layer covers the effective networking and end-to-end secure communication of devices. In the future, it's also possible to build the remainder of the framework using a simulation environment. The framework is functional and may be quickly built in the future, thanks to our insights. Table 1 lists the simulation parameters. Our suggested model has 13 IoMT sensor motes that are distinct from one another, with 8 sensor motes acting as a sensor (shown in yellow color), 1, 6, 11, and 12 acting as unicast senders to transmit data to the broadcast sensor mote 14 using RPL routing [31, 33–40]. Anybody may read the value of the sensor with the aid of any digital device, such as a mobile (M) or PC, by forwarding this information to the internet via an IPv4 or IPv6 connection through the broadcast router. It supports both unidirectional and bidirectional correspondence and works with the dynamic output of the Destination-Oriented Directed Acyclic Graph (DODAG). Higher growth results in unusual events and constrained behavior. Each device node in RPL has the option of sending packages up to their root or down to the child

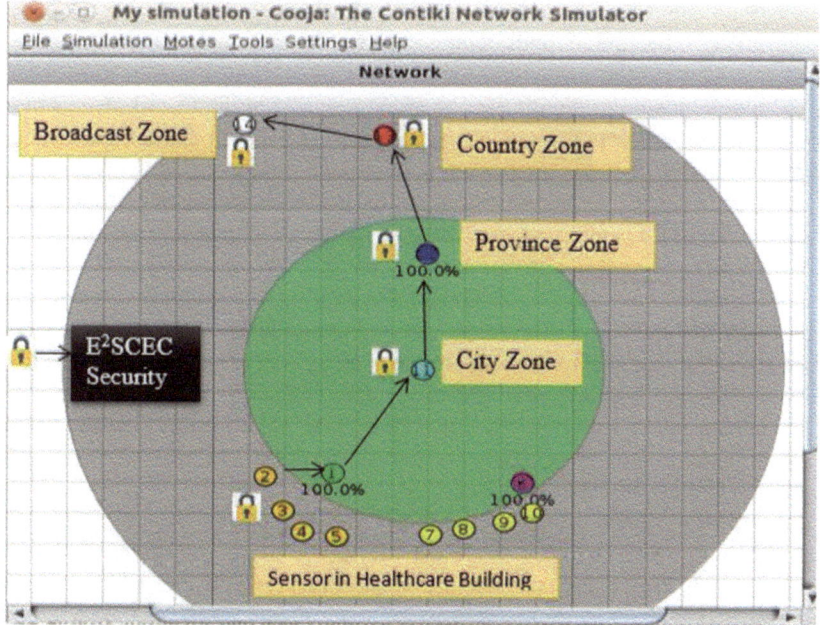

Fig. 3 Cooja simulation setup with different sky motes in a secure healthcare building

Table 1 Simulation parameters

Parameter	Value
Operating system	Contiki 2.7
Simulator	COOJA
Computer	RAM8GB,i5 Pro
Transmission range	60 m
Interference range	65 m
Simulation time	99 min
Routing protocol	CoAP
Number of nodes	15
Topology	Random
MAC layer	802.15.4
Node type	Skymote
Node distribution	Random
Networking	Mesh
Packet size	56 Byte
Packet rate	5P/s

node. IoT security sections must be built with the usage of IoT in various domains by ensuring that the protected packages are intended to prevent disruptions and that all communications may be properly secured [41–48].

The suggested technique will demonstrate how data is gathered using a mathematical model in accordance with the guiding principle. Drag and drop all nodes into various topologies according to the suggested scenario in the Contiki Cooja simulator. The CoAP protocol is intended to be on the IoT network application layer and offers security services for trust, authorization, and data integrity for transmitting data between service agents throughout the IoT network. In the area of the internet of things, Sky Mote is an excellent sensor-supported IPv6 protocol that offers very fast data rates for very low-power devices. Some of the characteristics include interoperability that is supported, and the sky mote uses very little current. Slumber and quick wake-up modes, USB Data Interface, light humidity, and temperature sensors are quite helpful. Top networking assistance X.509 v3 certificates issued by a shared Mutual Certificate Authority are used by the communicating parties to authenticate DTLS using a Public Key Infrastructure (PKI) (CA). DTLS establishes the right of access to certain themes and general security rules over the whole domain using permission and governance guidelines.

The following graph compares the power consumption of the proposed stable E2SCEC2 and RSA when the key sizes for both algorithms change. In contrast to RSA, which uses more power as the main size grows (as seen in the graph), the suggested E2SCEC2 uses less power. Figure 4 illustrates how RSA's power consumption becomes linear when the main size increases above 512 bits.

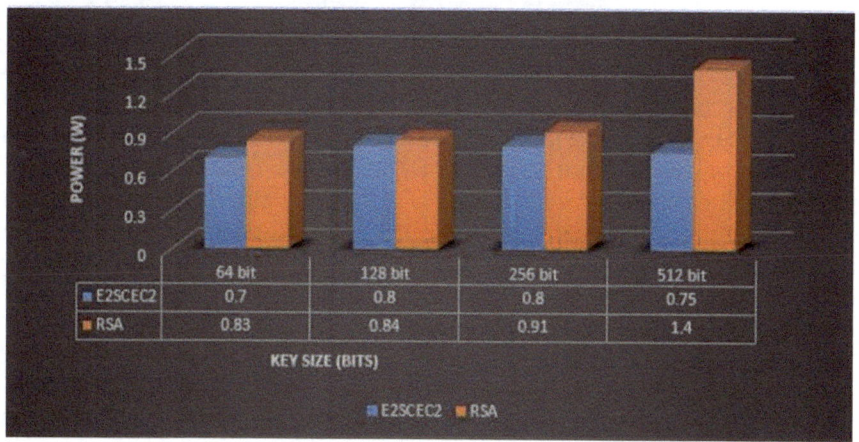

Fig. 4 Power consumption of E^2SCEC^2 and RSA

Due to the fact that E2SCEC2 does not need to expend as much energy on intensive prime number production as RSA does, it can construct private/public key pairs more quickly and with equal key lengths as RSA. The stunning triumph of E2SCEC2 against RSA is seen in the graph below. Figure 5 illustrates how key generation time grows exponentially for RSA and linearly for ECC as key sizes increase. E2SCEC2 does not need any resources to generate keys, however, both RSA and E2SCEC2 spend some time calculating the hash function of the message in order to generate signatures.

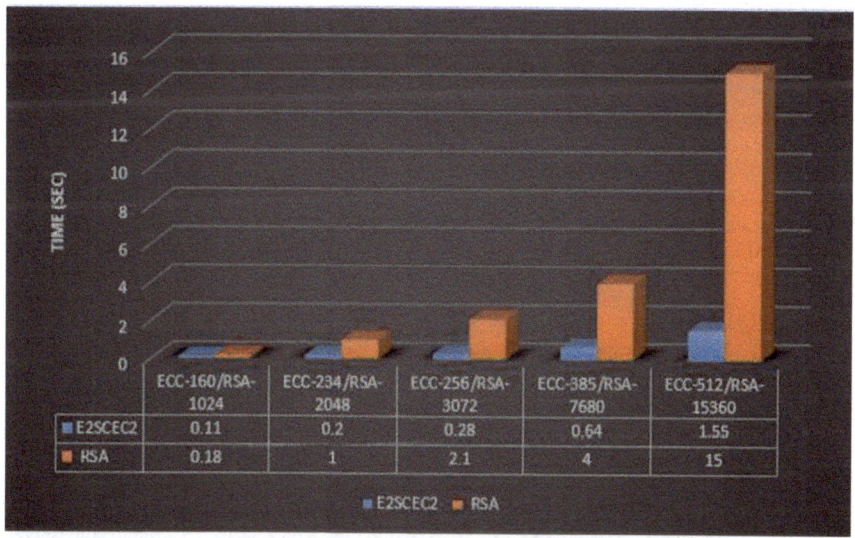

Fig. 5 Key generation time of E^2SCEC^2 and RSA

6 Conclusion

The development of IoT-based smart solutions has changed how people live. This study suggested using E2SCEC2 to create a secure system for IoT-based medical sensor data authentication and encryption. The recommended authentication approach combines user credentials and biometric characteristics to increase system security. The proposed approach will be utilized to address key management and associated protection issues in resource-constrained IoT devices and enable their secure operation across weak networks. To show off the suggested scheme's great resilience against pertinent cryptographic assaults, it is mathematically assessed. Since E2SCEC2 provides the same level of safety for significantly smaller keys as RSA, it requires less key storage than the latter. Our simulation findings allow us to draw the conclusion that, in networks with constrained computing, power, and storage resources, E2SCEC2 performs better than the RSA method. As a consequence, client-side CoAP solutions based on IoT may use E2SCEC2 at a reasonable price. Operational Cooja simulations are another important aspect that can be readily carried out while employing sensors to provide the best performance. Although we cannot promise perfection, we anticipate having expanded the area of interest for academics and practitioners in the IoMT system.

References

1. Kamarudin A, Yap NK, Ali MA, Phang LS (2020) A preliminary component model for IoT. In: IOP Conference series: Materials science and engineering, vol 769, no 1. IOP Publishing, p 012003
2. Shakib KH, Neha FF (2021) A Study for taking an approach in Industrial IoT based Solution. In: Journal of Physics: Conference Series, vol 1831, no 1. IOP Publishing, p 012007
3. Hussain A, Draz U, Ali T, Tariq S, Irfan M, Glowacz A, Daviu JAA, Yasin S, Rahman S (2020) Waste management and prediction of air pollutants using iot and machine learning approach. Energies 13:3930
4. Hammood DA, Alkhayyat A, Rahim HA, Ahmad RB (2020) Minimum power allocation cooperative communication based on health-care WBAN. In: IOP conference series: materials science and engineering, vol 745, no 1. IOP Publishing, p 012051
5. Rana AK, Krishna R, Dhwan S, Sharma S, Gupta R (2019) Review on artificial intelligence with internet of things-problems, challenges and opportunities. In: 2019 2nd international conference on power energy, environment and intelligent control (PEEIC). IEEE, pp 383–387
6. Abdullah WANW, Yaakob N, Badlishah R, Elobaid ME, Yah SA, Zunaidi I (2019) Fragmentation in MAC IEEE 802.15. 4 to improve delay performance in Wireless Body Area Network (WBAN). In: IOP Conference series: Materials science and engineering, vol 557, no. 1. IOP Publishing, p. 012020
7. Kumar A, Sharma S, Goyal N, Singh A, Cheng X, Singh P (2021) Secure and energy-efficient smart building architecture with emerging technology IoT. Computer Commun, 176C:207–217
8. Hussain A, Wenbi R, da Silva AL, Nadher M, Mudhish M (2015) Health and emergency-care platform for the elderly and disabled people in the Smart City. J Syst Softw 110:253–263
9. Kumar A, Sharma S (2021) IFTTT rely based a semantic web approach to simplifying trigger-action programming for end-user application with IoT applications. Semant. IoT: Theory Appl.: Interoperability, Proven. Beyond, p 385

10. Yew HT, Ng MF, Ping SZ, Chung SK, Chekima A, Dargham JA (2020) IoT based real-time remote patient monitoring system. In: Proceedings of the 2020 16th IEEE international colloquium on signal processing & its applications (CSPA). Langkawi, MA, USA, pp 176–179

11. Sagduyu YE, Shi Y, Erpek T (2019) IoT network security from the perspective of adversarial deep learning. In: 2019 16th Annual IEEE international conference on sensing, communication, and networking (SECON). IEEE, pp 1–9

12. Almahdi EM, Zaidan AA, Zaidan BB, AlSalem MA, Albahri OS, Albahri AS (2019) Mobile patient monitoring systems from a benchmarking aspect: challenges, open issues and recommended solutions. J Med Syst 43:207. [CrossRef]

13. Shah STU, Badshah F, Dad F, Amin N, Jan MA (2019) Cloud-assisted IoT-based smart respiratory monitoring system for asthma patients. In: Applications of intelligent technologies in healthcare. Springer: Cham, Switzerland, pp 77–86

14. Rahmani A-M, Thanigaivelan NK, Gia TN, Granados J, Negash B, Liljeberg P, Tenhunen H (2015) Smart e-health gateway: bringing intelligence to internet-of-things based ubiquitous healthcare systems. In: Proceedings of the 2015 12th Annual IEEE consumer communications and networking conference (CCNC), Las Vegas, NV, USA, pp 9–12. Institute of Electrical and Electronics Engineers (IEEE): Piscataway, NJ, USA, pp 826–834

15. Nanayakkara M, Halgamuge M, Syed A (2019) Security and privacy of internet of medical things (IoMT) based healthcare applications: A review. In: Proceedings of the international conference on advances in business management and information technology. Istanbul, Turkey

16. Mohammed KI, Zaidan AA, Zaidan BB, Albahri OS, AlSalem MA, Hadi A, Hashim M, Albahri AS (2019) Real-time remote-health monitoring systems: a review on patients prioritisation for multiple-chronic diseases, taxonomy analysis, concerns and solution procedure. J Med Syst 43:223

17. Sun Y (2019) Securing body sensor networks and pervasive healthcare systems. Ph.D. Thesis, Imperial College London, London, UK

18. Selvaraj P, Doraikannan S (2019) Privacy and security issues on wireless body area and IOT for remote healthcare monitoring. Intell Pervasive Comput Syst Smarter Healthc:227–253. [CrossRef]

19. Yan Z, Zhang P, Vasilakos AV (2014) A survey on trust management for Internet of Things. J Netw Comput Appl 42:120–134

20. Fortino G, Messina F, Rosaci D, Sarné GL (2018) Using trust and local reputation for group formation in the Cloud of Things. Futur Gener Comput Syst 89:804–815

21. Qiu T, Chi J, Zhou X, Ning Z, Atiquzzaman M, Wu DO (2020) Edge computing in industrial internet of things: Architecture, advances and challenges. IEEE Commun Surv & Tutor 22(4):2462–2488

22. Qi J, Yang P, Min G, Amft O, Dong F, Xu L (2017) Advanced internet of things for personalised healthcare systems: A survey. Pervasive Mob Comput 41:132–149. [CrossRef]

23. Yasin S, Ali T, Draz U, Jung LT, Arshad MA (2018) Formal analysis of coherent non-redundant partition-based motif detection algorithm for data visual analytics. J Appl Environ Biol Sci 8:23–30

24. Logambal M, Thiagarasu V (2017) Healthcare monitoring systems: A WBAN approach for patient monitoring. Int J Future Revolut Comput Sci Commun Eng 3:26–31

25. Kim S, Lee I (2018) IoT device security based on proxy re-encryption. J Ambient Intell Humaniz Comput 9(4):1267–1273

26. Huang H, Gong T, Ye N, Wang R, Dou Y (2017) Private and secured medical data transmission and analysis for wireless sensing healthcare system. IEEE Trans Ind Inform 13:1227–1237. [CrossRef]

27. Hathaliya JJ, Tanwar S (2020) An exhaustive survey on security and privacy issues in Healthcare 4. Comput Commun 153: 311–335. [CrossRef]

28. Ali T, Masood K, Irfan M, Draz U, Nagra A, Asif M, Alshehri B, Glowacz A, Tadeusiewicz R, Mahnashi M et al (2020) Multistage segmentation of prostate cancer tissues using sample entropy texture analysis. Entropy 22:1370

29. Izza S, Benssalah M, Drouiche K (2021) An enhanced scalable and secure RFID authentication protocol for WBAN within an IoT environment. J Inf Secur Appl 58:102705
30. Dhawan S, Chakraborty C, Frnda J, Gupta R, Rana AK, Pani SK (2021) SSII: Secured and high-quality Steganography using intelligent hybrid optimization algorithms for IoT. IEEE Access.
31. Rana AK, Sharma S (2021) Contiki Cooja Security Solution (CCSS) with IPv6 routing protocol for low-power and lossy networks (RPL) in internet of things applications. In: Mobile radio communications and 5G networks. Springer, Singapore, pp 251–259
32. Vatsa SV, Mohan A, Kumar A (2021) Implementing QZMAC (a decentralized delay optimal MAC) over 6TiSCH under the Contiki OS in an IEEE 802.15. 4 Network. In: 2021 International conference on communication systems & networks (COMSNETS). IEEE, pp 145–147
33. Morales-Molina CD, Hernandez-Suarez A, Sanchez-Perez G, Toscano-Medina LK, Perez-Meana H, Olivares-Mercado J, Garcia-Villalba LJ (2021) A dense neural network approach for detecting clone ID attacks on the RPL protocol of the IoT. Sensors 21(9):3173
34. Rana AK, Sharma S (2021) Industry 4.0 manufacturing based on iot, cloud computing, and big data: manufacturing purpose scenario. In: Advances in communication and computational technology. Springer, Singapore, pp 1109–1119
35. Niu X (2021) Optimizing DODAG build with RPL protocol. Math Probl Eng
36. Koster LA, Meinardi JE, Kaptein BL, der Linden-Van V, der Zwaag E, Nelissen RG (2021) Two-year RSA migration results of symmetrical and asymmetrical tibial components in total knee arthroplasty: a randomized controlled trial. Bone & Jt J 103(5):855–863
37. Halder J, Acharya T, Chatterjee M, Bhattacharya U (2021) ES-RSM-RSA: A novel energy and spectrum efficient regenerator aware multipath based survivable RSA in offline EON. IEEE Trans Green Commun Networking
38. Tripathi SK, Gupta B, Pandian KS (2021) An alternative practical public-key cryptosystems based on the dependent RSA discrete logarithm problems. Expert Syst Appl 164:114047
39. Rana SK, Kim HC, Pani SK, Rana SK, Joo MI, Rana AK, Aich S (2021) Blockchain-based model to improve the performance of the next-generation digital supply chain. Sustain 13(18):10008
40. Rana A, Chakraborty C, Sharma S, Dhawan S, Pani SK, Ashraf I (2022) Internet of medical things-based secure and energy-efficient framework for health care. Big Data 10(1):18–33
41. Yogesh S, Chinmay C (2021) Augmented reality and virtual reality transforming spinal imaging landscape: a feasibility study. IEEE Comput Graphics Appl 41(3):124–138. https://doi.org/10.1109/MCG.2020.3000359
42. Dhawan S, Chakraborty C, Frnda J, Gupta R, Rana AK, Pani SK (2021) SSII: secured and high-quality steganography using intelligent hybrid optimization algorithms for IoT. IEEE Access 9:87563–87578
43. Kumar A, Sharma S, Singh A, Alwadain A, Choi BJ, Manual-Brenosa J, Goyal N (2021) Revolutionary strategies analysis and proposed system for future infrastructure in internet of things. Sustainability 14(1):71
44. Kumar A, Sharma S, Goyal N, Singh A, Cheng X, Singh P (2021) Secure and energy-efficient smart building architecture with emerging technology IoT. Comput Commun 176:207–217
45. Rana A, Sharma S, Nisar K, Ibrahim A, Ag A, Dhawan S, Goyal N (2022) The rise of blockchain internet of things (BIoT): Secured, device-to-device architecture and simulation scenarios. Appl Sci 12(15):7694
46. Kumar Pandey R, Gandomkar A, Vaferi B, Kumar A, Torabi F (2023) Supervised deep learning-based paradigm to screen the enhanced oil recovery scenarios. Sci Rep 13(1):4892
47. Karupusamy S, Mustafa MA, Jos BM, Dahiya P, Bhardwaj R, Kanani P, Kumar A (2023) Torque control-based induction motor speed control using anticipating power impulse technique. Int J Adv Manuf Technol:1–9
48. Ragab M, Ashary EB, Aljedaibi WH, Alzahrani IR, Kumar A, Gupta D, Mansour RF (2023) A novel metaheuristics with adaptive neuro-fuzzy inference system for decision making on autonomous unmanned aerial vehicle systems. ISA Trans 132:16–23

A Robust Remote User Authentication Scheme for Supply Chain Management Using Blockchain Technology

Inderpal Singh, Balraj Singh, and Arun Kumar Rana

Abstract For a company's product life cycle to be faster, more efficient, and more effective, supply chain management (SCM) is critical. Nevertheless, present SCM systems must be improved to ensure genuine products and transactions are private and secure. As a result, this study proposes a safe SCM system depending upon blockchain and the Internet of Things to ensure that the products are genuine. All SCM stakeholders can initiate an encrypted, secure transaction for their goods or services due to the Quick Response (QR) scanner that works with the Internet of Things (IoT) and the blockchain-integrated distributed system. Finally, a genuine product from the manufacturer will be sent to the customer. To make authentication quicker and easier for scattered IoT devices, a lightweight asymmetric key encryption mechanism known as Diffie-Hellman key exchange and Hyperledger Fabric-based blockchain technology with on-chain smart contracts are deployed. The service provider registers each SCM stakeholder and provides them with their own public and private keys, which will be used to authenticate participants and IoT devices. Examining security and scalability shows that the recommended solution is more trustworthy and secure than existing methods.

Keywords Internet of Things · Supply chain · Blockchain · Asymmetric encryption · And security and identification are some words that come to mind

I. Singh · B. Singh (✉) · A. K. Rana
School of Computer Science and Engineering, Phagwara, India
e-mail: balraj.13075@lpu.com

I. Singh · B. Singh
Lovely Professional University, Phagwara, Punjab, India

A. K. Rana
Galgotia College of Engineering, Greater Noida, Uttar Pradesh, India

© The Author(s), under exclusive license to Springer Nature Singapore Pte Ltd. 2024
S. Namasudra et al. (eds.), *Data Science and Network Engineering*, Lecture Notes in Networks and Systems 791, https://doi.org/10.1007/978-981-99-6755-1_38

1 Introduction

Because of the lightning-fast expansion of the Internet and other forms of technology, the entire planet is currently undergoing the fourth wave of the industrial revolution, also known as Industry 4.0 [1]. The Internet of Things (IoT) [2] is a significant factor in how this occurs in a wide variety of settings. The Internet of Things (IoT) is a network of linked physical items (that is, machinery, appliances, and other devices) with sensors, software, and electronics placed on them. These objects also have identifiers that are unique to them. Sensors connected to the Internet of Things are also capable of communicating with one another through the Internet without the intervention of a human. It can look at the related objects, generate information about them, and decide what to do with the generated data. It has a significant potential to deliver exciting services in various areas, such as business, medical services, smart buildings, green infrastructure, social networking sites, and more. These are some of the locations that it may serve. The supply chain management (SCM) system has been altered due to the network of Things [5].

SCM is for supply chain management, which refers to managing the flow of goods through several stakeholders, such as retailers, distributors, and customers [3]. It simplifies monitoring the flow of goods and information throughout the system. Several stages in the supply chain must be completed before a good or service can be delivered to the end user. During these processes, raw materials are moved and transformed into completed goods, which are transported to their final destination and handed to the customer.

Connected gadgets from the Internet of Things can be attached to a product to determine its authenticity, origin, and quality. In addition, Internet of Things devices can guarantee a product's real-time tracking, traceability, and visibility at every stage in the supply chain. According to a recent poll, Internet of Things (IoT) gadgets have begun to make their way into the supply chains of Australian merchants. It incorporates barcode technology based on the Internet, smartphones, tablets, mobile apps, navigation position awareness, Internet-based monitoring and safety system, and detectors and analyzers [6].

2 Related Studies

There is clarity that there are benefits to utilizing the IoT in the supply chain. Despite its advantages, there are several concerns regarding the integrated supply chain that uses IoT. IoT devices generate significant data saved as plaintext on a centralized server or in the cloud. Consequently, the centralized server may be dishonest and improperly use the sensitive information provided by the users. When an Internet of Things infrastructure is centralized, there is a significant threat to the confidentiality and integrity of user information [7]. The majority of supply chains presently exist are not IoT-integrated [8, and as a result, the safety and confidentiality of users and product data are seriously threatened].

Additional research also looks at how the Internet of Things (IoT) and blockchain [9] can be employed in the supply chain. These studies are in addition to the publications that we have just discussed. Nevertheless, research must investigate how elliptic curve cryptography (ECC), an asymmetric key encryption method, can be utilized with the Internet of Things and supply chain systems. In addition, none of the prior studies discussed in Sect. 2 center their attention on. How to deploy key agreements and key distributions to authenticate internet-connected devices. A blockchain is a decentralized and distributed workgroup in which all participants share the same ledger of transactions. In this particular network, there is no central server. Because the records of transactions in the blockchain ledger can't be changed, it assures authenticity, transparency, traceability, security, and visibility between the many actors in the supply chain. The immutability of the blockchain platform ensures the truthfulness and integrity of the data relating to SCM transactions, but it does not guarantee the confidentiality of the information. Therefore, it is necessary to prevent sensitive information about users from becoming public. Traditional PC-based cryptography solutions are incompatible with most Internet of Things devices [10], as these devices need more memory, battery, or computational power. Because of this, the system requires a straightforward protocol for performing cryptographic operations.

This research uses lightweight asymmetric key cryptography (ECC), the Internet of Things (IoT), and Hyperledger Fabric to enable safe and trusted transactions in supply chain environments. This will assist in putting a stop to the issues that have been plaguing the supply chain. To ensure the reliability of internet-connected devices, a lightweight key agreement method based on ECC has been implemented. Hyperledger fabric ensures that financial transactions in the supply chain are carried out more quickly and anonymously than in the past. When they leave the factory, all finished products and services are tagged with a rapid response (QR) code. To read the codes, the suggested system will use a QR code scanner that is compatible with the Internet of Things (IoT). The information about the transactions will be automatically saved and encrypted within the blockchain. A simple authentication procedure will be used to register the QR scanners of all participants in the blockchain network (which might include a manufacturer, distributor, or store, for example). The information on the goods is saved in the blockchain via the QR scanner when two Internet of Things devices have been registered and mutually validated. The suggested method creates a workgroup, trustworthy distributed supply chain that enables real-time tracking and traceability of products, guarantees the authenticity of product information, and maintains its secrecy through an authenticated Internet of Things device. The Internet of Things (IoT) will make a blockchain-based supply chain more flexible, traceable, transparent, autonomous, and audible in real-time, and private during transactions. Moreover, IoT will make the supply chain loud in real-time.

- Internet of Things and blockchain technology is utilized to record transactions in the supply chain with much less assistance from individuals;
- The SCM protocol, an asymmetric key encryption method based on the ECC method, is where key distribution and key agreement are worked on. ECC is employed in managing cryptology procedures and in the straightforward authentication of entities.
- Blockchain technology, which is based on Hyperledger fabric, will ensure the confidentiality and safety of transaction data;
- Both security and privacy evaluations demonstrate that the procedure in question is effective.

3   Research Analysis

In this section, a robust authentication system is suggested for use in supply chain management systems. The system is intended for use by distant users. The plan is resistant to common security flaws and is divided into five stages: (a) the precomputation phase; (b) the registration phase; (c) the login phase; (d) the authentication phase; and (e) the password-changing phase. In Table 1, you'll find a list of the symbols that will be utilized in the scheme that has been proposed.

Figure 2 shows how this study thinks a data-driven supply chain ecosystem based on blockchain and IoT could work. We carefully investigate this system's authentication mechanism, consensus process, and registration protocol. Those taking part in this system include M, D, R, C, and SP.

There are two parts to the whole system. These are:

- Registration
- Authentication

Mi asks SPi to register it during the registration process, and SPi says yes. This finishes the procedure of registry for Mi. Likewise, Di and Ri complete the registration process. During the registration process, Mi, Di, and Ri each follow the steps in Sect. 5.1 and get their public and private keys from SPi. Smart contracts handle all interactions, and all the exchange is documented in chain 1. Anyone in the network can see the public key of anyone else. People use protocol 5.2 to verify each other and their IoT devices during the authentication process. Consider a scenario where and want to participate in the authentication. Using asymmetric encryption, they prove they are who they are. All the contracts and transactions are handled by their intelligent control, which is all recorded in the chain. All people who participate in the authentication part will also use their smart contracts to verify that the public key from chain 1 is confirmed.

Table 1 Notations

Symbol	Meaning
M_i	ith Manufacturer
SP_i	ith service provider
R_i	ith Retailer
D_i	ith Distributor
C_i	ith Customer
V	Secret parameter known only to the SP_i
Id_M	Unique identity of manufacturer
PW_M	Strong user password of manufacturer
BI_M	Biometrics of manufacturer
SC	Smart card
R_1, R_2, R_3	Secret random nonces
P, q	Large distinct primes such that $p = 2q + 1$
ΔT	Permissible time interval for the allowed delay
S_{Key}	Shared session key
$h(\cdot)$	Hash operation
$H(\cdot)$	Biohash operation
$\|$	Concatenation operation
$\oplus$	XOR operation

3.1 Phase of Precalculation

In offline mode, the parameters are set by the server. SPi creates a prime U, an integer V such that $U \times V \equiv 1(mod(p-1)(q-1))$ and calculates $V \equiv U - 1 \ (mod(p-1)(q-1))$. SPi considers U as the public key and V as the private key. The public key is shared, but the private key is kept secret. The proposed scheme uses the lightweight cryptographic hash function: "h: 0, 1* 0, 1 l, where l is the output length of h()."

3.2 The Sign-Up Phase

The registration process of Mi, Di, and Ri with SPi. The steps for registering M with SP are shown below, and the exact steps are used for registering D and R. Figure 2 shows the registration process. Mi takes the following steps to register successfully.

Step R1: The manufacturer Mi picks their own unique identity IdM and password PWM. Then, Mi makes a random number called R1. Further, Mi calculates a masked password MPWM = h(R1 ∥ PWM) and A1 = h(IdM ⊕ R1). Then, it uses a secure channel to send the message "IdM, MPWM, A1" to the remote SPi.

Step R2: After receiving the message, SPi at registration time T, computes A2 = h(IdM ∥ V ∥ T), A3 = h(A1), A4 = A3 ⊕ MPWM, A5 = h(V) ⊕ A1 and A6 = V ⊕ A2. SPi stores the parameters "A4, A5, A6, h()" on a smart card called SC, which is then given to Mi through stable connections.

Step R3: When the manufacturer Mi gets the smart card SC, they print their biometric BIM on the sensor and figure out K = R1 H(BIM) and L = h(IdM ∥ R1 ∥ PWM). Mi puts K and L into the smart card SC's memory, which now holds "A4, A5, A6, h(), K, L."

3.3 Phase Login Phase

In this phase, the manufacturer submits a login request to the server. The steps for M to log in to SP are listed below. D and R also use the same measures. To start a session, the steps below are taken.

Step L1: The manufacturer Mi inserts their smart card into the card reader, inputs IdM, PWM, and imprints biometric BIM at the sensor.

Step L2: Smart card SC extracts R1 = K ⊕ H(BIM) and calculates L = h(IdM ∥ R1 ∥ PWM). Then, it checks to see if condition L = ? L holds. If the condition fails, the session is terminated. If not, the process goes to the next step.

Step L3: SC also figures out that MPWM = h(R1 ∥ PWM), A1 = h(IdM R1), A3 = A4 MPWM, and h(V) = A5 A1.

Step L4: SC provides a random unique id R2, calculates Auth1 = h(V) ⊕ R2, Auth2 = IdM ⊕ h(R2 ∥ A3) and Auth3 = h(IdM ∥ R2 ∥ A3). SC sends SPi A6, Auth1, Auth2, and Auth3 login requests.

3.4 Authentication Phase

In this phase, the service provider and the manufacturer mutually authenticate each other. A shared secret session key is created after successful authentication. The workflow of the authentication phase is shown in Fig. 4. The authentication process consists of the following steps:

Step A1: as soon as the login message appears, the SPi extracts A2 = V ⊕ A6, R2 = Auth1 ⊕ h(V), and IdM = Auth2 ⊕ h(R2 ∥ h(A2)).

Step A2: SPi calculates Auth3 = h(IdM ∥ R2 ∥ A3) and validates if Auth3 = ? Auth3. If the condition doesn't work, the session ends. If not, the process moves on to the next step.

Step A3: SPi makes a random number R3, makes a session key SKeySU = h(IdM ∥ SIdS ∥ A3 ∥ R2 ∥ R3) using identity SIdS, Auth4 = R3 h(IdM ∥ R2), and Auth5 = h(SKeySU ∥ R2 ∥ R3). It then sends Mi an authentication request message {SIdS, Auth4, Auth5}.

Step A4: after receipt of message{SIdS, Auth4, Auth5}, the SC extracts R3 = Auth4 ⊕ h(IdM ∥ R2) to calculate session key SKeyUS = h(IdM ∥ SIdS ∥ A3 ∥ R2 ∥ R3). Further, SC calculates Auth5 = h(SKeyUS ∥ R2 ∥ R3) and verifies if Auth5 = ? Auth5. If the condition is met, then the service provider has been verified. If that doesn't happen, the process stops.

Step A5: Next, SC determines that Auth6 = h(SKeyUS ∥ R3 ∥ R2) and sends SPi the reply message Auth6.

Step A6: Upon receiving a reply, SPi validates the authenticity of Auth6. SPi determines that Auth6 = h(SKeySU ∥ R3 ∥ R2) and then checks to see if Auth6 = ? Auth6. If the condition does not hold, the session is terminated. If not, the manufacturer Mi is real, and the session key SKeyUS has been checked. After both sides have proven their identities, the session key encrypts all the messages.

3.5 Change Password Phase

The proposed scheme permits authorized manufacturers to update their passwords. Before changing a password, the smart card verifies that the user is who they say they are. The user can choose their own password. This step is important because changing the password often is an excellent way to keep things safe. To change the password, you need to do the following:

Step P1: Mi inserts his smart card SC, enters their identity IdM old password PWM, and imprints biometric BIM at the sensor.

Step P2: SC figures out L = h(IdM ∥ R1 ∥ PWM) and R1 = K H(BIM). Does it then verify if condition L = ? L holds. If the condition fails, the session is terminated. If not, the process goes to the next step.

Step P3: Mi picks a new PWMnew password. SC figures out that MPWM = h(R1 ∥ PWM), MPWMnew = h(R1 ∥ PWMnew), A4new = A4 MPWM MPWMnew, and Lnew = h(IdM ∥ R1 ∥ PWMnew).

Step P4: When SC gets the new password, it replaces the parameters A4 and L already stored with the new one.

4 Analysis of Security

The proposed plan has been carefully analyzed based on the threat model. The analysis shows that the proposed method can withstand all the most common network attacks.

4.1 Attack Forgery

Our scheme is secure against forgery attacks. Let's say that enemy A gets the login message A6, Auth1, Auth2, and Auth3 sent by Mi. Adversary A fails to forge SPi with a fake login request, as should know parameters PWM, biometric BIM, and nonce R2. These parameters are not unknown and never transmitted. So, A needs help to come up with an original request.

Our system is protected against assault. An adversary A may be able to read messages while they are being sent. Let's say an adversary records A6, Auth1, Auth2, Auth3, where Auth1 = h(V) R2, Auth2 = IdM h(R2 ∥ A3), and Auth3 = h(IdM ∥ R2 ∥ A3). To send a request successfully, an adversary must know the exact values of several parameters. But the one-way hash function h() has been used to protect all parameters. You can't figure out how to decrypt them, and the manufacturer still needs to send some of the parameters.

4.2 Attack on the Password Attack Forgery

Our scheme is secure against password-guessing attacks. Adversary A intercepts request messages A6, Auth1, Auth2, Auth3, and response messages SIdS, Auth4, and Auth5, to decrypt password PWM. To breach a password, one must know about PWM, which has never been transmitted over insecure channels between entities. Also, our scheme uses MPWM = h(R1 ∥ PWM) to figure out the masked password, and this MPWM is sent over channels. A can never decrypt it because it is encrypted with a secure hash and masked with a random nonce.

4.3   Attack on the Session Key

Our scheme is secure against attack on the session key disclosure. The proposed method makes two session keys, SKeySU and SKeyUS, where SKeySU = h(IdM ‖ SIdS ‖ A3 ‖ R2 ‖ R3) and SKeyUS = h(IdM ‖ SIdS ‖ A3 ‖ R2 ‖ R3). However, adversary A will never be able to get access to any of these session keys because these keys have never been transmitted between involved entities. These are sent by encrypting using the hash function, Auth5 = h(SKeyUS ‖ R2 ‖ R3). SKeySU and SKeyUS are always shipped with encryption for SKeySU and SKeyUS. Even if A intercepts Auth5, they won't be able to figure out what SKeySU and SKeyUS are because they are encrypted with a hash.

5   Formal Verification of the Proposed Scheme Using AVISPA

This part shows how the AVISPA was used to simulate the proposed plan. The Automated Validation of Internet Security Protocols and Applications (AVISPA) [2] "focuses on evolving a push-button for the evaluation of wide-ranging internet security-sensitive protocols and applications. High-Level Protocol Specification Language is the name of the language used to write the code for these protocols (HLPSL). The results show that the proposed plan is safe and can withstand significant network attacks (Figs. 1 and 2).

6   Conclusion

To ensure that items are authentic and transactions are confidential and safe, this research shows a secure SCM system based on the Internet of Things (IoT) and blockchain technology. Participants and IoT applications are verified via QR scanners, Diffie-Hellman key exchange, and blockchain technology based on Hyperledger Fabric. Analysis of security and scalability shows that the suggested system is more trustworthy and secure than other current methods. The Internet of Things (IoT) is a network of actual objects connected by detectors, software, and technology known as the Internet of Things (IoT). It has the potential to offer valuable services in various sectors, including business, healthcare, smart homes, smart cities, social media, supply chains, and more.

```
role server (Mi, SPi : agent,
        h : hash_func,
        Bh : hash_func,
        SKus : symmetric_key,
        SND, RCV : channel(dy))

played_by SPi
def=
local State : nat,
IdM, PWM, SP, MPWM, BIM, A1, P, Q, U, V : text,
Auth1, Auth2, Auth3, Auth4, Auth5, Auth6 : text,
R1, R2, R3, SIdS, T : text

const alice_bob_r1, alice_bob_r2,
bob_alice_r3, bob_alice_t,
subs1, subs2, subs3,sub4,sub5 : protocol_id

init State := 0
transition
% receive registration request message from the manufacturer
1. State = 0 ∧ RCV({IdM.h(PWM.R1').h(xor(IdM.R1'))}_SKus) =| >
State' := 1 ∧ request(Mi, SPi, alice_bob_r1, R1')
        ∧ T' := new()
        ∧ secret({V', P', Q'}, subs1, {SPi})
        ∧ secret({PWM, BIM, R1'}, subs2, {Mi})
        ∧ secret({IdM}, subs3, {Mi,SPi})
        ∧ SND({xor(h(h(IdM.V.T)).h(PWM.R1')).xor(V).
          h(xor(IdM.R1'))).xor(V.h(IdM.V.T))}_SKus)
        ∧ witness(SPi, Mi, bob_alice_t, T')

2. State = 1 ∧ RCV({xor(V. h(IdM.V.T)).xor(h(V).R1').xor(IdM.h(R1'.
        h(h(IdM.V.T)))).h(IdM.R1'.h(h(IdM.V.T)))}_SKus) =| >
   State' := 2 ∧ R2' := new()
        ∧ secret({R2'}, subs5, SPi)
        ∧ SND(xor(V.h(IdM.V.T)).Auth1'.Auth2'.Auth3')
        ∧ witness(SPi, Mi, bob_alice_t2, T2')

3. State = 2 ∧ RCV({SIdS', Auth4', Auth5'}) =| >
        ∧ request(Mi, SPi, alice_bob_r1, R1')
end role
```

Fig. 1 Role specification of service provider SPi

```
% OFMC
% Version of 2006/02/13
SUMMARY
SAFE
DETAILS
BOUNDED_NUMBER_OF_SESSIONS
PROTOCOL
/home/avispa/web- interface- computation/
./tempdir/workfilevOpMGm.if
GOAL
as_specified
BACKEND
OFMC
COMMENTS
STATISTICS
parseTime: 0.00s
searchTime: 0.33s
visitedNodes: 103 nodes
depth: 11 plies
```

Fig. 2 Simulation results

References

1. Lee I, Lee K (2015) The Internet of Things (IoT): Applications, investments, and challenges for enterprises. Bus Horizons:431–440. https://doi.org/10.1016/j.bushor.2015.03.008.
2. Kemp S (2019) Digital 2019: Global Digital Overview—DataReportal—Global digital insights. In: DataReportal—Global Digital Insights [Internet]. DataReportal—Global Digital Insights; 31 Jan 2019 [cited 3 Feb 2020]. Available https://datareportal.com/reports/digital-2019-global-digital-overview.
3. Saif I, Peasley S, Perinkolam A (2015) Safeguarding the Internet of Things: Being secure, vigilant, and resilient in the connected age. https://dupress.deloitte.com/dupus-en/deloittereview/issue-17/internet-of-things-data-security-andprivacy.html [Retrieved: 2015-07–27]
4. Jin J, Gubbi J, Marusic S, Palaniswami M (2014) An information framework for creating a smart city through the Internet of things. IEEE Internet of Things J 1(2): 112-121
5. Haber S, Stornetta WS (1991) How to time-stamp a digital document. In: Proceedings of the 10th annual international cryptology conference on advances in cryptology, pp 437–455
6. Alonso F, Fernndez Marco L, Salvacha J (2017) Iaacaas: Iot application-scoped access control as a service. Futur Internet 9(4)
7. Hummel R, Shafagh H, Raza S, Voigt T, Wehrle K (2014) Delegation-based authentication and authorization for the ip-based internet of things. In: 2014 Eleventh annual IEEE international conference on sensing communication and networking (SECON), pp 284–292
8. Andrea I, Chrysostomou C, Hadjichristofi G (2015) Internet of things: Security vulnerabilities and challenges. In: 2015 IEEE symposium on computers and communication (ISCC), pp 180–187

9. Tang B, Kang H, Fan J, Li Q, Sandhu R (2019) IoT passport: a blockchain-based trust framework for collaborative internet-of-things. In: Proceedings of the 24th ACM symposium on access control models and technologies, pp 83–92

10. Rashid MA, Pajooh HH (2019) A security framework for IoT authentication and authorization based on blockchain technology. In: 2019 18th IEEE international conference on trust, security and privacy in computing and communications/13th IEEE international conference on big data science and engineering (TrustCom/BigDataSE). IEEE, pp 264–271

11. Moussaoui D, Kadri B, Feham M, Bensaber BA (2021) A distributed blockchain based PKI (BCPKI) architecture to enhance privacy in VANET. In: 2020 2nd international workshop on human-centric smart environments for health and well-being (IHSH), pp 75–79. IEEE

12. Lohachab A, Karambir (2019) ECC-based inter-device authentication and authorization scheme using MQTT for IoT networks. J Inf Secur Appl 46:1–12

13. Ma W, Ma J, Zhang Q, Xue H, Li Y, Dang X, Zhao M, Zhang J, Han C, Wu J (2020) Attribute revocable data sharing scheme based on blockchain and CP-ABE. In: Proceedings of the 4th international conference on computer science and application engineering, pp 1–7

14. Ahmad A, Din S, Paul A, Jeon G, Aloqaily M, Ahmad M (2019) Real-time route planning and data dissemination for urban scenarios using the Internet of Things. IEEE Wirel Commun 26:50–55

15. Wu M, Wang K, Cai X, Guo S, Guo M, Rong C (2019) A comprehensive survey of blockchain: From theory to IoT applications and beyond. IEEE Internet Things J

16. Hassija V, Chamola V, Saxena V, Jain D, Goyal P, Sikdar B (2019) A survey on IoT security: application areas, security threats, and solution architectures. IEEE Access 7:82721–82743

17. Li D, Peng W, Deng W, Gai F (2018) A blockchain-based authentication and security mechanism for IoT. In: 2018 27th International conference on computer communication and networks (ICCCN), pp 1–6

Secure User Authentication Protocol for Roaming Services in Mobile Networks Using Blockchain

M. Indushree and Manish Raj

Abstract Increase in wireless devices made mobile communication pervasive. Global Mobile Networks (GLOMONET) provision the roaming service to accomplish this, where mobile users must experience secure and seamless roaming services over multiple foreign agents. The main objective of network providers is to have mutually authenticated, secured, and lightweight service to guard mobile user's data and privacy. Many interesting roaming authentication protocols have been proposed to achieve the security and privacy of mobile users in traditional communication networks. But they all suffer from one or another known security attack with the fact that current mobile networks are prone to attacks. Blockchain technology offers its advantages to establish a secure connection and authentication by safeguarding mobile user information and privacy with its immutable nature. The study shows that limited work has been done in space protocol design for GLOMONET using blockchain technology and the main goal of the protocol is to maintain security for transactional data and privacy of the mobility users along with anonymity property. In this article, soulbound tokens are used to issue credentials between the mobile user and Home Agent (HA) by serving as a secure and decentralized form of digital identity. The idea behind using soulbound tokens for issuing credentials is to create a tamper-proof and easily verifiable system that reduces the reliance on centralized authorities for identity verification. In addition, the smart contracts for user authentication have been implemented through solidity programming and the security strength of the proposed protocol is verified through a formal verification tool called AVISPA (Automated Validation of Internet Security Protocols and Applications).

Keywords Mobility network · Security · Smart contract · Soulbound tokens · Ethereum

M. Indushree (✉) · M. Raj
School of Computer Science Engineering and Technology, Bennett University, Greater Noida, Uttar Pradesh, India
e-mail: indushree.june1@gmail.com

M. Raj
e-mail: manish.raj@bennett.edu.in

© The Author(s), under exclusive license to Springer Nature Singapore Pte Ltd. 2024   511
S. Namasudra et al. (eds.), *Data Science and Network Engineering*, Lecture Notes in Networks and Systems 791, https://doi.org/10.1007/978-981-99-6755-1_39

1 Introduction

GLOMONET is becoming one of the most important developing network environments for providing flawless roaming services in foreign networks. Mobile devices have become an integral part of our daily life on the GLOMONET, with applications such as commerce, social networking, communication, and information exchange, among others. In a mobility network, the mobile user is allowed to pair their devices with alternative widgets utilizing Bluetooth, GPS, and WiFi in order to receive broad internet services and other region-based services systematically [4].

In network contexts, various fundamental challenges such as security and privacy issues such as mutual authentication, secrecy, and user anonymity are also raised. As a result, in network environments, user anonymity becomes a vital component. The current situation requires the development of reliable network architecture for global mobile environments. Furthermore, GLOMONET provides roaming services to mobile subscribers. These services let a user obtain access to the services that are offered by HN in an FN [7]. The well-built secure authentication techniques will be used to secure roaming services in GLOMONET between users HN and FN being overtaken. When a mobile user moves from one location to another in a global mobility network, roaming service is used to ensure the continuity of the mobile network service. Constructing a user model in a mobility network, the Home Agent (HA), Foreign Agent (FA), and Mobile User (MU) are all involved in authentication [11]. An intruder can eavesdrop on the communicated message across the public channel between MU and FA, which is the reason for this environment's security breach. As a result, secure data communication between HA, MU, and FA is required as indicated in Fig. 1.

1.1 Blockchain Importance in GLOMONET

In general, designing a secure architecture for providing roaming services requires proper organization, collaborations that can significantly benefit from the design of a decentralized network, and consideration of any use cases that necessarily require an accurate combination of network services provided by the network environment [2]. In particular, blockchain is being utilized to address a variety of challenging tasks across the entire network, and it is providing support in the continuous growth of a number of wired devices, such as data traffic and data services. A blockchain is a decentralized, immutable record that ensures network confidentiality, integrity, privacy, and authentication. Blockchain is a P2P network that runs on the IP protocol [20]. It is transparent, decentralized, and provides the security to the users. Using cryptographic hash, the previous blocks are interconnected to each other once the new block is added to a blockchain network, it makes sure that the chain is never broken since all the blocks are subsequently connected and each activity is stored permanently in the network. The capacity of blockchain plays a vital role in acquir-

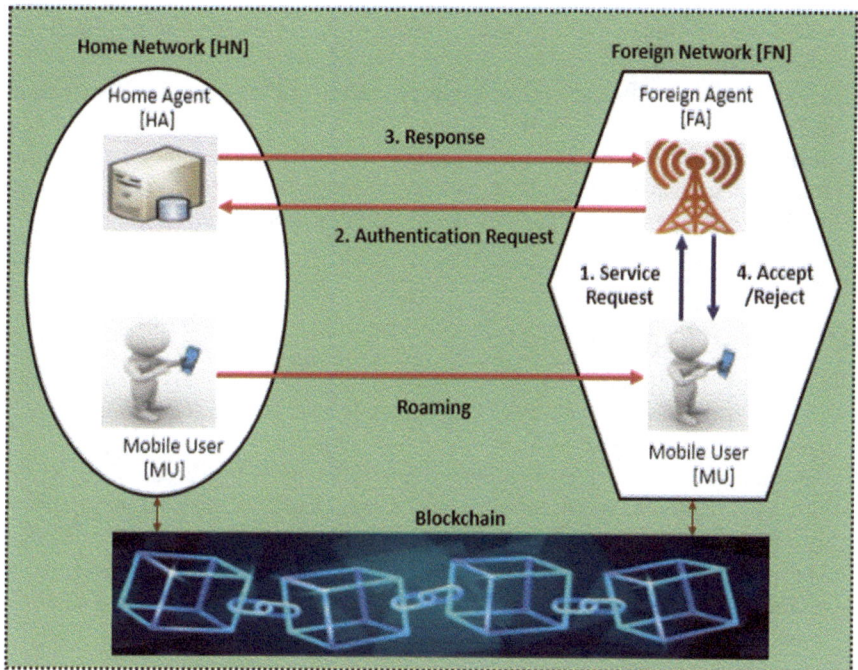

Fig. 1 Authentication scenario for Mobile Users in GLOMONET

ing noticeable attention in telecom industries. With the advancement of Soulbound Tokens (SBT), the user could then share this digital identity with other parties, such as employers or educational institutions, as proof of their identity and qualifications. Because the SBT cannot be replicated or transferred, the digital identity created using the token would be secure and easily verifiable, without the need for a centralized authority to validate the identity [15].

1.2 Motivations

Several privacy concerns have arisen between network customers using service providers and mobile phones in GLOMONET. We explored various MU authentication approaches that have been established to provide subscribers with roaming services in this section. The following are some of the limitations of the previous authentication techniques.

1. The majority of authentication protocols in global mobility networks are vulnerable to well-known network security flaws.
2. HA, FA, and MU use an unfair key agreement mechanism to distribute secret information among mobile user authentication entities. If any entity's static secret

values are exposed over the network, that network's entire security mechanism is compromised [18].

3. Allocating resources and consumption produced by communication operations and processing is a fundamental challenge in the global mobile network. Mobile devices have limited resources in terms of processing power, memory, bandwidth, and computational capacity. To implement existing authentication methods, there is a larger communication and computation overhead is expected [12].

1.3 Research Contributions

1. We propose a novel blockchain-based mobile user authentication mechanism using a non-transferable SBT.
2. Exploiting and testing various security attacks by using security tools and blockchain technology is used to provide the confidentiality, transparency, privacy, security as well as authentication to the protocols using decentralized network.
3. The proposed system is also verified using a formal verification tool called AVISPA (Automated Verification of Internet Security Protocols and Applications).

1.4 Organization of the Paper

The remaining portion of the article is organized as follows: Background information covered in Sect. 2. Security attributes in GLOMONET are described in Sect. 3. The proposed authentication framework is described in Sect. 4. Formal security verification and the implementation details, and experimental results are provided in Sect. 5. Finally, the article is concluded in Sect. 6.

2 Literature Review

With the rapid advancement of communication technology, mobile users can now move throughout the world and utilize the mobile network's ubiquitous services. GLOMONET has recently emerged as one of the most promising venues for providing flawless roaming services in foreign networks. However, the wireless and mobility environments, on the other hand, are well known for being more vulnerable to attacks. The attacker has the ability to eavesdrop, manipulate, or prevent sensitive data sent via the radio channel. As a result, in the mobility environment, the mutual authentication procedure between communication entities is critical.

In 2015, Karuppiah et al. [5] examined the Rhee et al.'s method, observing that it is susceptible to Perfect forward secrecy, impersonation, password guessing, and user anonymity attacks, as well as the point that Rhee et al.'s technique does not notice incorrect passwords immediately. After that, Karuppiah et al. [5] also presented a novel secure authentication mechanism with user anonymity for roaming service in GLOMONET. However, Fan et al. [16] shows that Gope and Hwang's authentication mechanism is insecure for GLOMONET. Thus, Fan et al. [16] come up with a novel mutual authentication system using a two-factor authentication scheme proposed to address the flaws in all of the above-mentioned recent schemes in 2016. Recently, in 2021, Shashidhara et al. [10] demonstrated Xu et al. authentication's and key agreement technique had some security weaknesses and Shashidhara et al. [10] stated that Xu et al. protocols are exposed to some common known attacks. Further, Shashidhara et al. established a secure and robust mutual authentication scheme for mobility networks as a solution. The proposed protocol is written in HLPSL and validated with AVISPA. Furthermore, the implemented security scheme is lightweight, secure, and computationally efficient, according to the performance evaluation. In 2020, Nikooghadam et al. [8] reviewed and illustrated the flaws in the work presented by Ghahramani et al. Then, for roaming users in GLOMONET, Nikooghadam et al. presented a more secure and efficient authentication and key agreement technique and also verified the proposed scheme's security in both a descriptive and formal manner using Scyther, a formal verification tool.

To design proper authentication protocols using blockchain in GLOMONET, very little research work has been done to date, and it is becoming a popular and well-supported technology. Most of the work in this field is currently taking place in the blockchain space, and some of the existing work that is published for authentication protocol using blockchain is highlighted as follows.

In 2022, Paul et al. [14] published a paper titled "A systematic literature review of Blockchain cyber security" in which they provide a detailed review of blockchain for cyber security and the author created a novel blockchain application for networks, public key cryptography, certification schemes, web applications, and machine visualization.

In 2021, Xu et al. [6] proposed an authentication and dynamic group key agreement protocol based on blockchain. This protocol helps every user's belonging to each group. Users only need to authenticate their lift neighbor once they complete the authentication which improves the efficiency. It also reduces the computation and communication costs. In addition, there are several mutual authentication protocols have been proposed for mutual authentication systems [1, 3, 13, 19].

3 Security Attribute in GLOMONET

The implementation authentication protocol must comply with the following security requirements:

R1 *Mutual authentication*: Every session establishes secure communication between FA and MU. Firstly, the MU, FA, and HA must mutually authenticate each other, and after the mutual authentication, it generates the SK for further processing.

R2 *User untraceability*: An Intruder $\mathscr{A}$ either detects roaming subscribers during authentication sessions or connects authentication activities that are associated with the same party.

R3 *Session key (SK) security and fairness*: Because the interactions between MU and FA are encoded by the SK, an SK (session key) negotiation is essential for maintaining a more secure connection. Furthermore, all communication agents contribute to the generation of the session key in a certain way.

R4 *Robustness against attack*: Even if an attacker has access to all of the data stored on the SmartCard, the protocol can withstand many threats.

R5 *Computational efficiency*: The authentication technique should be simple in both computation and communication.

R6 *No Time Synchronization*: Because transmission delay is unpredictable in existing networks, remote user authentication systems that use time stamps to ensure message freshness may still be vulnerable to replay attacks. Furthermore, in existing network systems, clock synchronization is complex and costly.

4   Proposed Authentication Protocol for GLOMONET

The proposed protocol consists of the following phases such as (i) initialization phase, (ii) registration phase, and (iii) mutual authentication phase. The following is a brief summary of the purpose of these phases:

1. **Initialization phase**: The system parameters are chosen by the HA.
2. **Registration phase**: Registration phase before providing any healthcare services to the patient user has to register with the health server so the registration phase is between the user and sever is established. So, once the registration is completed user can access the services from the health server this process is done in mutual authentication phase.
3. **Mutual authentication phase**: Here, the registered user should be authenticated first so after that services should be provided that is authentication in this phase the user has to authenticate the server even if the server has to authenticate the user so the mutual authentication will be established. Once the user and health server are agree or authenticated each other so then session key agreement will take place also SK is order to encrypt the information in the secure way.

4.1 Blockchain-Based Soulbound Token (SBT) for Authentication

Soulbound tokens can be used to issue an identity token between a user and a server by serving as a secure and decentralized form of authentication.

Here's how SBT could work in the GLOMONET:

1. The MU would first obtain an SBT, which is unique to them and cannot be replicated or transferred.
2. The HA would generate an identity token and send it to the mobile user.
3. The MU would use their SBT to sign the identity token, which would create a tamper-proof digital signature that proves their identity.
4. The user would then send the signed identity token back to the HA, which would verify the digital signature using the MU's SBT.
5. If the digital signature is verified, the HA would authenticate the MU and grant them access to the requested resources or services.

Overall, using SBTs for issuing identity tokens between users and servers could help to create a more secure and decentralized system of authentication, which could be used in a wide range of applications, such as online banking, e-commerce, and social media.

4.2 Minting and Issuing the Soulbound Token

An HA can mint an SBT to a MU by generating a unique key pair consisting of a private key and a public key. The private key is securely stored on HA, while the public key is shared with the MU.

Steps in SBT Minting and Issuing:

1. The HA would generate a key pair consisting of a private key and a public key.
2. The HA would securely store the private key, which would be used to sign and verify transactions involving the SBT.
3. The HA would send the public key to the user, which would be used to create the SBT.
4. The MU would use the public key to create a new SBT, which would be uniquely bound to their identity.
5. The MU would store the SBT securely on their device, such as a smartphone or a hardware wallet.
6. The MU would then use the SBT to authenticate themselves to the server, which would verify the digital signature using the private key stored on HA.

The authentication protocol has been designed using SBT smart contracts, which are lines of programming code that are stored on a Blockchain network and are only executed when specific requirements are met. It is referred to as smart since it is

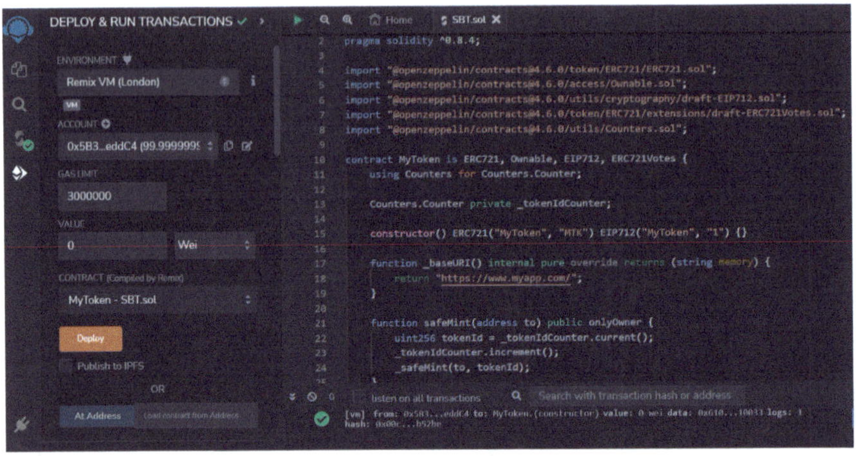

Fig. 2 Smart contract compilation and deployment process using Remix

capable of independently verifying and executing a contract. The contract, which contains all of the details of a specific agreement, is included in the decentralized blockchain network. The SBT has been implemented using a solidity smart contract and deployed to the Ethereum blockchain network.

Accuracy, openness, independence, security, and uniformity are all intended benefits of smart contracts. On the blockchain, the Ethereum nodes execute smart contracts written in the solidity programming language. Every 10 seconds, at least two additional network nodes must validate each node in the blockchain. Written contract functions may then be triggered and carried out after that.

The SBT contract has been compiled through Remix and deployed on the Goerli test network. The contract compilation and deployment as shown in Fig. 2. In addition, the contract deployment process, minting and issuing an SBT to the MU address, and transaction details on the blockchain as shown in Figs. 2 and 3, respectively.

5 Formal Security Verification of the Proposed Protocol

The AVISPA toolkit is a collection of tools for testing and implementing formal security protocol models. The role-based language HLPSL is frequently used to construct scheme models. AVISPA (Automated Validation of Internet Security Protocols and Applications) is a widely used simulation tool for determining if a newly proposed approach authentication technique or protocol is more secure against numerous of commonly known security vulnerabilities. The high-level protocol specification language is used in the AVISPA tool to implement communication.

The proposed scheme is used AVISPA tool to validate the informal analysis and automated security. To demonstrate that the recommended protocol scheme is resis-

Fig. 3 Minting and issuing
an SBT to the mobile user

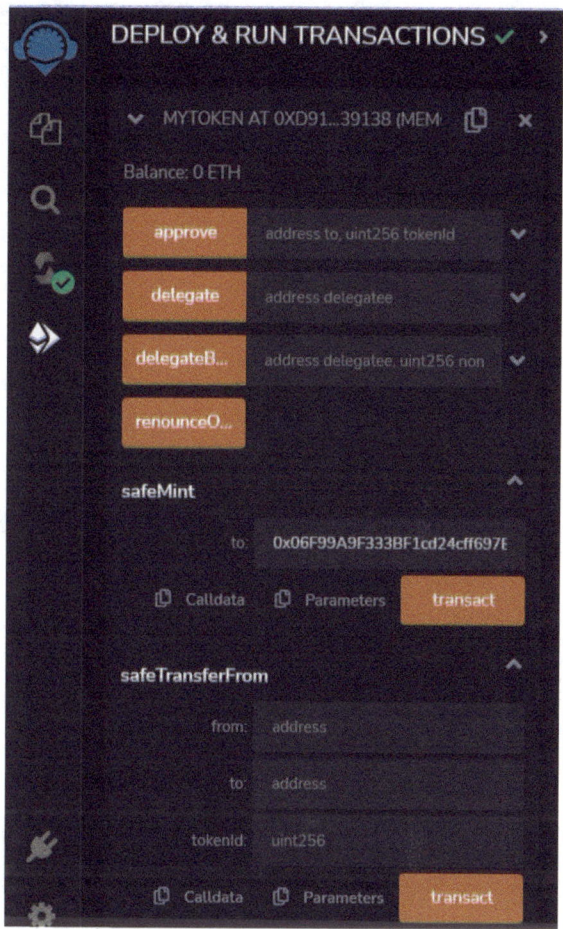

tant to various common security vulnerabilities, including replay, forgery, smartCard, and MITM attacks. The main goal of this scheme is to accomplish mutual authentication, establish the session key, and also implement de-synchronization, resist the common security attacks, and finally, reduce the computation and storage burden. Further to check the correctness of the formal security authentication properties, it uses a tool called AVISPA.

AVISPA tool contains backends that are divided into four categories. (i) On-the-Fly Model Checker—OFMC: a figurative strategy for exploring state space in a demand driven manner. (ii) SAT-based Checker—SATMC, (iii) Tree Automata based on automatic approximations for analysis of Security Protocol—TA4SP, (iv) Constraint Logic-based Attack Searcher—CL-AtSe. This input is first converted to IF—Intermediate Format, then to OF—Output Format.

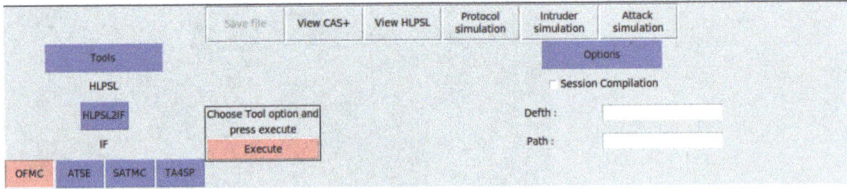

Fig. 4 Autentication protocol using OFMC

Each role is self-contained, receiving initial input via arguments and communicating with other roles via channels. The adversary is depicted using the DolevYao paradigm, with the intruder having the capacity to play a legitimate role in a protocol execution. The role system also describes a number of sessions, principles, and roles. OF is formed based on the four backends, and after successful protocol execution, OF defines whether the protocol is secure or insecure as well as under what conditions the output is acquired. The HLPSL language is used to specify the responsibilities of the MU, FA, and HA, the environment, and the sessions in the newly proposed scheme. Figure 4 shows the output of the proposed protocol AVISPA. The designed protocol is written in HLPSL and we have executed using one of the backend OFMC. The result of the proposed protocol is SAFE. If the result is UNSAFE, we can use the intruder simulation and attack simulation that show what information is revealed to the attacker that information helps us to rebuild the protocol.

5.1 Performance Evaluation

Table 1 clearly demonstrates that the proposed security framework meets all the necessary security properties required for the mobile user roaming service. Moreover,

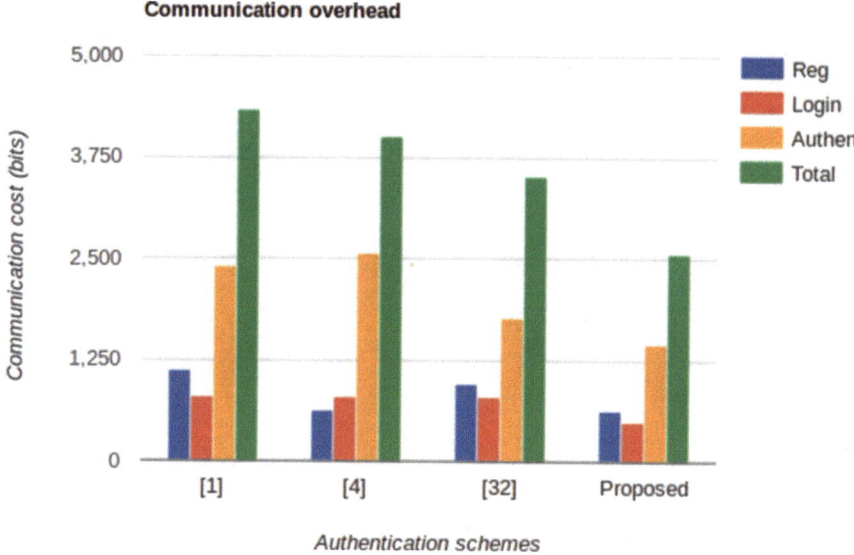

Fig. 5 Comparison of communication cost

it is crucial for the authentication protocol to maintain reasonable levels of both communication and computational complexity. To this end, we have estimated the efficiency of our proposed protocol based on its computation and communication costs in Fig. 5, which are key metrics for evaluating its performance. By conducting this evaluation, we have ensured that the protocol is efficient in terms of both its processing speed and its use of network resources while still maintaining the required security properties. As mobile networks continue to expand, the number of mobile users accessing online services and applications is increasing rapidly. This growth in user demand can put pressure on authentication systems and lead to performance issues, such as slower response times, increased latency, and reduced system availability. To address these challenges, it is important to develop authentication protocols that can scale to accommodate large numbers of users and high network traffic. In the proposed protocol, we have implemented a blockchain-based authentication framework that utilizes zero-knowledge proofs (zk proofs) for increased security and privacy. We have executed this protocol on Polygon's zk-EVM (Ethereum Virtual Machine), which is a high-performance platform that can handle more transactions per second than traditional Ethereum networks. This implementation can provide faster and more efficient processing of authentication requests, allowing the system to support more users without sacrificing security or performance.

Table 1 Security requirements and functionality comparison

Security requirements and Functionalities	Protocol [5]	Protocol [9]	Protocol [17]	Protocol [10]	Proposed
Mutual authentication	✓	✓	✓	✓	✓
Mobile user privacy	✓	×	✓	✓	✓
Prevents insider attack	×	×	✓	✓	✓
Withstand impersonation attacks	×	×	×	✓	✓
Withstand stolen-verifier attack	×	×	✓	✓	✓
prevent password-guessing attacks	×	✓	✓	✓	✓
Prevent replay attacks	✓	✓	✓	×	✓
Perfect-forward secrecy	✓	×	×	✓	✓
Anonymity and untraceability	×	×	✓	✓	✓
Fair session-key negotiation	✓	✓	×	✓	✓
Security against DoS attacks	×	✓	✓	✓	✓
Clock-synchronization problem	×	×	×	✓	✓
Decentralization	×	×	×	×	✓
Local password verification	✓	✓	✓	✓	✓

6   Conclusion

In this article, We proposed a novel blockchain-based protocol for roaming service in GLOMONET. To communicate between the MU, HA, and FA, the protocol was developed to take advantage of blockchain's features, such as authentication and its preservation of user privacy. Using a formal verification process, we were able to demonstrate the proposed protocol security. The confidentiality of sensitive information and participant authentication have both been successfully validated by this protocol. Overall, using soulbound tokens for identity verification and authentication can help to create a more secure and decentralized system that reduces the reliance on centralized authorities for identity management in the Global Mobility Network (GLOMONET).

References

1. Chaffer TJ, Goldston J (2022) On the existential basis of self-sovereign identity and soulbound tokens: an examination of the "self" in the age of web3. J Strat Innov Sustain 17(3):1
2. Gangwani P, Perez-Pons A, Bhardwaj T, Upadhyay H, Joshi S, Lagos L (2021) Securing environmental iot data using masked authentication messaging protocol in a dag-based blockchain: Iota tangle. Futur Internet 13(12):312
3. Goldston J, Chaffer TJ, Osowska J, Goins II CV (2023) Digital inheritance in web3: a case study of soulbound tokens and the social recovery pallet within the polkadot and kusama ecosystems. arXiv:2301.11074
4. Indushree M, Raj M, Mishra VK, Shashidhara R, Das AK, Bhat V (2023) Mobile-chain: secure blockchain based decentralized authentication system for global roaming in mobility networks. Comput Commun 200:1–16

5. Karuppiah M, Saravanan R (2015) A secure authentication scheme with user anonymity for roaming service in global mobility networks. Wirel Pers Commun 84(3):2055–2078
6. Li J, Qiao Z, Peng J (2022) Asymmetric group key agreement protocol based on blockchain and attribute for industrial internet of things. IEEE Trans Ind Inf
7. Madhusudhan R (2016) An efficient and secure authentication scheme with user anonymity for roaming service in global mobile networks. In: Proceedings of the 6th international conference on communication and network security, pp 119–126
8. Nikooghadam M, Amintoosi H, Kumari S (2020) A provably secure ecc-based roaming authentication scheme for global mobility networks. J Inf Secur Appl 54:102588
9. Reddy AG, Das AK, Yoon EJ, Yoo KY (2016) A secure anonymous authentication protocol for mobile services on elliptic curve cryptography. IEEE Access 4:4394–4407
10. Shashidhara R, Bojjagani S, Maurya AK, Kumari S, Xiong H (2020) A robust user authentication protocol with privacy-preserving for roaming service in mobility environments. Peer-To-Peer Netw Appl 13(6):1943–1966
11. Shashidhara R, Indushree M, Sneha N (2022) Design of a secure blockchain based privacy preserving electronic voting system. In: Emerging research in computing, information, communication and applications: ERCICA 2020, vol 1. Springer, Berlin, pp 1–9
12. Shashidhara R, Nayak SK, Das AK, Park Y (2021) On the design of lightweight and secure mutual authentication system for global roaming in resource-limited mobility networks. IEEE Access 9:12879–12895
13. Sumithra V, Shashidhara R, Mukhopadhyay D (2022) Design of a secure and privacy preserving authentication protocol for telecare medical information systems. Secur Priv 5(4):e228
14. Taylor PJ, Dargahi T, Dehghantanha A, Parizi RM, Choo KKR (2020) A systematic literature review of blockchain cyber security. Digit Commun Netw 6(2):147–156
15. Weyl EG, Ohlhaver P, Buterin V (2022) Decentralized society: finding web3's soul. Available at SSRN 4105763
16. Wu F, Xu L, Kumari S, Li X, Khan MK, Das AK (2017) An enhanced mutual authentication and key agreement scheme for mobile user roaming service in global mobility networks. Ann Telecommun 72(3):131–144
17. Xu G, Liu J, Lu Y, Zeng X, Zhang Y, Li X (2018) A novel efficient maka protocol with desynchronization for anonymous roaming service in global mobility networks. J Netw Comput Appl 107:83–92
18. Yu S, Park Y (2022) A robust authentication protocol for wireless medical sensor networks using blockchain and physically unclonable functions. IEEE Internet Things J 9(20):20214–20228
19. Zhan Y (2023) Research on synchronising soulbound token between blockchain and local database
20. Zhao W, Aldyaflah IM, Gangwani P, Joshi S, Upadhyay H, Lagos L (2023) A blockchain-facilitated secure sensing data processing and logging system. IEEE Access 11:21712–21728

Author Index